FROMELLES

100 YEARS OF MYTHS AND LIES

Analysing the background to, the treatment of, and setting the record straight for The Battle of Fromelles.

3RD EDITION, REVISED 2026

GEOFFREY BENN

Published 2026 by Geoffrey V Benn
3rd Edition

ISBN: 978-0-6452470-3-9 - Hardback Edition

---ooo000ooo---

Cover Imagery:
Main picture: Pierre Seillier © 13.02.2025. Prisoners of war captured at Fromelles being marched through the village of Hallennes-lez-Haubourdin, en route to Lille to be gaoled in the citadel before being sent to Germany, overseen with figures of 68 Pte Wesley Choat, 32nd Battalion (who escaped captivity), and Captain Charles Mills, 31st Battalion also POW and who did so much for the POWs and missing soldiers with Australian Red Cross after Armistice.

In revising the contents of this book, I again dedicate it to the passion,
and now the memory of a great Australian:
Neville Thomas Kidd.

So much has been revealed from historical records
that Neville was not privy to during his life.
If there is a heaven, I am sure he is smiling broadly there,
with what has been revealed since he left us.

---ooo000ooo---

Geoffrey Benn

February, 2025

FROMELLES: 100 YEARS OF MYTHS AND LIES

Analysing the background to, the treatment of, and setting the record straight for The Battle of Fromelles.

---ooo000ooo---

[Picture 2: William Polding Ryan, from John Ryan family collection]

WILLIAM POLDING RYAN

Lost So Far From Home

And

Found So Close to Our Hearts

Foreword

19 July, 1917 Tweed Daily, Murwillumbah, NSW

ROLL OF HONOR

RYAN.

In loving memory of our dear son and brother,

SERGEANT WILLIAM POLDING RYAN,

killed in action in France, July 19, 1916;

aged 20 years and 3 months.

A young life nobly ended,

In Fleurbaix they have laid him,

Our soldier so brave and so true,

He fought for freedom and liberty,

And died 'neath the Red, White and Blue.

- Inserted by his sorrowing parents, brothers and sisters,

Crystal Creek.

So was published the heartfelt rending of the loss of a family member, a year after the Battle of Fromelles, which claimed this soldier's life. He was according to official accounts, one of one thousand, nine hundred and seventeen (1,917)[1] Australian and five hundred and nineteen (519) British soldiers who lost their lives during the events of 19th and 20th July, 1916 at Fromelles in French Flanders.

Many others died in the days, weeks, even months after, due to wounds sustained in the Battle of Fromelles.

But to this day, no family member can point to a location at which they can see the name of the place where this soldier was killed, on *any Australian Capital City memorial within Australia.*

[1] You will discover in later chapters, that this number is understated.

The only memorial with the name Fromelles inscribed on it to this author's knowledge is on the shores of Lake Macquarie in Toronto NSW. The place at which he died is not inscribed elsewhere in Australia on *Australian* memorials to those who died in the service of their country.

Ironically, "Fromelles" is inscribed on a column located in the grounds of the French Embassy in Canberra, erected *to honour the memory of Australians who fell on French soil in the 1914-1918 War*. It was unveiled on 28 July, 1961, in a function attended by then Prime Minister, Sir Robert Menzies.

To see 'Fromelles' inscribed on a war memorial otherwise, one needs to travel overseas to London, England, to Fromelles in France, to Zonnebeke in Belgium, and to Ploegsteert, also in Belgium.

In London you can attend the Australian War Memorial at Hyde Park Corner.

At Fromelles, France, you can visit the VC Corner Military Cemetery, and the Fromelles (Pheasant Wood) Military Cemetery, near Fromelles

Near Zonnebeke in Belgium, you can visit the A.I.F. 5th Division's own monument at Polygon Wood. Elsewhere in Belgium, you can visit the Berks Cemetery Extension Memorial at Ploegsteert.

Despite suffering the largest loss of life by the Australian soldiers in a single day, Fromelles was not then, and has not since, been formally classified as a "Battle" by Australia's military, because of a lie initiated by the British High Command following the Battle of Fromelles.

As if to excuse this, we see the informal, wide-spread use of the word "Battle" by governments, media, and the public generally since then.

The continuation of the lie claims that Fromelles is 'recognised' as part of an action in an area of France some 80 kilometres away, in another province, on another river system, three days' march from where this action took place.

This book is written so that you can understand how the Battle of Fromelles has been wrongly classified in the past, why the soldiers who fought and died there never received the recognition of a Battle Honour, and why we need to redress this in the present, so that future generations can know the truth.

William Polding Ryan, and all of those heroes killed and maimed, their families both then and now deserve nothing less.

---ooo000ooo---

Acknowledgements

There are so many people who I must acknowledge, for without them there would be no book at all. First of course are the Australian and British soldiers who fought in the Battle of Fromelles in 1916, whether they were killed in action, were wounded and later died, became prisoners, or went on to fight in other battles, and lived to tell their tales. They were all dishonoured by having the action in which they fought classified as an 'Attack', when it should always have been classified as a Battle. Their service would again be dishonoured when their efforts to secure a Battle Honour for Fromelles were 'omitted' by Australia's Commanders in Chief following the Great War. And similarly in Britain where such attempts were unsuccessful.

To the people who believed with all sincerity that there were many who had not been accounted for. To my friend Lambis Englezos, and his friends Ward Selby and John Fielding, and all who joined with them to convince authorities that there was a mass grave next to Pheasant Wood near Fromelles.

To the thousands who have registered with the Unrecovered War Casualties Unit in the hope that an ancestor's place of burial can finally be placed with certainty.

To those within the Fromelles Association of Australia who have made linkages with people all around the world in the hope of tracing a family member who can donate DNA for cross-matching with a recovered soldier's DNA. To those persons within the Association who sought out every piece of material they could find referring to 'Fromelles' for the last one hundred years, which led to the contents of my original book, and now for this Revised Edition. I am forever in the debt of Royce Atkinson, and Marg O'Leary OAM, for their discoveries.

To Jim Munro and Russell Curley from the Families and Friends of the First AIF group, who have provided me with much valuable material in my search, and for Jim in particular who ensures there is a meaningful commemorative service on 19 July each year at the Anzac Memorial, Sydney.

To the members of Unrecovered War Casualties Unit – Army, who continue to sort through submissions in the hope that a soldier's remains can be conclusively linked to those who have donated their DNA for cross-referencing in the quest for an identification.

I also wish to thank Tim Lycett, whose abiding interest in Fromelles saw that those who were originally listed as having been killed on 21 July, 1916 are now properly listed as having been killed on 19-20 July, 1916. While a formal acknowledgement of the success of his submission has not been forthcoming, we know that it was successful because all their headstones now reflect 19-20 July as their date of death. His work assisted me greatly in compiling my chapter on 'Cause and Effect - the Real Numbers'.

In the UK, I thank Paul Cobb, Chris Coade, and Paul Humphriss, who have all assisted me to identify British Prisoners of War, and to make sense of some historical material I would not have located otherwise. The numbers and names for British soldiers have come about due to their assistance.

Finally, I give my thanks to the man this book is dedicated to, Neville Thomas Kidd, who served in the 2nd AIF in WW2 (NX94329). Despite him having no familial connection with Fromelles, in his time he became its greatest champion. His book *An Impression Which Will Never Fade* detailed the story of Major Roy Harrison who survived Gallipoli but fell leading the charge at Fromelles.

I met with Neville for the first time at his home in 2013, and became absorbed into his quest for inscription of memorials where places of battle are honoured within Australia. His attempts to convince authorities, particularly at the Anzac Memorial in Sydney of this need were rejected time and again, because in the past the trustees of these memorials were not fully informed of the details that this book reveals. I hope that changes.

His quest and my book are forever connected. I was with him in his last days, and he passed away on the day I had written my final draft of what was to become my first book's Chapter 19, but before I had tackled the information which would become then Chapter 13, which was inserted afterwards, and in many ways has validated his quest.

In writing this book, I again dedicate it to the memory of Neville Thomas Kidd, a great Australian. He was right about Fromelles in so many ways, but didn't have the details that would have proven him to be so.

---ooo000ooo---

Geoffrey Benn

February, 2025

1. 'raids', 'battles', 'Battles', and Battle Honours

'When I use a word,' Humpty Dumpty said, in rather a scornful tone, 'it means just what I choose it to mean—neither more nor less.'

'The question is,' said Alice, 'whether you can make words mean so many different things.'

'The question is,' said Humpty Dumpty, 'which is to be master — that's all.'

LEWIS CARROLL (Charles L. Dodgson), Through the Looking-Glass, chapter 6, p. 205 (1934). First published in 1872.

---ooo000ooo---

William Polding Ryan, the soldier who is featured in the Foreword, was killed at Fromelles, France, during the evening of 19 July, 1916. He was killed in what was the worst day of casualties in Australia's military history. He had crossed No-man's land, and reached the German trenches, where he was killed by an exploding shell.

His death was recorded initially as having taken place during a '*raid*'. That word grossly understates the extent of the fighting that took place, and the purpose for it taking place that evening.

In later years, the fighting at Fromelles was classified as being part of an '*Attack*'.

But the fighting at Fromelles on the night of 19/20 July, 1916 should have always have been classified as being part of a '*Battle*'.

The meaning of the terms '*raid*', of '*Attack*', of '*battles*' of '*Battles*' and of '*Battle Honours*' along with others, will be expanded on in this chapter, and it is essential that we do so at this stage of this book, so that as a reader you are not confused by terminology in the course of reading what transpires throughout this book.

What Is a 'Raid'?

When reading background details for the Battle of Fromelles, you will frequently see reference to the term 'raid'. Raids are important in the context of this work, so much so that we can see that a 'raid' is likely to be classified as a lower level action in terms of significance than a 'battle' or a 'Battle'. We should therefore firstly consider what is meant by the term of 'raid'.

My everyday form of looking at the meaning of a word is to grab my Collins Australian Pocket Engish Dictionary[2], and look up the word. From this source, a 'raid' is described thus: -

[Picture 3: Collins Dictionary]

raid (rād) *n.* [< ROAD, in obs. sense "a riding"] **1.** a sudden, hostile attack, as by aircraft, bandits, etc. **2.** any sudden invasion by police, to discover violations of the law —*vt., vi.* to make a raid (on) —**raid'er** *n.*

[2] Collins Australian Pocket English Dictionary, William Collins Sons & Co., 1981, Reprinted 1986.

In the militaristic sense, a raid is also known as depredation, a military tactic or operational warfare mission which has a specific purpose and is not normally intended to capture and hold terrain, but instead finish with the raiding force quickly retreating to a previously defended position prior to enemy forces being able to respond in a co-ordinated manner or formulate a counter-attack.

A raiding group may consist of combatants specially trained in this tactic, such as commandos, or as a special mission assigned to any general troops. Raids are often a standard tactic in irregular warfare, employed by warriors, guerrilla fighters, or other irregular military forces.

The purposes of a raid may include:

- to demoralize, confuse, or exhaust the enemy;
- to ransack, pillage, or plunder;
- to destroy specific goods or installations of military or economic value;
- to free Prisoners of War;
- to capture enemy soldiers for interrogation;
- to kill or capture specific key persons;
- to gather intelligence.

Accounts of raids fill the history books, due to their use of small groups, in action behind enemy lines, often at night. Raids were the most common form of warfare between tribal societies. So it was normal progression for raids to be merged into military use with the formation or opposing armies. The largest raids in history were the series undertaken during and following the Mongol invasion of Central Asia.

What Is the Difference Between A 'battle', and A 'Battle'?

For many people, these terms may be referred to in speeches honouring the service of soldiers past and present, but for the most part their individual significance has not been explained, nor has their application when it comes into the realm of classifying the manner in which soldiers' actions are recorded for posterity, mourning, and respect.

It is necessary before we examine the content of this book to look at the definition of each term. In doing so, we will clarify those two different terms, their roles and their purposes. We will also in due course, be referring to 'Battle Honours' separately.

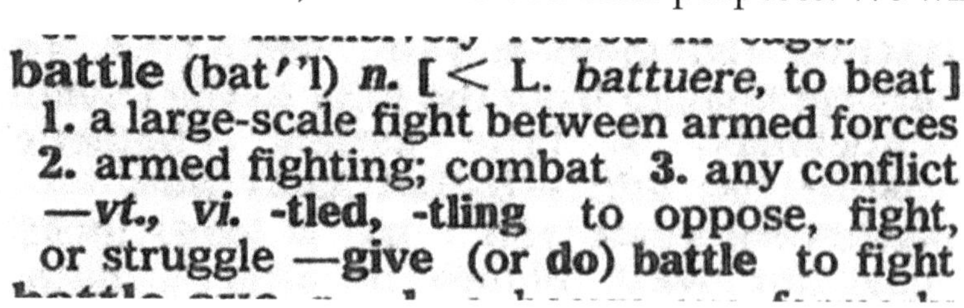
battle (bat′'l) *n.* [< L. *battuere,* to beat] **1.** a large-scale fight between armed forces **2.** armed fighting; combat **3.** any conflict —*vt., vi.* -tled, -tling to oppose, fight, or struggle —give (or do) battle to fight

[Picture 4: Collins Dictionary]

Again, using my same ever faithful Collins dictionary, you find that a 'battle' is defined thus: -

What are 'battles' (lower case 'b')?

We can consider the meaning of the simple word '**battle**', by reference to: -

(a) The ordinarily accepted layman's definition; and

(b) the Military definition.

In doing so, your author might seem to slip into using the capitalized form of 'Battle' from time to time. This is not accidental. This is specifically to signify that the particular incident being described has been *formally* described or classified for military historical purposes as a significant event warranting elevation above minor incidents. More on that in due course.

It thus appears, perhaps unexpectedly, that the ordinary form of 'battle' also embraces the military application of the term.

What Then, is a 'Battle' (upper case 'B')

This term, capitalized, is the end result of having been classified in accordance with various criteria. In this instance, classified by the Battles Nomenclature Committee, (the BNC) an Imperial creature, which was created originally to, and continues to, classify military engagements. The BNC's Report was prepared as a result of the committee being appointed in 1919, to tabulate the actions fought in The Great War 1914-1919, and to classify these actions with a definite system of nomenclature which would denote their relative importance. The work of the Committee will be examined critically in Chapter 12.

The BNC presented its report in 1922, with the following at clause 6: -

> *'The extension of the period covered by the so-called battles in Europe from days to weeks, and even to months, has introduced an entirely new factor, and the term 'battle' in the ordinary pre-war application of that word is obviously no longer sufficient. From the dispatches it is plain that the Commander-in-Chief's conception of such operations as the so-called 'Battle of the Somme' (1916) was a series of great blows each delivered with the full force available, after the necessary pause for preparation. The Committee have endeavoured to present them in this form by resolving them into their successive stages: - first the main attack – then the succeeding general attacks marking the opening of new phases. Each of these constitutes in itself a 'Battle' of the first magnitude, and the whole operation thus comes to form a Group of Battles.'*

The committee went on (in Clause 7) to state that: -

> *… the rank of 'battle' has been confined as a general rule to engagements of primary importance fought out between forces <u>not smaller than the corps</u> [emphasis added].*

Other than the Collins Dictionary referred to above, there are numerous other sources we can check for the definition of a battle. In some instances, however, the ordinary meaning of 'battle' and the militaristic meaning of 'Battle' appear to become blended. But the two are quite different in their usage, something which most people will not be familiar with.

Whereas a battle is defined as combat between opposing forces, in the military usage, battle is one where the major components of total forces committed to a military campaign are used *to achieve specific military objectives*.[3]

A **battle** is a combat in warfare between two or more armed forces, or combatants.

A war sometimes consists of many battles. Battles generally are well defined in duration, area, and force commitment.[4]

A series of planned battles directed to achieving a particular result are often called an **operation**.

A victory in battle is achieved when one of the opposing sides forces the other to abandon its mission, or to surrender its forces, or routs the other, i.e., one forces the other to retreat or renders the other militarily ineffective for further combat operations.[5]

While Battles can be found on land, sea, or in the air, this work will focus on the work carried out resulting in the Battle of Fromelles.

Battles are usually named after some feature of the battlefield geography, such as the name of a town, forest or river, commonly prefixed. Following the First World War, the British Battles Nomenclature Committee (the BNC) was formed to decide on standard names for all battles and subsidiary actions. There were instances whereby the descriptor given for a particular Battle did not appear to have a readily seen connection to the actual fighting. For example, a soldier fighting at Beaumont Hamel on November 13, 1916 was probably unaware he was taking part in what the committee would call the 'Battle of Ancre'.

This manner of referencing has unfortunately found the location of Fromelles, about which this book is based, on different geographic descriptors in three different years. So geographic descriptors alone are not sufficient.

There also arises a series of classifications whereby 'Battles' are not warranted, so other terms are used, such as 'action', 'skirmish', 'firefight', 'raid' or 'offensive patrol'. These are used to describe small-scale battle-like encounters. These combats often take place within the time and space of a battle and while they may have an objective, they are not necessarily 'decisive'.

---ooo000ooo---

Before World War 1 was even finished, articles appeared all over Australia such as the one published in *The Age, Melbourne on 3rd August, 1918*, under the heading '*Australia's War Record. Four Years' Work Reviewed.*'[6] In that article, a summary of the deceased, missing, prisoners of war, and other classifications of soldierly outcomes was posted, along with the number of those

[3] Dupuy, Trevor Nevitt (1992). Understanding war: History and Theory of Combat. London: Leo Cooper. ISBN 0-85052-293-5, p.67

[4] Dupuy, p.65.

[5] Dupuy, p.67.

[6] https://trove.nla.gov.au/newspaper/article/155139205?searchTerm=Australia%27s%20War%20Record.%20Four%20Years%27%20Work%20Reviewed%20%20%20%20%20%20%20%20%20%20%20%20&searchLimits=

by rank. There would be more to come once the war was over. It included reference then to what would become an early: -

RECORD OF PRINCIPAL BATTLES.

The following are the principal battles in which Australian troops have fought during the last four years: -

IN GALLIPOLI.

1915.

Landing at Anzac, 25th April

Krithia, 8th May.

Turkish attack, 19th May.

Lone Pine, 6th August.

Sari Bair, 6th - 31st August.

IN FRANCE.

The Somme Series.

1916.

Fromelles (sometimes called Fleurbaix,

sometimes called Laventie), 19th - 20th July.

Pozieres, Mouquet Farm, 23rd July - 6th

September.

Flers, 5th-14th November.

1917.

Ligny, 27th February.

Bapaume, March.

Capture of Boursies, 10th April.

Capture of Hermies, 10th April.

Capture of Dermicourt, 10th April.

Capture of Riencourt, 11th April.

Capture of Hindcourt, 11th April.

Capture of Beaumetz, Noreuil, Doignes.

March.

Lagnicourt, 15th April.

Bullecourt, 3rd - 10th May.

Messines, 6th-13th June.

About Ypres.

Ypres, 24th August.

Polygon Wood, 20th September.

Broodseinde and Passchendaele, 4th October.

1918.

Dernancourt, 28th March.

Villers-Bretonneux, 25th April.

IN PALESTINE.

1916

Katia, May.

Ogratina, 17th August.

Maghdaba, 23rd December.

1917.

Rafa, 7th-8th January.

Gaza, 23rd-29th March and 19th-22nd

April.

Beersheba, November.

Jerusalem, 9th December.

1918.

Jericho, 22nd February.

Amman, 26th - 31st March.

Note that even in 1918, two years after Fromelles, and three years before the BNC formulated its catalogue of classifications of battles and offensives, that Fromelles was shown in Australian newspapers as a separate battle, but at the same time it was part of 'The Somme Series' in 1916. The classification of the various Principal Battles would come only after the end of hostilities, in 1921, and will be dealt with in more detail later in this book. The related, but consequential issue of Battle Honours would not be finalised for Australian units, until 1926 and 1927.

What Are Battle Honours, and Why Are They So Signfificant?

> *A Battle Honour is a public commemoration of a battle, action or engagement, of which not only past and present, but future generations of the Regiment can be proud.*
>
> *Army Council Instruction 58, 28 January, 1956*[7]

Another way of expressing it is to say that a **battle honour** is an award of a right by a government or sovereign **to a military unit** to emblazon the name of a battle or operation on its flags ('colours'), uniforms or other accessories where ornamentation is possible.

We will look at the origin of Battle Honours in detail, in Chapter 13. But for now, a brief outline will assist.

In European military tradition, military units may be acknowledged for their achievements in specific wars or operations of a military campaign. In Great Britain and those countries of the Commonwealth which share a common military legacy with the British, battle honours are awarded to selected military units as official acknowledgement for their achievements in specific wars or operations of a military campaign. These honours usually take the form of a place and a date (e.g. 'Cambrai 1917').

Battle honours, theatre honours, honour titles and their ilk form a part of the wider variety of distinctions which serve to distinguish military units from each other.

According to Rodger[8], on 16 July, 1760 near the small German town of Emsdorff, a regiment formed only the previous year –Elliott's or 15th Light Dragoons – so distinguished itself in its first engagement that six years later the regiment inscribed its helmets with: -

> *Five battalions of French defeated and taken by this Regiment with their Colours and nine pieces of cannon at Emsdorff 16th July 1760.*

This inscription was confirmed by Royal Warrant dated 19 December 1768. According to Rodger, Emsdorff, although possibly not the first honorary distinction, is generally recognized as the first officially-named **battle honour**.

But it was not to be until in 1784 that infantry units were entitled to bear battle honours on their 'colours', i.e. their flags and insignia.

The need to develop a centralised system to oversee the selection and granting of battle honours arose in the 19th century following the increase of British military engagements during the expansion of the Empire. Thus in 1882, a committee was formed to adjudicate applications of battle honour claims. This committee, later called the Battles Nomenclature Committee, still maintains its function in the British Army today. In dealing with the classification of Battles, and ensuing Battle Honours, the following points have emerged: -

1. **Entitlement to Battle Honours**

---

[7] As quoted in Rodger: Battle Honours of the British Empire and Commonwealth Land Forces 1662-1991, p.9.

[8] See again, Rodger, p.9.

Battle honours are usually presented in the form of a name of a country, region, or city where the unit's distinguished act took place, usually together with the year when it occurred.

2. **Not every battle fought will automatically result in the granting of a battle honour.**

Conversely, a regiment or a battalion might obtain more than one battle honour over the course of a larger operation.

3. **<u>Nor does a Unit have to successfully defeat their adversary to earn a battle honour</u>**. (emphasis added)

The Hong Kong Volunteer Defence Corps received the battle honour 'Hong Kong' despite the defeat and capture of most of the force during the Japanese invasion of Hong Kong; and the cruiser HMAS Sydney was awarded the naval engagement honour 'Kormoran 1941' after being sunk with all aboard by the German raider Kormoran.

---oooOOOooo---

The purpose of these terms being explained at the beginning of this book is simply to show the differences, the distinctions, the qualifications involved, and the significance of the terms themselves. They are not interchangeable.

You will now be able to appreciate that *raids* are not considered to be as significant as '*battles*' and that not all '*battles*' are classified as '*Battles*'.

But Fromelles was indeed the site of a Battle. Not having been classified as a Battle has meant that Fromelles has not been able 'to be claimed' for a Battle Honour.

Inscriptions On Memorials Around Australia

What is generally not understood is that 'battles' do not normally appear inscribed on war memorials, but that 'Battles' regularly do. Because Fromelles was not ultimately classified as a 'Battle' it remains the case that it has not (with the exception of one memorial at Toronto NSW), been inscribed on the war memorials of any Australia Capital City.

Anyone travelling around Australia will invariably come across a war memorial in most suburbs and country towns, on which is inscribed the individual names of those who died, and who came from that town, district or city. Individual loss was measured on a local level, and not ascribed only to a place.

Somewhat differently, it is the custom of Australian authorities when erecting memorials to those lost in battle, to either make the inscription of such 'battles' as general and broad so that they can be all-inclusive, or to refer to specific fields of war, which appear to be referenced solely by the supposed Battle Honours as their means of honouring the fallen. Consequently, many other actions or battles are not included.

For example, the war memorial erected in Kings Park, Perth, Western Australia, refers to 'The Glorious Dead'. Taking the most inclusive approach, it does not refer to any individual place of battle. Thus, it is by default, inclusive in reference to any soldier who was killed.

Conversely, the Australian monument erected at Villers-Bretonneux Memorial features a long wall in four sections where the commemorated names are carved on stone panels in battalion order, and within each battalion the names are listed alphabetically by rank. This is to honour those who served in France in World War 1.

Taking the specifics approach, the long wall lists many places of battle. It has inscribed on the top of its walls, the following:

SOMME 1916-18	POZIERES	BAPAUME 1917	ARRAS 1917
BULLECOURT	MESSINES 1917	YPRES 1917	MENIN ROAD
POLYGON WOOD	BROODSEINDE	POELCAPPELLE	PASSCHENDAELE
AVRE	ANCRE 1918	VILLERS-BRETONNEUX	LYS
HAZEBROUCK	HAMEL	MARNE 1918	AMIENS
ALBERT 1918	CHUIGNES	MONT ST. QUENTIN	
HINDENBURG LINE	EPEHY	ST. QUENTIN CANAL	BEAUREVOIR

Note that it does not contain the name of every place which would have been called a significant place of battle such as was listed in The Age, Melbourne on 3rd August, 1918, e.g. Fromelles.

It is submitted in the course of this book that places of homage such as the Villers-Bretonneux memorial, and the ANZAC War Memorial, Sydney, which appear to honour by reference to Battle Honours, have by doing so, effectively excluded those who served in Fromelles.

It is trite to state that if one seeks to honour those who fell at Fromelles by honouring them under the Battle Honour of 'The Somme 1916', that this continues to uphold the absurdity that honour is being given to a Battle which occurred in another location (French Flanders), in a completely different geographic and administrative region of France, and which is some 80 kilometres from The Somme.

It is often said that all dachsunds are dogs; but not all dogs are dachsunds. And so it is with battles, Battles, and Battle Honours. For while all Battle Honours originate from battles, not all Battles are awarded Battle Honours. Nor it seems have appropriate battles become classified as Battles.

So it is with Fromelles and the Somme. While Fromelles can fairly be said to be part of the overall ***Operation*** or ***Campaign*** of The Somme, Fromelles is in no way geographically connected to or part of the Somme. Those who died at Fromelles did not die in or on 'the Somme', and nor will they ever be able to be properly or meaningfully classified by reference to geographical considerations.

The Somme and Fromelles are not part of the same province, the same region, the same river, or the same geographic system. While what I will be calling the Battle of Fromelles from this point on is not *officially* cited as a Battle, you will learn why it should always have been so called. Only a Command-level process of deliberate lying has allowed that to happen.

---ooo000ooo---

In coming chapters, you will read that the planning for the Battle of Fromelles underwent many revisions, and when finally commenced, it conformed to the normal terminology requirements of a Battle in that our soldiers were instructed ***to attack German positions and to hold them.***

A raid, demonstration or attack simply fails to fulfil this criteria. Simply put, the Battle was a disaster in planning and execution, with the lives of our soldiers sacrificed for no tactical gain.

The disparity in the treatment and classification of Fromelles was the subject of a motion, adopted unanimously on 25 July, 2015, at the Annual General Meeting of the Fromelles Association of Australia, held at Cessnock NSW, that: -

'Our Association calls on the Australian Government to formally record that those who served and were killed or wounded at Fromelles during the evening of 19 & 20 July, 1916 were in fact killed or wounded in a Battle during World War 1.

Having been so involved, that 'Fromelles' must be recorded and inscribed on war memorials honouring the service of soldiers who served in World War 1.'

---ooo000ooo---

It is controversially contended by various authors that Fromelles was or was not a success in its stated purpose. In the course of the research for this book, it became apparent that there were different understandings of just what the purpose of the Battle at Fromelles actually was. That will be discussed and revealed in coming chapters.

What hasn't been explained up until now is why Fromelles has not been classified as a Battle. In the coming chapters, I will seek to do that, and to point out the absurdity of that failing, because: -

- Those soldiers and their families were treated with disrespect because of it.
- Those soldiers died or were injured because of incompetent / disgraceful leadership.
- We now know about Fromelles in detail, whereas the previous generations were unable to obtain the degree of information which we are now privy to.
- The exhumations and identifications of those recovered in 2009 ensure its front page.
- The Australian Government through its members, agencies and servants constantly has, and continues to identify and confirm the wording: 'The Battle of Fromelles' without seeking to have it formally re-classified as a Battle.

- It is laughable that it is not now, nor has ever been officially designated as a Battle. Those who served deserve due recognition, and to not be consigned to some 'cutting room floor'. And:
- It is plainly and simply wrong to continue to pretend otherwise.

---ooo000ooo---

2. The First Battles at and Near Fromelles

– An Introduction.

In August 1914, Secretary of State, Lord Kitchener issued instructions to Sir John French. Something of a public declaration it had among its 500 words a few sentences applicable to the subject of this book.

> *It must be recognized from the outset that the numerical strength of the British Force and its contingent reinforcement is strictly limited, and with this consideration kept steadily in view it will be obvious that the greatest care must be exercised towards a minimum of losses and wastage.*

And further on:

> *In minor operations you should be careful that your subordinates understand that risk of serious losses should only be taken where such risk is authoritatively considered to be commensurate with the object in view.*

And:

> *...officers may well be reminded that in this their first experience of European warfare, a greater measure of caution must be employed than under former conditions of hostilities against an untrained adversary.*

As the events of the Great War unfolded on the Western Front, he might just as well have kept his thoughts to himself. His leaders do not seem to have paid attention to his recommendations at all.

---ooo000ooo---

The Battle which took place at Fromelles in 1916 was not the first time in which that location, or that area, had been the subject of a military conflict during WW1. Indeed, the area has been the subject of conflicts over many years preceding the war itself.

We need to look into the area surrounding Fromelles in historical terms before we can fully appreciate the impact of what took place there during the events of 19/20 July, 1916. In doing this, we will begin by looking at the area from 1914 and up to the preparations for the actual engagement in July, 1916.

If you walk along the road sometimes called the Rouges-Tommybrucke Road, but in reality the Rouges Bancs Road, near Fromelles, you will find a quiet agricultural area, with farms neatly tended. But the area referred to in Barton's book, The Lost Legions of Fromelles, as being 'astride the Rouges-Tommybrucke road' was the scene of a slaughter of British soldiers during the action known as the *Battle of Aubers Ridge, 9 May, 1915*. That ground is today partly occupied by a memorial park within which lie the rumbled concrete remnants of the old German front-line defences and a larger-than-life bronze representation of an Australian soldier carrying a wounded comrade from the field of battle.[9]

Neither the park nor the sculpture, however were designed to commemorate the 1915 action, but one that was fought at the same location more than a year later: the encounter of the 19/20 July 1916 known to the Australian nation as the Battle of Fromelles. The neatly kept patch of lawns and paths is the Australian Memorial Park, and the statue is the celebrated 'Cobbers Memorial'.

Some 300 metres away, to the north-west and located in the old No-man's land, lies the VC Corner Cemetery and Memorial. Within its walls lie the remains of 410 unidentifiable soldiers. The cemetery's name has its origin in a yet earlier British enterprise: what one might call the *'first' battle of Fromelles* which took place at the birth of trench warfare, in December, 1914.

After the war, during the period of official recovery between 1918 and 1921, the bodies of soldiers who died in all three actions were gathered from the fields and reburied. In preparation for the VC Corner Cemetery, however, each body was carefully identified by nationality and equally carefully separated to make certain that *only* Australians were interred there. Engraved upon a screen at the cemetery's rear are the names of 1300 missing Diggers – men with no known grave. The battlefield upon which the cemetery and screen were established has since become *their* battlefield – an Australian battlefield.

To learn the names of missing British soldiers who died there between August 1914 and November 1918 one must visit the Loos or Ploegsteert Memorials, each more than 20 kilometres from the site.

At neither the memorial park nor the cemetery is there reference to any pre-July 1916 or post-July 1916 actions, nor indeed mention of the endeavours of any nation but Australia. These omissions mean that few Australian visitors are aware of the many other events that took place precisely where they stand – and the numerous chilling similarities between each one. If they were, their appreciation and understanding of what happened to their forefathers in July 1916, and indeed *why* it happened, would be greatly enhanced.

The three major encounters at Fromelles – those of December, 1914, May 1915 and July 1916 – were fought in exactly the same sector and upon exactly the same ground.

In **1914** French, Indian and British troops fought side by side here, stemming the German advance during the *Battle of Armentieres*, and thus playing their part in the creation of the impasse that was fully established during that December. There then followed the earliest set-piece encounters of the period, plus an extraordinary and well-documented Christmas Truce.

The **9 May 1915** battle provided a bloody and bizarre rehearsal for 19 July 1916. In late September 1915, a string of diversionary assaults were launched at nearby Bois-Grenier, Neuve Chapelle and Givenchy to assist the grand but disastrous British push at Loos, itself designated

[9] See Barton p.1

to support an even grander but equally ill-fated French offensive in Champagne. These assaults offered more valuable learning opportunities for the Germans. [GVB: But sadly the British do not appear to have taken advantage of the learning opportunities.]

The latter two actions unfolded in almost the same way, and on each occasion the final German *coup de grace* took place at precisely the same location, astride the Rouges Bancs road and within the boundaries of today's Australian Memorial Park. Even more uncannily, the number of missing as a result of the British attack in 1915 is almost identical to the Australian figure for July 1916: approximately 1300 men. The most important factor of all, however, is how the lessons learned in 1915 influenced the catastrophe of the following summer.

The British appear to have learnt nothing. The methods of war did not change. The Germans however, put their lessons to good use. General Erich von Falkenhayn, the German Chief of the General Staff had ordered a construction programme on the Western Front in January 1915, to make it capable of being defended indefinitely by a small force against superior numbers. An elaborate, carefully sited and fortified front position was built behind fields of barbed wire, with camouflaged concrete machine-gun nests and a second trench close behind the front trench to shelter the trench garrison during bombardments. To evade Allied artillery-fire intended to obstruct the movement of reinforcements from the new rear defences, communication trenches were built. The front position was to be held at all costs as the main line of resistance but in May 1915, Falkenhayn ordered a reserve position to be built along the Western Front, 2,000–3,000 yds (1,800–2,700m) behind the front position, out of range of enemy field artillery. To contain a breakthrough, the second position was to be occupied opposite a sector broken into and serve as a jumping-off point for by counter-attacking troops. If counter-attacks failed to recover the front line, the rear position could be connected to the remaining parts of the front line on either side.

The construction programme was a huge undertaking, which took until autumn 1915 and had several opponents, notably the 6th Army commander Crown Prince Rupprecht, who claimed that a rear position would undermine the determination of soldiers to stand their ground.

After Falkenhayn's defensive building program, there would be 75 shelters with 9 – 12 inches (230 – 300mm) of concrete protection.

The last allied blood to be shed at Fromelles was that of the 47th (London) Division, when they finally liberated the village in October, **1918**. But before that the sector had been occupied by Portugese troops and consumed by the colossal German spring offensive known as Georgette, a firestorm that came perilously close to concluding the war in favour of the Central Powers. In other words, from the first moment of fighting in August/September 1914 to the day when the guns fell silent, there was sniping, shelling, gassing, tunneling, aerial combat, patrols, raids, a host of associated nearby actions, and of course the three key closely related encounters of 1914, 1915 and 1916. Amalgamation of casualties sustained in this sector throughout the war makes it one of the bloodiest stretches of battlefield on the entire Western Front. There is therefore a wider and greater tale to be told, a tale that complements 19 July, 1916 by adding not only colour but essential perspective to an action that – solely because of the Fromelles Project – has now become one of the highest profile episodes of the Great War.

At present the history of the Fromelles sector is at best truncated, at worst hidden completely. The danger of obscuring the wider context of conflict in favour of a single fourteen-hour event in 1916 is a real one. No battlefield that witnessed four years of positional warfare should be seen as frozen in time.

18 December, 1914 – The First Battle at Fromelles.

It was in October 1914 that the British found themselves in the Fromelles and Aubers region.[10] Having departed the now-stabilising Aisne and Marne battlefields, marching by night and resting by day they had headed north to try to arrest further German advances, on the Somme, in Artois, and then in French and Belgian Flanders. It was the period known as the 'Race to the Sea' when the Germans, baulked in their hopes of excising Paris from the east, sought to outflank their enemy and swing down upon the capital from the north. The fighting was to surge further and further northwards until it reached the coast. When Belgian engineers opened sluices at Nieuport to flood the great reaches of the polder plain with sea water, the Germans were forced back inland to the nearest high and dryer ground – at Ypres. Clashes in the neighbouring Lille region formed a part of this unpredictable and shifting area.

On 20 October, elements of the British 19th Brigade, alongside troops from General Louis Conneau's Mounted Division, some on bicycles, patrolled the ridge between the villages of Aubers and Radinghem, looking to push forward.

On 21 October German troops in field grey appeared *en masse*. Their approach was announced by gunfire, the heavies that many had yet encountered. Both Briton and Frenchman were forced into withdrawal.

On the morning of 23 October, a defensive line was established between the settlement of La Boutillerie and a small area encompassing a tiny hamlet on the flat plain below Fromelles village known as Rouges Bancs. Here, the troops dug primitive trenches and used roadside ditches for cover. To east and west, the Germans now owned all of the low, mean, but significant ridge in front of them as far as the eye could see. For the next two weeks, a sequence of hostile assaults was beaten off, enemy fortunes often being hampered by heavy mists, typical of the area. The enemy was digging in.

On 15 November, 1914, the 2nd Scots Guards took over the Cordonnerie Farm sub-sector and on the 27th a small party led by Lieutenant Sir Edward Hulse crept through unharvested turnips to make the first British *raid* in the neighbourhood.

The lines were now firmly fixed and the Germans were already installing a designed defensive infrastructure. The great siege had begun.

[10] Barton p.24.

The December 1914 action at Fromelles was the very first set-piece attack in the sector.[11] The battleground selected lay between the Sailly-sur-la-Lys-Fromelles road (the Rue Delvas, later christened the *Tommy-Strasse* by the Germans) and a track south-east of La Cordonnerie Farm. It was exactly the area that would be assaulted by the British 8th Division on 9 May the following year, and again by the 5th Australian Division in July 1916, over a year and a half hence.

The orders given to the 2nd Scots Guards and the 2nd Border Regiment (Borders) on 18 December 1914 were chillingly similar to those issued to the Australians in July 1916 – and so was the way the action unfolded. Their attack was also timed for 6.00pm, *but being European mid-winter, this made it nocturnal.*

The troops, carrying spades and sandbags to consolidate the expected gains, were required to assemble in No-man's land and upon the whistle rush and hold the German front trench.

Should it be unoccupied they were to push on to the second line.

The Scots Guards assumed responsibility for the right-hand sector facing Rouges Bancs, which includes the area today occupied by the Australian Memorial Park, while to their left the Borders attacked positions immediately west of Delangre Farm.

The action was unsuccessful. The cross fire from well placed German machine guns played a big part, accounting for very heavy casualties. Despite the menace which came from the use of machine guns, future allied leaders, in particular General Haig, were not inclined to regard them with appropriate significance at the time.

VC Corner

The battalion's casualties were approximately 180 killed and wounded – 50 percent of the force. The Borders were left with 128 casualties, several of whom still lay in No-man's land. *For voluntarily bringing in wounded 'by daylight on 19 December, 1914 under heavy rifle and machine-gun fire,' Privates Abraham Acton and James Smith, both of the 2nd Border Regiment, were awarded the Victoria Cross – the origin of the place name 'VC Corner'.*

On the morning of 20 December, soldiers from both sides by an almost tacit understanding as to no firing, arranged to bring in bodies of soldiers who had been left in No-man's land up until that time. Five days later, on 25 December, 1914, on exactly the same piece of ground one of most celebrated events of the war took place: the Christmas Truce.

How Is the Action Of 1914 Classified in The Battle Accounts?

The Report titled: "Army, The Official Names of the Battles and Other Engagements Fought by The Military Forces of the British Empire During The Great War, 1914 – 1919, and the Third Afghan War, 1919 [the BNC Report] contains **no reference** *to any action taking place on 18 December, 1914.*

The Attack on Wytschaete, on 14 December 1914 by II Corps, 3rd Division is listed.

[11] Barton p.32.

The Defence of Givenchy, 1914, 20-21st December by Indian Corps: 1st Lahore, and Meerut Divisions is listed.

But there is no reference to any Battle, Action, or Miscellaneous Incidents on 18 December, 1914.

Compounding this injustice, in "A Record of the Battles and Engagements of the British Armies in France and Flanders, 1914-1918", by Captain E.A. James, there is no reference to a battle of any description taking place on 18 December, 1914.

Was this an oversight? Or was it a case of just not thinking it was of any significance?

[Picture 5: Angus McBride, original artwork from Look and Learn No. 415 (27 December, 1969) © Bridgeman Images. While the image includes reference to a game of football, there is scant evidence to support the claim that any such game took place at Fromelles.]

10-13 March, 1915 – The Battle of Neuve Chapelle

The meeting of the British War Council of 13 January 1915 made 2 decisions that would have profound impacts upon the Australian Imperial Force and families in Australia.

The War Council established a sub-Committee, chaired by Lord Kitchener, to consider the employment of a military force in a new theatre of war, which focussed initially on the Balkans, but in response to an appeal from the Russian Tsar, Grand Duke Nicholas, eventually settled on a naval attack on the Dardanelles. The War Council also considered the Western Front campaign strategy for 1915.

One of the lessons of the fighting on the Western Front in 1914 was that while infantry positions could be improved by means of trenches and defensive positions, nothing could replace or compensate for the tactical advantage of high ground for steady and

continuous artillery observation of the enemy's position. Strategically, in the campaign planning for 1915, the capture of Lille, the fourth largest town in France (and identified on the map below), was regarded as a necessary prelude to any significant advance by the British Expeditionary Force (BEF).

As Aubers Ridge provided direct observation of the railway lines leading from the east and southeast of Lille and on the town itself, the capture of Aubers Ridge was of great strategic importance. In addition, an offensive against Aubers Ridge could be co-ordinated with and receive support from the prospective French 10th Army offensive on the Artois plateau further south (Vimy Ridge) and together the two attacks would threaten the rail, road and canal junction of La Bassee from north and south.

On 15 February 1915, Sir John French, Commander-in-Chief asked General Haig to draft schemes for an offensive by the First Army with La Bassee and Aubers Ridge as its objectives.

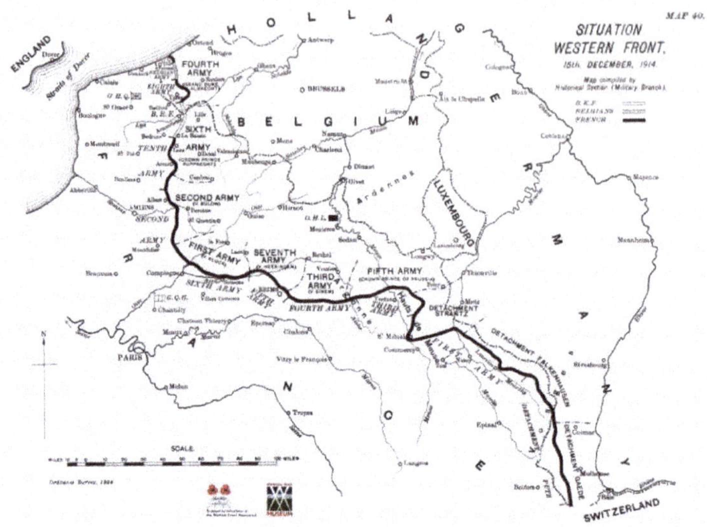

[Picture 6: Map 1: Location of the Western Front 15 December 1914[12]]

Reproduced in accordance with the Open Government Licence [13]

The Germans had been in possession of the small village of Neuve Chapelle since October, 1914. It lay some three kilometres south-west of Aubers, the lines protecting it forming a shallow salient. The prospective battleground was a narrow 3 kilometres, and the first enemy line of defence just a single lean breastwork protected by apparently derisory wire entanglements. Behind it lay further trenches, all unfinished and in the process of development.

With the breaking of the spring and the drying out of the water-soaked meadows of Flanders, an era of larger more ambitious operations had set in, involving, this time, little change of position, but for larger forces on the side of the British. The first hammer-blow of Sir John French was directed, upon 10 March, 1915, against the village of Neuve Chappelle which had … changed hands several times, and eventually remained with the Germans during the hard fighting … in the last week of October, 1914. The British trenches had been drawn a few hundred yards to the west of the village, and there had been no change during the last four months.[14]

Behind the village was the Aubers ridge, and behind that again the whole great plain of Lille and Turcoing. This was the spot upon which the British General had determined to try the effects of his new artillery.

The troops chosen for the assault were General Rawlinson's Fourth Army Corps upon the left and the India Corps upon the right, upon a front of half a mile, which as the operation developed broadened to three thousand yards. The object was not the mere occupation of the village, but

[12] Sources: *Official History of the Great War: Military Operations: France and Belgium 1914, Volume 2 Map 40* [extract from *The Western Front Association: Mapping the Front in association with the Imperial War Museum*].

[13] http://www.nationalarchives.gov.uk/doc/open-government-licence/version/3/

[14] Corfield p.23

an advance to the farthest point attainable, a line beyond Aubers that extended five miles across the ridge to the hamlet of Rouges Bancs below Fromelles.

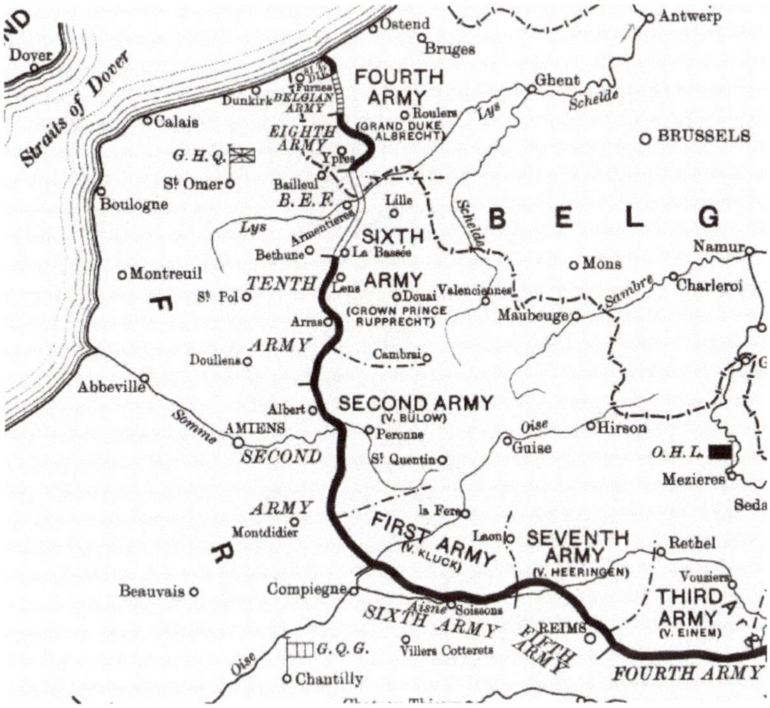

[Picture 7: Map 2: Location of the Western Front 15 December 1914[15]]

The obstacle in front of the Army was a most serious one. The barbed wire entanglements were on an immense scale, the trenches bustling with machine guns, and the village in the rear contained several outlying houses with walls and orchards, each of which had been converted into a fortress. Barbed wire represented a slowing down of any advance, and that made attacking soldiers vulnerable. Heavy artillery fire was designed to break up the wire entanglements, but was spectacularly unreliable in doing so.

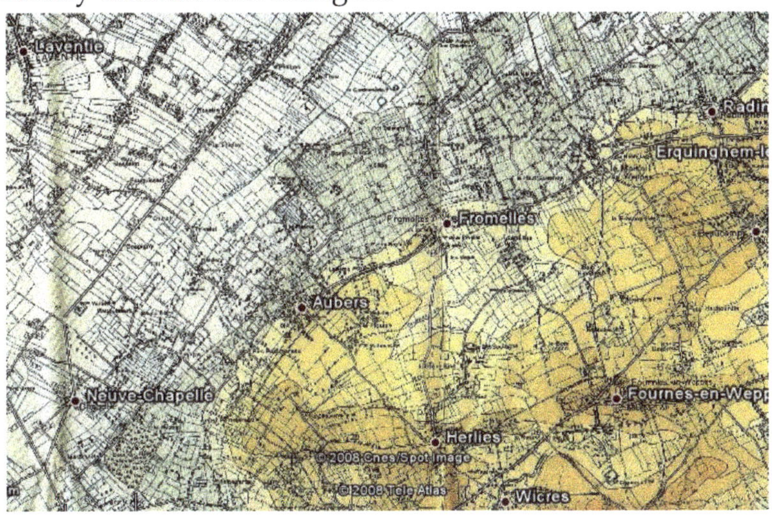

[Picture 8: Map 3: The Battle of Aubers Ridge – Location map showing topography with River Layes lowlands overlooked by Aubers Ridge.[16]]

This was the first time that an artillery "barrage" was ordered and the first time a timetable was issued giving each battery a purpose and target for each phase of the bombardment. It was also the first time "objective maps" were prepared with a "red line

---

[15] Sources: *Official History of the Great War: Military Operations: France and Belgium 1914, Volume 2 Map 40* [extract from *The Western Front Association: Mapping the Front in association with the Imperial War Museum*] Reproduced in accordance with the Open Government Licence. http://www.nationalarchives.gov.uk/doc/open-government-licence/version/3/

[16] In the *Official History of Australia in the War of 1914-1918 Volume 3, The AIF in France 1916*, the *Riviere des Layes* is presented as the *Laies River. Reproduced in accordance with the Open Government Licence. http://www.nationalarchives.gov.uk/doc/open-government-licence/version/3/*

Sources: Google Earth & *Official History of the Great War: Military Operations: France and Belgium 1915.*

objective" and "blue line objective" etc. These became standard practice for all subsequent battles of the war.

After a huge artillery barrage by some three hundred guns firing without pause for 30 minutes, the infantry went forward. As shell shortages limited the bombardment to thirty minutes, surprise was complete and Neuve Chapelle soon fell.

Those on the right got to the German trenches, and then to the village. Those on the left however found much of the wire intact and suffered horrendous losses, some battalions reduced to 150 or less, and most of the officers killed. The Indian Corps in their first large scale attack had no better luck, and yet it was 3rd Brigade of Gurkhas and the rifles of the 25th English Brigade that took the town by midday. But misunderstandings and delays in deploying reserves derailed the advance to the final objective. The British and Indians pushed on before being halted near the Laies River. They had gained a thousand yards.

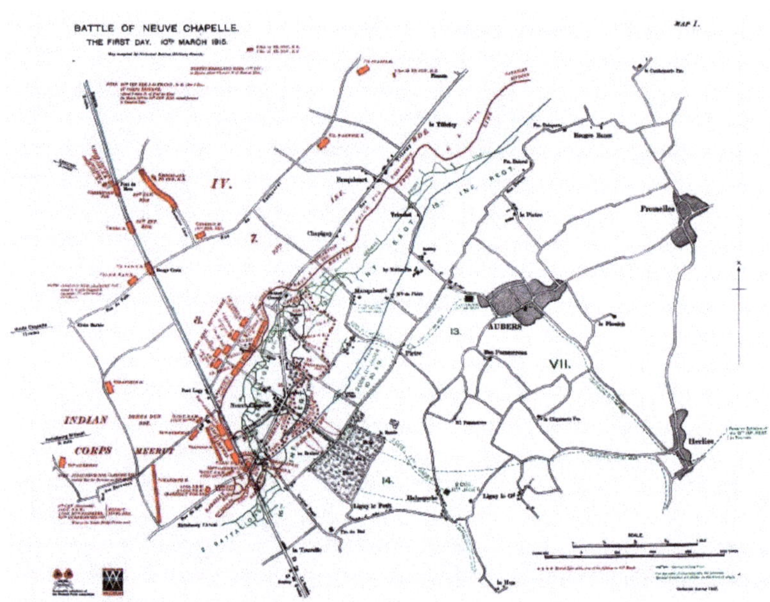

[Picture 9: Map 4: Battle of Neuve Chapelle, 10 March 1915[17]]

Shocked at how easily Neuve Chapelle had been lost and aware that the ridge might have followed but for British mix-ups, the Germans acted quickly. After the 13th and 14th Divisions sidelipped westwards, the 6th Bavarian Reserve (BR) Division, which had been rushed from Lille during the battle, took over the 8,000 yards of the line opposite the Fauquissart-Bois Grenier sector. It comprised 16 Bavarian Reserve Infantry Regiment (BRIR) – also known as the List Regiment after its short-lived first commander – and 17 BRIR from 12 BR Brigade and 20 and 21 BRIR from 14 BR Brigade. As three divisions now occupied an area previously held by two, the regimental fronts shrank from 3,000 to 2,000 yards, yielding the manpower needed to beef up the defences.[18]

On 11 March, a thick all day mist prevented any artillery work or infantry advance. It however provided the Germans with a very necessary opportunity to rebuild and reinforce. The 12th was a better day weather-wise and the Germans made several strong counter-attacks, but decided

---

[17] Sources: *Official History of the Great War: Military Operations: France and Belgium 1915, Volume 1 Map 1* [extract from *The Western Front Association: Mapping the Front in association with the Imperial War Museum*]. Reproduced in accordance with the Open Government Licence. http://www.nationalarchives.gov.uk/doc/open-government-licence/version/3/

[18] Pedersen p.17

to settle into new positions – *roughly the position that they were in when the Battle of Fromelles would begin on 14 July, 1916.*

Further to the north, opposite Fromelles, the line was held by the Canadian Division. Neuve Chapelle was captured and the British line advanced more than half a mile (1,000 yards) before German counter attacks stabilised the front.

On 13 March, 1915 at 10.40am, the German reinforcements made their first attempt to take back Neuve Chapelle. But they were stopped on the edge of the village, exhausted, starving, and all their officers dead.[19]

Haig now turned his artillery on Aubers Ridge and ordered the infantry to storm the ridge. The Germans had machine guns everywhere and as more British tried to get to Aubers they were cut down and then there were just no more reserves to call upon, and the battle came to an end.

The losses at Neuve Chappelle were for the British 583 officers and 12,309 men: 7,500 from the Fourth Corps, 4,000 from the Indian Corps. The Germans lost in the vicinity of 18,000. The reward was a slice of ground no larger than a moderate farm. Despite these casualties, the British Command concluded that success in trench warfare was dependent upon artillery fire. They achieved nothing except one would have hoped, a very healthy respect for the German machine guns and artillery on Aubers Ridge.

How Was The Battle Of Neuve Chapelle Classified In The Battle Accounts?

The BNC Report refers to the *Battle of Neuve Chapelle*, from 10 – 13 March, 1915. The geographical information refers to Road Richebourg-l'Avoue-Croix Barbee-Pont du Hem-Fauquissart-Aubers.

Captain E.A. James in his Record of Battles and Engagements follows the same battle reference.

9 May, 1915 – Another Attack On Aubers Ridge.

Less than two months after the Battle of Neuve Chapelle, on 9 May, 1915, there began another attack on Aubers Ridge.

The battle was part of the British contribution to the Second Battle of Artois, a Franco-British offensive intended *to exploit the German diversion of troops* to the Eastern Front.

French Commander Joffre's tactical objective for the Artois attack was to seize the heights 140 - 132 at Vimy Ridge.

The French Tenth Army was to attack the German Sixth Army north of Arras and capture Vimy Ridge, preparatory to an advance on Cambrai and Douai.

The plan was for a principal French attack between Carency and Roclincourt, supported by a flank attack to the north with the immediate objective of capturing the heights of Notre Dame

[19] Corfield p.22

de Lorette, then pushing forward into the Douai plain; and a flank attack to the south aimed at the heights at Points 96 and 93 east of Arras. (The latter being west of Bailleul Sir-Berthoult).

The British assault at Aubers would go in on the day after the main attack.

[Picture 10: Map 5: The Battle of Aubers Ridge, 9 May 1915.[20]]

The British First Army on the left (northern) flank of the Tenth Army, was to attack on the same day and widen the gap in the German defences, expected to be made by the Tenth Army and *to prevent German troops from being moved south of La Bassée canal.*

The British approached Aubers with optimism.[21]

The line was from Bois Grenier south to Festubert, with the greatest thrust at Rouges Bancs just in front of Fromelles. Beginning at 5.00am on Sunday morning with a bombardment which was much less than it should have been because ammunition was in short supply, it was an attack that never had a chance of success. Since the last attempt, the Germans had worked hard at improving their defence system.[22]

Detailed operational planning was placed into the hands of Sir Douglas Haig's First Army. Neither this or the overall strategic plan was notably affected by the German gas attack north of Ypres on 22 April, with the exception of various units of First Army being moved or placed on standby to move to Ypres. Indeed, Allied resolve was strengthened by the launch of the German Gorlice - Tarnow offensive in the east on 1 May and the sinking of the "Lusitania" on

---

[20] Sources: *Official History of the Great War: Military Operations: France and Belgium 1916, Volume 2 Map 2. Reproduced in accordance with the Open Government Licence. http://www.nationalarchives.gov.uk/doc/open-government-licence/version/3/*

[21] Barton p.42

[22] Corfield p.23

7 May. Plans began to take shape, and units began to move into position from early April. The British attack at Gallipoli was launched on 25 April 1915, opening another front and placing an additional (and growing) demand on already-stretched military resources.[23]

British Order of Battle: I Corps, 1st Division[24] and 47th (1/2nd London) Division. IV Corps

7th Division and 8th Division. Indian Corps, 3rd (Lahore) Division and 7th (Meerut) Division.[25]

There were to be two separate assaults, both pressed by First Army divisions under Sir Douglas Haig. The southern enterprise, once more in the Neuve Chapelle sector, involved a battlefront of 2.5 kilometres. Here four divisions were to break the line and surge eastwards.

At Fromelles, two British divisions prepared to assault the enemy trenches on a front of 1.5 kilometres.

The battleground was boundaried by two features: the Sugar-loaf Salient, a small outward bulge in the enemy line, and Delangre Farm, a moated farmstead some 100 metres within German territory.

At Fromelles, the British 8th Division under the command of General Rawlinson attacked the 6th Bavarian Reserve Infantry Division that had taken over part of the front line in March. The German Division included the 21st, 16th and 17th Bavarian Reserve Infantry Regiments (BRIR) that would oppose the Australian 5th Division and British 61st Division 14 months later.

Following a south-westerly angle of penetration, the plan was for two brigades of the 8th Division to swiftly seize and consolidate the ground up to and including Fromelles.

Through this breach the supporting 7th Division would drive beyond the village, swarm across the meagerly defended plateau beyond, and converge with the Indian Corps from the southern attack, thus enveloping Neuve Chapelle, Aubers and Fromelles.

In reserve, ready to exploit success at the very earliest opportunity, lay three infantry divisions and two corps of cavalry. The Royal Flying Corps were tasked to have machines constantly in the air, updating infantry commanders.

Zero hour for both attacks was set for 5.00am, and each was to be preceded by a forty-minute bombardment. A total of 190 artillery pieces had been allocated.

The barrage did little to undermine the German defences and there was concern over the quality and quantity of shells provided for the artillery (which contributed to the fall of the Asquith Government and Lloyd George becoming Prime Minister). But it did nothing to save the rest of

---

[23] See: http://www.1914-1918.net/bat11.htm

[24] The commander of the 1st Division at the time of the Battle of Aubers Ridge (also referred to as the first Battle of Fromelles) was Maj.Gen Richard Haking – later commander of the IX Corps, in command at the Attack at Fromelles 19/20 Jul 1916. At the Battle of Aubers 9 May 1915 the attacking brigades of his 1st Division lost over 50% of their fighting strength in little over an hour.

[25] See: https://en.wikipedia.org/wiki/Battle_of_Aubers_Ridge

the British and Indian Assault further south. There was a breakthrough at Festubert at very high cost to both sides, the Germans desperate to check any incursion onto Aubers Ridge.

Haig again was directing the attack. As a man he did not attach real importance to machine-guns, *'the machine-gun is a much over-rated weapon and two per battalion is more than sufficient'* (14 April, 1915). Haig was to find after two hours that the German General relied totally on them with spectacular success.

To the south of Neuve Chapelle, the 1st Division Commanded by Lt Gen Sir Richard Haking attacked the 13th and 14th Divisions.

The British I Corps, IV Corps and Indian Corps (Lahore and Meerut Divisions) attack was timed to coincide with a major French attack in the second battle of Artois, attacking Vimy Ridge. The attack was preceded by a significant artillery bombardment on defences that were then described as "well-developed".

So after desperate efforts throughout the day, at 5pm the offensive was called off and the British and Indians counted the cost – 11,600 officers and men. The abject failure was put down to a lack of high explosive shells and as a result of The Times correspondent reporting this fact, a great turmoil began in the British Government, culminating in a huge shake-up in munitions manufacture and supply, and the government itself.

The Bavarian narrative states that 'Burying the British dead took a week'. So it had after the action of December 1914, and so it would in July, 1916. British 8th Division losses in the attack at Rouges Bancs on 9 May 1915 were 4682. In their assault upon the Sugar-loaf the 2nd Northants lost 426 men.

The combined casualty count for both 9 May attacks was over 11,500 killed, wounded, or missing.

In the final analysis, this battle of 1915 was an unmitigated disaster for the British army-l No ground was won and no tactical advantage gained. It is doubted if it had the slightest positive effect on assisting the main French attack 15 miles (24 km) to the south. The British Official Historian, J. E. Edmonds recorded British casualties as 11,619 men. - Edmonds, J. E. (1928). *Military Operations France and Belgium, 1915: Battles of Aubers Ridge, Festubert, and Loos*. History of the Great War Based on Official Documents by Direction of the Historical Section of the Committee of Imperial Defence, pp. 37–41.

---ooo000ooo---

The 1915 battle concluded in exactly the same place and in exactly the same way as the 1916 encounter would: in and around the site now occupied by the Australian Memorial Park. [26]

The Battle of Aubers Ridge would especially influence the fate of the 5th Australian Division, for on 9 May, 1915 the Tommies of the British 8th Division attacked the same battlefront and indeed temporarily occupied the same German positions as would the Diggers of 1916 – almost to the

[26] Barton p.54

metre. In critical addition, although the strategic objectives were different, the tactics employed on both occasions were practically indistinguishable. As a result, the two battles unfolded in alarmingly similar fashion. [27]

Consequently, there is a tendency for some commentators to say that Lt Gen Haking commanded the same ground in 1915 and 1916 and therefore was doubly responsible.

Although British casualty figures were similar to those suffered by the Australians fourteen months later, there are several differences in the way the battle has been regarded, remembered and commemorated. Over 1300 men remain missing as a result of the May 1915 action; but no memorial marks their passing, or indeed the importance and legacy of the engagement. It has never been fully studied as a template or even as a hypothetical adjunct to the action of 1916. Although not one among their ranks would have known it, the Australians who stepped out on No-man's land on the evening of 19 July, 1916 and advanced over the sprawled mortal remains of hundreds of their British comrades were playing directly into the hands of Bavarian tacticians. The ravaged endeavours of the 8th Division on 9 May, 1915 formed the very foundation of the almost pre-destined ill fortune that the nation of Australia was at that moment embarking upon.[28]

How Was This Further Attack On Aubers Ridge Classified?

The BNC report details the <u>*Battle of Aubers Ridge*</u> (with date reference 9th May, 1915) with *Attack at Fromelles*, and *Attack at Rue du Bois*.

Under 'Attack at Fromelles', the geographical indicators are stated: Road Aubers – Fauquissart – Laventie -Rouge-de-Beut - Fleurbaix (exclusive) – la Boutillerie – Bas Maisnil.

Under 'Attack at Rue du Bois', the geographical indicators are stated: Road la Quinque Rue (exclusive) - le Touret – Lacouture – Crois Barbee : thence a line to the Bois du Biez (south west corner).

Captain E.A. James in his Record of Battles and Engagements follows the classification of Battle of Aubers Ridge, including "Attack on Fromelles".

A map drawn to scale using the same markers, would show that the battles occurred in the same place.

15th May, 1915 – The Battle of Festubert.

A few days later the Battle of Festubert commenced on 15 May about 3 miles to the south of Neuve Chapelle in the first night attack of the war and continued for 10 days with 16,000 casualties. The attack was again carried out by the British First Army commanded by Sir Douglas Haig and included the British 2nd and 7th Divisions and Indian Meerut Division.

---

[27]Barton p.42

[28] Barton p.57

It was 18 May, 1915 before the British were relieved by Canadians, who in five days' time were sent into attack but failed also to break the German resolve.

On 25 May, Sir John French ordered Haig to hold the line, so the French army could be supported elsewhere. The Battle of Festubert casualties now totalled 710 officers and 15,938 other ranks.

How Was the Action At Festubert Classified in The Battle Accounts?

It was recorded by the BNC as the Battle of Festubert, running from 15 to 25 May, 1915. The geographical information stated: The la Bassee Canal to Gorre; thence road to le Touret – Lacouture – Croix-Barbee: thence a line to the Bois du Biez (South West Corner).

25 September To 16 October, 1915 – The Battle Of Loos.

Later, commencing on 25 September, and ending on 16 October, the Battle of Loos took place. There were 6 Divisions of the British First Army (2nd, 9th, 7th, 1st, 15th and 47th Divisions between 5 and 10 miles to the south of Neuve Chapelle. Commanded by Sir Douglas Haig, the offensive commenced on 25 September and after counter attacks and renewed attacks, finished on 18 October with the loss of 800 British officers and 15,000 men, and with 1,200 officers and over 33,000 wounded.

If any battle of the British army in the Great War endorses the German officer's remark *'Lions being led by donkeys'*, the Battle of Loos was it. Regiment after regiment was sent to oblivion: new men, seasoned soldiers on both sides were wasted in huge numbers. The British 1st Army had another go at Aubers Ridge via Rouges Bancs and Fromelles, and the Indians were again sent forward at Neuve Chapelle, and despite initial small advances they could do no better than the earlier efforts and no ground was won along the front that concerns [this volume], but the British took Loos and an area to its north.[29]

It thus appears that units in the battle of Loos attacked via Fromelles. They crossed the same terrain, but instead of referring to Fromelles, references are to Loos. The contradictions set out in Chapter 1 in relation to naming of places of battle are immediately evident.

What is most important about the Battle of Loos in so far as the Battle of Fromelles is concerned, is Haking's role at Loos. Here we can see him for the first time in charge of a Corps – the XI consisting initially of the 21st and 24th Divisions, later the 12th and 46th plus the Guards and 3rd Cavalry. A considerable upgrade from just the 1st Division at Aubers Ridge 6 months earlier. What Haking did with his command is worth a couple of paragraphs because it clearly and without any ambiguity establishes what he brought to Fromelles and what was on the record.

Edmonds reported that arising out of the Neuve Chapelle and Festubert experiences one of the lessons was *'how quickly troops became exhausted and how soon fresh divisions were required to carry the attack forward; it had further shown the need for taking immediate advantage of the first success.'*

[29] Corfield p.25

[Author: There seems from this reporting, that there were debriefs. So where are the records of those debriefs for all Fromelles battles? Edmonds is reputed to have destroyed hundreds of thousands of documents in the course of preparing his 'History'. Is this an example of such documents?]

The general turn of events caused French, in Edmonds' words, to comment *'very strongly'* to Haig *'on the futility of pushing reserves through a narrow gap in the enemy's defences.'* Thus the two egos (of French and Haig) were on collision course: and Haking backed Haig.

The appalling destruction of the 21st and 24th Divisions in basic figures: 21st – 4051; 24th – 4178 was the outcome.

The 21st and 24th now destroyed and sent back, retired from battle, and not of sufficient strength to make a fighting unit. Haking got another two divisions, 12th and 46th, to form a *'reconstituted XI Corps'*: both had been at Ypres for some months. For the last part of the Battle of Loos, Edmonds never again mentions Haking's name, but instead unlike earlier accounts he makes a point of naming virtually every Brigade and Battalion commander under Haking's command.

The shambles that followed caused the Division 6,537 casualties. Overall Haking's command of XI Corps in this battle accounted for 16,830 casualties out of the 59,247 British losses. The Germans had 19,836 casualties. What more needs to be said? Except that with 1st Division losses at Aubers of 3,968, that gives Haking a total of 20,798 casualties in under six months.

As usual the Official History noted that the men were 'delighted' at the prospect of getting at the enemy, but as Alan Clark wrote in '*The Donkeys*', it forgets to point out, forgets to note that the men had no sleep for 48 hours and almost no food. Clark writes:

> *'Punctually at eleven o'clock the British rose out of the ground ... and formed up in a dense mass ... sheer weight of numbers must, it seemed, would carry the British through the thinly spread German outposts ... For fully ten minutes the Germans held their fire as the two divisions deployed in columns of extended line and started obediently off ... at the range of 1,000 yards the order to fire was given.'*

Thus, we see the 'Grand Old Duke of York' in the nursery rhyme being employed, instead of planning, thinking, and considering what was on the other side of the trenches. Nothing had been learnt from earlier engagements. Nothing.

Then follow descriptions of how the Germans could not grasp that the British would act with such disregard for the realities of the situation, then when the battlefield was covered with dead and dying and the attack was over:

> *'no shot was fired at them from the German trenches for the rest of the day, so great was the feeling of compassion and mercy for the enemy after such a victory.'*

The short sharp massacre of three and a half hours cost the British 385 officers and 7,861 men, whereas the Germans do not seem to have lost a single man.

Perhaps the most extraordinary consequence of this campaign, is that Sir John French was replaced by Haig, and with the advent of 1916, Sir Douglas Haig became the Commander-in-

Chief, loyally attended by such men as Haking who had been participants in, witnesses to, and planners of many of the disasters these last pages have dealt with. Perhaps their mutual association in the campaign against Sir John French and in the failures, gave them a special bond. [30]

---ooo000ooo---

[30] Corfield p.29

3. Events Around Fleurbaix Up Until 1 July, 1916, and The Somme Offensive

The Prelude to the Battle of Fromelles.

---ooo000ooo---

When this blanky war is over, Oh, how happy I shall be,

When I get my civvy suit on, No more soldiering for me.

---ooo000ooo---

The Allied war strategy for 1916 was for simultaneous offensives mounted by the Russians on the Eastern Front, the Italians in the Alps and an Anglo-French attack on the Western Front in the Somme Valley. However, a major German offensive at Verdun commenced in February and the French Government pressured the British Government for the attack in the Somme Valley to become largely a British attack to draw German forces away from Verdun. Australians were brought into the Western Front to assist with the British efforts.

Much is made of the Australian 5th Division being the first Australian fighting unit to be involved in a Battle on the Western Front. But there was quite a deal of activity in the region before the Battle of Fromelles took place. The 1st, 2nd, and 4th Divisions were all active in some way from April, 1916 onwards.

From late April, 1916 there were Australians in the Sailly[31] area, battalions of the 4th Division mainly, with others from the 1st and 2nd, together with New Zealanders, under General Birdwood and titled I Anzac Corps.

The whole area in this part of the line was called "The Nursery," as it had the reputation of being a quiet sector, and here new troops were introduced to the firing line before being sent off to more dangerous sectors. The actual sector was a frontage of about 1,000 yards at Cordonnerie Farm and the Ancient Abbey of Chartreux, but the name by which the boys of the 11th Battalion and others always referred to the sector was Fleurbaix, after a village to the left rear.

The first significant losses in the area were in 20th Bn (2nd Div., 5th Bde, NSW) with over 100 casualties, including some taken prisoner on 5 May. These were the prisoners who were paraded in Lille and, so the story goes, cheered and offered chocolates and cigarettes by the locals much to the Germans' dismay.

On 5 May, 1916, a German raid on the Bridoux Salient, a mile east of La Boutillerie, captured two of the still secret Stokes mortars, embarrassing the AIF for a long time. It was the loss of the new weapon, much valued by the British, which projected this German raid into the history books. The OC 20th Bn being sent back to Australia as punishment.

---

[31] The reference to 'Sailly' is generally a truncation of Sailly-sur-la-Lys. It became the headquarters of Lt Gen Haking during the Battle of Fromelles.

Accounts of the daily activities of various units appeared in newspapers after World War One. One such account was for the 11th Battalion AIF. The 11th Battalion AIF was among the first infantry units raised during World War. It was the first battalion recruited in Western Australia, and following a brief training period in Perth, the battalion sailed to Egypt where it undertook four months of intensive training. In April 1915 it took part in the invasion of the Gallipoli Peninsula, landing at Anzac Cove. In August 1915 the battalion was in action in the Battle of Lone Pine. Following the withdrawal from Gallipoli, the battalion returned to Egypt where it was split to help form the 51st Battalion. In March 1916, the battalion was deployed to the Western Front in France and Belgium where it took part in trench warfare until the end of the war in November 1918.

The activities of the 11th Battalion were initially described in an article in the Western Mail, Perth WA, on 28 April, 1938, titled: **CHAPTER XVII. 'FLEURBAIX'**[32]. The article later became part of the *Story of the Eleventh Battalion A.I.F., by Captain Walter C. Belford M.A.*

On 20 May, 1916, the 11th Battalion moved into the lines to take over from the 4th Battalion, which by that time was to move south to the Somme, and would be involved from July 23, in The Battle of Pozieres. As the battalion moved to take up position in Fleurbaix, the leading platoons fell into their swing, some of the boys started to sing with great feeling:

"When this blanky war is over,

Oh, how happy I shall be,

When I get my civvy suit on,

No more soldiering for me."

The soldiers were loaded up to their teeth with all equipment, helmets, gas masks, periscopes, bombs (grenades) and all the paraphernalia of the modern soldier. While an army may march on its stomach, this one very much carried its provisions on its back.

The battalion, less transport, moved into the firing line with 27 officers and 929 other ranks and occupied the left sector of the brigade frontage in the Petillon area.

It was the first time in the line for all but those who had experienced Gallipoli. There was a mixture of feelings, chiefly curiosity, and a desire to get to the real thing. There was occasional shelling, and machine guns could be heard hissing elsewhere, but the 11th Battalion was in the line for the first time in France.

The country was naturally extremely flat, except where the ground rose to a slight elevation at the famous Aubers Ridge. Maps of the region revealed that all of the country south of the Lys River, between Estaires and Armentieres, was a mass of streams, ditches and water channels. The Petillon sector and over the then German line towards Aubers and Fromelles was perhaps

[32] See the full newspaper article available via National Library of Australia via the following link: https://trove.nla.gov.au/newspaper/article/37847320?searchTerm=chapter%20XVII.%20Fleurbaix&searchLimits=

the worst bit of the whole area. The water table was never more than a foot or two below the surface of a very large part of the country.

In the area, there were also a number of mine shafts, leading to tunnels that had been under construction by a British tunnelling company, but which had been taken over by the 2nd Australian Tunnelling Company.

The ground in the vicinity was just drained and no more by the Laies river and tributary ditches, so that it was practically impossible to dig dugouts of the type used on Gallipoli or so well known later in France in the Somme area. 'Trenches', a misnomer in the first place, were built-up sandbag mounds, which had been battered out of all shape by shell-fire and weather.

All the shelters were surface huts, built of filled sandbags and timber, often with galvanised iron roofs covered with a layer of sandbags, just enough to make them splinter proof. Most of the men lived, or rather slept in burrows dug under the wide parapets.

It was wet weather when the 11th Battalion took over the trenches, so all the troops were issued with rubber thigh boots, called by the boys "gum-boots" and in these the troops used to slither up and down behind the ramparts. There was a duckboard track over the worst sections, but with the constant traffic, the slats soon broke and it was often more of a trap to walk on, especially at nights, than the slippery ground.

There was also a firestep right along the wall of the parapet and here the men would sit and yarn or smoke, as it was one of the few dry places, outside of the shelters at least, where the troops could sit with any comfort.

All the time there were two sentries in each bay, provided with periscopes and these kept an eye on the trenches in front. Very often the tinkle of broken glass would be heard, indicating that some German sniper had "rung the bell."

In the mornings after stand down, the Germans used to have the sun directly behind them, and consequently the Australian lines stood out clearly and distinctly. At this time the enemy snipers were most active, and fired at all periscopes and any movement that they could spot. There was no doubt about these fellows being marksmen. The boys of No. 11 Platoon, "C" Company, used to hold up on the end of sticks, emergency ration tins. These were about three inches by two. They were carefully pushed above the parapet and Fritz taking the shiny surface for mirrors, used to fire at them. It was computed that he put two bullets out of three through these tins. The nearest point of the German lines at this point was 350 yards distant.

In the days following the 11th Battalion's move into position, the troops had their introduction to the various projectiles that "Jerry" used to hurl over from his lines. The missiles consisted of large low velocity shells such as *Jack Johnsons*, and *Coal boxes*, big trench mortars called *Rum-jars*, *Minnies*, and *Flying pigs*, high velocity shells of small calibre, mostly 77 millimetre, referred to as *whizz-bangs*, large shrapnel shells bursting into a cloud of brownish looking woolly smoke, the famous *woolly-bears*, and small trench mortar bombs called *Pineapples*, and a great many more. It was not long before the boys were able to distinguish between the various missiles and to make the necessary reactions, which often meant a head-long dive for cover.

Gas was the most feared weapon of the Germans at this time and the trenches were plentifully supplied with all manner of gadgets to prevent the troops being taken by surprise. Gongs, klaxon horns, and iron bars were in great evidence, as were also empty 18-poundershell cases. Rattles were also used and a special SOS rocket, to be fired if gas were actually seen issuing from the enemy's line.

There were also numerous little wind vanes, some beautifully made, to represent aeroplanes, which the troops would religiously gaze at from time to time to see that the wind was not in Jerry's favour. A few gas scares occurred, the most important being on the night of May 30, when all precautions were used and the troops wore their helmets in the ready position all night, with the result that the acid in the flannel of the old P.H. helmets burnt many of the boys' foreheads and some had such bad sores that they had to be medically attended to.

Brigadier General Sinclair-MacLagan used frequently to make his rounds of the trenches in the early mornings. He was very particular as to waste, and used to make a great fuss if any empty cartridge cases were left lying about.

Although the boys would have been quite satisfied if he had come less often, these visits of the General at least showed that he desired to see for himself what the front line conditions were like, and this was one feature about commanders that the boys appreciated more than any other.

In contrast to the up-front activity of Brigadier General Sinclair-MacLagan, we will read in later chapters how for example, Brigadier General 'Pompey' Elliott was critical of the command of the XI Corps, in that they did not even reconnoitre the site before the upcoming Battle of Fromelles.

---ooo000ooo---

When 21 BRIR raided Cordonnerie Farm on 30 May, 1916, the bombardment, to which the *minenwerfer*[33] at Delangre Farm contributed tellingly, was the heaviest the Australians had known and caused most of their 116 casualties. The Commander of the 2nd Division, Major General J.G. Legge, admitted that the initiative lay with the enemy, who was, 'so far as he can be without actually attacking, somewhat superior in the offensive'.

---ooo000ooo---

The article published on 28 April, 1938 became the subject of a letter to the Western Mail, Perth, published just over two months later, on 2 June, 1938. The writer identified only as "F.G.H. (11th), Three Springs". He described the barrage that occurred on 30 May, 1916 as the ***'Fleurbaix Inferno'***[34]. In his letter he wrote: -

---

[33] Minenwerfer is the German name for a class of short range mortars used extensively during the First World War by the Imperial German Army. The weapons were intended to be used by engineers to clear obstacles including bunkers and barbed wire that longer range artillery would not be able to target accurately. Wikipedia

[34] See the full article via the following link: https://trove.nla.gov.au/newspaper/article/37848895?searchTerm=While%20reading%20the%20%22WM.%22%20recently%20I%20confess%20to%20an%20icy%20feeling%20down%20the%20spine%20%20%20%20%20%20%20%20%20%20%20%20&searchLimits=

While reading the "WM." recently I confess to an icy feeling down the spine, while what little breeze there was about seemed to concentrate about my person, finding an inlet at my trouser ends and travelling upward to the highest point of my anatomy. It was merely a recurrence of a malady contracted 22 years ago on May 30 [1916]. Memories of that awful night even affected my slumber as the whole scene came before me like a ghostly apparition in all its hideousness.

We had only been in the trenches in France a few days and although I had spent nearly eight months at Gallipoli, I had yet to learn the brutality of war as waged by the Germans.

Jack Woods (killed some months afterwards) and I were manning one of the breastwork bays held by "B" Company to which we both belonged, when it descended on us like a mighty thunderclap. Its intensity grew to such an extent during more than an hour that it lasted, that the different projectiles on the position of "B" and "A" Companies were almost as numerous as the grains of rice or particles of confetti that fall from well-wishers on the heads of a bride and bridegroom.

Jack and I stood to our gun till the bay on both sides of us was demolished and the end of our own, and we then crawled into two manholes under the parapet. None too soon, either, as a shell got the parapet, almost completely burying both of us a second or so later.

It was a difficult matter emerging from this predicament, particularly with our encumbering equipment getting hooked up with everything possible, but we eventually scratched our way out like a couple of reptiles that had finished their winter's hibernation.

I then encountered Corporal Stanley (nick-named Splinter), who commanded our section, a brave lad not yet 21, but very emotional, and he enjoined all his section to come into a tin shanty just by, and these were his words: "We have been a very happy family and we'll all die together."

That we didn't all die at that time was due merely to bad workmanship by some German manufacturer, as a shell came through the roof, bounced off the hard ground (which had been protected from rain for perhaps a couple of years), hit the roof again and then fell harmlessly into the corner.

Daylight revealed a ghastly sight of carnage; even the water in all the gutters around was crimson tinged, but we hadn't time even to take a last look at our dead cobbers as the line had been seriously breached and we worked like Trojans to repair it. An unfortunate lieutenant was blamed for provoking this frightful onslaught and from then onwards was about as popular as a black snake. He certainly disturbed a sleeping enemy in the "Nursery" with his "Chinese Cracker Artillery," but that was his job.

It is my considered opinion that the Germans on that memorable night were grimly determined to exterminate every living creature in that part of the line - not even excluding the rats that found sanctuary there.

---ooo000ooo---

In response to a directive issued by General Haig on 28 May, 1916, 1 ANZAC itself carried out fourteen ***raids*** between 5 June and 2 July 1916. Now the BEF's Commander-in-Chief, General Haig called on his other armies to launch ***raids that would distract German attention from the Fourth Army's preparations for its offensive on the Somme and wear down divisions the Germans might send there after it began on 1 July***.

The British 61st Division relieved the 38th (Welsh) Division early in June. Within a month, the 61st Division[35] had launched eight raids, more than any of the neighbouring 1 ANZAC divisions carried out. The results were mixed and the cost high.

---ooo000ooo---

Raiding activities became the subject of a further article by Captain Walter C. Belford M.A. of the 11th Battalion, as published in The Western Mail, Perth, WA, on 12 May, 1938, and which would again later become part of the *Story of the Eleventh Battalion A.I.F.*, titled: -

IN SUPPORTS AT FLEURBAIX.[36]

The companies of the battalion were very much scattered, but the troops were mostly quartered round La Croix Marechal and the Rue des Lombards, two localities about a mile behind the sector of trenches that had lately been occupied. The area was divided into a number of posts, and each company was responsible for one or more.

Some of the quarters in the posts were pretty fair, but some on the Rue des Lombards were very much knocked about by shellfire.

The battalion strength was now 31 officers and 870 other ranks. The total casualties for the first tour of the line, certainly a long period compared to future service in the actual front line, were 43 killed, 115 wounded and 6 missing, a total of 164. This was a fairly heavy total for a quiet sector, but the 11th Battalion had been very unfortunate in having had to suffer a raid while in the line. That could only be regarded as a piece of bad luck and might have happened to any battalion holding the sector at the time.

On June 9, Lieut. Colonel Roberts asked Captain Medcalf to take charge of and train a party of the 11th Battalion, in order to make a raid on a point called The Tadpole, opposite to the trenches held by "D" Company while it had been in the line. There were sixty men chosen as raiders.

During the time the battalion was in the support line, nearly all the work was done at night, as this minimised the movement during the day around the billets, and also because much of the work was done on redoubts, emergency roads to the communication trenches and other such work of a tactical nature.

All day long the troops rested, or went picking strawberries, which were now ripe and plentiful round the deserted farms.

It was now summer and most of the days were warm and fine. There were still occasional showers and the ground in this low-lying part of Flanders was still muddy, and after the showers invariably sloppy. The previous wearing of gumboots for long periods in the trenches, and the constant wet feet of these days in support, tended to soften the feet of the troops, and the effect of this was very noticeable later on when the battalion had heavy route-marching to do.

---

[35] The Division being the same as which would go on to fight at Fromelles on 19 July.

[36] See the full article at: https://trove.nla.gov.au/newspaper/article/37847963?searchTerm=In%20supports%20at%20Fleurbaix%20%20%20%20%20%20%20%20%20%20%20%20%20%20%20%20%20%20%20&searchLimits=

Several gas alarms occurred while the battalion was in support, and on several occasions the troops had to stand to until the command "All clear" came through. On June 14, the date that the Daylight Saving order came into force, in support of the decree of the French Government, there was a gas attack on the front near the 11th Battalion and the troops were on the "qui vive" all night: the effect of the gas was felt in Sailly, about three miles in rear, and it was reported that some civilians were suffering from the gas, but no one in the 11th Battalion was affected.

On 12 June, 182 Brigade (UK) supplied 268 men to the Australian tunnellers. 2/1 Bucks contributed 510 of the 2,500 men who carried 1,500 gas cylinders dumped on the Rue Masselot to the front line on 18 and 19 June.

Elsewhere on 15 June, Lt Gen Haking had suggested thirteen operations that XI Corps might undertake to support the Somme offensive.

A great artillery battle took place on the night of June 16 and the troops in support saw the tremendous picture of a big barrage and heavy shell fire, but luckily only from the onlookers' point of view. The scene was terribly magnificent, and everything stood out clearly outlined in the vivid light caused by the bursting shells. Then, from far and near came the sounds of bells, gongs, rattles and beaten shell-cases, indicating that another gas attack had been launched by the enemy. Again the gas was felt in the rear areas, but there were no casualties in the battalion.

Now that the weather was generally fine and clear, an increased amount of aerial activity was noticed. As this form of warfare was new to most of the troops, very little having been seen on Gallipoli, it was a source of great interest, and generally all hands would stop whatever they were doing to follow the wonderful evolutions of the fighters high aloft. In fact right to the end of the war an aerial scrap held the attention of the troops more than any other form of warfare.

Owing to this increased aerial activity on the part of the enemy, there was considerably more shelling of billets and back areas. On one occasion the Germans shelled and set fire to a billet which was situated next to an estaminet.

The latter soon went on fire also. As the liquid assets of the estaminet were in serious danger of being destroyed, the troops managed to rescue them FIRST.

The great Somme offensive *was now being planned, and for some time the commanders of the armies in the north had been under instructions to make as many demonstrations as possible in order to retain as great a force of the enemy in the northern sectors as possible, and to prevent him transferring any divisions down to the Somme area. To this end, and also for the purpose of identifying the units opposed to them, and so fixing the position of as many German divisions as possible, a great number of* ***raids*** *were ordered, not only by the Australian division, but also by British units in the northern sectors.*

For one of these raids, the party of the 11th Battalion under Captain F. G. Medcalf, was training in quarters behind Sailly, not far from Doulieu. Maps and plans of the Tadpole, for such was the name given to the portion of trenches to be attacked, were distributed to all the raiders, and these were carefully studied, and a model of the plan to be adopted in the actual attack was practised dally.

---ooo000ooo---

German shelling savaged a raiding party from 2/4 Ox and Bucks just before it went over on 21 June and smashed a raid by 2/8 Warwicks five days later. These same units would become part of the 61st Division at Fromelles.

Soldiers and Officers alike were being sent for training in musketry, in sniping school, and for signallers.

The 4th Australian Division had arrived and was billeted in the neighbourhood of Sailly, and on the night of June 24-25 "B" Company was relieved by troops of the 50th Battalion. The 51st relieved the 10th Battalion, and there was no chance of old comrades meeting, as the relief took place late at night on June 28-29.

For the next four days there was a little company drill, but as the battalion had a considerable amount of night fatigues to perform the men were generally resting through the day, and so were able to visit their French friends in the village and surrounding country. The 3rd Brigade was going to carry out two raids early in July, and the raiders were hard at work training for the job. There seemed to be increased activity on the Fleurbaix front, but as the various batteries had to register for the raids there was a good deal of covering fire to prevent the registering from being too obvious.

The battalion Lewis gun establishment was increased to six guns on June 30, and reserve crews of 24 men were immediately selected and instructed in the use and care of the weapons.

As part of its programme, 1 Battalion hit 20 BRIR east of Petillon on 28 June. Three nights later, 150 raiders from 9 Battalion inflicted 79 casualties on 21 BRIR east of the Sugar-loaf salient.

---ooo000ooo---

Elsewhere on 1 July, 1916, the great **Somme Offensive** had begun. But the work of the 11th and other Battalions had not finished in this area.

The 9th Battalion, made a very successful raid on the night of July 1-2, in which 21 prisoners were captured.

The raids were the subject of planning and execution by Lt. Col. Maurice Wilder Neligan, as to which, more detail will be set out in relation to the events of 1 July in the following chapter.

---ooo000ooo---

Snipers, shelling, bombings in many forms, artillery concentrated on the soldiers.

Already there had been gas used in the area.

Explosions, collapsing and exploded breastworks, and underlying wet soil.

And the Battle of Fromelles on 19 and 20 July, 1916 was yet to come.

One has to wonder if this Battalion and the 9th Battalion had been so successful in the use of raids, why they were not permitted to remain and harass the German defenders opposite, or even to move into the area of Fromelles, instead of having untried battalions which had never fought together under fire, go into what would become the Battle of Fromelles, in less than three weeks time.

4. Day 1 of the Great Somme Offensive

From Beaumont Hamel to Fromelles, 1 July to 18 July, 1916.

A Comparison of a Battle and some Raids.

---ooo000ooo---

On 1st July, 1916, the **great offensive** or the ***operation*** which came to be described as the 'Battle of the Somme' began. The Fourth Army (commanded by Sir Henry Rawlinson) and the Reserve Army (later renamed the Fifth Army and commanded by Sir Hubert Gough) attacked on a front of 15 miles. Fighting and battles conducted on such a distance could only realistically be categorised as a 'campaign' but for reasons which will be discussed later, the reference of 'Battle' was retained. In the coming pages, we will be looking at events of the days between the start of what should rightly be called the 'Campaign' of The Somme, and that at Fromelles.

The scale of the damage inflicted on soldiers in the British and French contingents can be appreciated from the following extracts:

They had, as instructed, formed into a line, rifles held across their chests, bayonets at the ready; they advanced slowly in a disciplined way – they could do little else carrying six stone packs. They walked and watched. They presented a facile target. The German gunners laughed and beckoned them on. As men fell their comrades still carried on in the same way, slowly, walking in line. There was almost no panic. They felt terror and anger, but they were numbed. The survivors of the first wave, the first of six tragic waves, carried on walking in a line ….[37]

By this time, the British had been at war since August, 1914, so 22 months had not changed their method of engaging in deadly battle, from a style of military tactic better suited to the Middle Ages.

After a week's bombardment the great assault was delivered at 7.30 am on July 1st – a day of fiery heat. Thirteen British Divisions, with six more close up, began to advance at a stately pace which gave the enemy time, where ever No-Man's Land was wide, to bring out their machine guns from shelter after the barrage lifted. In the furnace of their fire British lines melted with tragic swiftness. Sixty thousand men fell on that one day, the heaviest day's loss in the history of the British Army. And for little result.[38]

Actual figures: 21,392 dead, 35,493 wounded. It took seven days and 58 packed hospital trains to clear the wounded.

The Battalion with the highest casualties on 1 July was 10th Bn West Yorkshire with 306 killed, 404 wounded. In comparison with the upcoming Battle at Fromelles, the 60th Battalion, the most damaged at Fromelles, had 398 killed and around 360 wounded.

[37] M. Middlebrook: *First Day on the Somme* (1971)

[38] B.H. Liddell Hart: *Foch* (1937):

British casualties were 57,500 on the first day, while German casualties were approximately 8,000, including 2,200 taken prisoner.

---ooo000ooo---

Before we move on to an examination of the events that unfolded at Fromelles, it is worthwhile looking at one of the battles fought on the Somme, simply to see what happened on The Somme, and to focus on the work of the Newfoundland (Canadian) contingent, at the Battle of Beaumont-Hamel. It is being included in order to highlight the appalling inability of the British High Command to plan, to perceive, and to properly portray what took place on The Somme.

There are a number of parallels in the way activity took place at Fromelles, as we will see later. [39]

By way of comparison to the methodology adopted in the fighting of Battles on the Somme, we will also examine the methodology adopted by Australian Battalions at Fromelles, in the use of raids, fought on 1-2 July, 1916.

The Battle of Beaumont-Hamel: 1 July, 1916

If you drive from Fromelles (Pheasant Wood) Military Cemetery to the Beaumont-Hamel Newfoundland Memorial Park you will take slightly more than an hour to drive approximately 71 kilometres from point to point in a generally more south than south-west direction. The park is approximately half way between Amiens to the south, and Arras to the north.

The Beaumont-Hamel Newfoundland Memorial site is situated 9 kilometres north of Albert, France near the village of Beaumont-Hamel, in an area containing numerous cemeteries and memorials related to the Battle of the Somme. It is a memorial site dedicated to the commemoration of Dominion of Newfoundland forces members who were killed during World War I. The 74-acre (300,000m^2) preserved battlefield park encompasses the grounds over which the Newfoundland Regiment made their unsuccessful attack on 1 July 1916 during the first day of the Battle of the Somme.

It was the regiment's first major engagement, and during an assault that lasted approximately 30 minutes the regiment was all but wiped out. It is the largest site dedicated to the memory of the Newfoundland Regiment, the largest battalion memorial on the Western Front, and the largest area of the Somme battlefield that has been preserved.

---

[39] For a more expansive article dealing with Beaumont-Hamel, consider the following material at:

http://www.heritage.nf.ca/first-world-war/articles/beaumont-hamel-en.php.

And further material on Wikipedia: "Beaumont-Hamel Newfoundland Memorial." at

https://en.wikipedia.org/w/index.php?title=Beaumont-Hamel_Newfoundland_Memorial&oldid=835134339 (accessed September 22, 2018).

The site is one of the few places on the former Western Front where a visitor can see the trench lines of the First World War and the related terrain in a preserved state.

The Newfoundland Regiment's Pathway to Beaumont-Hamel

During the First World War, Newfoundland was a largely rural Dominion of the British Empire with a population of 240,000, and not yet part of Canada.

The regiment trained at various locations in the United Kingdom and increased from an initial contingent of 500 men to full battalion strength of 1,000 men, before being deployed.

After a period of acclimatization in Egypt, the regiment was initially deployed at Suvla Bay on the Gallipoli peninsula with the 88th Brigade of the British 29th Division in support of the Gallipoli Campaign.

The Plan

After its removal from Gallipoli, the Newfoundland Regiment was informed on February 25, 1916 that it would be part of the impending Somme offensive. It departed Egypt on March 14, 1916 and arrived at France eight days later. In France, the regiment went into the line in April 1916 at Beaumont-Hamel.

For the next three months, between April and July, in readiness for combat, the men trained rigorously, did tours of duty on the front line, dug trenches, strengthened defences, and observed the enemy. Five casualties were suffered on 24 April, when German bullets wounded four men and killed 19 year old Private George Curnew. More casualties followed in the coming weeks. Already, it was clear that the area was dangerous.

Facing the Newfoundlanders, and the 29th British Division were the defences manned by experienced troops of the German 119th (Reserve) Infantry Regiment of the 26th (Wurttemberg) Reserve Division. These same German opponents had been involved in the invasion of France in August 1914 and had been manning the Beaumont-Hamel section of the line for nearly 20 months prior to the battle. The German troops were also spending a great deal of their time not only training but fortifying their position, including the construction of numerous deep dugouts and at least two tunnels. They were entrenched and were not going anywhere.

Seven Days of Preliminary Action

Commencing 24 June, the Allied powers began their bombardment of the German front lines along the Somme with artillery. It was planned to last for a week, ahead of what was to take place on July 1. It was both intended and expected to weaken the enemy defences in advance of the July 1 ground attack. The entire battleground along the River Somme was a 34-kilometre ribbon of land. Allied trenches stretched along one side, and the Germans along the other. In between lay No-man's land. Beaumont-Hamel was a small part of the overall larger set piece.

At 9:00pm on June 30, the Regiment departed Louvencourt and marched three hours to its trenches on the battlefield.

The Germans held much of their front in great strength. They had created a three-tiered system of forward trenches. The first line, 300 to 500 metres away, was was well dug in, shielded with extensive protective wire, and capable of surviving sustained bombardment.

The second German line of trenches was constructed further behind, at ranges of between 2,000 and 5,000 metres. Further back still, the Germans were working on yet a third line of trenches. This network of heavily defended lines presented a formidable obstacle to any attacking force.

The Newfoundland Regiment was stationed in trenches near the French village of Beaumont-Hamel, which lay behind German lines. It was a strategically difficult position. The German front lines were about 300 to 500 metres away, down a grassy slope and heavily guarded by barbed wire entanglements. The German 119th Reserve Regiment was tough and experienced, and had turned the natural defences of a deep Y-shaped ravine into one of the strongest positions on the entire Somme front.

[Picture 11: Map 6. British Objectives at the Somme, 1 July 1916[40]]

The Newfoundlanders' assignment (along with the rest of the 88th Brigade) was to seize control of the German trenches near the village of Beaumont-Hamel. The Regiment would be part of a third wave of attackers to leave Allied trenches.

---

[40] Wikimedia Commons contributors, "File:British plan Somme 1 July 1916.png," *Wikimedia Commons, the free media repository*, https://commons.wikimedia.org/w/index.php?title=File:British_plan_Somme_1_July_1916.png&oldid=142044454 (accessed January 6, 2019).

Zero Hour

It was believed that a local high point – Hawthorn Ridge – should be the site for an underground explosion to precede the ground-based attack. The ridge was an important German stronghold on its frontline, about 700 metres to the west of Beaumont-Hamel.

In the weeks preceding the attack, 18,000 kilograms of explosives had been buried under Hawthorn Ridge. It was to be set off prior to the ground attack. But just how much before the attack?

Debate ensued as to whether it should be four hours before the ground attack, two minutes before the attack, or at the very moment the soldiers left their trenches. A compromise set the detonation to be 10 minutes before zero hour.

The offensive began on the morning of July 1, 1916, at 6.00am Allied Forces bombarded the Germans with artillery for about an hour. And then at 7:20am, the explosives under Hawthorn Ridge were detonated. The blast turned the area into a giant crater, measuring 40 metres wide and 18 metres deep. The first wave of the attack was to take place 10 minutes later at 7.30am

But the timing was a terrible mistake. The explosion of the mine successfully destroyed a major enemy strong point but also served to alert the German forces that a land attack was imminent, while the 10-minute delay gave them just enough time to strengthen their defences and makle ready their machine guns.

Following the explosion, the German troops immediately deployed from their dugouts into the firing line, even preventing the British from taking control of the resulting crater as they had planned.

As the first wave of Allied troops left their trenches at 7:30am, they were greeted by a devastating barrage of enemy artillery and machine gun fire. It was far stronger than anyone had anticipated. Most men were killed or wounded in minutes.

The second wave of troops left their trenches soon after and met with the same fate.

The Newfoundland Regiment was still in its trenches, awaiting orders to go over the top as part of a third wave of attack.

By now it was obvious that more soldiers were being killed than progressing. The troops from the 86th and 87th Brigade of the 29th British Division were quickly stopped. With the exception of the 1st Battalion of the Royal Inniskilling Fusiliers on the right flank, the initial assault foundered in No-man's land at, or short of, the German barbed wire.

By 8:00am, there was considerable confusion along the Allied lines. The first two waves of attack had been devastating failures, but at divisional headquarters, Maj Gen Beauvoir De Lisle and his staff were trying to unravel the numerous and confusing messages coming back from observation posts, contact aircraft and the two leading brigades. One such report for example, suggested that what turned out to be a German flare (to indicate that shells were falling short of target) was mistakenly identified as a British flare used to indicate the first objective had been taken by the 87th Brigade.

There were also indications that some troops had broken into and gone beyond the German first line. In an effort to exploit the perceived break in the German line De Lisle ordered the 88th Brigade, which was in reserve, to send forward two battalions to support the attack.

And then at 8:45am, Brigadier General Douglas Edward Cayley ordered the Newfoundland Regiment, which was to be the third wave of the attack, to advance "as soon as possible".

The Newfoundlanders were situated at 'St. John's Road', a support trench 230m behind the British forward line and out of sight of the enemy. However movement forward through the communication trenches was not possible because the trenches were congested with dead and wounded men and under shell fire.

Lieutenant Colonel Arthur Lovell Hadow, the battalion commander, decided to move immediately into attack formation and advance across the surface, which involved first navigating through the British barbed wire defences.

The Newfoundland Regiment Advances

The men left their trenches nearly two hours after the start of the attack, at 9:15am, with orders to seize the first (at 300 to 500 metres) and second lines (from 2000 metres away) of enemy trenches. But as they moved down the exposed slope towards No-man's land, no friendly fire covered their advance.

As they breasted the skyline behind the British first line, they were effectively the only troops moving on the battlefield and clearly visible to the German defenders. Modern soldiers know the problem with being silhouetted against a backdrop is that you are fully exposed in the absence of covering fire. The soldiers became sitting ducks.

Subjected to the full force of the German 119th, most of the Newfoundland Regiment who had started forward were dead, dying or wounded within 15 to 20 minutes of leaving St. John's Road trench.

The Danger Tree

There were other flaws in the Allied strategy. The weeklong bombardment that preceded the attack did not weaken German defences as much as Allied commanders had hoped. Instead, it had the unexpected effect of leveling the landscape of No-man's land, which deprived the advancing men of any cover from enemy fire.

About halfway down the slope between the British and German trenches, however, a skeleton of a blasted apple tree did survive. It had been used as a landmark by a Newfoundland Regiment trench raiding party in the days before the Battle of the Somme.

Now known as the Danger Tree, it had been part of a clump of trees located about halfway into No-man's land. British and German artillery bombardments eventually stripped the tree of leaves and left nothing more than a shattered tree trunk.

During the Newfoundland Regiment's infantry assault, the tree was once again used as a landmark, where the troops who had made it into No-man's land were ordered to gather.

But the tree provided meagre cover and the men became silhouetted against the sky as they approached it, making them easy targets for the German gunners.

The tree was also a highly visible landmark for the German artillery and the site proved to be a location where the German shrapnel was particularly deadly.

As a result, most reached no further than the Danger Tree, and many men died near it.

The few men who did reach the German lines were horrified to discover that the week-long artillery barrage that preceded the attack had not cut the German barbed wire.

Allied leaders had already obtained this information, thanks to a series of raids that reconnaissance teams had carried out in the nights before the attack. However, commanders had dismissed the reports, convinced that the raiders were inexperienced and the artillery bombardment much more effective than it actually was. As a result, the majority of the soldiers who reached the enemy trenches were killed, tangled in the uncut wire.

The Attack Was A Failure

The attack was a devastating failure. So far as can be ascertained, 22 officers and 758 other ranks were directly involved in the advance. Of these, all the officers and slightly under 658 other ranks became casualties.

For all intents and purposes the Newfoundland Regiment had been wiped out, the unit as a whole having suffered a casualty rate of approximately 80%. The only unit to suffer greater casualties during the attack was the 10th Battalion of the West Yorkshire Regiment, attacking west of Fricourt village.

[Picture 12: Barbed Wire at Beaumont-Hamel, 1916. © Imperial War Museum. Image Q 1530]

At 9:45am, the Regiment's commanding officer, Lieutenant-Colonel Arthur Hadow, reported to headquarters that the attack had failed. He received initial orders to gather any unwounded men and resume the offensive, but wiser counsel prevailed and the order was countermanded.

Throughout the day, survivors tried to make the long and dangerous journey back to their own lines, dodging enemy snipers and artillery fire. Private James McGrath got to their barbed wire, where he was first shot; then went to jump into their trench when he got a second in the leg. He lay in No-man's land for fifteen hours, and then crawled a distance of a mile and a quarter. He was fired on again, this time the sniper's bullet 'fetching' him in the left leg. He waited for another hour and moved again, only having the use of his left arm now. As he was nearing his own trench he was 'fetched' again, this time around the hip as he crawled on. He managed to get to his own line which was evacuated as his artillery was playing heavily on the German trenches. He was rescued from a hole after another hour, and taken to the dressing station a distance of two miles.

In a single morning on The Somme, almost 20,000 British troops died, and another 37,000 were wounded. The Newfoundland Regiment had been almost wiped out. When roll call was taken, only 68 men answered their names; 324 were killed, or missing and presumed dead, and 386 were wounded. The following days brought more fatalities.

[Picture 13: A Wounded Man is Brought in at the Battle of the Somme, 1916. © Imperial War Museum. Image Q752.]

The Rest of The War

After July 1916, the Beaumont-Hamel front remained relatively quiet while the Battle of the Somme continued to the south. Ultimately, in the final act in the Somme Offensive, Beaumont-Hamel was assaulted by the 51st (Highland) Division on 13 November 1916 on the opening of the Battle of the Ancre. Within two days, all the 29th Division objectives of 1 July had been taken along with a great many German prisoners. However, in March 1918, the German Spring Offensive was here checked on exactly the same battle lines as before.

The Effect of Beaumont-Hamel On Newfoundland

Today if you meet someone from Newfoundland, you will learn that 1 July is their special day of remembrance. It was the first great conflict experienced by that dominion. Beaumont-Hamel became to them what Gallipoli became to Australians. The Newfoundlanders had been at both places.

Of all the battles that the Newfoundland Regiment fought during the First World War, none was as devastating or as defining as the first day of the Battle of the Somme. The Regiment's tragic advance at Beaumont-Hamel on the morning of July 1, 1916 became an enduring symbol of its valour and of its terrible wartime sacrifices.

The losses at Beaumont-Hamel plunged Newfoundland and Labrador into a period of mourning. Casualty lists appeared in local newspapers, but information came in slowly from the front lines and it was often incomplete. Many families had to endure a few tortured weeks of wondering whether their loved ones were alive or dead. Being closer to Europe than those in Australia, news was received in weeks instead of months.

British officers sent letters of condolence to the many families who had lost sons, husbands, fathers, and brothers, and Allied leaders publicly praised the Newfoundland Regiment. Among them was Sir Douglas Haig, Commander-in-Chief of the British Forces who [typically] portrayed the outcome as a success:

> *"Newfoundland may well feel proud of her sons. The heroism and devotion to duty they displayed on the first July has never been surpassed. Please convey my deepest sympathy, and that of the whole of our armies in France in the loss of the brave officers and men who have fallen for the Empire, and our admiration of their heroic conduct. Their efforts contributed to our success, and their example will live."*[41]

So as to perpetuate the memory of those who were lost on 1 July, 1916, in the 1920s, the Newfoundland government bought the ground over which the Newfoundland Regiment fought. The memorial park it established at Beaumont-Hamel has since become a place of pilgrimage for Newfoundland and Canadian people wishing to honour and mourn the Regiment.

Raids Undertaken In Fleurbaix And Fromelles on 1 July, 1916.

While 1 July, 1916 marked the beginning of the Somme Offensive, and numerous battles were being fought along that river system, raids were continuing to take place, directed at the German soldiers defending Fromelles. Two particular accounts have emerged of raiding activities undertaken by the 9th and the 11th Battalions during the evening of 1-2 July, 1916.

The Work of The 9th Battalion.

The actions of the 9th Battalion were directed by **Captain MAURICE WILDER NELIGAN**, later Lieutenant Colonel, C.M.G., D.S.O. and Bar, D.C.M., Croix de Guerre, M.I.D. (5 times)

It is worth taking time to examine the methodology adopted by Neligan, a leader who was systematic in planning, preparing, and executing raids on the German lines at Fleurbaix / Fromelles. A biography of Neligan has been the work of Peter Holmes in: *A Magnificent ANZAC. The Untold Story of Lt. Col. Maurice Wilder Neligan*, but it is the activity undertaken in Fleurbaix / Fromelles on the evening of 1-2 July, 1916 with which we are concerned here.

Neligan was a mercurial and eccentric Englishman who joined the Australian Army as a private soldier at the beginning of the First World War, lying about his identity, age and birthplace. He landed with the 9th Battalion at Gallipoli on 25 April, 1915 and was both decorated for gallantry, and commissioned as an officer, a few days later. No sooner had Neligan arrived at the western front than he suggested to the Commanding Officer of 9th Battalion on 21 May, that he be

---

[41] Evening Telegram July 21, 1916

allowed to lead a raid on German trenches; in fact he 'begged' to lead it. Neligan was soon to get his wish.

In June the Australians organised a series of raids; twelve in all from a variety of units. The 9th Battalion was selected to organise a raid on the German trenches in early July, and Neligan was to play a pivotal role in its success. Brigade Commander Brigadier General Sinclair-MacLagan was an early arrival at the C.O.'s office. He advised the C.O. that he had selected 9th Battalion to do the 3rd Brigade's raid, and that he would like [then] Captain Wilder Neligan to lead it.

The orders which Neligan prepared for the raid and issued on 24 June were thorough, well written and direct; they left nothing to chance. He spelled out precisely what each member of the party should take with them; this would depend upon which section they belonged to. These sections were composed of ladder men, bomb throwers, bayonet men, telephonists, messengers, tape men, blocking men, and salvage men. There were to be three parties, consisting of a right, left, and centre party. Neligan specified the type of weapons to be carried, the number of rounds of ammunition and bombs, and equipment such as tape and tools. Hands and feet were to be blackened, no papers to be carried, and white calico armbands would be worn, three inches thick, with a band of luminous paint around the centre of the band.

Detailed instructions were also given about the composition of lines for the raid and how the raid would be carried out. The men were told what to do when they entered the enemy trenches, what direction to take and what each section should be responsible for. He ordered that wounded men must be left, and collected on the return journey. He even threatened a court-martial for any man who left the attacking party to tend to a wounded colleague.

Neligan decided that the raid should be carried out without an artillery bombardment – it was to be a "silent raid". The only bombardment was to be trench mortars to assist in cutting the enemy's wire and a small bombardment in a flanking sector to deceive the enemy of the real location of the raid. No-Man's Land was about 300 yards wide at the point of entry and one reason for the selection of this particular point was the location of a German machine gun, which had caused much loss of life and was a threat to the Battalion.

The teams for the raid had been very carefully trained behind the allied lines, and on the evening of 1 July, they were taken by bus to an old house near the firing line. Their faces were blackened and they were given chewing gum to suppress any coughing. In the afternoon before the raid, trench mortars were used to cut the wire in front of the enemy trenches, and at 11.00pm, the pre-arranged artillery bombardment was directed at part of the enemy's line *as a feint*.

At 11.00pm the three parties began to cross No-man's land and by 12.45am they were all in position, 200 yards from the German front line trench. Two machine guns then opened fire to cover the noise of a further advance and the party moved another 100 yards toward the trench; they reached this objective by 1.30am. The order was then given to advance to 50 yards toward the trench. None of the raiders had yet been discovered. At 2.03am, Neligan gave the pre-arranged code of "gallop" and an artillery bombardment was commenced behind the German lines and machine gun fire swept the flanks. All was so far going to plan.

Shortly after the artillery bombardment had begun, the parties rushed the German trenches and by 2.12am, all had reported their entry into the trenches. Lieutenant Colonel Robertson, Commander of the 9th Battalion ordered a rocket signal to be fired signalling the men to return, and by 2.17am, all parties had returned to their own lines and the artillery bombardment had ceased.

The raid was quick and effective but there was much bitter fighting in the German trenches. The troops found the machine gun which had caused the Australians so much trouble, but they were unable to remove the gun from its stand, and so they carried it back – with its base – to the Australian line. The right flank party, led by Captain Claude Benson, entered their sector of the trench and killed many Germans and captured four. Captain Benson, the last to leave the trench was unfortunately killed as he was assisting two wounded men back through the wire. Eyewitnesses said he was shot just above the heart and was killed instantly. He was a popular officer. His death must have been a blow to Neligan.

As for his part in the raid, Neligan's actions were described in an article in *The Fighting Ninth* as follows: -

Neligan moved in behind the listening post, and having dropped quietly into the enemy position, he clattered down the duck boards towards the listening post, taking no pains to conceal his approach. He rightly calculated that whoever was occupying the position would suppose that the noisy approach was being made by some German Officer on a round of inspection. He found as he expected, 2 German soldiers peering into No-man's land, as though all their attention was centred upon their duty of watching and listening. They had their backs turned to Neligan, and fell victims to 2 rapid blows from the only weapon he used – a knobkerrie (a short stout stick with bolts attached to its end). Thereafter the raid on the German Trenches was carried out. Neligan came unexpectedly upon a party of wounded Germans in a trench. These threw a bomb toward him and he was seriously wounded in the head. The enemy occupying the post was killed; but badly wounded as he was, Neligan held on to his command. Finally, the managed to crawl back to the lines, where in the casualty station he refused attention till his men who had been wounded had been attended to.

The raid was a triumph, with C.E.W. Bean describing it as *probably the most brilliantly executed of this series of raids.* But war on the Western Front was a bloody affair and the Australians had lost 9 dead, 21 wounded and 5 missing, and according to a German source, the Germans had lost 14 dead, 40 wounded and 25 had been taken prisoners (although the Australians claimed to have killed 53 Germans and captured 21).

Neligan had impressed on his troops that *one prisoner is of greater importance to the Army Corps than that of three dead Germans.* The Australians had captured 21 German soldiers. Neligan would have been very happy with this result.

After Neligan returned to the lines he was immediately hospitalised. He had been struck in the head by a piece of shrapnel, which had fractured his skull and he was given emergency surgery on 2 July, and the depressed fracture was raised. Following Neligan's operation at a hospital in France, he was sent to the Empire hospital in London (which specialised in brain injuries) with a severely fractured skull and underwent several more operations. He was completely paralysed for some months. He had lost a large amount of bone over his right occipital parietal (the right side of his scalp) on his head and was very lucky to be alive. He had permanent damage to his skull and suffered from severe headaches for some time after his injury and possibly for the rest of his life. His recovery was long and painful.

While he was convalescing from his head wound in London, he was awarded his initial DSO for his planning and leadership of this raid, and was presented with his DSO by the King, at Buckingham Palace.

Neligan went on to re-join his unit and lead it in battles culminating in the end of the war. He would go on to become Lieutenant Colonel, and be awarded C.M.G., D.S.O. and Bar, D.C.M., Croix de Guerre, M.I.D. (5 times).

The Work of The 11th Battalion.

Separately, the action of the 11th Battalion would also show how prior planning, supported by trained artillery led to a result which was so unexpected, that the soldiers sent to finalise the raid on the German trenches, were able to walk back to their lines, instead of having to crawl on their stomachs.

---ooo000ooo---

On 2 July, 1916 the 11th Battalion received a sudden order to move, and about 9.00pm set out for Oultersteene, a small village close to Merris.

After a long march, which finished about midnight, the 11th Battalion arrived at its destination. There was a good deal of hunting round in the dark, but eventually the troops were settled in their billets.

On the following morning, 3 July, 1916, the troops heard the result of the raid by the 11th Battalion, which took place on the previous night. The actual date of the raid had been kept a secret until the day before the event. At the appointed time Medcalf led his party up to the front line. All watches had been synchronised with that of the artillery. At zero hour the guns opened with a crash, and the shelling was well on the target. The wire in front of the enemy trenches had been smashed by trench mortars earlier in the day.

Captain Medcalf's own report of the operation recounted as follows[42]: -

Under cover of our concentrated artillery fire, we crawled forward across No-man's land in our formation, and came to a halt within 50 yards of the German parapet.

Our artillery was bursting its shells right on top of the Hun's parapet, and there we lay in perfect safety only 50 yards short of it, waiting for it to lift, so that we might follow along and enter the German trench. One would have thought that the boys were on the parade ground — the only difference was that they

[42] From the article by Captain Walter C. Belford M.A., published in the Western Mail, Perth WA, on 12 May, 1938: titled: - **In Supports At Fleurbaix,** available via the following link: https://trove.nla.gov.au/newspaper/article/37847963?searchTerm=In%20supports%20at%20Fleurbaix%20%20%20%20%20%20%20%20%20%20%20%20%20%20%20%20&searchLimits=

were keener and steadier and better disciplined at that moment than ever they were on parade. By and by our guns lifted and put a curtain of fire behind the Hun's trench, in order to prevent him bringing up supports.

This was the moment we were so anxiously awaiting; so I sat up and gave the order to adopt our attacking formation and move forward. The boys doubled through the wire in silence; one or two were hooked up in the barbs, but most of the wire had been smashed previously, so it did not offer much difficulty.

Through the breach in the Hun parapet, we went, and spread out inside the trench, seeing many dead and taking numerous identification papers, helmets, and respirators. There was only one live Hun left in all that amount of trench, our artillery had been so deadly. Unfortunately, an over-zealous private bayoneted this man instead of taking him prisoner.

When we had been in the trench for the specified time. I gave the order to return through the place by which we had entered and make for our own trench again. The men just walked home across No-man's land. I was never so surprised in my life as to see our chaps walking in that particular area, where a man generally crawls gingerly on his stomach.

But our artillery and machine guns were so effective in silencing the Huns that it was quite safe. Every man in our party returned safely to our trenches.

The members of the raiding party were at once checked and sent to the rear, where motor waggons picked them up, and hurried them back to billets.

After a few hours' rest the raiding party followed the 11th Battalion to Oultersteene (approximately 12 miles, or 20 kilometres from Fromelles).

The identifications gained by this raid proved that the trenches were then held by the 21st Bavarian Regiment of the 6th. Bavarian Reserve Division, the same Regiment which was to be faced by the 5th Division in coming weeks, at Fromelles.

---ooo000ooo---

A church parade was held on Sunday, July 9, when both the 9th and 11th Battalions attended. General Birdwood attended the service, which was long remembered by the troops for the sermon preached by the padre, who told the boys that life was "a mere incident," and that its loss was nothing, and more to the same effect.

Rumours had been going round that the Australian divisions were going down to the Somme, where there had been such a colossal loss of life already. Consequently, the troops resented the padre's homily that life was a mere incident. It was rather a tactless sermon to give to troops who were about to be thrown into the terrible maw of the Somme battlefield.

After the service there was a short address by General Birdwood, who then definitely informed the troops that they were going to the Somme, where the "Big Push" had recently started. Birdwood told the boys that they would have a chance of open fighting, which, he said, he was sure that the diggers would prefer to the trench fighting that they had been engaged in up to this time in France.

The news was received with great joy by the boys, and they marched back to their billets entirely satisfied with themselves and the world generally.

Immediately after midday the 11th Battalion received orders to "fall in" and it was marched off to billets about a mile and a half north-east of Meteren. The next three days saw a march to The Somme, where Pozieres awaited.

---ooo000ooo---

The preparations for, and the execution of the raids was recounted by Sgt. H. ('Squatter') Preston, 9th Battalion in an article –

EARLY RAIDS IN FRANCE – 9TH BATTALION COUP AT FLEURBAIX

published in the Daily Examiner, Grafton NSW, on 17 January, 1935, and in which Preston confirmed the meticulous nature of planning which can precede the carrying out of raids.

Soon after the arrival of the Australians in France orders were issued to them for the capture of German prisoners, and, despite numerous tries, it was not until the end of May, 1916, that a prisoner was taken. After this several strong patrols and ***raids*** were organised to capture prisoners and inflict loss on the enemy.

Preparatory to the raids, No-man's land was well patrolled. During one such patrol on the night of June 18, the soldiers had reached a position right on the German barbed wire when they were spotted, and then after a flash. 'Bang!' and Captain Warren was dead.

The raid on the evening of 1-2 July was a "silent one," without preliminary bombardment.

For days previous the enemy barbed wire was cut by trench mortars. One day during one of these bombardments, where the trenches were separated by 350 yards, the butt of a German rifle was flung 100 yards behind the Australians' lines.

The raid was carried out in three formations, led by Captains Benson, Young and Ramkema, with 150 other ranks.

Most of the activity was centred on Preston's right. A ***sap*** was dug into No-Man's Land, so as to deceive the enemy of the exact point of entry.

The raiders, all volunteers, were trained behind the lines, in the vicinity of Steenwerk. All were kept in ignorance of the date of the raid until the night of July 1, when they were moved to the final rehearsal point — an old farmhouse just behind the lines.

Preston went on to recount: -

Our hands and faces were blackened, and we were given chewing-gum to prevent us from coughing. My wristlet watch with an illuminated dial was taken from me, and this was my first intimation that the actual raid was about to take place.

I was allotted to Lieutenant Ramkema's party, which was in the centre. After taking up our position about 12.30 in No-man's land, 50 yards from the enemy, word was sent to the artillery and machine guns.

The artillery laid down a barrage beyond the objective, while the machine guns swept the flanks. I will never forget that mad rush to the enemy lines; men falling on both sides of me. My gas helmet was torn from me by barbed wire and I fell into a deep shell hole and lost my rifle, but soon recovered it.

Shells were screeching overhead and bursting behind the enemy front line. The wall of flame resembled a bush fire.

I could see quite clearly the heads and shoulders of Germans above the parapet — the enemy shooting at us from only a few yards distant.

Just as I reached the parapet somebody alongside of me — I think it was Sergeant Barry — was shot, and fell into the trench.

I next heard someone call out 'gas!'

I reached for my helmet, but it was gone.

When I jumped into the trench, which was about 8ft. deep, I was confronted by a number of Fritzies in an overhead covered communication trench.

I had a good mate in Tom Horsington. He had a good supply of bombs, so we soon disposed of the Fritzies and cleared the trench.

Turning into the next bay I saw a dark object. Flashing my torch on to it (the torch was attached to my rifle), I noticed a big Fritz. All he had to defend himself with was a dagger. This Fritz was taken prisoner. I still have the dagger, as a souvenir.

Further down the trench an enemy, machine gun was collared. Dan Mahoney did good work here, and was rewarded with the. M.M. Then came the word "imshee" — the signal for all to leave the trench.

As I ran along to find a way out I came across Lieutenant Meyers wounded in the neck. I helped him out over the top and then turned in the direction where a lot of fuss was going-on.

Lieutenant Ramkema had discovered a dug-out with about 20 Germans in it.

He emptied his revolver into them and ordered the survivors (about 12) to come out. Sergeant Keyon and myself drove them towards our lines.

By this time No-Man's land was being swept by machine gun and artillery fire. Well on our way our prisoners, terrified by the bombardment, started to give trouble. Some were shot and the remainder came in without further resistance.

During the stay in the enemy lines all parties had heavy fighting. Captain Wilder Neligan found an outpost with three men in it. He shot two, the third threw a bomb and Neligan was wounded in the head, but he carried on in command.

Lieutenant-Colonel Robertson, who was watching operations from our lines, sent up the rocket for our recall at 2.30.

All parties returned and the artillery ceased firing. Sergeant Lucas and myself returned to No-man's land and recovered a lost telephone.

The prisoners, about 30, were blindfolded and taken behind the lines. The raiders — those who were left — were taken back by motor lorries, given a bath, and an issue of coffee and rum and were allowed to sleep all next day.

After we entered the German lines with faces and hands blackened, one Fritz who could speak English, on being taken prisoner, said: "Australia no black," and opened up the tunic of one man to see his skin.

Each raider who survived was promised a holiday in Blighty, but instead had a three days route march to the Somme and put into Pozieres.

The Rhondda Sap

The 'sap' referred to in the preceding article was called the Rhondda Sap. It was effectively a trench dug in No-man's land in order to be able to have soldiers come into it, ahead of any attack on the German lines. It was so called because it was started by Welsh men. Rhondda, or the Rhondda Valley, is a former coal mining valley in Wales, formerly a local government district, consisting of 16 communities built around the River Rhondda. The sap had been used much earlier in a manoeuvre to reduce the width of No-man's Land. It would be pushed out some 200 yards and then joined left and right with other trenches, which when well established would become the new British front line. But the water table, let alone the machine gunners of the Sugar-loaf had made the task impossible. That it should be started up again in the battle which was to be very dependent on its success, makes one wonder how intelligent were the Intelligence Staff? Much of the original work on the Sap was done by the 3rd Australian Tunnelling Company, and the sad fact that more Australians were lost on this pitiless task, and that many were buried in it, makes it of special importance.[43]

More information about the Rhondda Sap and the work of the 3rd Australian Tunnelling Company will be spelt out in Chapter 6.

---ooo000ooo---

While the 1st Divn. A.I.F. certainly knew about Maurice Wilder Neligan, it is doubtful that the results of his ability to plan, prepare and execute raids were passed on to the relieving 5th Divn. by the staffs of Generals Birdwood and Plumer. The confrontation with the Germans in the coming Battle of Fromelles later that July might well have proceeded differently if these results had been.

Neligan's orders were a textbook demonstration of patient and careful planning and this would not have been lost upon the men who served under him. To some extent his raid had been an experiment, but it was an experiment which had paid off in bounds. He certainly understood what a ***raid*** was. It is therefore not only a mystery as to why his methods were not adopted more and more, but whether or not they had even come to the notice of Haig and Haking so as to influence their thinking in the planning for the Battle at Fromelles on 19/20 July.

They probably weren't.

[43] Corfield, p.43

5. The Chain of Events from the Start of the Somme Offensive Until the Battle of Fromelles

One can find a great deal of historical accounting for the Battle of Fromelles in the *Official History of Australia in the War of 1914-1918 Volume 3, The AIF in France 1916* Chapters 12 & 13 The Battle of Fromelles. Compiled by C.E.W. Bean, the history devotes almost 120 pages to the Battle of Fromelles (pp 328-447). This is available on line through the Australian War Memorial website www.awm.gov.au. In this chapter, we will attempt to distill the events which occurred between 1 July, 1916 and the original *intended* eve of the Battle of Fromelles, on 17 July, 1916.

---ooo000ooo---

The British High Command sought actions to prevent the German Army reinforcing their forces on the Somme. On 5 July prospects of a break-through on the Somme appeared so promising to General Haig, now Commander-in-Chief of the British Expeditionary Force that he ordered the other armies to prepare attacks in case the enemy was thoroughly beaten there. In that event the Third Army, commanded by General Allenby, on the northern flank of the battle, would launch an offensive; but Haig on 5 July also ordered:

The First and Second Armies should each select a front on which to attempt to make a break in the enemy's lines, and to widen it subsequently.

He pointed out that the German armies on the Somme might contemplate withdrawal, and in that case the First and Second British Armies, by attacking, might '*turn the retreat on the Somme into a general retreat.*'

Responsibility For, and Plans For, An Attack At Fromelles:

In general terms, the command in relation to events around Fromelles, went from Haig to Butler, to Plumer, to Monro, to Haking; thence Haking to Mackenzie and McCay.

Haking suggested a First Army attack towards Fromelles, not towards Lille as originally planned.

John Bourne commented[44]: '*that (Fromelles) took place at all owed most to the ambition and willingness of Haking to carry it out, and his unshakeable confidence that it would work. Fromelles is difficult to justify as the point for an attack, even a feint attack as it was flat ground, broken up by water obstacles, and overlooked by Aubers Ridge.*'

Fromelles lay near the boundary of Second and First Armies, opening the possibility of participation by Second Army, whose GOC Plumer was reluctant to mount a diversionary attack at Ypres or Messines. Haking's plan did not take into account the earlier failure on the same ground in May 1915.

---

[44] As referred to by Beckett, Dr Ian F; Corvi, Steven J (editors) (2006). Haig's Generals. London: Pen & Sword. ISBN 9781844158928

General Plumer of the Second Army that held the line from north of Ypres to south of Armentieres, noted that where the Second Army joined the First Army opposite the Sugar-loaf Salient, the Germans held their front more lightly and he proposed a joint operation to General Monro, the commander of the First Army. General Monro on 8 July requested General Haking, an experienced and distinguished officer, formerly commanding the British 1st Division and now commanding the XI Corps of the First Army, to draw up plans for an offensive, and to assume that his Corps would be reinforced by a division from the Second Army, together with some of that division's artillery.

General Haking's scheme, aiming at the capture of the Fromelles-Aubers Ridge, a mile behind the enemy's front, was rejected by General Monro, being of the opinion that the capture of the Aubers-Fromelles Ridge, though of great advantage if the rest of the front was to remain stationary, would be of little assistance in the case of the advance on the Somme, for which he had been asked to prepare.

By then, however, the situation on the Somme had changed and Sir Douglas Haig's general staff were now looking into several operations recently suggested and concluded that the attack on Aubers-Fromelles, undertaken as 'an artillery demonstration,' would 'form a useful diversion and help the southern operations.'

Australian troops commenced moving to the Western Front from training camps in Egypt and England from March 1916 with the 1st and 2nd Divisions of the AIF taking over part of the line in the 'nursery' sector south of Armentieres. The 4th Division Unit Diary for 2 July states that the 4th and 5th Divisions were to relieve the 1st and 2nd Divisions so that they could be moved to a reserve area ready for a move to the south (Somme) or north (Ypres) if required. The 4th Division relieved the 1st Division in the Fleurbaix / Fromelles area on 3 July while the 5th Division was still arriving in France (from 26 June to 8 July), but on 6 July the 4th Division was also added to the reserve force to move to the south and was relieved by the 5th Division, which took charge of part of the line on 12 July as part of the First Army under the command of General Monro.

In the first week of July, there were a number of changes, principally 12th and 13th Bdes relieving 1st and 3rd Bdes. At this time some battalions were badly damaged by bombardments, not least the 51st Bn (13th Bde, WA) with 151 casualties: an event ignored by Bean. In another 13th Bde battalion, the 47th Bn Pte William Elsdale (No. 4484, Qld) was killed in action at Fleurbaix on 7 July. A full-blooded Aboriginal from Queensland he may have well been the first of his people to be killed on the Western Front.

Also little known is that the 2/4 Glosters actually got into the German trenches on 4/5 July.

5 July, 1916: -

On 5 July, 1916, prospects of a break-through on the Somme appeared to Haig so promising that he ordered the other armies to prepare attacks in case the enemy was thoroughly beaten there.

In that event the Third Army, on the northern flank of the battle, would launch an offensive.[45]

But on 5 July, Haig also ordered the First and Second Armies to select a front on which to attempt to make a break in the enemy's lines in the following terms:

The First and Second Armies should each select a front on which to attempt to make a break in the enemy lines, and to widen it subsequently.

He neither prescribed there be a feint, or an attack, but these few lines in the hands of various generals in due course gave rise to the Battle of Fromelles.

Farther south, where the Second Army joined the First opposite the Sugar-loaf Salient, the Germans held their front more lightly.

On 5 July, 1916, as the 5th Division's officers were visiting the line between Bois Grenier and the Sugar-loaf that they were about to take over from the 4th Division, 13 Jager Battalion, which had formed part of the Sugar-loaf garrison, was identified on the Somme. Plumer had already sent II ANZAC another appeal from Haig to prevent such transfers.

7 July, 1916: -

On 7 July, Godley told McCay 'that raids and all possible offensive (sic) should be undertaken at once'. General Godley issued to the II ANZAC Corps – of which the 4th Division was heading for the Somme, and comprised only the New Zealand and 5th Australian Divisions – the following order: -

> ***Raids*** *must therefore take place immediately, and must be on a larger scale than has hitherto been attempted – about 200 men or a company …*

8 July, 1916: -

On 8th July, 1916, the 4th Division, less its artillery, was ordered to follow the 1st and 2nd to the Somme and the 5th Division prepared to relieve it.

General Monro, at a conference of his corps, directed Haking to draw up plans involving the First and Second Armies.

By then however, the situation on the Somme had changed. The Germans had been able to bring up reinforcements, and a minor offensive elsewhere was required … for that of pinning his forces to their existing fronts and preventing their movement to the Somme.

Sir Douglas Haig received information that the Germans had transferred to that front from the Lille – Lens area some nine battalions as reinforcements. Nothing would be more likely to prevent further transfers than a threat, made by the First and Second Armies, of a British advance upon Lille. [46]

---

[45] Bean p.330

[46] Bean p.332

The general staff, now looking into the several operations recently suggested, concluded that the *attack on Aubers-Fromelles, undertaken as 'an artillery demonstration,' would 'form a useful diversion and help the southern operations.'*

It was a very different scheme from that which Haking had been asked previously to draw plans. It was, however, suggested that a scheme should be worked out for an infantry advance to the Aubers Ridge, in case this might 'at a later stage' become advisable.

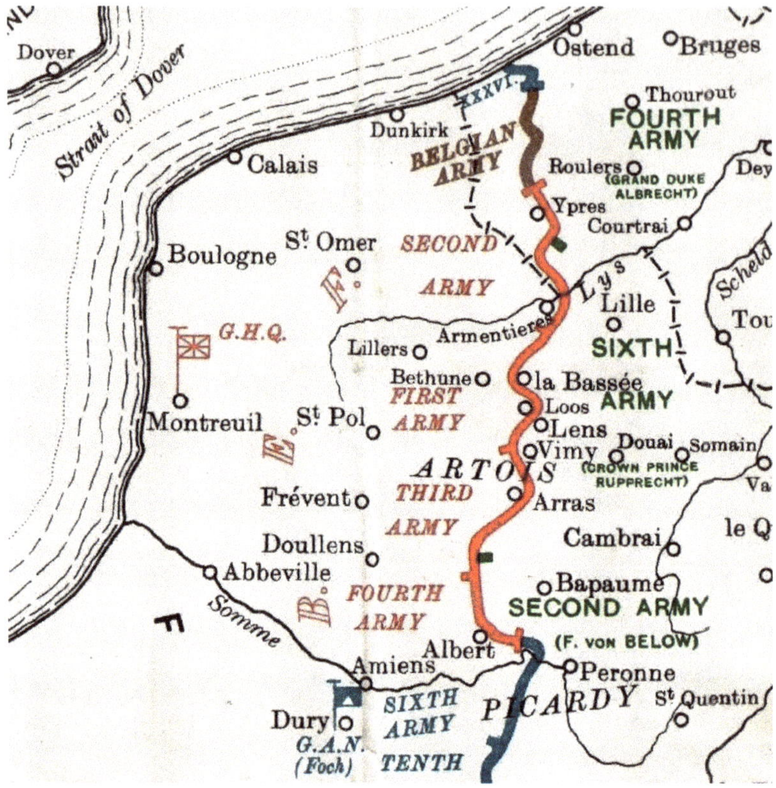

[Picture 14: Map 7: Location of Armies on the Western Front 30 June1916[47]]

That very day when Haking was instructed to draw up his plan, the 4th Australian Division had been suddenly ordered to follow the rest of the 1 ANZAC Corps to the Somme. The 5th Division, ordered up to relieve the 4th, began on the same day, its first march towards the front area.

9 July, 1916: -

On 9 July, Haking outlined a scheme for an attack by three divisions on a 4,600 yard frontage against Aubers Ridge, 'to include the two main tactical localities on the ridge, the high ground around Fromelles and the village of Aubers'. General Sir Charles Monro felt that a thrust from Loos held better prospects in the event of a Somme success. He turned Haking down.

On that same day, the 2/5 Warwicks attacked in daylight.

Bean wrote, [48] outlining the effect that various raids had on German forces from 9 July, 1916 onwards: -

Raiding was therefore at this time no easy process. Moreover, it is difficult to see how it could cause any serious anxiety to the enemy, and German accounts now available make it quite evident that it did not.

---

[47] Sources: *Official History of the Great War: Military Operations: France and Belgium 1916, Volume 1 Map 3* [extract from *The Western Front Association Mapping the Front in association with the Imperial War Museum*]. Reproduced in accordance with the Open Government Licence: http://www.nationalarchives.gov.uk/doc/open-government-licence/version/3/

[48] At 328

Next day, July 10, orders were given to pack up and be ready to move off at a moment's notice. No move was made till 7.15pm, when the battalion was moved off to the north past Le Coq de Paille to Mont des Cats, where the troops bivouaced for the night. It was fine weather and it was no hardship to sleep out among the fields and trees that beautiful summer night.

10-11 July, 1916: -

On 10 July, the 8th Brigade relieved the 4th Brigade in the left section.

29 and 30 Battalions left Erquinghem for 8 Brigade's sector in front of Bois Grenier on the left, while 57 and 58 Battalions from 15 Brigade, co-located with the divisional headquarters in Sailly, moved in opposite the eastern face of the Sugar-loaf on the right. 14 Brigade filled the centre with 55 and 56 Battalions from Fleurbaix on 11/12 July. Each battalion held 1,000 yards of front and the 4th Division's artillery, deemed too inexperienced to go to the Somme, stayed behind in support.[49]

The 14th Brigade relieved the 12th in the centre section.

At 5.50am on 11 July, the Germans blew a mine in a tit-for-tat response to one the Australian tunnellers had exploded under them a week before.

12 July, 1916: -

General McCay at noon on July 12th in Sailly chateau took over from General Cox the command of the sector. [50]

His division, though last of the AIF to arrive in France, would be the first in serious action. Haking's plan as then explained to him, was to seize 6,000 yards of German front line – from the Fauquissart – Trivelet road to a point opposite the Boutillerie – with three divisions: the 61st and 31st of Haking's own corps, would assault the south-western and northern fronts of the Sugar-loaf; the 5th Australian Division would extend the front of attack as far as Boutillerie.

That day, a Stokes mortar fired without warning near D Company, 2/4 Ox and Bucks. Replying instantly, the Germans cleaned up the entire platoon.

13 July, 1916: -

2/4 Royal Berks had their turn on 13 July, but the Bangalore torpedo party was hit on the way across and only the first wave of eleven men got through the German wire.

Haig was informed that eight German battalions had followed 13 Jager to the Somme from the Lille area. As *the raiding programme had failed to stop them*, stronger action was necessary, especially with the second great attack on the Somme about to begin. You can easily conclude that if a raiding programme was seen as having failed to stop the Germans, that attacks, and indeed Battles would be needed.

---

[49] Pedersen, p.33

[50] Bean p.335

But at a conference held on the 13th July, these alternative plans seem to have got mixed, and a wretched, hybrid scheme, which might well be termed a '*tactical abortion*,' emerged from the Council of War.[51] At this conference, Major General Butler, Haig's Deputy Chief of the General Staff, met General Monro and his Chief of Staff, General Barrow. Plumer was not present, but was represented by General Harrington, his Chief of Staff.

This conference picked out the spot at which the 'demonstration' was to take place, viz., just at the junction of the two armies, i.e., around the Sugar-loaf Salient.

A new departure, however, was provided in that it was decided that in the 'demonstration' some infantry should take part. The 1st Army, under Monro, should provide two divisions, and the 2nd Army, under Plumer, one division.

---ooo000ooo---

Brig Gen Harold 'Pompey' Elliott of the 15th Brigade, 5th Division AIF had deep misgivings: -

At this stage it is just to mention that I had all along had the gravest misgiving as to this attack. When I moved into the line I had carefully reconnoitred it from every point of view. It had been the scene of an attack early in the war (viz., 9th May, 1915) by some 30,000 (three divisions) of Regular British troops with 25,000 more in reserve, supported by 500 guns who fired 80,000 rounds of ammunition. They suffered 10,000 casualties, and did not gain a single yard, and the attack was abandoned at nightfall on the ground that the continuance of the attack would mean a useless waste of life. Moreover, General Haking had a little farther south on the 29th June, 1916, carried out a similar attack, which was repulsed with heavy loss. Vide page 346, Bean's note.[52]

I finally persuaded Major Howard, one of General Butler's assistants on General Haig's staff, to accompany me from my H.Q. at Rouge de Bout to No-man's land. I showed him my orders. I pointed out the difficulties. I explained why I had placed machine guns in the space between my right and the British left so as to cover my flank in the event of the British failing to take the Sugar-loaf and to cover our advance until they had done so, a measure of which he warmly approved.

I asked him then if he could give me any suggestion to better my orders, and he admitted he could not.

Then I turned upon him and said, 'Now you know all I do. You have been on this front nearly two years; I've been here ten days. I give you as my fixed opinion that we have not an earthly chance of success. I want you to tell me straight out, as man to man, and strictly between ourselves, what you think of it all?'

His reply was: 'If you put it to me like that sir, I must answer you in the same way as man to man. It is going to be a bloody holocaust.' 'Well, then,' I said, 'if that is your real opinion, will you promise me that you will go to-night on your return to General Haig, tell him that is your opinion and that it is mine also.'

[51] Elliott p.2

[52] Elliott p.6

He was much moved by this appeal, and promised faithfully to do so. Though I have never spoken to him since or heard from him, I know that he faithfully observed his promise, as is shown by the events that followed. The subject is referred to in Captain Bean's History, see pages 346 and 413, Volume III.

---ooo000ooo---

At 11.00pm on 13 July, McCay arrived at La-Motte-au-Bois to be told that his division was temporarily joining XI Corps for the attack. The New Zealand Division and 60 Brigade from the 20th Division would take over the left half of its sector so that the Australians could concentrate on the narrower assault frontage.

14 July, 1916: -

At 2.00am on 14th July, Haking learned that the artillery he was receiving was not the amount he was expecting. Instead of sending him the guns of three divisions, Second Army was providing only the incompletely trained artilleries of the 4th and 5th Australian Divisions. Worse still, the ammunition allocation was slashed by a third to 215,000 shells. He immediately scaled down the operation by excluding the 31st Division, and wisely, decided to narrow his front of attack to 4,000 yards, and to attack with two divisions, the 61st UK Division striking at the Sugar-loaf, as already arranged, from the west, and the 5th Australian Division (instead of the 31st Division) from the north, with its left flank at Cordonnerie.

On the morning of 14th, McCay learned of the change in the plans. It was decided that an infantry attack should form part of the demonstration, the First Army probably providing two divisions, and the Second Army one. This coincided with the 5th Division receiving its final 5,000 trench helmets. The bombardment was to begin on 14 July with all the artillery then available, and was to last about three days. General Haking's scheme of attack was therefore approved, its object (according to the First Army order issued on 15 July) being:

'to prevent the enemy from moving troops southwards to take part in the main battle. For this purpose (it was added) the preliminary operations, so far as is possible, will give the impression of an impending offensive operation on a large scale, and the bombardment which commenced on the morning of the 14th inst. will be continued with increasing intensity up till the moment of the assault.'

Elsewhere, The Battle of Bazentin Ridge Took Place, from 14 July to 17 July, 1916

14 July, 1916 saw the start of what is now named the *Battle of Bazentin Ridge*. It was to serve the Capture of Longueval, the Capture of Trouee Wood, and the Capture of Ovillers.

That Action took place from the 14th July to 17th July. Geographical information in the BNC report states: Road Hardecourt – Maricourt – Fricourt – Becourt – Albert – (exclusive): thence to River Ancre. It was more than 80 kilometres from Fromelles.

The BNC Report attaches Fromelles to the Battle of Bazentin Ridge as a *subsidiary Attack at Fromelles on 19 July, 1916.*

Geographical information in the BNC report for Fromelles states: Road Aubers – Fauquissart – Laventie – Rouge-de-Bout – Fleurbaix (exclusive) – la Boutillerie – Bas Maisnil.

The BNC Report also attached subsequent *Attacks on High Wood* from 20-25 July. The geographical information for which states: Road Flers – Longueval – Bazentin-le-Grant – Bazentin-le-Petit – Martinpuick.

Captain E.A. James in his Record of Battles and Engagements adopts the same referencing of dates, and again refers to the 'subsidiary Attack at Fromelles (on the Aubers Ridge), 19th July, 1916'.

In her article: -

'YESTERDAY, 17TH JULY, 1916, THE BATTLE OF BAZENTIN RIDGE ENDED[53],

Nicki Giles wrote: In many ways, the Battle of Bazentin Ridge, part of the second phase of the Battle of the Somme, was a thing of tactical brilliance. The plan was conceived by XIII corps commander Lt Gen Walter Congreve VC in conjunction with Fourth Army Commander Rawlinson. They had learned from the slaughter of the Battle of Albert on 1st July 1916, when 20,000 Allied troops had perished. Not for them the week-long softening barrage, which would herald the imminent attack as clearly as a trumpet call when it finally ceased. Not for them the approach in broad daylight, for fear that green troops would be confused by low visibility, nor the slow, methodical approach on enemy lines. Never again would they make the same mistakes.

This time, the attack would be launched very early, under cover of darkness. Field Marshall Haig argued strenuously against this tactic, fearing the new soldiers would struggle, and at first over-ruled Rawlinson's wish. However, after a second application by Rawlinson at the urging of the corps commanders, Haig reluctantly came around.

The initial barrage, it was agreed, would this time be just five minutes long. The Germans, expecting a much longer assault, would be caught on the hop. Last time they had calmly retreated to their shelters to wait out the storm of missiles, then, in the delay between it stopping and the troops launching their somewhat slow attack, climbed back out at their leisure and settled themselves behind their guns. On this occasion, the troops would sneak out into No-man's land while the barrage was going on and lie as close as possible to the German defences, then, as soon as the coast was clear, a bombing party would leap up and rush them, with the rest of the infantry right behind to mop up the surviving German soldiers. A 'creeping barrage', a lesser attack that would take place in front of the troops, halting periodically to let them advance and moving up the field with them, would add to their protection. If they were lucky, many of the defenders would be caught still in their dug-outs.

[53] Published 18 July, 2014. See: - https://www.forces-war-records.co.uk/blog/2014/07/18/yesterday-17th-july-1916-the-battle-of-bazentin-ridge-ended

Although this attack used less guns than the Battle of Albert, it would take place over a narrower front for maximum impact. Having already captured Mametz Wood, the Allies were in a prime position to launch their plan stealthily.

At 3.20am on 14th July 1916 the plan kicked off - and it went without a hitch. At 3.25 exactly, when the artillery bombardment stopped, the bombing party and infantry rose up from No-man's land and rushed the Germans. By mid-morning, all objectives had been successfully achieved and the Allies were ready to launch the next stage of the plan, an attack on High Wood. It had been planned as a cavalry charge, but it soon became clear that the wood was temporarily empty. The Germans, confused by the speed and ferocity of the first attack, were still regrouping and had left this vital chink in their line un-defended.

The attack having succeeded unexpectedly quickly, nobody had realised that the cavalry to be needed this early. According to Mark Adkin's 'The Western Front Companion' there were actually three Calvary Divisions on standby, the 2nd Indian Calvary Division, the 1st Calvary Division and the 3rd Calvary Division. Since the Indian Division, led by 7th Dragoon Guards and Deccan Horse, had been specifically assigned to attack High Wood, they were the only ones ordered forward from their position four miles south of Albert. The order didn't come until 7.40am, and by the time the Division moved off it was 8.20, and there were still miles of churned battlefield to get through. Permission was sought for the infantry to attack instead; it wasn't given. Here was where the Allies still had a lot to learn. They had not yet realised that, in a desperate war such as this one, any possible advantage had to be exploited to the full, and that the side that won would be the one to seize the moment.

The cavalry didn't arrive until early afternoon, then confusion as to whether or not Longueval had been cleared of Germans (the town was within shooting distance of High Wood) meant that the order to proceed was delayed still longer, first until 5.15pm, then again until 7.00pm to allow extra time for the infantry to organise themselves. Nine hours after the gap in the German front line had first been identified, the cavalry bravely charged. By now the Germans had spotted the weakness, and defenders had moved slowly back into the wood and adjoining fields. As the horses plunged into the wood, the machine guns opened fire. Many horses and men perished needlessly. It would not be until the next major offensive that the Allies finally secured the wood that could have been theirs without a fight, and the Battle of Bazentin Ridge ended in a stalemate rather than in victory.

---ooo000ooo---

It will be noted by any student of geography that the various sites of actions, ***except for Fromelles***, are in the Somme region, some 80 kilometres away from Fromelles. And in addition, the actions reported under this heading took place between 14 July and 25 July. But Fromelles was neither in the Somme, nor could it be the subject of a battle at Bazentin Ridge, as that Battle had concluded before the Battle of Fromelles had even begun.

A modern motor car can take approximately 1 hour and 10 minutes to travel from Fromelles to Bazentin. A soldier in 1916 was likely would take 3 or 4 days marching in good conditions.

---ooo000ooo---

Separately, for the period of 14 – 18 July, 1916, the 3rd Australian Tunnelling Company (3ATC) Unit Diary makes reference to work being done at Fauquissart i.e. near Fromelles: -

- The work of sinking the 2nd shaft at Fauquissart (M30.c.05.30).
- New defensive systems and extensions of the Red Lamp System were then started.
- At Red Lamp new work was done at (W.14a.10.30) & (W.136.35.00)
- At Fauquissart new work was carried out at N.136 Central.
- At Tillaloy new work was done at M.246 Central.
- At Chapigny new work was done at M.24 c & d.

Fauquissart is located behind the British defence breastworks, on D171 Rue Tilleloy. The 'Red Lamp' system was the work being done to expand on the Rhondda Sap as preparation for the forthcoming Battle of Fromelles.

So near Fromelles, and also at two other points much further south, the 3rd Australian Tunnelling Company, which worked in Haking's area, had been ordered to prepare to break the surface of No-man's land by thrusting forward from Rhondda Sap, beneath the surface, a pipe filled with explosive.

15 July, 1916: -

Having fulfilled his mission – to ascertain whether all staffs were in agreement – Major Howard reported the arrangements to G.H.Q. But it may be presumed that his grave fears were also at least verbally represented, because Sir Douglas Haig, before whom the report was laid, noted at its foot:

Approved, except that infantry should not be sent in unless an adequate supply of guns and ammunition for counter-battery work is provided. This depends on what guns enemy shows. D.H. 15 July, '16.

A bomb-fight occurred in the trenches of the 58th Battalion during the night of 14 / 15 July, and three members of an Australian Lewis-gun team with their gun were afterwards found to have been captured. As usual, the thoroughness of the bombardment caused heavy loss in the thickly-garrisoned front area (5th Division: 40 killed in action, 6 died of wounds, 118 were wounded, 4 missing; and for the 6th Oxfordshire: 10 killed, 19 wounded). One of those killed was Sgt. George Challis, who will be featured in more detail in later chapters.

In the German front line, however, the impression at first existed that the bombardment was merely retaliation for the previous night's raid.

---ooo000ooo---

And on this day, Major General G. de S. Barrow of General Staff, First Army issued his instruction that:

1. The XI Corps, assisted by portions of the Second Army, will carry out an offensive operation as early as possible **with a view to seizing the enemy's first line system of trenches on the front between the Fauquissart-Trivelet road and La Cordonnerie Farm**. (emphasis added)

2. The object of the operation is **to prevent the enemy from moving troops southwards to take part in the main battle**. (emphasis added) For this purpose the preliminary operations, so far as is possible, will give the impression of an impending offensive operation on a large scale, and the bombardment which commenced on the morning of the 11 inst, will be continued with increasing intensity up till the moment of the assault.

3. For the period during which the operations last, the XI Corps will **continue to hold** the front as far north as point N.10.c.7.8. (emphasis added)

We might just pause to note here that the suggested course of action was going far beyond what has been seen defined in chapter 1 as a 'Raid'. The actions being contemplated were to take ground **and hold it**. Planning for a battle is being foreshadowed, one which we will see should be classified as that: a Battle. However:

> *This order appears to cast aside any intention of secrecy or surprise (the supreme element in war). On the contrary (although an infantry attack was now contemplated), the operation was to be widely advertised by the bombardment — a policy which might conceivably have been a wise one had no infantry attack been required, but was simply suicidal if the intention was to follow it up by an immediate infantry attack.*[54]

That afternoon, gas was mistakenly released from 2/7 Warwicks line, causing several friendly casualties.

At 9.00pm a gas discharge for an hour long was commenced. 17 BRIR lost 25 men. The Germans retaliated with a bombardment on A and B Companies of 58 Battalion. As the shelling reached a crescendo at 11.00pm, 1/21 BRIR showed up. 96 raiders went into the shattered lines, capturing a Lewis gun team after a sharp bomb fight. 58 Battalion lost another 42 killed and 118 wounded and 6/Oxfordshire suffered 29 casualties.

On the night of 15th July for two hours from 9.30pm the Germans put down a heavy bombardment on the Cellar Farm and Mine Avenue area where A and B Coys of 58th Bn were about to be relieved by the 6th Oxfords from the incoming 60th British Brigade, who were going to hold the sector on the left of the 5th Division. The barrage wrecked the dugouts, killed many and was cover for a German raiding party (2 officers, 95 ORs of 21st Bavarian RIR). They lost 10 killed and 22 wounded and left the 58th with one officer and 41 ORs dead, and one officer and 102 ORs wounded. Four men were missing, three being a 58th Lewis gun team complete with weapon: all were taken prisoner – the main purpose of the raid. The Germans thus learned that the 5th Australian Division was in front of them and had been there three days. The German

[54] Elliott

report on the raid said they got no indication of a forthcoming attack either from their excursion into the allied trenches or from their prisoners.

In relation to Elliott's comments on the loss of the element of surprise, the comments set out following, are interesting in themselves, for on the 15th July, the 2/1st Buckinghamshire Regiment, part of the British 184th Brigade, were arriving in the area. Captain G. Christie-Miller wrote the following material in his memoirs: -

> *My recollections are that from the moment we reached Laventie North area again, the troops were practically engaged every night in the line or on working parties, in preparation for an attack to be launched on the morning of the 17th (Monday).*
>
> *I may be asked what should necessitate all this work and why should the assaulting troops be worn out doing it? The second question I cannot answer. The first I should like to be able to answer in detail. Suffice it to say that some hundreds of preserved rations and tins of water, thousands of Mills Bombs, wire and picquets in profusion, and last but not least, an abundance of trench mortar ammunition, heavy, medium and light, if prepared on such a scale can provide a respectable carrying load for a couple of Battalions for a couple of nights.*
>
> *In addition, extensive revetting arrangements for the consolidation of captured positions were made, and endless TM* (Trench Mortar) *Emplacements were built.*
>
> *The details are wanting, but it is a fair general statement that practically none of the troops detailed for the attack, had a decent night's rest from that at Richebourg on the Wednesday (a clear week) before the attack, and then only had one after coming out of a tour in the line.*
>
> *The Battalion were accommodated on the road to the right of and parallel with the railway running from Laventie Station.*
>
> *(It was a curious coincidence that September 9th, 1918, I went up the same road with the Gloucesters to attack Junction Post not 1000 yards from the scene of the Bucks attack of July 19, and that the men killed that day were buried beside their comrades of July, 1916.*
>
> *.... Every road towards the line from Laventie was packed with wheel transport, mostly artillery limbers, and columns of traffic often stood for hours without movement.*
>
> *The concentration must have been obvious, and why it was not interfered with and why the roads were not searched by enemy artillery I have never understood.*

But the movements were indeed very obvious to the onlooking German sentries, who knew that something was about to take place, and were making ready for it.

16 July, 1916: -

On 16 July, Haking issued his final instructions. His 'plan' – in all essentials the same that he had originally devised – was that each division should attack with all three brigades in line, and each brigade with two battalions. They were to go no farther than the enemy's support lines, except to take two ruined farms, Ferme Delangre and Ferme Delaporte, which formed strong-points just beyond the 5th Division's objective.

[The order to take these was eventually countermanded but on the afternoon of the actual attack.]

At this stage the attitude of the British G.H.Q. towards the projected offensive underwent a remarkable change. It is evident from the records that Haig's staff, far from pressing for the demonstration to be made, regarded its probable results with deep misgivings, and it seems certain that some members of the staff would gladly have seen the orders cancelled. [55]

The 'purely artillery' demonstration which some of the staff had favoured, was a very different affair from the projected infantry attack now ordered. It is true that the general on the spot, Haking, was confident of the success of his plan, but the staff must have been acutely conscious of the fact that, three weeks before, Haking had launched a strikingly similar operation on a smaller scale with disastrous results.[56] The local attacks made in the same area and for much the same purpose in May and September, 1915, had signally failed.[57]

This assault, suggested by Haking in the same document which contained his original proposal for the larger operation, had been delivered on June 29 by the 116th Brigade of the 39th Division. The intention was to attack, cut off, and permanently hold the 'Boar's Head Salient', two and a quarter miles south-west of the Sugar-loaf salient. The right battalion reached and held for a time the enemy's support line; the left penetrated the enemy's trench at only a few points. All were subsequently forced to withdraw with heavy loss. Major the Hon. Neville Lytton, who was present, concluding an account, naturally coloured by the feelings of a participant, says (The Press and the General Staff, p.42): 'The Divisional General was ungummed, but it seemed to us that there were others who were responsible, and if they had lost their commands after this failure, possibly greater disasters might have been avoided, for a similar experiment was made a little later on with two divisions and the result was exactly the same. Naturally in the Communique, our attack appeared ***as a successful raid...'***

Incompetence seems to be its own reward. Haking by modern standards would have been relieved of his command. To continue with such stupendous exercises in the absence of any success in the past, using the same methods reminds of what Albert Einstein has erroneously been credited as having later said:

It is insanity to do the same thing, expecting a different result.

Haking's subordinates, like McCay's, found him difficult. Shortly after the 46th Division arrived in XI Corps, its senior staff officer, Lieutenant Colonel Philip Game, was calling Haking 'a vindictive bully'. In February 1916 Game, who would later become Governor of New South Wales, went further, describing him as 'really impossible, untruthful and a bully, and not to be trusted'[58]. Haking also had a fixation about Aubers Ridge. He was 'always very keen' to take

---

[55] Bean p.345

[56] (in footnotes to) Bean p.346

[57] Bean p.346

[58] Pedersen p.36

the feature, remembered a brigade major in XI Corps, and 'always told us if we behaved ourselves, we should be allowed to attack it'[59].

In a conference at Chocques on the day, General Butler pointed out that Haig did not wish the infantry to attack at all unless the commanders were satisfied they had sufficient artillery and ammunition *not only to capture, but to hold and consolidate, the enemy's trenches.* [60]

A report of the conference states that General Haking: -

> *was most emphatic that he was quite satisfied with the resources at his disposal; he was quite confident of the success of the operation; and considered that the ammunition at his disposal was ample to put the infantry in and keep them there.*

All the commanders present, however, were unanimously against a postponement.

When Haking asked whether, in the event of great success, he might push on to Aubers Ridge, the answer given on behalf of Haig was 'No'; the objective was a strictly limited one, and the Commander-in-Chief did not intend to embark in more extended operations, 'however inviting'.

Haking's Memo on 16 July, 1916 was addressed to: 31st Divn, 39th Divn, to 51st Divn, and to 5th Australian Divn, in the following terms:

It has been ascertained that the enemy is moving his troops from our front to resist the attacks of our comrades to the south. The Commander-in-Chief has directed the XIth Corps to attack the enemy in front of us, ***capture his front system of trenches***,[emphasis added] *and thus prevent him from reinforcing his troops to the South.*

[To illustrate this, the 184th Infantry Brigade Order No. 16 on 16 July, 1916 stated at 3. (1) that: *The 61st and 5th Australian Division have been ordered* ***to capture and hold*** *the German Front Line and Support Trenches from the Fauquissart-Trivelet Road to South of Cordonnerie Farm.*]

The memo from Haking went on, and finished with: -

'The objective will be strictly limited to the enemy's support trenches and <u>no more</u>.'

The order then set out in detail, finishing with: -

This order to be read to all troops taking part in the attack...

Heavy rain fell in the afternoon, so heavy that the artillery 'heavies' were unable to be fired to register.

This plan contained a major command and control error over the Sugar-loaf Salient, one of the 61st Division's objectives, but just a stone's throw from the inter-divisional boundary. Commanders regarded boundaries as unfortunate necessities because the co-ordination measures operations across them require can easily break down. Hence, they should not be sited

---

59 Game and E.C. Jepp quoted in McMullin pp205-6

60 Bean p. 347.

on or near places critical to operational success, which the Sugar-loaf certainly was. As the 'vital ground' for XI Corps, it had to be taken for Haking's attack to succeed. [61]

At least four machine guns had been identified in the position. They enfiladed both sides of the salient, where No-man's land, 420 yards across, was at its widest. As the ground was so flat, they were also capable of grazing fire, in which the centre of each burst does not rise above the height of a standing man, for hundreds of yards along No-man's land. This combination, enfilade grazing fire with a long dangerous zone, maximizes a machine gun's lethality by enabling it to catch an assaulting flank, with no dead ground for cover. Because the Sugar-loaf's machine guns could devastate both the Australian and British assaults in this way, the commanders on either side of the boundary near it had to be able to act instantly without compromising each other. Not easy when they were out of touch in the heat of battle. Shifting the boundary so that only one formation was affected would have been beneficial.

As in May, 1915, everything depended on the artillery suppressing the German line until the infantry reached it. At 258, the number of field guns and howitzers at Fromelles in 1916 was about the same as it was in 1915. From McCay to the brigadiers[62]: -

> *The mode of taking the trenches should be as follows – first wave stays at and clears enemy out of first row of enemy trenches … Then advances further. Meanwhile second wave passes first wave to next enemy row … and so on till all works of enemy first line system … are taken.*
>
> *It is the rearmost row of enemy's first line that is to be at once fortified and held when it is taken.*
>
> *Clearly understand that each wave, so soon as it has cleared of enemy the work it gets into, goes on to the prescribed limit of attack, i.e. the rearmost enemy work in his front line system, 100 yards to 150 yards behind his parapet.*

Clearly this was not to be a 'raid' in the defined sense set out in previous chapters. When a unit is sent to ***attack, seize and hold*** it is no longer a raid, but a design for other purposes. Clearly the preparations for Fromelles in 1916 were more elevated than any 'raid' could ever hope to be.

C. E. W. Bean reports that General Haking, in a letter read to all troops on 16 July, the eve of the day appointed for the assault (it was later deferred 3 days) explained, firstly, the reason for the operation, and then the methods. In describing the latter he said that the feint bombardments in the south would be continued on the morning of the offensive …

> *'…whilst our guns along the front of our real attack will be getting the exact range of the enemy's trenches without attracting undue notice. When everything is ready, our guns, consisting of some 350 pieces of all descriptions, and our trench mortars, will commence an intense bombardment of the enemy's front system of trenches. After about half-an-hour's bombardment the guns will suddenly lengthen range, our infantry will show their bayonets over the parapet, and the enemy, thinking we are about to assault, will come out of his shelters and man his*

[61] Pedersen p.40
[62] Pedersen p.43

parapets. The guns will then shorten their range, and drive the enemy back into his shelters again. This will be repeated several times. Finally, when we have cut all the wire, destroyed all the enemy's machine-gun emplacements, knocked down most of his parapets, killed a large proportion of the enemy, and thoroughly frightened the remainder, our infantry will assault, capture, and hold the enemy's support line along the whole front. The objective will be strictly limited to the enemy's support trenches and no more.'

General Haking believed that the rear-most trench of the enemy's front system would probably be found at from 100 to 150 yards beyond the German front line and that the two allotted battalions of each brigade would suffice for an advance so limited. But the third line of trenches that he had coveted just did not exist.

17 July, 1916: -

At 4.00am on 17th, the time when the final seven hours' bombardment should have started, a heavy mist lay upon the country. The hour was accordingly put off, first until 8.00am, and then till 11.00am. At 8.30am, as the air was still too misty, Haking wrote to the First Army Commander advising with great reluctance that the operation should be postponed.

News of the postponement was received with intense relief by both divisions, whose men were well-nigh worn out with the hurried preparation. The army commander decided that the assault should not be undertaken for at least two days.[63]

Monro, however, in agreeing to the postponement, had decided to cancel the whole operation. In an urgent dispatch he informed Haig of this decision, and asked for leave to inform Plumer. In answer he received the following:

The Commander-in-Chief wishes the special operation mentioned in the above letter to be carried out as soon as possible, weather permitting, provided always that General Sir Charles Monro is satisfied that the conditions are favourable, and that the resources at his disposal, including ammunition, are adequate both for the preparation and execution of the enterprise.

If the Battle was a success, no doubt Haig would have accepted credit for it. If it was not a success, then he could lay blame at Monro instead. As Monro had already given his opinion that the resources were sufficient, it was a foregone conclusion that the operation would now take place.

---ooo000ooo---

The work of 3ATC for the installation of push pipes in No-man's land was still under way.

In his report dealing with the push pipes being installed near the Rhondda Sap, Lt B. PRIESTMAN, O.C. No. 4 Section, 3 Australian Mining Company R.E. (3ATC), wrote that the pushing **was started on the evening of 17/7/16** and reached a distance of 91ft.

[63] Bean p.349

18 July, 1916

The work of 3ATC in installing push pipes in No-man's land near the Rhondda Sap had not been delayed. Prior to and during the battle 3ATC under Major L.J. Coulter were working, continually under fire. Carter had members of the 1/5th Duke of Cornwall's and some engineers to assist. We know that there was a section from 58th Bn also there, and that they worked all night in darkness and under bombardment from both sides, but their task was unachievable.

In his report dealing with the push pipes being installed near the Rhondda Sap, Lt. O.R. Howie of 3ATC stated that On the evening of the 18th last I placed the detonating canister in the pipe and carried the leads (electric & instantaneous fuses) over our parapet. The leads were in good order.

---ooo000ooo---

6. Wednesday 19 & Thursday 20 July 1916

--- The Battle of and at Fromelles ---

Yesterday evening, south of Armentieres, we carried out some important raids on a front of two miles in which Australian troops took part. About 140 German prisoners were captured.

Official Communique released by British High Command, at 2.50pm on 20 July, 1916

---ooo000ooo---

The village of Fromelles was captured in 1914 by the Germans during the Race to the Sea, when the German advance across northwest Europe was halted by the British and French. The opposing armies had tried to outflank each other, until their frontlines reached Belgium's North Sea coast.

Fromelles had remained in German hands ever since, nestled behind a long and well-entrenched front line, held in this sector by the 6th Bavarian Reserve Division, and further supported by strategically positioned units along Aubers Ridge to the northwest of the village.

The Allied aim for the battle that is about to begin is to deceive German high command into keeping their reserves in place, rather than sending them south to reinforce their units on the Somme, where the mighty Allied offensive is taking place. While the 'official' line is that it is intended as a feint, the sheer scale of the weaponry and the manpower involved will show that it is really only able to be classified as a Battle.

Before this day, British forces have launched two unsuccessful attacks in the area in front of Aubers Ridge. The first, in December, 1914 only two months after German forces occupied the area, was intended to capture the Ridge itself.

Then, in May 1915, less than a year after this first attempt, the British 8th Division sought to seize the German positions before the Ridge. Now, some 14 months later, the Battle of Fromelles is beginning to unfold on a hazy Wednesday morning, as the British 61st Division and the Australian 5th Division move into position to launch their attack.

Between May 1915 and July 1916, General Erich von Falkenhayn, the German Chief of the General Staff had ordered a construction programme on the Western Front, to make it capable of being defended indefinitely by a small force against superior numbers. An elaborate, carefully sited and fortified front position was built behind fields of barbed wire, with camouflaged concrete machine-gun nests and a second trench close behind the front trench to shelter the trench garrison during bombardments.

To evade Allied artillery-fire intended to obstruct the movement of reinforcements from the new rear defences, communication trenches were built. The front position was to be held at all costs as the main line of resistance but in May 1915, Falkenhayn ordered a reserve position to be built along the Western Front, 2,000–3,000 yds (1,800–2,700m) behind the front position, out of range of enemy field artillery.

The German salient at Fromelles contained some higher ground facing north-west, known as the Sugar-loaf, resembling the shape of a local cake-loaf. The small size and height of the salient gave the Germans observation of no-man's-land on either flank. The 5th Australian Division (under Maj Gen Sir James Whiteside McCay) was to attack the left flank of the Sugar-loaf salient by advancing south as the 61st Division attacked on the right flank from the west.

Aubers Ridge itself provided a good screen for German artillery positions but is not high enough for effective, direct observation of the enemy. As a result, the Bavarians had constructed platforms in trees, observation posts in church towers and chimneys, and concrete structures within taller farm buildings.

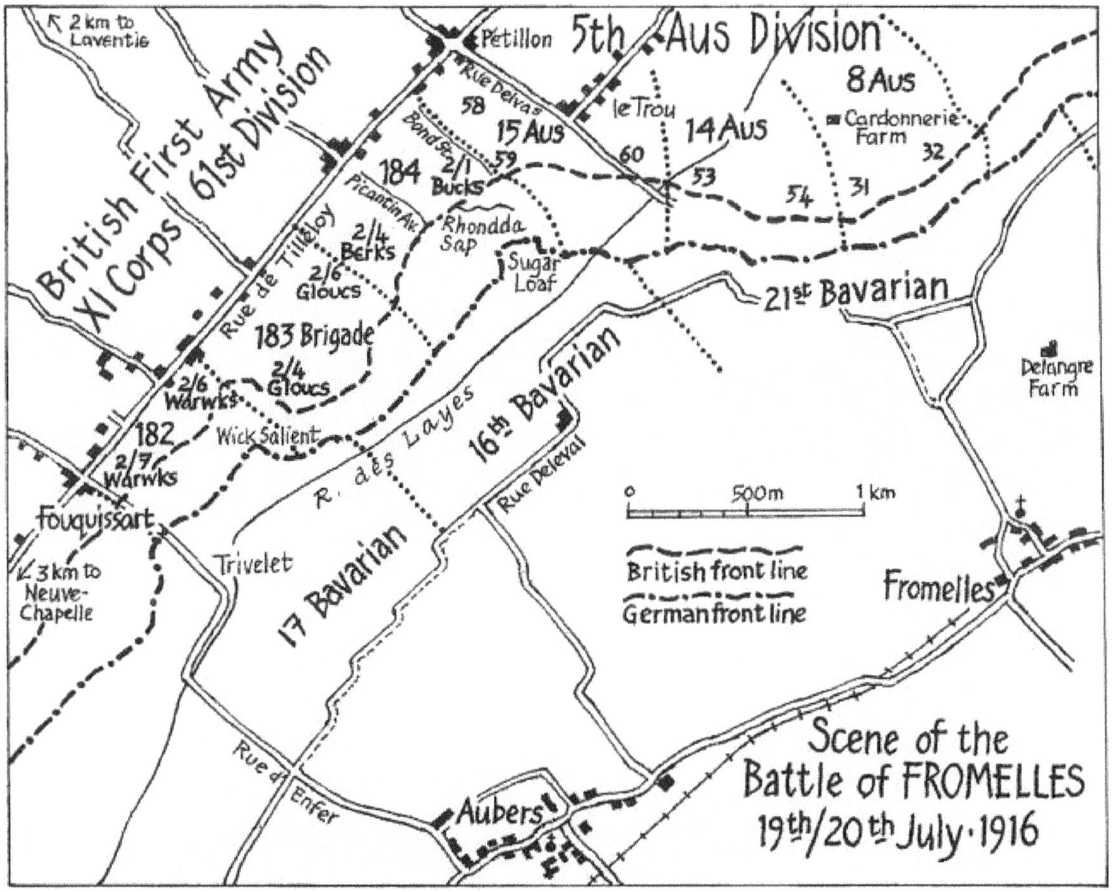

[Picture15: Map 8: The Battle of Fromelles – Order of Battle for British and German forces.][64]

And there was another salient on the southern end of the line of attack, the Wick Salient, facing the British soldiers. The Wick salient was in front of the 2/7 Warwicks and the 2/6 Warwicks of the 182nd Brigade. To their left, the 2/4 Glosters and the 2/6 Glosters of the 183rd Brigade were between the Wick and the Sugar-loaf salients.

The Sugar-loaf was directly in front of the 2/1 Bucks and the 2/4 Berks of the 184th Brigade.

[64] Michael Senior, Fromelles 1916, Pen & Sword Books, Barnsley England. Reproduced with permission.

Both the Wick Salient and the Sugar-loaf salient were heavily fortified and from where machine gunners overlook no-man's-land and the Allied lines beyond. It was a sandwich of death just waiting for the German machine guns.

Along this, the German line, there were about 75 solid concrete shelters. From their front-line trenches, the Bavarians had dug tunnels under the wire and out around thirty metres into no-man's-land, which they'd use as listening posts. At night searchlights, flares and star shells illuminate the frontline.

A trench tramway network, known as a 'push line', had also been constructed along the entire length of, and parallel to, the trenches. The supply lines are in some locations connected to tracks for horse-drawn or petrol-driven vehicles, and in turn meet up with the standard-gauge French railway network.

A buried telephone network and electricity supply was also in place, powering lighting, water pumps, concrete-mixing stations and other amenities. It is by far a more sophisticated and solid defensive system than that occupied by the British and Australians.

Having occupied this part of the line for about two years the Bavarians also know the terrain intimately, in contrast to the newly arrived Allied units.

From 10/11 July, 1916, until 19 July, various units moved into the area ahead of the impending battle. Villagers in Laventie, La Gorgue and other billets are reputed to have openly talked about an impending operation, and told 2/4 Glosters of such before their own orders arrived.

Lt Gen Haking, believed that the objective, the rear most trench of the enemy's front line, lies about 140 metres beyond no-man's-land. Haking is based at XI Corps HQ in Sailly-sur-la-Lys, 9 km away, and is convinced the objective is achievable by a smaller force than previously considered appropriate for the task.

He decided that one division from XI Corps, under Maj Gen Colin McKenzie, fight alongside the 5th Australian Division, commanded by Gallipoli veteran Maj Gen Sir James Whiteside McCay. Regrettably, the British 61st Division is generally (and unjustly) considered to be under-equipped, and the 5th Australian Division have not been involved in fighting on the Western Front previously.

Preparations for the attack were rushed, the troops involved lacked experience in trench warfare and the power of the German defence was significantly underestimated, the attackers being outnumbered 2:1. Of the Australian assaulting battalions, the 60th had not yet been in the front line on the Western Front; the 32nd and 54th had been there for part of a day, and the 59th somewhat longer; the 31st and 53rd for two days.

Thus, the 5th Division of the AIF which had recently arrived in the Fromelles area, was committed with short notice, to attack a defensive system. A system which was described as "well developed" 14 months earlier. Their opponents, the troops of the 6th Bavarian Reserve Infantry Division had been holding these defences for all of this time.

The ground to be attacked was waterlogged, flat and easily overlooked from Aubers Ridge, behind the German front line to the south. The British 39th Division and 31st Divisions moved their boundaries north as the 61st Division concentrated along its attack front from the Fauquissart–Trivelet road to Bond Street. The 20th (Light) Division moved its boundary south to Cordonnerie Farm to take over the left of the 5th Australian Division, which concentrated the divisional frontage from Cordonnerie Farm to Bond Street.

The Build Up

Preparations for the battle had begun four days earlier, providing a series of signals to the Germans that a larger offensive may be about to occur. The Germans had seen them clearly from their observation positions, and from balloons above it, where they could see the roads behind the British line clogged with men, vehicles, and horses.

Adding to these concerns, the distance to be covered across no-man's-land is highly variable across the line of attack, and in some places is up to 400 metres wide.

Much of this ground was in clear view of the German positions. From the Sugar-loaf salient alone, machine guns were able to fire forward and to each side.

The 6th Bavarian Reserve Division held more than 7 kilometres of the German front line. Each of the Division's regiments had been allocated a sector, each in turn manned by individual infantry companies.

At their strongest, the German trenches are protected by sandbagged breastworks over two metres high and six metres deep, which makes them resistant to all but direct hits by artillery. This line is further protected by thick bands of barbed wire entanglements.

It is now dawn on Wednesday 19th July, 1916, northwest of the village of Fromelles in northern France. Elsewhere it is now nineteen days into the Somme Offensive on the Western Front.

Brigadier General "Pompey" Elliott's misgivings about the battle to come are well founded.

On the 19th July, the 3ATC with the 61st Division Infantry was ordered by Chief Engineer XIth Corps to undertake push pipe operations prior to the assault to enable communication trenches to be quickly constructed, & to blow up crater lips to block machine gun fire.

The three sites chosen were: -

No. 1 at N 5 D 10.50 From Rhondda Sap, resulting in a crater 220′ long, giving 6′ of cover.

No. 2 at N.14a.10.20 from Red Lamp salient, resulting in a crater 100′ long, giving 6′ of cover.

No. 3 at N.13d.48.90 from Red Lamp salient, resulting in a crater 150′ long, giving 5′ of cover.

'Red Lamp' salient was the code used by 3ATC to refer to the work being done to install push pipes in No-man's landm near the Rhondda Sap.

The Bombardment

The main bombardment was to begin at midnight on 17/18 July and run for seven hours, but rain forced a postponement. Over the final three hours, the artillery was to lift and the infantry show bayonets and dummy figures several times, to simulate an infantry advance, then the artillery was to resume bombardment of the front line to catch the German infantry out of cover.

The bombardment finally began when the mist cleared at **11 o'clock** in the morning of Wednesday, 19 July, with the infantry attack re-scheduled to 6.00pm, instead of 11.00am as originally planned.

The Allied artillery averaged one gun for every eight metres of German front line. With support from First Army artillery to the south, 296 field guns and 78 heavy guns were ready, which gave a greater concentration of heavy artillery than that of the Fourth Army on the first day of the Somme Offensive.

The artillery has two decisive objectives. Firstly, to put enemy machine guns and their crews out of action or at least keep them from firing until the infantry reaches the German trenches.

Secondly, to destroy the barbed wire and collapse the German parapet, depriving the German infantry of their defences and preserving the lives of the attacking troops.

As the bombardment begins, the quiet landscape is transformed into a deafening, shuddering tumult. Machine Gun Sergeant Les Martin, recalled later in a letter to his brother: *'From about 11.00am till 6.00pm there was not a space of a second's duration when some of our guns were not firing, the row was deafening. I put wadding in my ears while we were down in the supports waiting to go forward'.*

At 11.30am, half an hour into the bombardment 5 British aircraft fly over the Wick Salient area to observe the effects of the shelling on the German positions, including the front line, the wire entanglements and the rear areas, including the hamlets of Aubers and Herlies. They report many German shelters and dugouts appear buried and wire defences cut.

At 12.00 midday, the bombardment increases. German commanders send up two aircraft to locate the British artillery positions.

By 1.00pm, two hours after the Allied bombardment began, German artillery is responding. The German 32nd Battalion's commanding officer reports 'Hostile trench mortars and 18-pounders [are] smashing our parapets and trenches to pieces. [There are] Many casualties.'

By 2.00pm onwards, in response to the bombardment that commenced at 11 o'clock, the enemy's artillery began to answer the increasing British bombardment by shelling the communication trenches and reserve and support lines of both the attacking divisions. In the Australian area the ammunition and bomb-dump of the 31st Battalion was blown up and the battalion commander, Lieutenant Colonel Toll, and most of his signallers, messengers, and the medical staff of the battalion were wounded.

At about 2.30pm, Allied artillery observers report their bombardment on one of the prime targets, the Sugar-loaf salient, had been ineffective and the wire entanglements remained uncut. An order is sent to increase the barrage in order to cut the wire but the message is not received until after 5.00pm by which time it's too late to prevent the planned attack, due to start at 5.45pm.

God was not on the side of the gunners supporting 184 Brigade. Inexperienced, their observations obscured by dust and smoke, and taking heavy casualties, they made so little impression on the Sugar-loaf that a special concentration was ordered at 2.35pm by which time a German counter-bombardment was falling all along the attack front, causing casualties to the Australians and the field artillery crews of the 61st Division at Rue Tilleloy. Several ammunition dumps were exploded and the decoy lifts by the British artillery failed to deceive the Germans.

It made no difference: 'no effective destructive or neutralising of the Hun infantry, artillery or MGs took place. The total effect of our artillery preparation on the Hun resistance was Nil'.[65]

The German artillery, though, handed out to 184 Brigade the same treatment it had given to the 183rd. Packed behind the breastwork hours before their own bombardment began, 184 Brigade's assault battalions had to weather the German reply from start to finish. 2/4 Berks on the right lost forty men and 2/1 Bucks on the left almost a hundred, which came on top of their casualties in the accidental gassing the night before. Each battalion reinforced its assault companies with a platoon from its reserve company in the 300 Yard Line, forcing a hurried reorganisation an hour before the attack.

At around 3.00pm, the Germans had observed troops, opposite the Sugar-loaf, in the forward saps know to him as the "Australian Trench", and had turned his bombardment on them, thus cutting the electric leads to two of the push pipes.

Action was undertaken almost immediately by 3ATC to restore the leads to the ammonal charges and fuses, under difficult conditions and enemy fire.

The commander of 3ATC, Major L.J. Coulter (of Grenville, Victoria), and six of his men were wounded in attempting to fire them. The Second in Command, Captain A. Sanderson (of Perth W. Aust), fired the charges in one case.

Lt O.R. Howie (of Collie W. Aust), assisted by Sergeant M. J.M. Kerby (of Ballarat, Vic.) and Lance-Corporal W.A. McKay (of Kalgoorlie W. Aust), under difficulties, exploded one of the push pipes at **4.38pm**. The crater formed was about 100ft long, 30ft wide at top, average depth about 10ft.

Lt B. Priestman, O.C. No. 4 Section, 3ATC, assisted by Corporal W.A. Bayes (of Queenstown, Tas.), and Lance Corporal, E.E. Jackson (of Underwood, Tas.). exploded the last of the push

[65] Pedersen p.58

pipes at **4.39pm.** The resultant crater was approximately 150ft long and an average over about five (5) ft.

As the Germans seemed to be getting a far worse battering, morale held. The Australians cheered on seeing ragged gaps being torn in the enemy's breastwork. 'Wire netting and woodwork was observed flying through the air' after a large explosion opposite 15 Brigade and even the erstwhile gloomy Elliott, sharing the danger in the front line, was sufficiently impressed to assure his men:

> *'Boys, you won't find a German in the trenches when you get there.'* [66]

ORDER OF BATTLE AND FIGHTING UNIT ARRANGEMENTS.

ATTACK AND FIGHTING UNITS

The units involved in the attack were to be (from north east to south west):

The Fifth Australian Division	The 61st British Division GOC Major-General Colin Mackenzie CB
The 8th Brigade, made up of:	**184th Brigade** made up of
Assaulting battalions: - 32nd (SA & WA) and 31st (Qld, Vic) Reserve battalions: - 29th (Vic) and 30th (NSW)	Assaulting battalions: - 2/4th Royal Berkshire (right) and 2/1st Buckinghamshire (left) 2/1st Royal Berkshire (support) 2/4th Oxford and Buckinghamshire (support) and Reserves: - 2/5th Gloucestershire And also: - 184th MG Coy, and 184th TM Battery
The 14th Brigade (NSW) made up of	**183rd Brigade** made up of
Assaulting battalions: - 53rd and	Assaulting battalions: - 2/6th Gloucestershire (right) and

[66] Pedersen p.66

54th Reserve battalions: - 55th and 56th	2/4th Gloucestershire (left) 2/6th Worcestershire (support) Reserve battalions: - 2/7th Worcestershire and 2/8th Worcestershire
The 15th Brigade (Victoria) made up of	**182nd Brigade** made up of
Assaulting battalions: - 59th and 60th Reserve battalions: - 57th and 58th	Assaulting battalions: - 2/7th Royal Warwickshire (right) and 2/6th Royal Warwickshire (left) 2/5th Royal Warwickshire (support) Reserve battalion: - 2/8th Royal Warwickshire and also 2 Coys 2/8 Warwicks and 1 Section 182nd MG Coy 182nd TM Battery

Assistance Would Be Provided from the Following Units: -

AUSTRALIAN SUPPORTING UNITS

5th Division HQ

5th Division Artillery HQ

5th Divisional Sanitary Section

5 Divisional Signal Company

10th (HQ) Coy Aust Army Signal Corps

5th Division Train

757, 758, 759 760 Coy, Aust Army Signal Corps

5th Divisional Trench Mortar Battery

5th Divisional Machine Gun Company

5th Divisional Engineers

5th Divisional Traffic Control

5th Division Pioneer Battalion

5th Divisional Ammunition Column

5th Division Salvage Company

8th Brigade HQ

8th Machine Gun Company

8th Australian Field Ambulance

8th Light Trench Mortar Battery

8th Field Coy Engineers

18th Coy Aust Army Signal Corps

14th Brigade HQ

14th Machine Gun Coy

14th Austrralian Field Ambulance

14th Light Trench Mortar Battery

14th Field Coy Engineers

28th Coy Aust Army Signal Corps

15th Brigade HQ

15th Machine Gun Coy

15th Australian Field Ambulance

15th Light Trench Mortar Battery

15th Field Coy Engineers

29th Coy Aust Army Signal Corps

4th Divisional Artillery

11th Brigade AFA

49th Battery AFA

51st Battery AFA

13th Brigade AFA HQ

50th Battery AFA

113th Howitzer Battery AFA

13th Brigade Australian Field Artillery

14th Brigade Australian Field Artillery

53rd Battery AFA

55th Battery AFA

54th Battery AFA

114th Howitzer Battery AFA

15th Brigade AFA HQ

57th Battery AFA

59th Battery AFA

58th Battery AFA

25th Howitzer Brigade AFA HQ

52nd Battery AFA

60th Battery AFA

56th Battery AFA

115th Howitzer Battery

2 Australian Tunnelling Company

3 Australian Tunnelling Company

Australian Army Medical Corps

Australian Army Veterinary Corps

5 Divn Mobile Veterinary Section

Australian Army Pay Corps

Australian Provost Corps

Australian Army Ordnance Corps

Chaplains' Department

BRITISH SUPPORTING UNITS

61st Dision Artillery (RFA)

305 Brigade RFA

306 Brigade RFA

307 Brigade RFA

308 Brigade RFA

Royal Engineers

1/3 Field Coy (South Midland R.E

2/2 Field Coy (South Midland) R.E.

3/1 Field Coy (South Midland) R.E.

476, 478, 479 Field Coys R.E.

Field Ambulances

2/1 Field Ambulance

2/2 Field Ambulance

2/3 Field Ambulance

Divisional Signal Company

Divisional Train

61 Division Sanity Section

Mobile Veterinary Section, 61 Veterinary Section

Pioneer Battalions: -

A Coy 1/5 Duke of Cornwall Light Infantry

B Coy 1/5 Duke of Cornwall Light Infantry

Hampshire Yeomanry

61 Division Cyclist Company

AERIAL SUPPORT

On 19 July, 5th Division Artillery would also be assisted by No. 16 Squadron Royal Flying Corps, whilst reconnaissance and liaison work would be in the hands of No. 10 Squadron Royal Flying Corps.

Little wonder there were over 17,200 men involved, with units contained in approximately four square miles.

The twelve attacking battalions were supported by more artillery than at the Battle of Aubers Ridge in May 1915, when a similar number of battalions attacked in the same area. More ammunition was available than in 1915 and trench mortars were added to the artillery for wire cutting. With support from First Army artillery to the south, 296 field guns and 78 heavy guns were ready, which gave a greater concentration of heavy artillery than that of the Fourth Army on the first day of the Somme.

And as the last hours of 19 July ticked away, many men wrote to loved ones, 'just in case'.

Beginnings of The Assault

By **5.00pm** the German artillery was pounding the Allied front line where Australian and British troops prepared for the charge. The German artillery was supported by heavy machine guns unsilenced during six hours of bombardment.

Those in the 14th Brigade had already witnessed shells bursting everywhere, mostly high explosive; and saw machine-guns knocking bits off the trees in front of the reserve line and sparking against the wire. When men looked over the top they saw no-man's-land leaping up everywhere in showers of dust and sand...rather confirming their fears that the Germans knew something.

The next hour would become a perfect demonstration of what is often called 'the fog of war' as plans and orders are overtaken by the tide of battle.

But **at 5.10pm**, too late for the artillery to respond, Brigadier General 'Pompey' Elliott, received a warning. The entanglements west of the Laies River, which protected the side of the Sugar-loaf facing Elliott's men, were uncut. Observers in 60 Battalion saw three intact machine gun emplacements there. Adding to Elliott's problems, part of his force was not in position because of the ferocity of the German artillery.

At 5.15pm, to the right of the Australians, the British soldiers begin to leave doors called sally ports in the trenches and enter no-man's-land. The effect is very different to a line of soldiers going over the parapet - the small doorways naturally funnel the soldiers, causing congestion and making them an ideal target for German machine gunners who shoot them down as they try to fan out from these narrow exits. The effect is horrific as bodies pile up around the doorways.

Elsewhere along the line, British and Australian soldiers begin a series of movements from the rear trenches to the forward line.

The Australian 14th Brigade, having inadequate communication trenches, sent its third and fourth waves over the open fields between the "300 yards" and front lines at **5:25pm and 5:31pm**. So congested are the Australian trenches that many of these men need to go out over open ground to reach their starting point. They come under intense artillery and machine gun fire as they do so.

The Bloody Holocaust Has Begun

At the same time the infantry of the 61st Division began to file out from its front line through sally-ports leading into No-Man's-Land.

It was with this manoeuvre that the infantry operation really began.

Jumping off on their 300-yard frontage, the 2/4 Berks were to assault the line held by (German) 12 Company 316 BRIR. At **5.40pm**, they debouched from their two sally ports in full view of the German lookouts. The Germans were roused, and their machine guns began clattering away.

A Company on the left was annihilated and the survivors tried to retain their own breastwork. Emulating Brigadier General Lowry Cole the year before, Lieutenant Colonel John Beer leapt up onto the parapet to rally them and was quickly cut down. Mostly leaderless, the two platoons from B Company that got clear on the right started out for the German wire at 6.00pm but found it uncut. Unsupported and under heavy machine gun fire, they too fell back. The attack cost 2/4 Berks 163 casualties.

[Picture 16: AWM No. AO3042. [Soldiers of the 53rd Battalion prepare for the attack. Only Three returned; all wounded.]

Led by Captain Ivor Stewart-Liberty, D Company on the right, like the 2/4 Berks platoons alongside them, was mown down. Only one man, Lance Corporal Stevens, reached the Sugar-loaf. But a few men from A Company on the left, including their commander, Captain Harold Church, were seen on its parapet, which may have prompted the despatch of Captain Henry Buckmaster and C Company at 6.10pm. The 2/1 Bucks War Diary tersely remarks:

"again, the enemy's machine gun fire prevented any advance without extermination". 184 Brigade's Vickers guns, three of which were knocked out, expended 30,000 rounds trying to suppress it.

A and C Companies in 31 Battalion's first wave went over at **5.58pm** followed by its second wave on the barrage lift two minutes later. Losses were high as C Company fought hard initially but when the first wave closed with its objective, the 250 yards of breastwork, it met only a few men covering a hasty retreat. Going around a traverse, Lieutenant Drayton bumped into an officer clutching a grenade and shot him. Private Percy Weakley saw a machine gun crew about to leave its gun and killed four before shrapnel bursting overhead killed him.[67]

The leading waves had moved on when Lieutenant Colonel Toll arrived in the German front line with the combined third and fourth waves. Leaving some men to help the Lewis gunners get established, he set out after them and struck the same problem that had confounded Cass. He also looked for and was unable to find the German second line. Toll later reported:

[We] *swept on with the intention of capturing the second- and third-line trenches in the first line system, but we went on and on but no trace could be found of same. It now appeared evident that the information supplied as to enemy defences and aerial photographs was incorrect and misleading. The ground was flat, covered with fairly long grass, the trenches shown on aerial photos were nothing but ditches full of water…*

Leaving his deputy in charge, Toll and some of his staff braved the Australian barrage to confirm that the German support line did not lie ahead.

While the maps showed the system as it had been, the aerial photographs taken before the battle gave a clue as to how it actually was. Lined with earth-filled ammunition chests, the Kastenweg, (the 'Chest Way') stood out as having been built over the old trenches, which would not have happened were they still in use. The Australians lacked photographic interpreters but they might have drawn the right conclusion had here been time for close study of the pictures before the attack. But Haking's headquarters did not pick up the mistake either. As 2nd Lieutenant Reginald Hugh Knyvett, 59 Bn remarked afterward, *'a great number of men were lost by going too far into enemy territory, seeking the supposed third line'*.[68]

As the only defensible position seemed to be the old German front line, Toll withdrew 31 Battalion to it.

6.00pm: British and Australian troops have in some sectors reached the German front line and are beginning to break into the trenches as Allied artillery lifts its range to target areas behind the enemy's front. The fourth wave of infantry is advancing. Two and a half hours of daylight remain.

Major Geoff McCrae, a Gallipoli veteran, leads the fourth wave of the 15th Australian Brigade, including the 59th and 60th Battalions - about 2000 men including 60 officers. His men include signallers with wire and telephones to keep headquarters informed of the battle's progress. 75 metres into No-man's land, McCrae is shot in the neck and killed, as are many of the soldiers including all McCrae's signallers, who also fall dead or wounded. Messages now have to be to be carried back across No-man's land by runners.

6.10pm: Private Walter 'Jimmy' Downing of 57th Battalion is watching the attack unfold:

> *"The 60th climbed on to the parapet, heavily laden, dragging with them scaling ladders, light bridges, picks, shovels and bags of bombs. Scores of stammering German machine-guns spluttered*

[67] Pedersen p.77
[68] Pedersen p.81

violently, drowning the noise of the cannonade. The air was thick with bullets, swishing in a flat lattice of death. There were gaps in the lines of men - wide ones, small ones. The survivors spread across the front, keeping the line straight... The bullets skimmed low, from knee to groin, riddling the tumbling bodies before they touched the ground. Still the line kept on. Hundreds were mown down in the flicker of an eyelid, like great rows of teeth knocked from a comb, but still the line went on, thinning and stretching. Wounded wriggled into shell holes or were hit again. Men were cut in two by streams of bullets. And still the line went on... It was the charge of the Light Brigade once more, but more terrible, more hopeless."

It is to become one of the most poignant statements about Fromelles ever written. And a mere ten minutes have passed since 6.00pm.

From the Regiment Report we learn that at **6.15pm** Tupman knew that the 184th Brigade assault had failed. Next to them, of course, was the Australian 15th Brigade, at that time still sending men 'over the bags' with 400 yards of No-Man's Land to cross. If the adjacent British Brigades were aware of each other's positions, how come the British and Australians never had such liaison? The historian of the 2nd Buckinghamshire Regiment explained it simply: -

'...the Australians on the left of the Bucks, but owing to the salient unable to keep in touch'.[69]

With the two assaulting British Brigades effectively disposed of in the first 15 minutes of the Battle, the German gunners at the Sugar-loaf could concentrate on the Australians, which they did to terrible effect.

A little before **6.30pm** 53rd and 54th Battalion soldiers who are now occupying the German front line from Rouges Banc to near Delangre Farm press forward in an attempt to locate the key objective, the German second line.

Some of these men are further than 300 metres but only find occasional enemy soldiers taking cover in shell craters, many of whom retreat on sight. The Australian soldiers are under orders to resist the chase, instead some Germans are taken prisoner and sent back to Australian lines.

Out of touch in 184 Brigade's battle headquarters, 1½ miles rearward on the Rue Verte, Carter instructed Lieutenant Colonel Williams at **6.30pm** to attack again two hours later, sparking an angry exchange between them. The 2/1 Bucks war diary pulls no punches in conveying the frustration Williams, who was in the breastwork, felt on receiving Carter's orders 'at a time when every man, save a few telephone operators, orderlies and wounded, was in No-man's land'. Christie-Miller recalled: 'it proved next to impossible to persuade the authorities that there was nothing left to attack with'.[70]

The Regiment Report for 2/4th Gloucester Regiment, by Lieutenant Colonel J.A. Tupman CO is also of immense importance. The 2/4th Glosters were part of 183rd Brigade, which were to the left of 184 Brigade. The 184th was charged with taking the Sugar-loaf. To the left of 183rd Brigade

[69] Corfield p.51
[70] Pedersen p.61

were the troops of the Australian 15th Brigade, whose assaulting battalions were the 59th and 60th.

With parts of the German front line trenches being captured by both British and Australian units, the soldiers press their attack and move on towards the German second line.

53rd Battalion commander Lt. Col. Ignatius Norris arrives with battalion headquarters in the fourth wave, calling 'Come on lads! Only another trench to take!'

As Norris moves forward he is cut down by an 'unsuppressed machine gun'. Norris' last words are reported to be, 'Here, I'm done, will somebody take my papers?' With him is Private Leslie Croft, a battalion signaller, who attempts unsuccessfully, with others, to bring his wounded leader back to safety.

Others with Norris are also killed and wounded by German machine guns, including his young adjutant, 32-year-old Lieutenant Harry Moffitt.

According to Sergeant Patrick Lonergan, when he sees his colonel hit, Moffitt instantly calls for four men to carry him back to the Australian lines. As he does so, he too is hit and falls dead across Norris' body.

6:45pm: The Australian and British advance is stalled and turns into a retreat under fire from their own artillery as the guns continue to hit targets behind the enemy's front line, unaware of the progress of the infantry units.

German artillery also begins to open up on No-man's land, preventing other Allied units from advancing.

By around **7.20pm**: Brig Gen Elliott sends a report to Headquarters that the attack is failing in part because of a lack of support.

In response, Maj Gen McCay sends about 500 men, half of the reserve 55th Battalion, into the battle.

The German record of battle reports that "the forward trench was once again in our hands - 47 unwounded and 14 wounded Englishmen were captured, as well as seven machine guns and sundry material. A great number of wounded and dead English lay in front of our position. Bandages were ordered forward, the platoons which had been pushed out moved back into their positions ... two captured machine guns were turned against the enemy."

7.45pm: On receiving word that 15 Brigade was stalled, McCay authorized Elliott at 7.45pm to use half of his third battalion, the 58th, to reinforce the line in No-man's land and hopefully carry it forward.

But as **8.00pm** approaches, the Germans are reoccupying much of their front line. The Allied attack is failing everywhere.

Night Falls

8.00pm: Daylight begins to fade over the battlefield.

At XI Corps headquarters, General Haking orders a fresh assault to take place in an hour's time. Having received divisional reports on the progress of the battle so far, his intention is to support the limited successes where they can be found.

Haking is under the misapprehension that the Sugar-loaf salient has been dealt more damage than in fact it has. Plans are drawn up to launch an assault against it. But within half an hour these plans are cancelled as aerial reconnaissance reveals the position to still be intact.

Haking also receives news of the high number of casualties. He orders the three remaining British brigades to wait until morning before resuming the attack.

Whilst the order to delay does eventually make its way to McCay, Elliott who has been preparing his reserves to assault the Sugar-loaf is never informed of the change in plan and continues to prepare for the assault.

The headquarters of the Bavarian 6th Regiment are at the same time planning for a determined counter-attack. Equipped with a more realistic assessment of the situation their plans involve a combined artillery and infantry assault with clear lines of command and communication.

8.13pm: Brigadier General Carter's request for support on 184 Brigade's left flank when the 61st renewed its attack at 9.00pm reached Elliott at **8.13pm**. He tasked D and C Companies of 58 Battalion to advance through 59 Battalion against the eastern face of the Sugar-loaf while 184 Brigade assaulted on the western side. [71]

8.20pm: A renewed bombardment continued as preparations were made to attack all along the front at 9.00pm. But at 8.20pm Haking was given a first-hand report on the beating the 61st Division had taken. He directed Mackenzie to abort the 9.00pm attack and ordered that all troops were to be withdrawn after dark. 183 and 184 Brigades were to withdraw any men left in No-man's land and reconstitute the front line with their respective third battalions, 2/7 Worcesters and 2/4 Ox and Bucks. Shortly afterwards, 182 Brigade advised that 2/7 Warwicks had been bombed out of the German line, eliminating the 61st Division's only toehold. '*The men who went over first are no more*', Lieutenant Colonel Nutt was told. Just in time to prevent A Company 2/7 Warwicks going forward, Haking decided that all three brigades should pull back and attack again next day.

Reinforcements for the 182nd Brigade received the order in time but the troops in the German line were overwhelmed, only a few wounded and stragglers returning. Troops pinned down in No-man's land withdrew under cover of the bombardment and parties went out to rescue wounded. More discussion between Mackenzie and Haking led to a decision that the 184th Brigade would attack the Sugar-loaf overnight, after a ten-minute hurricane bombardment; German shelling on the British front line then caused a postponement until the morning.

---

[71] Pedersen p.89

Breakdown of Communication

9.00pm: Elliott orders the Australian 15th Brigade reinforcements to advance and assault the eastern face of the Sugar-loaf, with two companies of the 58th Battalion and some survivors of the 59th. They are led by 21-year-old Major Arthur Hutchinson. Rising from cover, 200 metres from the Sugar-loaf, they are immediately met by machine gun fire from the emplacements ahead of them. The result is a slaughter.

At the same time fresh reserves of the Bavarian 6th Division gather at Delangre Farm in large numbers, in preparation for the counter-attack.

8th Australian Brigade headquarters receives reports that the 32nd Battalion is under great pressure: the "front line cannot be held unless reinforcements are sent. Enemy machine gunners are creeping up... The artillery is not giving support. Sandbags required in thousands. Men bringing sandbags are being wounded in the back. Water urgently required."

The signallers who kept 8 Brigade's telephone link intact for most of the battle – another tremendous feat – had a terrible time as well. Writing in the third person and calling himself Ted, seventeen-year-old Corporal Rowland Lording of 30 Battalion was tracing a cable run out at **10.00pm** recorded:

> *God! What sights they see out there. Huddled and stretched out bodies, khaki heaps that were once men – some of A Company digging a trench – others like themselves, making short crouching runs and flinging themselves down before anything that will afford the slightest cover.*
>
> *Crump! Bang! Crash! The shells fall, Zipzipzip – zipzip! Machine-gun bullets kick up the dirt around them. A lull and they are off again. Zipzip! Bang! Another twisted heap of khaki hits the ground.*
>
> *It is Ted. He does not move. His cobbers crawl over to his side. 'Where d'you get it?' they ask him. His lips move, but they do not hear his reply. His arm is shattered and blood is gushing from his side. He cannot last much longer – they think he is going west. His eyes ask them to do something. Stan rolls him on to a groundsheet and drags him yard by yard towards the trench. Shell splinters tear through the sheet. The ground rocks from a nearby shell-burst which almost covers them with mud. Stan drags him on. Ted is in mortal fear of being hit again. At last they come to the sally-port and he is carried on a duckboard into the trench.*
>
> *They give him the worst possible thing. He gulps down some rum, chokes, coughs blood, loses his breath; blood bubbles from his side, he is in the throes of death. He quietens. They give him water. If they can stop the bleeding he might survive. With a bandaged lead pencil, they probe back his lung and plug the wound with a field-dressing and pieces torn from a greatcoat. They fix a tourniquet and bind his arm to a piece of duckboard. This completes their rough but honest first aid.* [72]

[72] Pedersen p.95

Lording lost six ribs and his right arm and underwent many more operations before dying in a mental hospital in 1944.

Morning of 20 July, 1916 – The Withdrawal.

12.20am: By twenty past midnight, Brig Gen Elliott has learnt that the attack on the Sugar-loaf has failed disastrously and relays the information to McCay.

A third attempt is now deemed impossible. McCay's response is to order the attack abandoned and for the survivors to withdraw to their old positions and form a defensive line against an expected German counter-attack.

Meanwhile General Haking's headquarters prepare plans for another attack in daylight.

At the front, fighting continues as different units defend their positions and others attempt to retreat.

Soldiers of the Australian 8th Brigade still occupy some German frontline trenches but come under increasing pressure from the German infantry counter-attack and artillery fire.

Over the next several hours British and Australian soldiers will be pushed back to the Allied frontline through the confusion and darkness.

The postponement failed to reach the 58th Australian Battalion, which attacked with some of the 59th Australian Battalion and was stopped in No-man's land with many casualties, survivors from three battalions finding their way back after dark.

Despite reinforcements, the situation of the 14th Australian Brigade troops in the German lines became desperate as artillery fire and German counter-attacks from the open right flank, forced a slow withdrawal in the dark.

2.00am: On the left flank, more troops were sent forward with ammunition to the 8th Australian Brigade at dusk and at 2:00 am every soldier who could be found was sent forward. Consolidation in the German lines was slow as the troops lacked experience, many officers were casualties and there was no dry soil to fill sandbags, mud being substituted.

3.15am: German counter-attacks on the front and flanks, with machine-gun fire from Delangre Farm, Mouquet Farm and The Tadpole, began at 3:15 am on 20 July, forcing a retirement to the German first line and then a withdrawal to the original Australian front line. During the withdrawals some troops managed to fight their way out but many were cut off and captured.

4.00am: Just before 4.00am, orders to retreat reach elements of the 8th Brigade still holding positions on the German frontline. Whilst some units refuse to leave, remaining locked in hand-to-hand combat against the Germans, others make a break for the Allied line across No-man's land. Many are cut down by German machine gun fire as they try to reach the Allied line.

Captain Charles Arblaster and the remnants of the 53rd Battalion positioned behind the German frontline trenches come under heavy attack from German units and contemplate retreat to their own frontline which is now dangerously far away. As Arblaster leads the retreat his men are struck by a hail of German gunfire. Captain Arblaster is shot through both arms, and falls wounded. He and other survivors make their way back to the German front line. They seek cover in the trench and are taken prisoner.

5.00am Onwards

At **5.00am**, McCay and Mackenzie had met Haking at his headquarters in Sailly (some 8-9 kilometres away) to plan the 61st Division's fresh attack. Monro was also present. They had no sooner started than McCay's chief of staff telephoned with news from Pope that Cass's situation was critical and from Tivey that 8 Brigade was back in its own line. He wanted to know whether 14 Brigade should be reinforced or withdrawn. Monro opted to abandon the attack and withdraw. [73]

5.10am: The order for a general retreat is finally given by General Charles Monro, Commander of the British First Army. But retreat is at least as perilous as attack, and many soldiers fall as they try to get back to the Allied line.

Word of the retreat does not reach all units at the same time, and soldiers are still coming back as late as 8 o'clock in the morning, a good hour after German forces have recaptured much of their original frontline positions. Pockets of soldiers, such as those from the 53rd Battalion are trapped in their forward positions, forced to surrender and become prisoners of war.

The Australian 14th Brigade is one of the last units to receive the order to withdraw, as German forces press their counter-attack. 54th Battalion commander, Lieutenant Colonel Cass' report stated that "these attacks caused the 53rd to break and crowd in on my right. Many of the men were demoralized and a mixed party surrendered. It would appear that at least six Germans in Australian uniforms went over to the Germans from the vicinity of the right of the 53rd and others seeing this followed."

14th Brigade's retreat is assisted by the make-shift ditches and tunnels known as saps that have been dug by fellow soldiers through the night to assist in traversing No-man's land. These have even been laid with duckboards and Cass sends his machine gunners back one at a time, carefully establishing a rear guard during this orderly withdrawal.

News of the 8th Australian Brigade retirement reached McCay at a meeting with Mackenzie, Haking and Monro to plan the new 61st Division attack. Monro ordered the 14th Australian Brigade to be withdrawn.

At **5:40am**: A box-barrage around the brigade began, to protect the retirement.

Pope had warned Cass at 5.15am that a withdrawal was likely but that he should wait for the order. Because of another breakdown at McCay's headquarters, Pope was not given the order for over an hour. He sent it to Cass at 6.30am. But messages were taking 1½ hours to get through.

[73] Pedersen p.101

7.00am Onwards:

The withdrawal order finally reached Cass at **7.50am** – eight runners were needed to get it through. He spread the word to use the communication trench across No-man's land, the head of which Gibbins's rearguard was holding. Captain Norman Gibbins leads several assaults with grenades brought out through the 14th Brigade's communication trench across No-man's land.

German troops had got well behind the right flank and fired at every sign of movement, forcing the Australians to withdraw along the communication trench dug overnight.

Many were shot down while charging towards it. Those who succeeded could only move slowly along its congested course and the rearguard was hard pressed to keep the Germans at arm's length while they got away.

Gibbins deploys two Lewis guns to cover the withdrawal as his men hold the head of this trench. On receiving orders to retire, Cass was among the last to leave; the inspirational Gibbins was the last. Finding the trench blocked with the dead and wounded near the Australian line, he is forced to climb over the parapet. As he looks back at the German line, at that instant, he is shot in the head and killed. Half of his rearguard were lost.

Sergeant White, a 55th Battalion Lewis gunner recalled the order to retire and described the last moments of the battle:

> *"At that time Captain Gibbins was but a few yards from me and most of the infantry had retired. I was suddenly brought back to my senses by hearing Captain Gibbins call out 'come on, all you gunners.' I immediately picked up my spare parts and followed him...I saw him reach the top of our trenches where he turned his head around sharply and was immediately struck in the head by a bullet and killed instantly."*

From the comparative safety of the British frontline trench Corporal Williams observed:

"We were powerless to assist them, and had to watch them being shot down at point blank range... It seemed an eternity of time until the lucky ones reached our parapets, to be pulled in by willing hands. No sooner was our field of fire clear than we blazed into the Germans who had lined their parapets to punish the retiring troops."

In the meantime, the rest of 1/16 BRIR had streamed through Teufelsgraben, a strongpoint between Delangre Farm and Rouge Bancs, and linked up with 1/21 BRIR's attack.

Just before **8.00am**, German forces launch another attack. As Sergeant Archibald Winter of the 55th Battalion recalled: -

> *'God knows which way he came—we don't. He appeared to come from every direction. We were unsupported, consequently Fritz could come in on our own flanks. They had snipers everywhere and our own men were falling fast. Then we got into close quarters with the bombs but we were only a handful and Fritz was there in his thousands.'*

14th Australian Brigade are the last men out and are ordered to hold the Germans back. Their retreat is haphazard. Some men hold their positions to the end, whilst others are sent back by their commanding officers.

By **9:00am** the remnants of the 53rd, 54th and 55th battalions had returned; many wounded were rescued but only four of the machine-guns were recovered. Artillery-fire from both sides diminished and work began on either side of No-man's land to repair defences; a short truce was arranged by the Germans and Australians to recover their wounded.

At **9.20am** on the 20th July, the Bavarian 6th Division Command considers the battle won. An Intelligence Officer notes that the last of the British and Australians left defending themselves have been captured.

Prisoners of War

As to allied Prisoners of War, there remains some lack of precision as to just how many were captured.

In the case of Australian soldiers, Colonel A.G. Butler, in his book[74] provided a chart in which 470 is stated. Elsewhere, Paul Cobb[75] has written that there were 492, but in the research for this book, there have been 485 identified by name. Another 5 have been referred to in German Prisoner records, but do not appear to have corresponding information in the soldier files held with National Archives of Australia.

In the case of British soldiers, the only place at which any would have been captured is in the area of the 182nd Brigade, and particularly the 2/7 Royal Warwickshire Regiment and the 182nd Machine Gun Corps. In her book, Carole Wilkinson[76] suggests there were 51, but in the research for this book, only 37 names have been found to be referred to in German Prisoner records. Those do not have any serial numbers recorded against them, and the loss of British records subsequently in World War II may have seen those destroyed.

In the 182nd Brigade's Diary following the battle, on 21 July, 1916, Brigadier General Gordon signed off on a return indicating that in the 2/7 Royal Warwicks and the 182nd Machine Gun Coy, there were 180 missing following the Battle. But these numbers were precedced with the qualification that the exact proportion of missing and wounded cannot be determined until receipt of returns from the medical authorities.

In the case of German soldiers captured by the Allies, there has not been found a discrete set of data to identify these. It was estimated that after the Battle, that there were about 140. To identify those by name would require an examination of the Bavarian records piece by piece in order to do so.

---

[74] Official History of the Army Medical Services, Vol. 2, Western Front 1914-18

[75] See 'Fromelles 1916' ISBN 978 0 7524 5601 0, p. 118

[76] See Australia's Bloodiest Day at War, ISBN 978 1 742031 76 7, p.99

The Medical Services and The Wounded

The casualties presented at 14th Field Ambulance are charted in their War Diary. For the time preceding the attack at 4.00pm on 19 July, until 10.00am on 21 July, in two hour periods, the flow of casualties appears to be:

4.00pm 19/7	5	4.00am 20/7	115	4.00pm 20/7	35	4.00am 21/7	5
6.00pm 9/7	40	6.00am 20/7	240	6.00pm 20/7	65	6.00am 21/7	20
8.00pm 19/7	35	8.00am 20/7	100	8.00pm 20/7	35	8.00am 21/7	5
10.00pm 19/7	90	10.00am 20/7	170	10.00pm 20/7	30	10.00am 21/7	5
12,00 m/n 19/7	130	12.00 m/d 20/7	185	12.00 m/n 20/7	15		
2.00am 20/7	125	2.00pm 20/7	35	2.00am 21/7	25		

Captain Frederick Collier, the Regimental Medical Officer of the 60th Battalion, wrote:

> *'We worked all that afternoon, that night and all next day without ceasing. We could not show a light and when we came to a wounded man we would ask him where he was hit and feel for his wound with hands covered with dried blood and mud. There was no time and no water to wash hands.'*

On 21st July, Colonel C H Hardy, 5th Division Assistant Director of Medical Services, wrote to the General Officer Commanding the 5th Division Infantry Brigade:

> *Would you kindly convey to Medical Officers, Regimental Stretcher Bearers and A.A.M.C. Details my hearty thanks and congratulations for the excellent work done by them.*
>
> *Returns show that the evacuation of* **3,075** *in the period 2pm 19th July to 6am 21st July was an accomplishment of which they can justly be proud. I cannot too highly praise the work done by your Medical Officers.*

On 22nd July, Alfred Langan the Medical Officer of the 30th Australian Infantry Battalion, wrote to his father....

> *'For sixteen hours we worked with blood up to our elbows on the poor battered wrecks that were brought to us ... Men with shattered arms staggered in carrying or dragging men with battered legs and begged us to attend to their more unfortunate mates first. Not one of them murmured or complained. God, it made you humble and brought the tears to your eyes ... If ever there was a living Hell, that night and morning was it ... If there is a God in Heaven, may he strike the bastard Kaiser and his crew dead. We lost exactly half the battalion.'*

Australia's Official Historian Charles Bean wrote: -

> *'The scene in the Australian trenches, packed with wounded and dying, was unexampled in the history of the AIF', and was burned into the mind of everyone who saw it.'*

As 2nd Lt. Reginald Hugh Knyvett of the 59th Battalion remembered:

> *'if you had gathered the stock of a thousand butcher-shops … it would give you a faint conception of the shambles those trenches were',*

Not bothering with descriptive metaphors, Sergeant Martin saw

> *'…dead bodies lying in all directions, just as they had fallen, some without heads, other bodies torn about minus arms or legs, or pieces cut clean out of them by shells.'*

The Outcome

The battle had cost the British 61st Division 1,547 casualties, which represented a loss rate in the assault battalions of 46 percent. As nearly all of the 5th Australian Division had been sucked in to the battle, its casualties have almost continuously since said to be 5,533 men. The casualty rate in its assault battalions averaged 76 percent, while the flanking battalions suffered over 80 percent casualties. And then there were 485 men taken prisoner, in addition to the 3 captured on 15 July, 1916.

Casualties in the (German) 6th Bavarian Reserve Division totalled 1,582, of which 17 BRIR lost 274 men against the British attack, 21 BRIR lost 775 men against the Australians, and 16 BRIR lost 377 men against both.

From the German standpoint, the result was outstanding, and not just because of the highly favourable casualty ratio. Searching the dead and the prisoners, they found a copy of the general order from Haking that was read to all troops on the eve of the assault. By stating quite plainly that the objective was 'strictly limited to the enemy's support trenches and no more', it told them that a feint had been intended. They could now move troops from the Aubers/Fromelles area to the Somme or anywhere else, defeating the whole purpose of the operation. Yet there still remains some dispute as to whether or not the Germans actually did move troops from the area around Lille and Fromelles to the Somme.

Shells and bullets do not distinguish between veteran and novice, a point made the previous year when three tested divisions were shattered in the attack on Aubers Ridge. One of them, the 8th (UK) Division, assaulted over the same frontage and the same ground as the 5th Australian Division in this assault. Indeed, the two battles were so similar as to make Fromelles a re-run of Aubers Ridge on a smaller scale. In the 1915 battle, the Germans lost 1,551 men and the three British divisions principally involved 11,279, proportionally the same as their two divisions in 1916.

If the chances of success in May 1915 were proven to be nil, what is to be said of the prospects in July 1916? Using the attack as an example of how not to do things, Sir Basil Liddell Hart decried it as '*the final link of an almost incredibly muddled chain of causation*'. The planning process at the start was a dog's breakfast that Bean described in one of the most famous passages in the Australian Official History: -

> *Suggested first by Haking as a feint-attack; then by Plumer as part of a victorious advance; rejected by Monro in favour of attack elsewhere; put forward again by G.H.Q. as a "purely artillery" demonstration; ordered as a demonstration but with an infantry operation added, according to Haking's plan and through his emphatic advocacy; almost cancelled - through weather and the doubts of G.H.Q. – and finally reinstated by Haig, apparently as an urgent demonstration – such were the changes of form through which the plans of this ill-fated operation had successively passed. It was now definitely ordered. Haking arranged that the seven hours' bombardment should be begun at 11 o'clock in the morning of Wednesday, July 19th, and the infantry attack at 6.00pm. Thus, the assault originally planned to be delivered before noon, was now to be made three hours before dusk.*

The Australians were also angered by the Official Communique released by the British High Command, at 2.50pm on 20 July, 1916: -

> *Yesterday evening, south of Armentieres, we carried out some important raids on a front of two miles in which Australian troops took part. About 140 German prisoners were captured.*

It was plainly and simply a lie, made to conceal the severity of the disastrous reverse. Notwithstanding this purpose, Charles Bean questioned 'the good of deliberate lying like that', because the scale of the casualties and the loss of over 400 men as prisoners made the truth obvious.

But the truth was uttered by the men who survived that Battle, in honour of those who died, or were scarred for life.

The Australian public saw through it as soon as the facts were known and from then on British official statements, which had usually been accepted at face value, were treated with great scepticism.

It would become an embarrassment that so many leaders then, and for so long since, have by action or inaction, dishonoured so many brave Australian lives.

---ooo000ooo---

7. The Aftermath

'Good God Bill, what's happened to my brigade?'

Brig. General Harold (Pompey) Elliott to Captain Bill Trainor.

---ooo000ooo---

"To throw good money after bad is foolish. But to throw away men's lives where there is no reasonable chance of advantage is criminal. In the heat of battle mistakes in the command are inevitable and amply excusable. But the real indictment of leadership arises when attacks that are inherently vain are ordered merely because if they succeed, they would be useful. For such manslaughter, whether it springs from ignorance, a false conception of war, or a want of moral courage, commanders should be held accountable to the nation."

Liddell Hart "The Real War" 1930, p.202 (remarking apropos of the fighting at Ypres).

The Immediate Comments on The Results

Following the disastrous events of 19 and 20 July, the various troops were given the following messages from their superiors: -

Please convey to the troops engaged last night my appreciation of their gallant effort, and of the careful and thorough preparedness made for it. I wish them to realise that their enterprise has not been by any means in vain and that the gallantry with which they carried out the attack is fully recognized.

Gen. Sir Douglas Haig to Lt. Gen Haking

I wish to convey to all ranks of your Division [British] my appreciation of the gallant attack carried out yesterday by them. Although they were unable to consolidate the ground gained, the effect on the enemy will be far-reaching, and will prevent him from moving troops away from our front to the South.

I wish you all in your next attack a more complete and permanent victory, and that you may reap the full fruit of the energy and skill displayed by all Commands and their Staffs in the execution of their task.

R. Haking, Lt Gen Commanding XI Corps to Mackenzie, 20 July, 1916

The Australians were also lauded for their gallantry and achievement by the Commander-in-Chief, the Army Commander, the Corps Commander and the General Officer Commanding the 61st British Division. They were the first Australian troops to receive these accolades in France.

The Division not only fought gallantly but all ranks of every arm and service have carried out in a most exemplary and devoted manner, working day and night, an amount of labour which has highly tested their endurance and discipline, and merits my unqualified praise.

Colin Mackenzie CB, Commanding 61st Division to his Division 20 July, 1916

But in comparison, their own Division leader, had this to say: -

They'll get used to it.

Major General J.W. McCay to Brigadier General E. Tivey in the trenches, dawn 20 July, 1916.

---ooo000ooo---

But even after the reports of the Australians' gallantry at Fromelles, General Haig was less than effusive about the abilities of the other Australian soldiers under his command. The extent to which Fromelles, and the loss of almost half of an Australian Division concerned Haig is revealed in his diary entry of July 22nd, prior to the Battle of Pozieres:

> *'I visited General Gough after lunch to make sure the Australians had only been given a simple task. This is the first time that they will be taking part in a serious offensive on a big scale against the German forces.'*

None of the soldiers who served at Fromelles formed part of the units which were to fight at Pozieres. Soldiers who served at Fromelles would however, join in actions on The Somme later in the year.

The soldiers he was referring to had not been involved previously in large scale battles, but had been involved in a number of raids previously. Given that at Fromelles, Haig had just permitted the slaughter of 1917 Australian soldiers under his command, with more to be added in the course of the coming days, and the wounding or taking of 3616 others, he obviously had concerns about the Australians' general preparedness for battles.

The 'Official' Casualties

The battle caused one of the greatest-ever numbers of Australian deaths and casualties in action. Between 5.45pm on 19 July, and 8.30am on 20 July, a period of less than 15 hours, the Australian 5th Division suffered 5,533 casualties. The 8th Brigade suffered casualties of 1,547 men; the 15th Brigade suffered casualties of 1,776 men; the 14th Brigade suffered casualties of 1,717 men; and there were also 88 men from the divisional engineers. As we will see in Chapter 9, the official number of deaths of 1917, was in reality more than 2000.

Two battalions had so many casualties that they had to be rebuilt. Of 887 personnel from the 60th Battalion, only one officer and 106 other ranks survived unwounded. The 31st Battalion had 544 casualties. And the 32nd Battalion suffered 718 killed and wounded.

The 61st Division was already understrength before the battle, engaged half as many men as the 5th Australian Division and lost 1,547 casualties.

German casualties in the 6th Bavarian Reserve Division were 1,582 men.

Allied soldiers killed in the area that was re-taken by the Germans, were buried shortly after the battle. The burial pits were photographed from a British reconnaissance aircraft on 21 July but marked as dugouts or trench-mortar positions.

The Australian and British soldiers killed in the vicinity of and behind the German lines were gathered for burial by the Germans. It was high summer and they had to clear their lines so they dug a collective burial ground. They recorded the burials, removed and documented the ID discs and personal items, individually bagged them and sent them back to the soldiers' families through the Red Cross.

A Burial Order was issued on 21 July, by Julius Ritter von Braun, Regimental Order 5220 of the Bavarian Reserve Infantry Regiment No. 21. It specified that the 'English' bodies were to be buried in mass graves to the south of Pheasant Wood, and that H Company was to excavate mass graves for approximately 400 bodies. The bodies were taken by narrow gauge trench railway from 22 July, and progressively buried in eight 10 x 2.2 x5-metre (32.8 ft × 7.2ft ×16.4ft) pits.

---ooo000ooo---

No territory was gained or lost. The scale of casualties in one day would remain unsurpassed even by the first and second Battles of Bullecourt in 1917.

On 11 April, 1917, in the *first* battle of Bullecourt, the 4th Australian Division carried out a costly and unsuccessful attack at Bullecourt. Of the 3000 casualties, 1170 were prisoners of war – by far the most prisoners taken in a single World War 1 battle.

The *second* battle of Bullecourt, over a period of two weeks in May 1917, claimed a further 7,000 Australian casualties from three Divisions - the 1st, 2nd and 5th Divisions. The casualties included more than 1,100 captured.

---ooo000ooo---

According to Barton[77], an aggregation of losses in 1914, 1915 and 1916 makes the sector astride Rouges Bancs a candidate for the title of the bloodiest battleground of the Western Front. It is little wonder that human remains are still recovered from these fields. An observation such as this makes a mockery of the title "nursery sector". Perhaps we should now call them 'the killing fields' instead?

---ooo000ooo---

A graphic account of the action, the losses, and the aftermath of the battle was published three years later, as part of an article by Sgt. W.H. Downing, which appeared in The Geelong Advertiser, Geelong on 26 July, 1919: -

THE MASSACRE AT FLEURBAIX WHERE MANY GEELONG HEROES DIED.[78]

"OVER THE BAGS IN FIVE MINUTES, over the bags in five minutes": so it passed along. There were looks of sharp inquiry, but no word spoken. Then "Over you go." The 60th climbed on the parapet, - heavily laden, dragging with them scaling ladders, light bridges, picks, shovels, and bags of bombs. There was wire to go through, and sinking ground: a creek to cross, more marsh and wire, then the German line. Stammering scores of German machine guns spluttered violently, drowning the noise of the canonade. Hundreds were mown down in the flicker of an eyelid, like great rows of teeth knocked from a comb, but still the line went on thinning and stretching. Wounded wriggled into shell-holes, or were hit

[77] Barton p.57

[78] See: https://trove.nla.gov.au/newspaper/article/165750121

again. Men were cut in two by streams of bullets, and still the line went on. Fifty-six remained of a full thousand. It was over in five minutes.

And then the Fifty-ninth rose, vengeful, with a shout — a thousand as one man. The chattering metallic staccato of the tempest of hell burst in nickelled gusts. Sheaves and streams of bullets swept like whirling knives. There were many corpses hung inert on our wire, but the Fifty-ninth surged forward, now in silence, more steadily, more precise than on parade. A few yards and there were but two hundred marching steadily on. The rest lay in heaps and bloody swathes. They began firing at the German line as they advanced. Lewis gunners dropped into shell-holes and fired burst after burst, dashing from cover to cover. A hundred men broke into a wild but futile charge, determined to strike if possible one blow, but ironical enemies pressed their red-hot thumb pieces with blistered hands, spraying death from the tortured muzzles. That hundred lay flat in the attitude of sleep.

It grew quiet again. A few wounded crawled in the grass, sniped at by riflemen. Then there was silence. Eighty came back that night. Two companies from the Fifty-eighth rose from the breastwork. The remainder were carrying ammunition, and advanced by rushes with covering fire. In banks of battle reek the sun went down as red as blood. As darkness grew on, the Fifty-seventh went forward, but most were recalled almost before they left, for there was nothing to be gained by further loss of life. A few reached the creek.

It was the Charge of the Light Brigade once more, but more terrible, more hopeless — magnificent, but not war — a volley of death filled by some body's blunder or the horrid necessities of modern war.

The handful in our trenches stood to arms all night, held inactive by orders, since the line was now dangerously weak, for there were no supports, and no reserves, and many enemy corps d'elite were in front.

The bays were jammed with dead and wounded lying head to foot for two miles, in a treble row, on the fire steps, beneath them and behind the blockhouses. Wounded came crawling in, rolling over the parapet and sprawling to the bottom. A white-faced boy, naked to the waist, was being led along the trench, a hole in his side. He cracked some joke, then, "I think I'll spell -- a minute; it's all going dark." He sat down. An hour later someone shook him, but he was stiff and cold. By dawn, in spite of strenuous labours only half the wounded and a few of the dead had been brought in. For five nights this work was unremitting. Parties went out under fire in broad daylight. Some of the wounded were never found. A few crawled in three weeks later, with shattered limbs and gaping wounds. All flight the Eighth and Fourteenth Brigades were fighting for their lives, almost surrounded, up to their breasts in water with which the enemy had flooded their trench. For three days hundreds of wounded lay, uncomplaining in their torment, in our line.

For days an officer, blind and demented, wandered near the German lines, never fired on, but used as a decoy to attract his friends, to their death. These were shot on attempting to reach him. He wandered up and down the line till he died, avoiding friend and foe alike. The remnants of the 57th and 58th held the front-line system for a further fifty days, making fifty-nine in all, without relief.

Such, were they who builded for Australia — and such the men who laid the corner-stone of her renown.

Counting The Casualties. Killed, Wounded, or Prisoners.

Australian Soldiers

The shattered assault-battalions were collected on the afternoon of July 20th in the roads near their respective brigade headquarter. Of the 60th Battalion, which had gone into the fight with 887 officers and men, only one officer and 106 men answered the call [82% casualties]. The 32nd, 59th, and 53rd had suffered almost as heavily. A few of the missing at the roll-call came in afterwards; but the 5th Division had lost in one night's sharp fighting 5,533 officers and men [incl more than 400 POW], and the 61st Division (which was of less strength) 1,547 [incl about 80 POW]. The German loss appears to have been under 1,500 [including about 140 POW].[79]

Among the Australians, by far the heaviest casualties had occurred on the flanks, the attacking battalions from left to right suffering as follows (Official History, Vol III AIF France 1916): -

Unit.	Officers.	Other Ranks.	
32nd Battalion	17	701	
31st Battalion	16	528	
54th Battalion	19	521	
53rd Battalion	24	601	
60th Battalion	16	741	
59th Battalion	20	675	

The losses of officers and men respectively in the other battalions were:

29th Battalion	7	209	
30th Battalion	9	343	
55th Battalion	11	330	
56th Battalion	2	149	
57th Battalion	2	33	
58th Battalion	11	237	
8th MG Coy	5	49	
14th MG Coy	7	46	
15th MG Coy	2	34	
Div Troops		*168*	
Total	**168**	**5365**	**5533**

[79] Extrapolated from Bean: The Official History, Vol. III, AIF France 1916 – Chapter XIII, p.441 & following.

Adding up the figures provided in the Official History identified a deficit of 168, presumed to be from other Div Troops (e.g. Bde HQs, Pioneers, Engineers, Artillery, Medical, Services etc) – all shown here as Other Ranks.

At its inception, there were 1299 names of soldiers unaccounted for, recorded on the VC Corner Australian Cemetery Memorial.

There were 410 bodily remains buried at VC Corner Cemetery itself.

There were 266 unidentified and buried at Rue David Military Cemetery.

There were 142 unidentified and buried at Ration Farm Military Cemetery.

There were 120 unidentified and buried at Aubers Ridge British Cemetery.

There were 72 unidentified and buried at Y Farm Military Cemetery.

There were 52 unidentified and buried at Le Trou Aid Post Cemetery.

There were 27 unidentified and buried at Rue du Bois Military Cemetery.

There were 22 unidentified and buried at Rue-Petillon Military Cemetery.

There were 10 unidentified and buried at Anzac Cemetery at Sailly-sur-la-Lys.

There were 10 unidentified and buried at Sailly-sur-la-Lys Canadian Cemetery.

There were thus 168 bodies unaccounted for as well as unidentified. They are likely to be within the ground of No-man's land, and where they are likely to remain.

But in spite of the number of those previously unaccounted for, there are now 250 in the process of being identified, at Pheasant Wood Military Cemetery, Fromelles.

In July 1916 the 5th Divisions' war establishment was 19,360. According to its War Diary, on 18 July 1916 its effective strength of was 18,727. The 5,533 casualties suffered at Fromelles equated to 29.5% of the Division.

In accordance with the policy at this time adopted by G.H.Q., the severity of this reverse, though of course well known to the German Army and people, was concealed, some say 'covered up', from the British public in the official communiqué: -

Yesterday evening, south of Armentieres, we carried out some important raids on a front of two miles in which Australian troops took part. About 140 German prisoners were captured.

---ooo000ooo---

British Soldiers

At the Battle of Fromelles, on 19 July, 1916, some of the British 61st Divn did manage to penetrate the German first and second lines behind the Wick salient.

Those that fell and whom were killed in action were transported to a German Cemetery at Fournes, approximately 6km behind the German front lines, south east of Fromelles itself. The

Germans created several mass graves in the area, with Pheasant Wood being a well-known example.

In June, 1920, in Fournes, one mass grave was exhumed. Six sets of remains were discovered. Only one, Pte Edward Scudamore of the 182nd Machine Gun Company was successfully identified. He and his five comrades were reinterred at Cabaret Rouge Cemetery. Pte Scudamore had his own grave whereas the others were 'Known Unto God'.

In July, 1923, a unit returned to the German Cemetery in Fournes, exhumed a further 176 burials that had various dates of death. A large plot was excavated, a large trench grave very similar to those at Pheasant Wood in terms of size and shape etc. was revealed. Fifty-one (51) soldiers were found here, buried between the 23rd & 25th July, 1916. Fifty (50) of them were British, of whom five were named – Pte W Wale 2/7 Warks, Pte A Fisher 182 MGC (wrongly recorded as being of the 122 MGC by CWGC), Pte Harry Berry 2/7 Warks, L/Sgt J R Whateley 2/7 Warks and Pte Herbert Tipple 182 MGC. All the men were reinterred in graves 'Known Unto God' at Cabaret Rouge Cemetery. The five named men had Special Memorials numbered 7, 8, 9, 10 and 11 respectively indicating that they were either buried or believed to be buried in the cemetery. The one other successful identification was Pte Patrick Gearing of 53rd Battalion AIF.

Of the 813 British soldiers listed in Appendix I, there were

396 Memorialised at Loos Memorial,

4 Memorialised at Ploegsteert Memorial,

7 Memorialised at Thiepval Memorial, and

406 who have a headstone in a cemetery at which they can be honoured.

Reports and Recriminations Followed Soon After, and Into The Decades Ahead

Four days after the battle, Haking's four-page report accompanied Mackenzie's paper to First Army Headquarters. He wrote that 61st Division was *"not sufficiently imbued with the offensive spirit to go in like one man at the appointed time"*.

He conceded that the Australian Division had *"attacked in the most gallant manner and gained the enemy's position"* but added that they *"were not sufficiently trained to consolidate the ground gained"*, a phrase which caused angst in the Australian press and was quoted by Bean in the Australian Official History.

As was seen in preceding chapters, the task given was to capture and hold the German trenches. In this sense, the Battle up to that point was a tactical success. But the German counter-attack whittled that away to a defeat.

Haking claimed that "The artillery preparation was adequate. There were sufficient guns and sufficient ammunition" and that *"the wire was properly cut and the assaulting Battalions had a clear run into the enemy's trenches"*, omitting to mention that on 17 July he had mentioned that some gunners had never before fired on the Western Front, and that many of the casualties resulted from machine guns fired from positions that the bombardment had failed to suppress. The lies continued.

He claimed that: -

- *"... the lessons to be learnt from the attack apply more to the Divisions which took part than to ordinary trained Divisions"* …
- *"... the attack, though it failed, has done both divisions a great deal of good"* … and that
- *"… with two trained Divisions the position would have been a gift after the artillery bombardment; with these two new Divisions there was a good chance of success but they did not quite attain it"*

omitting to mention that he had had a good Division (the 1st) at Aubers Ridge in May 1915. He also wrote that: -

- *"… the attack … has done both Divisions a great deal of good"*

and also listed the chronology of dates and postponements, claiming that he had been *"directed to attack"*.

By 26 July Haking had received McCay's report and was more generous in his praise of the Australians, although he blamed 61st Division for failing to take the Sugar-loaf.

Field Marshal Sir Henry Hughes Wilson was one of the most senior British Army staff officers of the First World War. Wilson (diary 30 July 1916) described Fromelles at the time as "a botch job".

Captain Philip Landon testified in 1938 to Brigadier General Sir James Edward who was responsible for the post-war compilation of the 28-volume *History of the Great War*, that it had been:

> *"… as good an illustration as there was of the reckless extravagance in expenditure of life which ruled the minds of some of the subordinate commanders, like Gen Haking, at this stage of the war"* …
>
> *"… the weakness of GHQ lay in not seeing that a Corps Commander, left to himself, would also be tempted to win glory for his Corps by spectacular success, and would be prodigal in using the Divisions which passed through his hands for this purpose"*.

Colonel E.R. Clayton later testified to Edmonds that Haking's "undue optimism was one of the direct causes of failure" of the attack at Fromelles. In the Official History, Wilfrid Miles later wrote scathingly of the lack of preparation and the pointlessness of an attack by inexperienced troops, to seize a position which they could not possibly have held against counterattacks, and blames First Army for not cancelling the operation.

Brig Gen Harold 'Pompey' Elliott (in 1930) criticised Haking for exaggerating the amount of artillery that would be available, and for attacking without surprise. He exonerated McCay and argued that Haking, after Major Howard's report (which Haig had annotated to permit the attack only on condition that sufficient guns and ammunition were available) had persuaded Monro, who in turn persuaded Butler, and that Haking had been keen to win glory for himself. He also stressed how Haking had ignored suggestions from Monro that the attack be postponed because of the rain, and was scornful of Haking's after-battle report. Andy Simpson[80] wrote that Fromelles was "a costly failure" but "although the Australian Official Historian blamed Haking for the affair, fault also lay with the First Army and the 5th Australian Division itself".

The Disgraceful Treatment of Colonel Harold Pope.

Perhaps the most malicious example of how senior staff were used as scapegoats was the sacking of Colonel Harold Pope by his Commanding Officer McCay.

Harold Pope81 was born at Ealing, Middlesex on 16 October, 1873, the son of a solicitor. He was educated in England, joining the clerical staff of the Great Northern Railway when 16. He emigrated to Perth, Australia in 1895 to work for the railways. Married in 1896 to Susan Slater at Albany, he joined the Western Australian Military forces four years later and was a Lt. Colonel in 1908.

At the Landing, 25 April, 1915, he led the 16th Bn, and at nightfall they executed an important move in an area later called Pope's Hill. His conduct on Gallipoli throughout the campaign was spoken of with unreserved admiration: Monash thought highly of him, the men swore by him, 'a popular figure … with confident bearing, strong face and friendly eyes' was one description. Near the end of the campaign, he was appointed Companion of the Order of the Bath (CB) and in October was evacuated ill.

Appointed by General McCay to command 14th Brigade, Pope, now temporary Colonel, was blessed with experienced battalion commanders and at Fromelles was the most successful brigade commander in the battle.

Sent into attack at 6.00pm on 19th and ordered to withdraw at 5.40am on the 20th, the 14th suffered over 2000 casualties, including 496 dead.82 It was the last Brigade to withdraw.

From Bean's diary of the day, we learn that on hearing of the battle that Bean drove to Sailly-sur-la-Lys, arriving at 1.20pm on 20 July. He interviewed various people including Pope, about whom he then wrote: -

Pope of the 14th Bde rather disgusted me by the boastful way he talked – I think he had been refreshing himself after the strain. 'Well, we were really the only Bde who didn't come back till we were told to,' he said this morning. He was rather contemptuous about Elliott and 15th Bde.

[80] Simpson, A. (2006). Directing Operations: British Corps Command on the Western Front 1914-18. Stroud: Spellmount. ISBN 978-1-86227-292-7.

[81] Corfield p. 102

[82] Corfield p.260

When he went to sleep after lunch on 20 July after some sixty hours of high tension and little rest, Pope could not be awakened.

Pope's diary of the following day recorded:

Alleged to be drunk after lunch on Thursday. Went to sleep 3.00pm. Woke at 8.00 am 21st. Saw Pope (Captain) off. Had bath etc. Everything quiet. When starting to prepare report on operation at 11.10am got memo from McCay relieving me of command – as about 4.30pm he had found me incapable of duty.

McCay sacked Pope unceremoniously on 22nd July, and had him shipped back to Australia.

From Pope's diary again: *22 July: Pope (Captain) saw McCay 9.00am. He says on Thursday afternoon 4.30pm I was drunk and he knows it.*

McCay, despite appeals from many influential people never changed his mind. Pope asked Birdwood for a court-martial so he could clear himself, but it was refused.

When McCay was out of the way, Birdwood who had dodged the affair earlier on, backtracked and with Hobbs brought Pope back to France and in command of 52nd Bn. On 7 June, 1917 Pope was wounded at Messines. Later Mentioned in Dispatches, Pope was invalided home in February 1918. From then until the end of the war, he worked on organising troopships.

Pope appears to have been victimised by his divisional commander, who was himself out of his depth, and by his own Brigade Major who, like so many British officers, did not appear willing to endeavour to accept and understand the citizen soldiers of Australia. Pope was a successful soldier in the Gallipoli campaign, and on the Western Front, and his strenuous efforts to restore his reputation must be acknowledged so that in some part the record is corrected. Furthermore, he was held in high regard among his men and his peers, and that is a valid test of an officer's worth. Conversely, McCay was despised by many '19 July men', and that too is a reasonable gauge of his leadership.[83]

Was The Strategic "Objective" Realised?[84]

The strategic objective of the attack was contained in the First Army Operational Order 100 (15 July, 1916),

> ***...to prevent the enemy from moving troops Southwards, to take part in the main battle.***

In addition, Haking had ordered that the troops due to attack were to be told that the Commander in Chief [Haig] had directed XI Corps

> ***to attack the enemy in front of us, capture his front-line system of trenches, and thus prevent him from reinforcing his troops to the South.***

---

[83] Cobb p.172

[84] See Wikipedia: Battle of Fromelles. https://en.wikipedia.org/wiki/Battle_of_Fromelles.

Historian Michael Senior wrote[85] that historians generally judged the attack to have failed to prevent German troops being moved south to the Somme. Wilfrid Miles, the British official historian writing that the IX Reserve and the Guard Reserve corps had been sent. Peter Pedersen wrote that the Germans knew that Fromelles was a decoy and sent the reserves to the Somme. The Australian official historian, Charles Bean, wrote that the attack showed the Germans that they were free to withdraw troops. In 2007, Paul Cobb wrote[86] that the Germans were not deterred from sending troops to reinforce the Somme. Senior wrote that there was evidence that the transfer of troops to the south was delayed. A German intelligence officer of the 6th Bavarian Reserve Division wrote on 20 July, 1916:

> *There are no signs of any immediate repetition of the enemy attack .... However, judging by the general situation, a new push is not impossible.*

A Bavarian document discovered in 1923 related that:

> *An order was captured declaring that the object of the attack was to keep German troops engaged in the sector so as to keep pressure from the Somme...a repetition of these attacks is therefore to be expected.*[87]

And Bean wrote in 1930 that the Bavarians might have doubted that the British would sacrifice 7,000 men as a decoy.

But there is also evidence to show that the IX Reserve and Guard Reserve corps had been moved from the Souchez–Vimy area, 20 miles (32 km) from Fromelles, well outside the XI Corps sector. Troops which were kept in the Loos–Armentières sector opposite the corps for four weeks after 19 July, remained *as a precaution*.

German records show that eight divisions were between Loos and Armentières on 1 July and that two were sent to the Somme by 2 July, long before the Fromelles attack and the other six divisions stayed opposite XI Corps *for five to nine weeks after 19 July*. Had divisions moved earlier, the Battle of Pozières (23 July – 3 September), might have cost I Anzac Corps far more than 23,000 casualties.

Senior concluded that German troops **had** been retained opposite the XI Corps, because of the attack at Fromelles.

So, while many commentators have described Fromelles as a *tactical* failure, there is also compelling evidence, available from contemporary records and accounts, that its strategic objective had been partly successful.

Evaluating The Battle and What Took Place

---

[85] Michael Senior (2012): Haking: A Dutiful Soldier: Lieutenant General Sir Richard Haking: XI Corps Commander 1915-1918: A Study in Corps Command (hbk ed.) Barnsley: Pen & Sword Military. ISBN 978-1-84884-643-2.

[86] Cobb, Paul: Fromelles 1916. 2007, The History Press.

[87] Senior, M. (2012). Haking: A Dutiful Soldier: Lieutenant General Sir Richard Haking: XI Corps Commander 1915–18: A Study in Corps Command (hbk. ed.). Barnsley: Pen & Sword Military. ISBN 978-1-84884-643-2. p.132

One of the significant leaders involved in the lead up to and the execution of the Battle of Fromelles, was Brig Gen Harold (Pompey) Elliott. Following the cessation of hostilities, Elliott went on a post-battle inspection accompanied by Lieutenant J.D. Schroder, who reported:

"Ordinary sandbagged trenches were now heaps of debris, and it was impossible to walk far without falling over dead men, [but] Pompey went from battalion to company headquarters and so on right along the line. A word for a wounded man here, a pat of approbation to a bleary-eyed digger there, he missed nobody. He never spoke a word all the way back to advanced brigade [headquarters] but went straight inside, put his head in his hands, and sobbed his heart out."

Other survivors were struck by the intensity of his anguish. At one stage Elliott called out to Captain Bill Trainor,

> *'Good God Bill, what's happened to my brigade?'*
>
> *'Practically all my best officers are dead.'* And
>
> *'I presume there was some plan at the back of the attack but it is difficult to know what it was.'*

In a talk given in 1930 and reprinted from "The Duckboard" September 1, 1930, Elliott issued a number of critical observations of the Battle, and the manner in which it was carried out. The talk was titled '**The Battle of Fleurbaix or Fromelles**', and was a sensation at the time. It was portrayed as an accusatory exercise in which Elliot blamed the British for all that went wrong. The following is an edited account of what is covered in the full article.

Elliott outlined the history leading up to the Battle at Fromelles, which has been covered in the previous chapters. In doing so, he outlined what he described as the significant "Blunders" made in the planning of and execution of the battle.

Blunder No. 1: The large-scale bombardment which would take place, ahead of an infantry attack would cast aside any intention of secrecy or surprise (the supreme element in war). On the contrary, the operation was to be widely advertised by the bombardment—a policy which might conceivably have been a wise one had no infantry attack been required, but was simply suicidal if the intention was to follow it up by an immediate infantry attack.

Blunder No. 2: Insufficient artillery preparation, resulting in the wire being uncut and enemy front line and batteries not being silenced. The planned attack continued in the knowledge that only 2/3 of the original amount of munitions would be available. Haking was to have had 300,000 shells for his field guns and 30,000 for his 4.5 howitzers, but would be allocated only 200,000 and 15,000 respectively, and no medium trench mortars at all.

Blunder No. 3: Faulty division of the front. The proposals for the attack upon the Sugar-loaf could not have been closely studied.This formidable bastion, which required for success the closest possible liaison and co-operation between the forces engaged upon the task, should in accordance with all the known rules of war, have been entrusted to a complete unit—battalion, brigade or division.

In the outcome it was found that this bastion lay just on the Junction of two different units — notoriously the weakest point in an attack or defence, and Haking and his staff appear to have been so entirely oblivious of this fault that not the slightest effort that is discoverable was made to overcome this obvious difficulty.

Blunder No. 4: No arrangement for liaison between these divisions or to co-ordinate the advance itself or methods to be employed.

General McCay was only advised of the change of plan on the 14th July, and on the 16th he received his final instructions, viz., to attack with all three brigades in line, but to use only two battalions of each brigade for the assault.

This instruction again appears to have been badly conceived, for neither the brigadiers nor the divisional commanders had any forces in reserve to repair initial errors. Had this been done it might have been possible to push a company or two across into the space that had been gained, and by attack outwards, to have cleared the front of the 15th Brigade.

Sending battalion staffs forward with the last wave before any lodgment had been gained in enemy's line, was sacrificing valuable officers uselessly.

Blunder No. 5: Preparations had to be unduly hurried, two and a half days only being allowed, the men were tired out by nights without rest. Some of the bombs were sent up without examination in the 14th and 8th Brigade were found not to be detonated.

Blunder No. 6: Failure of the first duty of a commander — personal reconnaissance — resulting in selection of unsuitable ground commanded everywhere by the enemy observation and fire, and which gave no position upon which to consolidate.

The objective was limited to the capture of the enemy's front line and to his support lines, which were supposed to exist some fifty to one hundred yards in rear of the front line. Inasmuch as our own support lines situated in a relatively similar position were at this time utterly useless, and we were now developing a second line some 300 yards in rear, one would have thought it would have occurred to someone in authority to doubt if the supposed support line of the enemy really existed.

Careful reconnaissance, if not mere inductive reasoning, would have disclosed this to be the case. And a wide No-man's Land confronted us at a vital point.

Blunder No. 7: The reason why no trenches existed in the ordinary sense was simply that the water level rose to within a foot of the surface of the soil, and any trench cut below that level became at once waterlogged, and could not on our side at least be drained at all. Hence if no support trenches were to be found in the enemy's country, the problem of getting shelter for the men when they reached their final objective should have been faced, studied and provided for. No such preparations were made.

Adding to that, gas cylinders were kept in our line when it was certain that line would be bombarded. They had been embedded in their parapet by Haking's orders, in the hope that they could be used if the wind was favorable. These were exploded by the enemy bombardment which added to the confusion and loss.

Blunder No. 8: About a mile to a mile and a half inland behind the enemy's line rose the long ridge of the Aubers-Fromelles, of no great height, but in such country, it completely dominated our lines. From a protected position, an observer, with a telescope could, with perfect safety to himself, count every sentry in our lines. He had also an extensive view across our back areas, and could at once detect any preparation for attack. A more unsuitable site from which to launch a well-advertised attack could hardly have been found on the western front.

Blunder No. 9: The British (opposite the Sugar-loaf, and so next the 15th Brigade) had, contrary to my own personal advice, based on experience in Gallipoli in August, 1915, cut dozens of sally ports in their parapet. They had thus further advertised the intended attack, and at the same time weakened their parapet to a great extent, and they in particular suffered severely from this fire.

They would take longer to get out in that way than we would by going over the top deployed in line. The British decided to start to file out as early as 5.30pm We didn't start until 5.45pm The attack of the British battalion which was detailed to seize the Sugar-loaf, and which, as I have said, lost 140 men before they started, failed completely, the wire there, which was very strong, not being cut. The heavy artillery detailed to subdue this salient could hardly hit it, much less subdue it, their guns being the oldest in the front. They were really afraid, as we learnt later on, of aiming at the Sugar-loaf at all for fear of hitting our own lines.

---ooo000ooo---

Elliott then proceeded to bring the shortcomings he had observed, together.

Elliott recapitulated after his dissertation, saying that this operation has no single redeeming feature except the wonderful bravery shown by the officers and men in the line. The Duntroon men, Hutchinson, Elliott, Arblaster, Hopkins and Robinson, who were killed; Legge, and others engaged, did splendid work.

The whole operation was so incredibly blundered from beginning to end that it is almost incomprehensible how the British Staff, who were responsible for it, could have consisted of trained professional soldiers of considerable reputation and experience, and why, in view of the outcome of this extraordinary adventure, any of them were retained in active command.

Elliott proceeded to point out the comments by General Haking, in his first report of the action (set out earlier in this chapter), and recounted Haking's comments in a later report in which he said,

> *"The attack of the Australians was carried out in an exceptionally gallant manner. There is no doubt that the men advanced with the greatest determination. Their difficulties on the right flank were caused by the failure of the 61st Division to carry the Sugar-loaf."*

Haking thus endeavoured to throw the whole blame of the failure upon the infantry, particularly the British.

Captain Bean's observations were placed before Elliott's audience, in which Bean was quoted as having stated: -

The blame for our losses was in Australia and amongst the soldiers thrown exclusively, but quite unjustly, upon General McCay, for which he was, in fact, no more responsible than the humblest private in his force.

His whole after life was embittered by these unjust reproaches.

His security in his command was not in any way shaken by the popular error—the authorities in the A.I.F. being well aware of the truth.

But in accordance with the tradition of the British Regular Army, that no criticism must be made of a senior officer (in this case Haking), they uttered not one word or sentence before or since to raise the ban against him, and he suffered in admirable silence.

Bean also went on to say: -

Another particularly unfortunate, but almost inevitable result of the fight was that the Australian soldiers tended to accept the judgment, often most unjust, that the 'Tommies' could not be relied upon to uphold a flank in a stiff fight.

Elliott asked his audience: "*General Haking had cast the first stone against them, and who could blame those who followed his lead? The result of the action was to cripple the 5th Division for months to come, and the loss was not in numbers only, for our very finest officers, N.C.O.'s and men, many of them Gallipoli veterans, perished there. And what did it accomplish?*"

Elliott went on to observe that in its final form the Battle was intended as a feint to distract the attention of the German staff from Pozieres [on the Somme], and inasmuch as German records show that for a few days afterwards the Germans expected a renewal of the attack, the feint cannot with certainty be judged to have been completely ineffective. Yet the value of the result (if any) it can be said with certainty, was tragically disproportionate to the cost.

Comparing the successes of August, 1918, Elliott noted that we are apt to overlook the immensity of care and preparation that preceded it and so ensured its brilliant success, but in a failure such as Fleurbaix we see the result of the neglect to make such preparations.

He recalled that *in [our] Field Service Regulations it used to be taught that no general should launch his troops to an attack without a careful personal reconnaissance, but where, from the number of troops engaged such personal reconnaissance was impossible, he might delegate this duty in part to responsible staff officers. Here there were only two divisions engaged, and there was nothing to prevent such personal reconnaissance, yet there is not the slightest sign that Monro or Haking recognised any such responsibility.*

Elliott recounted that the only officer belonging to the higher formations who came near the place to his knowledge was Major Howard, whom Elliott '*captured at my own headquarters and insisted upon him accompanying me over my own sector*'.

According to Elliott's now well-publicised account of the meeting, Major Howard at once admitted the impossibility of the task and did his best, Elliott presumed, to prevent the slaughter that followed.

In great contrast, according to Elliott: - *General Harrington, Plumer's Chief of Staff, never failed to consult the brigade and battalion commanders to see if they were satisfied.*

---ooo000ooo---

Elliott gave Haking the chief blame for this debacle:

'The plan was his, and but for his insistence would have been abandoned almost at its birth. He can however, shelter behind the weakness displayed by Monro, by Butler, and even by Haig himself at this critical juncture.'

Elliott further suggested that: -

..if G.H.Q. had insisted on the 'demonstration' taking place at that particular spot at all costs with infantry, the frontage of the attack should have been slightly changed and the final objective made the line of the Laies River from opposite Trivolet to its junction with our lines. The new line to be established to be a little beyond that river.

Elliott believed that this would have given a definite line to hold on to.

The new trenches could have been drained into the river. The new line also would have tended to make a large proportion of the rest of the German line on the front (attacked during the actual operation by the 14th and 8th Brigades), untenable, thus gaining by manoeuvre what could only otherwise be gained by direct frontal attack at heavy cost.

Moreover, reserves would have then been available.

Look at it how we will, however, it is impossible to see how to make of it a satisfactory job, but the methods actually chosen seem to have been the worst possible.

With the benefit of having worked with tanks in later campaigns, Elliott theorized that probably without tanks to help, the attack must in any event have failed. But this comment needs to be put into context. Tanks were not yet a part of battle equipment on the Western Front, so this should be seen as the result of hindsight rather than what would have been available at the time.

---ooo000ooo---

We may well ask: - what became of the officers responsible? Elliott had no reluctance in giving his assessment of that:

'Haking sheltered himself behind Monro's weakness, and having laid the blame upon the infantry's alleged shortcomings, particularly upon his own British Tommies, got away with it and retained his command till the end of the war.'

'Monro was apparently quietly shelved by being sent away to India to be Commander-in-Chief there.'

Elliott recounted a passage from a "recent brilliant work" [first published in 1930], "The Real War," by Liddell Hart, vide page 202, quoting Hart's remarks apropos of the fighting at Ypres, but which could just as readily be attributed to Fromelles:—

"To throw good money after bad is foolish. But to throw away men's lives where there is no reasonable chance of advantage is criminal. In the heat of battle mistakes in the command are inevitable and amply excusable. But the real indictment of leadership arises when attacks that are inherently vain are ordered merely because if they succeed, they would be useful. For such manslaughter, whether it springs from ignorance, a false conception of war, or a want of moral courage, commanders should be held accountable to the nation."

And to add more dimensions than there had been before, Elliott continued, to criticize the military culture of shielding senior officers from blame for their disasters: -

'I have said that General McCay was blamed for the failure. It would appear that General Haking successfully unloaded the blame for his previous failure on a divisional commander. Major the Hon. Neville Lytton, who was present, concluding an account, naturally colored by the feelings of a participant, says (The Press and the General Staff, p. 42):

"The divisional general was ungummed, but it seemed to us that there were others who were responsible, and, if they had lost their commands after this failure, possibly greater disasters might have been avoided, for a similar experiment was made a little later on with two divisions, and the result was exactly the same. ***Naturally in the communique our attack appeared as a successful raid****."*

'The tradition about shielding senior officers from blame appears to have come down from Wellington's Day. At page 218 of Sir John Fortescue's "Life of Wellington," the following passage appears: —

This example shows how far Wellington was prepared to go in shielding a senior officer from public disgrace ... and more than once he visited the sins of careless generals upon good regimental officers. This was hard, sometimes cruelly hard; and the harder because when once he had pronounced judgment upon an individual, or even upon a regiment, he would never revoke it, no matter what facts might subsequently be brought to light. He would admit that, if he had known earlier of all the circumstances, he would have softened his censure, perhaps have spared it altogether; but, once declared, he would never recall it. ... Protests were useless. With him it was, 'What I have written, I have written,' and there was no more to be said.'

The Medical Services and How They Coped

We read in the previous chapter of the efforts of the Medical Services, as recounted from the War Diary of the 14th Field Ambulance, and the comments of Captain Frederick Collier, the Regimental Medical Officer of the 60th Battalion.

For the time preceding the attack, commencing at 4.00pm on 19 July, until 10.00am on 21 July, in two-hour periods, the flow of casualties appears to be as set out in the chart of casualties recorded over the course of the Battle, and which is again reproduced below.

4.00pm 19/7	5	4.00am 20/7	115	4.00pm 20/7	35	4.00am 21/7	5
6.00pm 9/7	40	6.00am 20/7	240	6.00pm 20/7	65	6.00am 21/7	20

8.00pm 19/7	35	8.00am 20/7	100	8.00pm 20/7	35	8.00am 21/7	5
10.00pm 19/7	90	10.00am 20/7	170	10.00pm 20/7	30	10.00am 21/7	5
12.00 m/n 19/7	130	12.00 m/d 20/7	185	12.00 m/n 20/7	15		
2.00am 20/7	125	2.00pm 20/7	35	2.00am 21/7	25		

The work of Captain Collier and his colleagues was acknowledged in Australia's newspapers after the war had come to an end, and one such item was the following one from Col. Reuter E. Roth, published in The Sydney Morning Herald, Sydney on 22 July, 1919, three years after the Battle:

> Sir, - I should like to remark that the R.AM Corps had nothing to do with the wounded at the battle of Fleurbaix. I happened to be the D.D.M.S. of the 2nd A.N.Z.A.C, and was responsible for all the medical arrangements. The grand work and sacrifice of the officers and other ranks of the Australian and New Zealand Army Medical Corps helped much to retrieve the want of success of the engagement.

---ooo000ooo---

On 24 July, 1919, James Green wrote to the Editor of the Sydney Morning Herald that: -

> … the splendid work, amongst the finest organisation I saw during the war, was solely the work of the three field ambulances of the 5th Division. Casualties were evacuated at the rate of one every two seconds during the night. The service of motor ambulances, protected by steel mesh from shrapnel and bullets, came right under fire. In addition to the splendid work of the Australian medical officers, may I now mention the fine work of the Australian Y.M.C.A. at Steenwerck. Fifteen hundred wounded men went through that town, and the Y.M.C.A. gave hot food and coffee to each. Thirty-five pounds' worth of cigarettes alone were distributed and a postcard describing condition etc., was sent for every man to his people in Australia.

---ooo000ooo---

Another commentary on the work of the Medical Officers, from J.R. Campbell, late Captain, A.I.F., appeared in The Register, Adelaide on 30 July, 1919 as follows:

> Lieut. Col. Toll pays a deserving tribute to the officers and men of the 5th Australian Division, who so valiantly upheld the traditions established at Gallipoli by the Australians.
>
> I wish now to draw attention to the work of the medical officers, of which no mention is made in the colonel's article. Perhaps no other division ever suffered as heavily as the "5th" did in that action, and thousands of wounded were passed through the C.C.S. in a few hours. Within 24 hours there was not a sign of casualty, and the credit of this is due,

to a large extent, to the organization and administration of Col. Hardy, with whom were assisted Lieut.-Col. Shephard and Major Clayton (now Lieutenant-Colonel), both of South Australia. The stretcher bearers worked for 24 hours without relief, and, although it was the first stunt for many, there was neither confusion nor "letting up."

Newspapers Carried Reports Dealing With The Lost and The Missing.

For example, from The Bendigonian, Bendigo Vic. 6 September, 1917

CAPTAIN H. M. PLOWMAN, AND LIEUT. J. H. STERLING, M.C.[88]

After the fighting near Fleurbaix and Fromelles, in France, on 19th July, 1916, a long list of missing soldiers was published. All hope of finding the majority of these men alive has apparently been abandoned, and relatives are being notified of their death.

Amongst the number was Captain H. M. Plowman, who was well known in the Bendigo Camp. He was 27 years of age.

Another soldier who was reported missing on 19th July was Lieut. J. H. Sterling, B.A., LL.B., M.C., son of the Rev. J. H. Sterling, who officiated at St. Andrew's Presbyterian Church for several months last year. Lieut. Sterling is now officially reported killed in action.

In the attack at Fleurbaix or Fromelles, some miles south of Armentieres, the Australians advanced about 450 yards and captured the enemy position, but the Germans diverted a stream of water into the captured trenches, and the Australians found them untenable and had to retire. The Australian losses were very heavy. Many Bendigo soldiers died in this fight on the 19th July. In a few cases those missing were afterwards found to be prisoners, but, as already stated, the great majority were killed in action or died subsequently from wounds received in the fight.

---ooo000ooo---

Families Sought Information on Their Soldiers

The work of those in Army Base Records, the conduit for information from Army to families, and for families to direct their enquiries to Army, has recently been the subject of a book by Carol Rosenhain. The book is titled "The Man Who Carried The Nation's Grief", and details the work of Maj. James Lean MBE who was in charge of Base Records. Anyone who has researched a soldier's records will at one point have seen a memo or letter signed by him, as - J M Lean.

[88] See: https://trove.nla.gov.au/newspaper/article/90855076?searchTerm=CAPTAIN%20H.%20M.%20PLOWMAN%2C%20and%20LIEUT.%20J.%20H.%20STERLING%2C%20M.C%20.%20%20%20%20%20%20%20%20%20%20%20%20&searchLimits=

Chapter is 5 titled "Fromelles and The Agony of The Missing Sons" and is 64 pages in length. It states that despite having 192 staff in May 1916, they were swamped in the aftermath of Fromelles with over 5500 casualties.

And The Conscription Debate Raged At Home.

One of the most significant effects following on from the Battle at Fromelles, and shortly after, the Battle at Pozieres, would be how Australia came to deal with the question of military service, and in particular, conscription.

Shortly after the outbreak of war there was a general election in Australia. The Labor Government under Andrew Fisher declared that Australia should support Great Britain with her last man and her last shilling. Senator Pearce was made Minister of Defence.

[Picture 17. AWM ART V08942. A WWI recruitment poster.]

Fromelles, and shortly after it Pozieres, served as a cruel introduction to the harshness of the war on the Western Front, and as word of the gruesome battles got out volunteer numbers began to fall. Australians back home were very aware of how terrible the casualties in Fromelles and Pozieres were.

Professor Joan Beaumont wrote: "The scale of losses are essentially what triggered the subsequent conscription debates, and the nature of those battles made the debates extremely emotional."

In early September, Australian divisions were taken out of the main battle in Pozieres to recover and rebuild their strength.

With volunteer numbers dropping, Australian Prime Minister Billy Hughes moved to hold a referendum on whether or not overseas conscription should be mandatory. Until that point, overseas service was purely voluntary, and those who served in Australian units were all therefore, volunteers.

When McCay returned wounded, to Australia from Gallipoli, Sir George Pearce appointed him Inspector General, a position he retained until Pearce overrode General Birdwood and made McCay a Major General and gave him the 5th Division. To the suspicious, Pearce's censorship of the Fromelles disaster, deliberately mixing it up with Pozieres, might have had something to do with his commitment to McCay.

It also had a lot to do with the forthcoming Conscription Referendum of which Pearce was the chief architect. It was announced a week after Fromelles and was voted upon and lost as the Battle of the Somme came to an end. Hughes had long wanted to introduce conscription, rather than voluntary recruitment, but could not get the support in the Government, so on October 28, 1916, Australians were asked:

> **"Are you in favour of the Government having, in this grave emergency, the same compulsory powers over citizens in regard to requiring their military service, for the term of this War, outside the Commonwealth, as it now has in regard to military service within the Commonwealth?"**
>
> **Result: CONSCRIPTION REFERENDUM OF 1916 DEFEATED**

The referendum was narrowly defeated - 51 per cent against versus 49 per cent for - with many attributing reports of the Battles of Fromelles and Pozieres as the reason.

But Australians remained bitterly divided over the issue.

Indeed in 1927, the Labor Daily, Sydney, on 23 June, carried an article advising of the technical issues involved in the referendum, and informing the populace that the government of the day could have carried legislation even without a referendum, if the Hughes Government had a majority in both houses of parliament. It didn't, hence the referendum.[89]

"There were lots of reasons why Australians voted 'no', ranging from labour unions fearing conscription would hand jobs over to immigrants, to libertarians outright saying it's not right to force a man to kill," Professor Beaumont says.

"But the meetings and debates became quite violent. Everyone knew who had served and who hadn't in their communities, and people would get on tables and name names.

It was by far one of the fiercest debates in Australia's history. Sir George Pearce persevered with the conscription idea and tried again in 1917, but its timing was equally bad: the Battle of Passchendaele with tremendous losses occurred just prior to that referendum. The referendum was lost.

According to Professor Beaumont, one of the consequences of the debates that lingers in Australia today, 'is the entrenched belief that you could not conscript men to fight overseas' - an issue that was touched upon again in subsequent wars.

---ooo000ooo---

One lasting impression regarding the Battles of Fromelles and next at Pozieres in 1916, is that they are widely seen as the first time that Australians played a 'full' role in WWI, and although many of that year's offensives did not achieve their goals, they helped deal significant blows to German forces.

---

[89] See: https://trove.nla.gov.au/newspaper/article/236623430?searchTerm=CONSCRIPTION%20REFERENDUM%20OF%201916%20DEFEATED%20%20%20%20%20%20%20%20%20%20%20%20&searchLimits=

But in the aftermath of WW1, Fromelles did not attain the heady recognition given to Gallipoli or Pozieres, despite it being singularly the largest loss of life (in a day, and in relative terms) for an Australian action in the war.

We know, for example, that during WW1, 61,720 soldiers were killed, and 137,013 were wounded.

In the Aftermath of the Battle of Fromelles, and from the Great War itself, one can only speculate on the follow-on effects on soldiers who returned to Australia, on themselves, on their families, and on the country.

We can only speculate on how many suffered from long term disabilities. We can only speculate on how many became alcoholics, or drug dependents, or whose families suffered. There was little support, if any for most. Many ended their lives only to be buried in unmarked graves.

And we can only speculate on how many soldiers' families' members (mothers, fathers, siblings and wives) died prematurely from grieving over their dead sons, brothers and husbands. The impact of coping, or not coping, with such loss should not be underestimated.

The presence of 'shell shock' spread among communities. Meanwhile veterans were supposed to just get on with their lives and go back to where and how they were before the war. That simply could not, and did not happen automatically.

But what we also know is that in the 10 years following the end of WW1, another 60,000 of those who returned, died.

The losses which had been 1 in 5 of those who served up to the declaration of the Armistice, had in 10 years become 2 in 5 of those who served.

---ooo000ooo---

After the disaster, the excuses. After the excuses, the reality.

And after everything else, we are forever left with the numbers,

the staggering blood-soaked numbers.

8. Comments and Reflections by Those Who Served in the Battle of Fromelles

'... once begun, myths under construction require not only continual supplementation, but protection. It was far from easy, for in the final analysis, history and legend do not make comfortable bedfellows.'[90]

---ooo000ooo---

It is clear from previous chapters that this battle could have been delayed, or averted, but it is equally clear that it was pursued with some vigour by Lt. Gen. Haking.

After the end of active fighting, and the withdrawal by the allied soldiers, the Germans – our opponents - offered ours a truce to gather in the wounded and the killed. It was mid-summer. It was hot. It was humid after rain earlier in the week. Ideal conditions for flies to feast on the flesh of wounds and dead bodies. But our illustrious Australian leader, General Sir James Whiteside McCay would not permit one. Result: our valiant wounded or deceased soldiers could not be officially recovered.

Our leader refused, not for humanitarian reasons, but because he and our Commanders were following "orders".

But despite the orders, or in spite of them, many were recovered under cover of nightfall. The soldiers themselves were not prepared to abandon their mates in No-man's land.

Captain R. Hugh Knyvett, in his book *Over There With The Australians, 1917*, recorded various incidents encountered in the course of soldiers and bodies being recovered. One famously quotes a delirious soldier days after the battle end as saying:

> *"I've been talking to dead men for days. There was two men came up to speak to me who carried their heads under their arms."*

But the one incident which was recorded as an extraordinary act of bravery during the rescue of soldiers describes an unnamed sergeant (most likely Alexander Gordon Ross of the 57th Battalion, a former horse tamer from Geelong) who was with him when they found a badly wounded man on the German barbed wire:

> *"...when we tried to pick him up, one by the shoulders and the other by the feet, it almost seemed that we would pull him apart. The blood was gushing from his mouth, where he had bitten through lips and tongue, so that he might not jeopardise, by groaning, the chances of some other man who was less badly wounded than he.*
>
> *He begged us to put him out of his misery, but we were determined we would give him his chance, though we did not expect him to live.*

[90] Barton p.381

But the sergeant threw himself down on the ground and made of his body a human sledge. Some other joined us and we put the wounded man on his back and dragged them thus across No-man's land, through the broken barbed wire and shell-torn ground, where every few inches there was a piece of jagged shell, and in and out of shell holes.

So anxious were we to get to safety that we did not notice the condition of the man underneath until we got to our trenches; then it was hard to see which was the worse wounded of the two. The sergeant had his hands, face and body torn to ribbons, and we had never guessed it, for never once did he ask us to 'go slow' or 'wait a bit'. Such is the stuff that men are made of."

Obviously, some of those killed or wounded and still lying in No-man's land were recovered under darkness in the coming days. But many more were left to die, to rot, and to disintegrate in the open, fed on by flies, for the rest of the summer, and the rest of the war. Only then, after the war, were the remains of 410 recovered and interred in two communal graves at VC Corner Cemetery. There are many whose remains still lie in the grounds which made up No-man's land, and in all likelihood will never be recovered. Their names and those of the unrecovered casualties have since been recorded on a memorial wall at VC Corner Cemetery.

It is reasonable to conclude, even with the benefit of hindsight, that if the truce was accepted, we could have gathered in almost every single soldier, tended to and saved many of the wounded, and buried the dead with dignity under a headstone of their own. Not agreeing to a truce ensured that wounded and heroic rescuers undoubtedly died, and our dead were left to rot.

Unlike our own illustrious leader and the British commanders, the Germans - our opponents - treated our dead with a degree of respect that was denied to them by our own at the time. In dealing with our soldiers, the Germans catalogued those who were recovered from the trenches and close quarters. They identified articles found with the bodies, and recorded those articles so that they could be passed on to family through the auspices of the Red Cross. And they arranged for a burial of the 250 soldiers who were later found in those mass graves.

So, by the end of the war, various cemeteries would become the final resting place of many Australian and British soldiers. Another was created by our enemy, but remained unmarked for the next 94 years, waiting to be found. That mass grave was found because of the efforts of retired Melbourne Arts teacher, Lambis Englezos against opposition at the time. The latter could have been avoided if a truce was upheld. We are often called on to honour our fallen, but in this case, who honoured who?

Was There A Cover-Up?

On 24 July, 1916, four days after the Battle, in a piece for the London Times and the Daily Mail, Official Historian Charles Bean labelled Fromelles an 'attack' that had resulted in 'severe' casualties. The powers that be were not pleased, for the British official communique described it as 'some important raids on a front of two miles in which Australian troops took part'. Raids … not attacks.

The Germans translated and circulated both communique and article, themselves noting the contrast. To Digger and Tommy, this was a distinction of considerable significance: their 'battle' had been downgraded, and yet all the participants looked upon the clash as something far greater than the word 'raid' could ever articulate.[91]

History, especially military history, has always been and always will be commandeered by agendas. During the Great War and for many years afterwards, the public was told what the powers-that-be wished them to know. It is the traditional approach, and continues to this day. It generally takes the form of soothing statements frequently repeated that build into a positive narrative that both justifies actions and protects individuals. But when dusty files finally emerge from murky vaults, opportunities generally arise to filter the finer reality. Combining the results of all archival sources, however, does not produce the truth: that remains forever beyond the reach of the military historian, because every man and woman who bore witness to the war experienced an individual and differing daily reality, and thus often recall and recount diverging impressions of the self-same event. As some of the evidence in this book illustrates, there cannot be a trustworthy 'common view'. Indeed, it is perhaps prudent to mistrust all those who declare themselves certain, for we can neither study nor know it all.[92]

It is one thing for us to read the official portrayal of what the Battle at Fromelles involved. We will invariably read the histories and biographies of the leaders installed to supervise the various Battalions, and the Divisions which took part in the fighting. After all, it is often quoted that *history is written by the victors.*

The Soldiers Themselves Revealed What Had Gone On.

What is usually not seen in texts given over to historical accounts of what took place, are the comments from the individual soldiers, who bore the brunt of the bullets coming at them, the artillery, the weather, the poor planning, the callousness of their 'leaders' and in many cases the social isolation afterwards. For example, refer again to the newspaper clipping of 24 July, 1919 from James Green writing to the Sydney Morning Herald, quoted in the previous chapter:

'For every man was issued with hot food and coffee, and … a postcard describing conditions etc. *was sent for every man to his people in Australia.*'

It is difficult to say there was a cover-up when individually thousands of soldiers were issued with postcards to be sent home, in which they would be describing the fighting in which they had been wounded, so soon after their embarkation from Egypt to France.

This is just one illustration of why it is not true to say in this modern era that the public was not aware of the full extent of the fighting at Fromelles. Major General Mike O'Brien CSC (Ret'd), supervisor of the archaeological dig at Fromelles in 2008 and 2009 has commented that it is in

---

[91] Barton p.381-382.

[92] Barton p.397.

fact 'manifestly untrue'.[93] The communique after the battle, classing it as a raid with some success, was insensitive and misleading. But let's be charitable, General Haig did not want to give succor to the enemy. But the casualties – all of them – were regularly announced in newspaper lists, untagged with locations or battles.

It has also been commented in various sources, that the people in Australia were not being given the full extent of the fighting until some time later. From reading these reports, it is hard to justify the suggestion that there was a '*cover-up*' about the event, especially when so many families were learning of the loss of their favourite son or father, or brother, or uncle.

Charles Bean issued a prompt heavily-censored dispatch to our newspapers which did not total the casualties. Their extent was not revealed accurately until A.D. Ellis' history of the 5th Division was published in 1921. Brig Gen Pompey Elliott was not the first to publicly criticize the blunders in the battle, though his words had the most authority and effect. The historian and chaplain of the 53rd Battalion, the controversial Father J.J. Kennedy, DSO, could well be awarded the title of first critic.

At the same time, it is also easy to understand why Australians, being the cynical lot that they are, would see a 'cover-up', if that was the myth being served up to them by official sources. But it is apparent that information was not so much covered up, as not passed on in a timely manner in a way which would make it apparent just what had taken place, until well after the event. But even so, they were reported in the newspapers of the day, as a search of files in the Trove repository of the National Library of Australia will abundantly reveal.

It is sometimes commented that reporting about Fromelles was delayed until after the Battle of Pozieres. There is no historical justification for this contention. Reports had already been issued about Fromelles before Pozieres was under way. The action that followed at Pozieres was in many reports mixed up, sometimes aggregated with that of Fromelles, but they were separate actions, fought by separate Divisions.

No good comes from trying to compare or aggregate them with each other. They each deserve respect.

Contemporaneous Reports From Those Who Were Involved.

What will be set out in this chapter is not an interpretation of what took place with the benefit of hindsight, but an insight from the written comments, included in newspapers, letters, and other information which is freely available for those determined to go looking.

It will deal with the comments made by soldiers who were involved in the fighting at Fromelles, as well as those who commented about it, as newspaper journalists, or any other 'authorized' source.

---

[93] In a speech to the Friends of the 15th Brigade, Melbourne, 19th July, 2017, on the 101st anniversary of the Battle of Fromelles.

One thing that becomes patently clear is that Fromelles was an unnecessary waste of lives, on a scale never before seen. Those who participated formed very firm opinions, and took little in the way of criticism in the years that followed. They were not prepared to let others diminish their efforts. Those who did offer criticisms were met with severe rebukes, as some of the following contributions will attest. Those rebukes deserve to be continued until Fromelles is properly classified as a "Battle".

TRENCHES WON AND LOST. ANZACS LACKED SUPPORT.[94]

Writing to his mother from France, Private **A. Cohen** described fighting at Fromelles, in which his division took part. He wrote: —

We suffered heavy casualties. We took three lines of German trenches, lost them, and had a bad time in general because of lack of support. We had to mount the parapet with fixed bayonets at 2.00am to beat off the German counter-attack. The fight lasted two days and a night. It was hell, fire, and slaughter.

---ooo000ooo---

In an article published in The Argus, Melbourne, on 24 July, 1916, just four days after the Battle of Fromelles, and also referencing the earlier Battle on 9 May, 1915, headed

WHERE THE AUSTRALIANS ATTACKED[95]

the newspaper contributes the following assessment:

The difficulties which the Australians had to face in the attack at Fromelles, which is five miles south of Armentieres, and seven miles west of Lille, are indicated by the following account by "The Times" correspondent of the result of a similar British assault on this portion of the German line as long ago as May 9, 1915, when it is certain that the enemy were ***not*** *as strongly entrenched as they are today.*

The trouble in those days was that the British were short of munitions.

The results of our attack on May 9 in the districts of Fromelles and Richebourg, (said The Times correspondent referred to) *were disappointing. We found the enemy much more strongly posted than we expected. We had not sufficient high explosives to level his parapets to the ground, after the French practice, and when our infantry gallantly stormed the trenches, as they did in both attacks, they found a garrison undismayed, many entanglements still intact and Maxim guns on all sides ready to pour in a stream of bullets. We could not maintain ourselves in the trenches won, and our reserve were not thrown in, because the conditions for success in an assault were not present. The attacks were well planned, and valiantly conducted.*

The infantry did splendidly, but the conditions were too hard. The want of an unlimited supply of high explosives was a fatal bar to our success. We have had many casualties this week, but if we did not win

[94] From The Sun, Perth W.A., published 27 March, 1917.

[95] See article at: https://trove.nla.gov.au/newspaper/article/1605103?searchTerm=WHERE%20THE%20AUSTRALIANS%20ATTACKED%20%20%20%20%20%20%20%20%20%20%20&searchLimits=

all we hoped for we have detained on our front forces equivalent to our own, and have greatly facilitated the French success on our right".

At the very least, the Australian assault may be expected to detain an equally large force in front of the Anzac sector, and to that extent contribute towards the operations on the Somme.

---ooo000ooo---

Inadequate account had been taken of the German defences. Despite the infantry attack being 'well planned, and valiantly conducted', the want of an 'unlimited supply of high explosives was a fatal bar to their success'. The lessons from 1915 had not made 1916 any more likely to succeed.

---ooo000ooo---

Again, on 24 July, 1916, from the pen of C.E.W. Bean, Official Reporter with the Australian Imperial Forces, came the following article published in The Argus, Melbourne:

AUSTRALIA'S SHARE IN GREAT OFFENSIVE[96]

A Heavy Engagement. All is Well on Somme Front. Important Russian Gains.

BRITISH HDQRS., IN FRANCE, July 20.

After a bombardment, an Australian force on Wednesday evening attacked German trenches south of Armentieres. On the left, the Australians seized the German front line, and passed beyond it to further trenches of the enemy's first system. In the centre, the Australians carried the whole of the German first system, and reached the more or less open country behind it.

On the right, our troops had to cross a much wider stretch of ground between the trenches. Here the Germans held a very strongly-fortified salient. The Germans were ready for the attack, and had managed to save a number of machine-guns from our bombardment. In spite of their very brave efforts, our troops on this flank were unable to cross the ground between the trenches, and only managed to reach the German trenches at isolated points. From these they were driven out. This success enabled the Germans to concentrate fire of all sorts of artillery on the portion of the trenches which we had captured. The Germans battered down their own trenches, where they were occupied by our men. They also turned water from a channel down their trench on the left flank, and the Australians there, shortly after reaching the trenches, found themselves standing in water, which rapidly rose until it was waist-high.

Our men endured a tremendous bombardment until early in the following morning, when, after 11 hours in the captured position, such Australians as retained the small remaining portion of the German line were ordered to retire. By dint of very brave work, our engineers and infantry, in working parties, had managed to get communication trenches dug completely through to the German trenches. These trenches were dug under very heavy shell-fire, and they enabled our troops to carry out their retirement with loss which was slight when the extraordinary difficulty of the operation is considered. Among, the last who

---

[96] See full article: https://trove.nla.gov.au/newspaper/article/1605102?searchTerm=AUSTRALIA%27S%20SHARE%20IN%20GREAT%20OFFENSIVE%20%20%20%20%20%20%20%20%20%20%20%20&searchLimits=

returned to our trenches were eight men, who said that they had got lost behind the German trenches, and had been wandering about till daylight in country in the rear of the enemy's frontline.

Our troops in this attack had to face shell-fire more heavy and more continuous than was ever known in Gallipoli. Many of them were quite untried previously. But the manner in which they carried the operations through seems to have been worthy of all the traditions of Anzac.

At least 200 Germans were captured, and we secured several machine-guns. Many Germans were killed. The losses among our troops engaged were severe.

[Author: Captain Bean, the official War Historian, may be seen as putting a good light on the Battle at Fromelles. It was not as graphic as what other later reports made alarmingly clear.]

---ooo000ooo---

Immediately following the previous article was reported:

A GERMAN CLAIM.

LONDON, July 23.

The following official German communiqué was received in London on Friday night: -

> *"An English attack in the Fromelles region yesterday by two strong divisions was repulsed. We made prisoner 481 men, and counted 2,000 bodies in front of our lines."*

Enemy Experts Puzzled.

The Amsterdam correspondent of the Central News Agency report that German war writers are puzzled by the attacks made by the British at Fromelles on Wednesday evening. The Germans say that two strong British divisions (40,000 men) were engaged.

Some of the experts consider that the attacks were the prelude to a new offensive, and others that they represent a serious attempt to advance towards Lille. Others, again think that the attacks were merely a demonstration to prevent the Germans sending troops to the sectors menaced at Hooge (east of Ypres).

---ooo000ooo---

The Germans obviously, and correctly, had mixed signals as to what was the reason for the Battle.

---ooo000ooo---

A letter from Private **Joseph Stephenson** 14 MG CO who was wounded at the Battle of Fromelles on 20 July, 1916, was published less than two months later.[97] He wrote this letter home to his mother several days later. It was reproduced in the Goulburn Evening Penny Post on

[97] See the full letter which was originally published in the Goulburn Evening Penny Post, and reproduced on 18 July, 2016 via the following link: https://www.goulburnpost.com.au/story/4035112/a-private-battle-fromelles-1916/

September 9, 1916. The information in the letter had passed through the censors of the day. But the information contained in the letter was a patent depiction of the severity of the Battle itself.

I was wounded at Fromelles on July 20. We charged and captured three lines of German trenches the day previous, but I managed to go through that day without a scratch. It was a terrible charge and one I will never forget.

How anyone got through without being hit is nothing short of a miracle.

For two days before the charge we bombarded the enemy very heavily, and Fritz was put to a lot of trouble as all their parapets were blown away. All of a sudden Fritz started on us with his big guns and high explosive shells, they kept it going all day, and I tell you things were very lively.

After a few hours our trench was considerably damaged, but considering the very few casualties we came to the conclusion it would take a lot to shift us. Our artillery started again and paid them back.

The night before the charge we had to build a new "possy" for our machine gun, as Fritz had knocked the other one down with a shell. Anyhow, we got on the parapet and started our job with sand bags, but flares kept going up and we had to get down.

When all was quiet we started once more, and keeping on all night finished our job before the morning. All morning Fritz sent along shells and we sent 60 pound bombs. Everybody was anxious, and at 5.30pm the order came along to fix bayonets and stand to the parapet. This was done, and when the whistle blew at 5.40pm our gallant 5th Division hopped the parapet and were out to make a name for themselves.

We faced heavy fire and lost a lot of men before we had gone 20 yards, but we still pushed on, followed by two more lines. We got to the German first line. Some of our boys stood on the parapet and used the bayonet, while others preferred to get right inside the trench and throw the bombs which we all carried. Most of the Germans cleared before we could get to close range.

We pushed further on through thick scrub which was alive with snipers. It was here we had very heavy hand-to-hand fighting and took a lot of prisoners. All was well done, excepting that they shelled us and let loose the gas, but fortunately the wind was favourable and carried the gas back over their own lines. Needless to say we had our gas helmets ready to pull on. Then they flooded the trench and in a short time we were knee deep in water.

We pushed on further until we reached their third line of trenches, but they counter-attacked, and we returned to the second line amongst the water, which was now up to our elbows, and we stayed in this for 11 hours, and were than relieved by the New Zealanders.

Most of our gunners had been killed or wounded, but we went to our old front line and stayed there to be relieved the next night, but my luck was out, for at 1.30am on the 20th, shrapnel burst above our gun crew, and I was wounded in the right forearm and left shoulder.

The field dressing was no good, as the wound in my arm was so big, so I took off my puttees and another wounded man bound up my arm. I had lost a lot of blood and got very weak, so I laid on a stretcher for an hour, when my wound was dressed, and I was sent to the clearing station.

Will you please take this letter to the local papers and ask them to publish same, so as my friends will know my whereabouts. My address is No. 2904, Gunner J.V. Stephenson, Eleanor West Ward, Edmonton Hospital, London 5.

[Picture 18: AWM. No. EZ0037. Soldiers waiting to be moved to the front lines.]

---ooo000ooo---

Those who had been involved in the fighting at Fleurbaix / Fromelles, also became the subject of spirited reporting in the newspapers in Australia. A few of those reports are reproduced just so the reportage can be read over a hundred years later.

Fighting at Fromelles where the trenches were taken and then lost by the Australians is described by Private N. Batten in a letter to Mr. Butler, general secretary of the Try Society, South Yarra. He gives the most vivid account yet received of that action, which was reported in The Sun, Sydney, on 3 May, 1917, almost one year later. At the time, Private Batten was in hospital in England. Note that almost one year later, the details of the Battalions involved were still censored from the report.

FROMELLES - TRENCHES WON AND LOST - HIGH EXPLOSIVE SHELLS.[98]

Melbourne, Thursday.

"It was near the village of Fromelles," writes Private Batten, "that we participated in one of the bloodiest battles the Australians have yet taken part in. Preceding our attack the artillery bombarded them unceasingly for eight hours. The shells used by this section numbered 60,000. (Some 'Strafe' that!) and believe me it played up with their trenches.

At a signal the artillery extended their range to Fritz's supports and communication saps, and our chaps as one man mounted the parapets with bayonets fixed, rifles loaded with one in breech and ten in magazine and pockets filled with hand grenades.

We rushed like fury toward the Huns, and quicker than you could count, our poor fellows fell everywhere in bunches. Wave followed wave in quick succession. However, we took two lines of their trenches, and

[98] See full article at: https://trove.nla.gov.au/newspaper/article/221962570?searchTerm=FROMELLES%20-%20TRENCHES%20WON%20AND%20LOST%20-%20HIGH%20EXPLOSIVE%20SHELLS%20%20%20%20%20%20%20%20%20%20%20&searchLimits=

held them till morning, and in the meantime they were busy extending our communication trenches from our original front line to the newly captured one.

Our glory was short-lived, however, as, early in the morning the brutes, considerably outnumbering us, counter-attacked and recovered their lost ground.

Descending to figures, after the affair was over, there were only 400 left of the [censored] Battalion numbering about a thousand. The [censored] fared about as well, and the [censored] a little better. The other brigades of our division suffered about as badly. My old company could only muster 54 out of 220, and of my platoon of 55 only 7 answered the roll call after the charge. One officer remained out of 6 in C Company. That was the O.C. Of my intimate pals, one, Cecil Lavender, is left. Two of them are wounded and are now in hospital In England, but three are missing."

---ooo000ooo---

From a report one year and three months later, on 2 November, 1917: Young Witness (Young NSW): -

A RETURNED HERO WHO WON THE D.C.M. AT FLEURBAIX.[99]

Sergeant F. A. Saunders, referred to hereunder, is a nephew of Mrs. A. A. Menzies, of Brundah, Grenfell and also a nephew of Sergeant C. A. Saunders (formerly of Grenfell), now on active service in France.

Perhaps the proudest and happiest man of the 464 soldiers who returned to Sydney yesterday was Sergeant F.A. Saunders of the 53rd Battalion living at Enmore, who won the D.C.M. at Fleurbaix. Twenty nine years of age, he first saw active service at Rabaul, where he got malaria. Keen to join the first batch for Gallipoli, he was blocked through the after effects of the fever. He kept on plugging away at the recruiting authorities until, after five attempts, he was accepted. He went to Gallipoli, and was one of the last to leave the peninsula. He was one of the "Die Hards" as the Australians called those who hung to the famous slopes till the last. Thence he went to France, where he saw hard fighting — early morning attacks in the cold, and through the mud, with comrades falling around; in the night tornadoes of shell, leaving death in their wake. Then came the fearful fighting at Fleurbaix, where so many brave men went out to it.

Hero Saunders won his D.C.M. In this way:

On a July day the attack on the enemy was launched. Saunders, in charge of 120 bomb throwers, showed the way. Barbed wire pulled them up, and enemy 'machine guns did the rest'. Men fell in threes and fours, and only seven or eight of the original 120 got into the German first-line trench. Overhead went screaming shells, the heavens seemed to belch fire, and all hell appeared loose. Near and far men fell, some never to rise, and the groans of the dying went unheeded in the battle. Snipers got some of the attackers, and Saunders, incensed, rushed a German officer

---

[99] See: https://trove.nla.gov.au/newspaper/article/113645058?searchTerm=A%20RETURNED%20HERO%20WHO%20WON%20THE%20D.C.M.%20AT%20FLEURBAIX.%20%20%20%20%20%20%20%20%20%20%20&searchLimits=

and brought him down with a revolver. Saunders mounted the parapet and started bomb throwing. Urged to leave the parapet, he replied "Give me more bombs."

Saunders and his few were hard pressed when reinforcements came up. He then decided to charge the Second trench. Disaster followed. Hidden machine guns mowed the attackers down until a few only were left. Saunders, badly wounded, told his mates to "beat it." Private Stan Moore, of Sydney, refused, and carried Saunders to the first trench. The Germans, after an unsuccessful counter-attack, flooded this trench. Moore again refused to leave Saunders, carrying him across No-man's land through an enemy barrage, besides driving ahead two German prisoners.

---ooo000ooo---

On 19 July, 1924, eight years after the Battle, in The West Australian, Perth:

FLEURBAIX-FROMELLES BATTLE.[100]

First A.I.F. Engagement in France. Eighth Anniversary To-day.

Today is the eighth anniversary of the battle of Fleurbaix-Fromelles, the first in which Australian troops took part in France. Writing with regard to this memorable engagement, Mr. E. T. Crain, of Central-avenue, Inglewood, says: -

It was an attack carried out by the 61st British and the 5th Australian Divisions. The immediate intention was **to capture and to hold the front and support lines** [emphasis added] of the enemy on a front of about 4,000 yards. And was hoped that, if the attack succeeded, it would be the prelude to a further and more important offensive against vital enemy positions on the Aubers Ridge. There are few incidents of the war which deserve a closer attention from the Australian public.

It was an hour of intense silent excitement that preceded the first frontal attack made by Australian troops in France. A crowded front line of trenches, men armed with grenade, rifle and bayonet, machine gunners standing by, orders whispered, while the artillery were pounding the enemy stronghold. At last the signal came, and from our lines leaped battalions of the finest men that ever breathed, only to be met with a withering fire from machine guns and field artillery at point blank range. The waves of men pressed forward steadily but just as steadily the enemy fire grew hotter and hotter, and the enemy front line was seen to be thickly manned. The losses mounted rapidly as the men pressed forward gallantly in the face of terrific odds. One brigade lost no fewer than 35 out of 39 of the assaulting officers and with them most of the N.C.O.'s.

In the circumstances the survivors could only hold on determinedly to what had been won and await such further action as might be devised to meet the situation.

---

[100] See: https://trove.nla.gov.au/newspaper/article/31242896?searchTerm=FLEURBAIX-FROMELLES%20BATTLE.%20%20%20%20%20%20%20%20%20%20%20&searchLimits=

On the right sector the English Division met a somewhat similar fate, and the losses were enormous. To even attempt to describe the battle is beyond human power, but the following official casualty list for the period from noon, July 19 to noon, July 20 (24 hours) will suffice to show the intensity of the operation: —

8th Infantry Brigade. 54 officers; 1.846 other ranks;

15th Infantry Brigade. 51 officers; 1.725 other ranks;

14th Infantry Brigade. 65 officers; 1652 other ranks:

Div. Engineers. 3 officers; 85 other ranks;

making a total of 173 officers and 5,308 other ranks.

The infantry battalions that suffered most heavily were the 60th, the 32nd, and the 59th, with 757, 718 and 695 casualties, respectively. The 53rd, 31st and 54th also lost more than half their strength, and it is doubtful that any British division in the whole war suffered equally heavy losses in the same space of time. The casualties in the English brigades were 577, 336, and 400, in the 182nd, 183rd, and 184th brigades respectively.

It should not be forgotten that while the operation in itself was an unqualified local failure its main purpose in preventing the withdrawal of troops to the Somme was doubtless to some extent achieved. Such fine Western Australian officers as Lieut. J. Ion and Lieut. A. Paterson were among those who made the full sacrifice. There are today in Western Australia some of those few who survived the terrible ordeal, and we, on this 8th anniversary, again pay our humble tribute to the memory of those who fell in defending a call on our Empire.

---ooo000ooo---

Seventeen years after the Battle, on 30 November, 1933, from The Western Mail, Perth:

A MEMORY OF FROMELLES.[101]

Major A.R. White, M.C., one time of Mt. Lawley; and now resident of Malaya, is holidaying in Perth. He did splendid work in the battle of Fromelles, on July 19, 1916; when his battalion, the 32nd, was on the left flank - of the attack - and lost 17 officers and 701 men killed, wounded or captured, within 24 hours. At the finish, when all the other attacking troops were withdrawn or had been driven back, the 32nd, with White in command of the firing line, was completely surrounded. What happened then is graphically described by Captain Bean in the official history:

White and Mills were at once convinced that any attempt to drive out the Germans was now out of the question. The only possible course was to turn their men round, and endeavour to charge through to the original Australian trenches. White, therefore, directed the machine-guns to be collected and sent back to the Australian lines by a long detour towards the west, where

[101] See: https://trove.nla.gov.au/newspaper/article/38015701?searchTerm=A%20MEMORY%20OF%20FROMELLES.%20%20%20%20%20%20%20%20%20%20%20%20&searchLimits=

he thought they would be less likely to encounter the enemy. Lieut. S.E.G. Mills walked along the line explaining the order, telling the men to wait for a signal and meanwhile to keep up constant fire. He succeeded in warning most of them even on the extreme flank, which was now being attacked.

They made their way to a ditch in rear of that held during the night; and from here at 3.45, with machine-guns firing at them from front, flank and rear, some 150 men attempted almost simultaneously to charge over the top to the Australian lines. A considerable part - including White, who like many others had to fight the Germans hand to hand, and Mills, who immediately afterwards fainted of his wounds - crossed the German front trench and the fragment of the 8th Brigade streamed across no-man's-land towards the Australian position.

Recounting Of Treatment By Prisoners of War.

The actions of those involved in the fighting, and who did not have the good fortune to return to their trenches was also revealed after the war was over, as revealed in The Argus, Melbourne, 4 October, 1918:

REPATRIATED AUSTRALIANS.

EXPERIENCES IN GERMANY.[102]

One of the Australians who had been repatriated was ***Corporal P. Freirat*** *(53rd Bn). He was taken prisoner at Fleurbaix in July, 1916. A wound in the leg was not dressed for five days. A second operation became necessary, and it was performed brutally. No anaesthetic was given. Freirat being held down during the operation. Later he was operated on for appendicitis.*

---ooo000ooo---

From The Examiner, Launceston Tasmania on 18 October, 1918, came the following account: -

PRIVATE C.J. MITCHELL REPATRIATED[103]

BRUTAL GAOLERS.

LONDON, Oct. 17.

Private **C. J. Mitchell**, of the 53rd Battalion, who was repatriated, has arrived in London. He belonged to Rockdale. He was prisonered at Fleurbaix having been wounded in the head and a leg. He received considerable treatment from German doctors, but subsequently was compelled to work in a sugar factory, and later on railway construction at Broomberg. The prisoners at the latter place, including English, French, and Russians, refused to work in the munition factory,

---

[102] See: https://trove.nla.gov.au/newspaper/article/1408674?searchTerm=fed%20on%20horse%20flesh%20%20%20%20%20%20%20%20%20%20%20&searchLimits=

[103] See: https://trove.nla.gov.au/newspaper/article/51005119?searchTerm=BRUTAL%20GAOLERS.%20%20%20%20%20%20%20%20%20%20%20&searchLimits=

although threatened with machine guns. Mitchell attempted to escape from Stettin, but was punished with work in a lime quarry, under the most unhealthy conditions and brutal gaolers. He contracted poisoned feet, and was exchanged into Holland.

---ooo000ooo---

From the Daily Herald Adelaide SA. 5 December, 1918: -

HEROES OF FLEURBAIX. AUSTRALIAN SOLDIERS RETURNING

(Reuters Telegram) London, December 3, 6.55pm

Repatriated Australians are arriving at Hull in big batches daily. Some of these brave Anzacs were captured at Fleurbaix, and they testify to the brutality of the Uhlan escort, which marched them to Lille. They also describe the brutal treatment of French citizens, who tried to give the Australians food. Many of these martyrs of militarism give harrowing accounts of the bestial treatment of prisoners in dungeons at Fort Macdonald and Lille.

The Fleurbaix prisoners will include men of the 32nd Battalion (S.A.), who were captured on July 20, 1916, after the Germans had flooded the trenches between them and their supports when they were holding an advanced position which they had themselves taken.

---ooo000ooo---

And from The Moree Gwydir Examiner and General Advertiser on 6 May, 1919:

STARVED AND ILL-TREATED.[104]

Private **Allson David Stone**, (32nd Battalion) a returned soldier, who was made prisoner with 150 Aussies and Tommies at Fleurbaix in June (sic) 1916[105], relates the shocking treatment and starvation of prisoners by the Germans. Five hundred men were picked for work in Prussian Poland from Dulmen Camp. At Schneidemuhl Camp they met many of the old "Contemptibles." They worked at unloading sugar beet from railway trucks into vats in twelve-hour shifts.

The prisoners protested against the severity of the weather and treatment, and for this were marched into an open field and stood to attention from eight o'clock at night to 12 noon the following day, a blizzard blowing at the time. Pte. Stone witnessed the bayonetting in the head of one of the prisoners by a German non-commissioned officer.

They experienced much brutal treatment, and were subsequently sent to work in the coal mines at Bochum Westphalia, where conditions were equally bad. Finally, their Red Cross parcels began to arrive. If it had not been for the Australian Red Cross, says Pte Stone,"the majority of the prisoners would have died from starvation and overwork."

---

[104] See: https://trove.nla.gov.au/newspaper/article/115776959?searchTerm=STARVED%20AND%20ILL-TREATED%20%20%20%20%20%20%20%20%20%20%20%20&searchLimits=

[105] The reference to June, 1916 was incorrect, as the 32nd Battalion had not entered the area until the days before the Battle of Fromelles which was originally scheduled to begin on 17 July, but was deferred until 19 July, 1916.

Recounting of The Battle at Church Services.

Church services were organized after the end of the war, to mark the anniversary of the Battle of Fromelles, which coincidentally also fell on what was proclaimed in the UK to be *Peace Celebration Day*. From a Public Notice in the Sydney Morning Herald on 19 July, 1919, came the following announcement:

BATTLE OF FLEURBAIX.[106]

The battle of Fleurbaix - Fromelles, the first Australians' battle in France, was fought on July 19, 1916. Its anniversary thus falls on Peace Day, and at the request of the 5th Division of the A.I.F. Chaplain James Green, C.M.G., will conduct a commemoration church parade in the Paddington Methodist Church tomorrow evening. Special permission has been given for A.I.F. men to wear uniform, and General Lee, C.M.G., will be present, or represented by his staff.

Recounting of The Battle From Those In The Medical Services.

Just as there were services to pay respects to the fallen, there were also letters to the Editor, pointing out the opinions of those who were either wishing to point out some exaggerations in the reporting of events, or who were themselves mistaken in what they understood. One such contributor was the former Lt Colonel C. MacLaurin, of the Army Medical Corps, AIF, who wrote to the Sydney Morning Herald on 26 July, 1919. His letter drew a quick retort from others engaged in the Battle. The relevant extracts from his letter are set out below:

FLEURBAIX.[107]

I spent the terrible night of Fleurbaix in the safe if intensely busy seclusion of the Second Casualty Clearing Station, where we received 801 casualties in 24 hours. Chaplain Captain Green has given an interesting account, but I think I can supplement it with essential facts which should be known, and I certainly think the official report of that disastrous battle should be published.

It must be remembered that just before that time the sector in question was occupied by the Second Division, which had brought to a fine art the system of raids. Every night they were over, killing, plundering, and taking prisoners, with but trifling loss to them-selves; but as a result that division became extraordinarily competent, and brother Boche became peevish in the extreme, and at last most infuriated. To attack such men came the Fifth Division, composed of half-trained and half-disciplined soldiers. Why they were sent into action may be surmised, rather than said.

About 10 o'clock the first convoy came in, the wounded literally cock-a-hoop. "Glorious victory — hundreds of prisoners — stoush for Fritz," they cried as the ambulances rolled in, and we set to work. We

---

106 See: https://trove.nla.gov.au/newspaper/article/15861100?searchTerm=BATTLE%20OF%20FLEURBAIX%20%20%20%20%20%20%20%20%20%20%20%20&searchLimits=

107 See: https://trove.nla.gov.au/newspaper/article/15857028?searchTerm=I%20spent%20the%20terrible%20night%20of%20Fleurbaix%20in%20the%20safe%20if%20intensely%20busy%20seclusion%20of%20the%20Second%20%20%20%20%20%20%20%20%20%20%20%20%20&searchLimits=

were operating for about 48 hours on end; but in the early morning we noticed a change in the spirits of the incoming men. They became dull and apathetic; they would not talk; not a word could we get out of them as to how the day was going. Some of them muttered, "That blanky Sugar-loaf." Most of the wounds were in the right arm or leg or the back, and we could infer that our boys were being enfiladed; and, further, so many wounds were of such a trifling nature that it was obvious that they were being tried too severely for raw troops. So the night wore on, and by next morning we were able to piece together the whole awful story, which, so far as I know, has never yet been published. Now that it is all over, and the Fifth has wiped out its failure so gloriously many a time, there is no reason for further secrecy.

It must distinctly be understood that our artillery was not to blame. The fault was in the impetuosity of the untrained infantry, who rushed on beyond their true objective and in the horrible misunderstanding as to what that objective was. ... There can be no doubt that the effects upon the morale of the Fifth Division were bad, and that that gallant force did not recover for a considerable time.

The infantry showed amazing courage but their lack of training was their undoing. Had it been the Second Division at that time, no doubt the results would have been very different. The Australian Artillery did splendidly. The staff work all through could not be criticised, for it did not exist.

For our ten thousand casualties we captured a few machine-guns and a hundred prisoners, and it was rumoured that the Boche had temporarily diverted 13 troop trains from the Somme. His casualties appear to have been trifling.

---ooo000ooo---

Lt Colonel MacLaurin's letter drew quick responses, such as this one, from "One of the Fifth Division", published three days later on 29 July, 1919 in the Sydney Morning Herald:

FLEURBAIX. - TO THE EDITOR OF THE HERALD.[108]

Sir, - I cannot refrain from expressing my feeling aroused by the publication of the account of "Fleurbaix" by "C. MacLaurin, late Lt Colonel, A.M.C. of A.I.F." in your Issue of this morning. I wish emphatically to deny the statement that the Fifth Division at this time was composed of "half trained and ill-disciplined soldiers." I invite you to trace the history of the 8th Brigade, which, with the 14th and 15th Brigades, formed the division. It was formed in July, 1915, just a year prior to the "Battle of Pettillon" known as Fleurbaix. The four battalions of this brigade were trained principally in musketry for three months prior to embarkation, landed in Egypt in December, and with the lessons of Gallipoli, the great retreat from Moas and other big engagements in Flanders during 1914-15 to study, had six months of solid, hard training, bringing those battalions to a very high standard of efficiency. Of the other brigades about 60 per cent, were veterans of Gallipoli. Had I the power and a pen of flame I could not portray the magnificent bravery, fortitude, and discipline of the Fifth Division in its initial attempt to uphold the splendid traditions of our comrades of Gallipoli.

---

[108] See: https://trove.nla.gov.au/newspaper/article/15861228?searchTerm=I%20cannot%20refrain%20from%20expressing%20my%20feeling%20aroused%20%20%20%20%20%20%20%20%20%20%20%20&searchLimits=

Let me emphasize this. The statement that the remnants of the division were soon moved away from the line is not in accordance with facts, for that sector was held by some of those depleted battalions for three weeks afterwards, and it was one of those battalions that put up a record for the British army for a raid into one of the enemy's strongest posts.

The Fifth Division did its duty, and deserves credit for its sacrifice.

---ooo000ooo---

In another response to Lt Colonel MacLaurin's letter, the following came from J. Williams in a letter to the Editor, Sydney Morning Herald, on 30 July, 1919: -

FLEURBAIX.[109]

Sir. … The Lt Colonel states that at this time the 5th Division was composed of half-trained and half-disciplined soldiers. Yet 90 per cent of the officers and non-commissioned officers of the 14th and 15th Infantry Brigades were men from the peerless 1st Division, and had taken part in most of the stunts since the landing, and the senior officers included such soldiers as Elliott, Pope, M'Conachie, Scott, Simpson, Norris, Cass, and many others. Such leaders had been training the units of the 5th Division in Egypt for months, under almost as severe a course as that which made the original 1st Division such an effective fighting force. The men in the ranks were enthusiastic and of a very good stamp. Therefore, it is only natural to come to the conclusion that the division was not as much lacking in training and discipline as the Lt Colonel states.

Lt Colonel MacLaurin gives it as his opinion, that if the 2nd Division had been entrusted with the operation the result would have been entirely different. I beg to differ, and go as far to say that no division on the Western front could have succeeded where the 5th Division failed. Surely the Lt Colonel knows that men wounded in the first excitement of an attack always show very different spirits from their comrades wounded hours afterwards, whom the nerve-shattering effects of a modern bombardment have half stunned.

The 5th Division was not removed from the line, but held the sector (from which they attacked) until October, 1916. Then it was relieved by the N.Z. Division, and immediately sent to the Somme. It took over the line from the 11th English Division, in front of Flers on October 21, 1916, and with the exception of a 10 days' rest in November, 1916, was never out of the forward area until April 24, 1917, and was in the line again at Bullecourt in the early days of May, 1917. This record may be set against the "shattered morale" dictum of the Lt Colonel. The 51st Highland Division (including A. and S. Highlanders) relieved the N.Z. Division on the left of the 5th Division, some weeks after the battle, but certainly did not take over any line from the "remnants of the 5th Division," the latter still being in the frontline, between the Germans, the 2nd C.C.S., Channel ports, and other objectives.

---

[109] See: https://trove.nla.gov.au/newspaper/article/15856713?searchTerm=Lt%Colonel%20MacLaurin%20gives%20it%20as%20his%20opinion%2C%20that%20if%20the%202nd%20Division%20%20%20%20%20%20%20%20%20%20%20%20%20&searchLimits=

Fromelles (or Fleurbaix) was a ghastly failure, in which an Australian division was butchered for no purpose. The infantry paid the price, not so much of their impetuosity, but of the blunder of some "brass hat." The true tale of July 19-20, 1916, has yet to be told, and why so many valuable lives were wasted.

Let the readers of the Lt Colonel's story not get the idea into their heads that the 5th Division had any disgrace to live down through their conduct in this action. The attacking battalions were set a task made impossible by Fritz knowing the full details of the operation beforehand, a total lack of staff work, the position to be attacked being one of immense strength; and then, if our artillery outnumbered the German, the latter certainly used his with better effect.

I will venture the opinion that the many returned men who have read the statements of Lt Colonel MacLaurin's story will not feel any shame in belonging to any of the "half-trained and half disciplined battalions" which took part in the battle of Fleurbaix.

---ooo000ooo---

In what seems to have been the last word in the exchange, Lt Colonel MacLaurin wrote again to the Editor, Sydney Morning Herald, and on 30 July, 1919, his letter (following) was published: -[110]

Sir, - Your correspondents seem to have misunderstood my article. I cast no reflection on the brave Fifth Division. It was not their fault that they had not at that time been thoroughly trained in the difficult team-work which characterised attacks in Flanders; and God forbid that any man should cast a word against those poor gallant boys who behaved with such noble courage and gave their lives so cheerfully.

The Fifth Division was at that time inexperienced, but the men did their duty magnificently, and lost enormously. The fault was not with them, but elsewhere. Where was it? Who was responsible for the massacre?

---ooo000ooo---

By the following year, 1920, reporting on the events at Fromelles were not leaving any doubt as to its true military status. It was referred to regularly as 'the Battle' or 'the battle' of Fromelles (or Fleurbaix), without hesitation. Formal classification would come in the years following. But the following extract from an article in the Argus, Melbourne, on 10 April, 1920, goes a long way to revealing what the feeling was in regard to the status of the battle at Fromelles:

FROMELLES, 1916! A GLORIOUS FAILURE. WHAT REALLY HAPPENED.[111]

[110] See: https://trove.nla.gov.au/newspaper/article/15856714?searchTerm=Your%20correspondents%20seem%20to%20have%20misunderstood%20my%20article%20%20%20%20%20%20%20%20%20%20%20%20&searchLimits=

[111] For full article, see: https://trove.nla.gov.au/newspaper/article/1690351?searchTerm=FROMELLES%2C%201916!%20A%20GLORIOUS%20FAILURE.%20%20WHAT%20REALLY%20HAPPENED%20%20%20%20%20%20%20%20%20%20%20%20&searchLimits=

"The English attack in the region of Fromelles was carried out, as we have ascertained, by two strong divisions. The brave Bavarian Division, against whose front the attack was made, counted on the ground in front of them more than 2000 enemy corpses. We have brought in so far 481 prisoners, including 10 officers, together with 16 machine-guns." — German communique, July 21, 1916.

"Yesterday evening, south of Armentieres, we carried out some important raids on a front of two miles, in which Australian troops took part. About 140 German prisoners were captured." — British communique, July 20, 1916.

In this form, readers of "The Argus" received their first information of the great action by Australian troops on the Western Front, which has since come to be known as "the attack at Armentieres," or "the fight at Fleurbaix," but was really "the Battle of Fromelles." For a long time, the secrecy of war kept a veil drawn over the detail of this sad page in the history of the Australian Imperial Force, but closely censored letters from members of the 8th, 14th, and 15th Infantry Brigades soon began to indicate that something more serious had happened than was at first suggested. Since then, more or less accurate accounts of the battle have been in circulation, but the most interesting features had necessarily to be withheld until after the armistice. The first consecutive record, prepared from official documents and elaborated by commanding officers who actually took part in the engagement, has been prepared by Captain A. D. Ellis, M.C., of the 29th Battalion, and it is included in his "Story of the Fifth Division" (London: Hodder and Stoughton).

In this the movements of the various units from the day the intention to attack at Fromelles was first announced until the work of clearing up the battlefield had been completed are carefully followed with the aid of diagrams, and even the most unmilitary reader will find much to stir him in this account of "our most glorious failure."

---ooo000ooo---

Feelings about what took place in Fromelles were still raw four years later, when in The Australian, Perth W.A., the following was published on 23 July, 1920: -

THE FLEURBAIX HOLOCAUST.[112]

Monday last was the fourth anniversary, of the disastrous day at Fleurbaix, when the Fifth Australian Division was cut to ribbons.

Mr. H. T. Crain (formerly Lieutenant in the 32nd Battalion) writes: —

Never perhaps will the real truth regarding the dreadful happening at Fleurbaix on July 19, 1916, be told the people of Australia It was on this day, four years ago, that the infantry of the 5th Australian Division was hurled over the parapets in Flanders against terrible odds. Someone blundered! Yes, but should that

[112] See: https://trove.nla.gov.au/newspaper/article/210243951?searchTerm=THE%20FLEURBAIX%20HOLOCAUST%20%20%20%20%20%20%20%20%20%20%20&searchLimits=

be sufficient reason for withholding the facts from the people of Australia? On that night 8740 of Australia's finest sons made the great sacrifice in the cause of liberty.

Great deeds were done that night. Our glorious infantry, after facing one of the heaviest barrages of shell and machine gun fire ever known on the Western front, and fiercely fighting hand-to-hand for 11 hours, were forced to retire to their own lines. What a story! And what a terrible cost! In one Western Australian and South Australian battalion (the 32nd), out of 1000 men who went forward less than 70 were able to carry on the fight next morning. Yet what has Western Australia heard of these glorious deeds of her sons? Is it not time the relatives of those brave men who gave their all were told of the glorious way in which their sons fought and died? Never let this day fade in your memory, for on this day pages were indelibly written in the history of Australia.

---ooo000ooo---

It is by now abundantly clear just how those who were involved in the events at Fromelles felt about what they had endured, how they were treated, and how their service should be considered in the light of history.

---ooo000ooo---

9. Cause and Effect - Classifying Those Who Died and When – The Real Numbers

It has been said that figures will not lie. It is equally true that liars will figure. It is our duty to prevent liars from figuring in the interest of any theory, by presenting original data fairly.[113]

---oooOOOooo---

Anyone who owns and operates a business has to know how their business is functioning financially, otherwise they are not able to plan ahead, either for growth, or for recovery. So accounting functions have developed to "balance" the books, and to "reconcile" the books with the bank account. When the balance and reconciliation takes place it does so, '***as at***' a particular date or time. That simple phrase '***as at***' can give definition to what has taken place, what needs to take place, and what is in place.

In military and civilian life, leaders invariably plan for 'normal' activity levels, and importantly, for expected peak load periods, emergencies and coming commitments. This process is vital for ensuring that sufficient logistical support is available at the initial moment of need, and for the expected duration of the event. Thus a contingency factor is built into the data, to ensure that unforeseen circumstances do not derail operational performance.

In the case of armed forces, the leaders always need to know how many soldiers they have available at any given time, how much ammunition is needed for their next project, and what provisions need to be made for transporting the soldiers, as well as feeding them. It is therefore vitally important in the event of actual ongoing battle to know what is the availability of each of these items '***as at***' a given date and time so that plans can be made for the future.

Accounts are taken of the operational status of each unit on a regular basis and are known as Reports. Reports usually follow a set pattern, and can be daily, weekly, and/or monthly. So a report taken each day is known as the Daily Report.

It is also important so that arrangements can be made for burying the dead, attending to the wounded, and issuing appropriate records setting out the details of the loss of a soldier.

The Official Numbers Are Wrong.

David Noonan was reported in the Sydney Morning Herald on 30 April, 2014 as stating "new research shows Australia's official casualty statistics are seriously distorted and must be revised immediately." Noonan's book 'Those We Forget. Recounting Australian Casualties of the First World War' released 1 May, 2014, was the outcome of his work. Noonan makes the following significant points in his work:-

---

[113] By the statistician Carroll D. Wright giving the opening remarks at a Convention of the Commissioners of Bureaus of Labor Statistics, on 25th June, 1889, as furnished to the American Statistical Association.

Search for details of Australia's dead and wounded in the First World War and the figures thrown up are remarkably similar: of the 331,000 men who embarked from Australia with the Australian Imperial Force (AIF), 60,000 were killed and there were 155,000 admissions for wounding. These statistics are presented, with minor variations, on the websites of the National Archives of Australia, the Australian War Memorial, the Australian War Graves Commission, the Australian Parliamentary Library, and in the records of the British War Office and Australia's official history of the First World War by C.E.W Bean.

The final figure is actually 61,720 killed during World War 1, with 137,013 wounded. Seventeen died on the very last day, 11th November, 1918.

This means statistically around two out of three soldiers died or were wounded in the First World War. These figures have been quoted in every publication referring to Australia's casualties since fighting stopped in 1918. Alas, they are wrong. Some men were wounded more than once so the number of soldiers wounded is fewer than 155,000. And some of those wounded would subsequently die in later battles and so be counted twice in this inadequate statistic.

So how many men were killed or wounded? How many men were hospitalised? Australian casualty statistics don't include hospitalisations for injury and illness while other countries such as Britain, US, Germany and France do. Are the official statistics understated?

We will see later in this chapter how the 'official' numbers for Fromelles are also inaccurate, simply due to the manner of recording the date of death, and not relating it to the cause of death during the action itself. While the official tally records outcomes solely on 19 or 20 July, in taking a holistic look at the outcome of the Battle, far different figures emerge.

As part of his research, Noonan developed a method to examine these records based on polling techniques popularised by George Gallup in the 1930s. It involved taking statistically robust random samples of more than 12,000 individual soldier attestation papers, counting them page by page, entering the collected data into the latest spreadsheet software and statistically extrapolating it to produce the first accurate count of casualties of the AIF. The results are startling.

First, of the 376,000 men who were prepared to serve overseas, 308,000 men, or 82% of those who volunteered for the AIF served in a theatre of war.

Second, in this conflict Australia does not count as 'casualties' the non-action deaths and total hospitalisations due to illness and injury, in addition to the deaths, hospitalisations from wounding and prisoners that resulted from action, unlike our allies and even Germany.

This therefore leads to the recording of an inaccurate outcome. As compared with the manner in which Britain, US, Germany and France record their outcomes, it appears from looking at Noonan's work, that there are more precise methods that can provide more accurate numbers.

Australia only records 155,000 wounding admissions and omits illness and injury. Once the international practice is applied, Australia's total hospitalisations were five times greater than officially acknowledged: 750,000 admissions for approximately the 308,000 men of the AIF who served in a theatre of war. Hospitalisations due to wounding were higher than that officially acknowledged too, climbing to 208,000 admissions (+/- 500), 30 per cent of which were admissions due to shell shock. The men of the AIF were decimated.

As a proportion of its fighting force of men who were actually exposed to a theatre of war, Australia's army suffered more deaths, more hospitalisations for wounding and more hospitalisations for illness and injury than the armies of Britain, Germany, France, Canada or the United States.

We know that during WW1, 61,720 Australian soldiers were killed. Of those soldiers who survived, more than half of them were discharged medically unfit. Of those who were not discharged medically unfit, 60 per cent of them applied for pension help in the post-war period; so four out of five servicemen survivors were damaged or disabled in some way.

One can therefore only speculate on the follow-on effects on soldiers who returned to Australia, on themselves, on their families, and on the country. We can only speculate on how many suffered from long-term disabilities. We can only speculate on how many became alcoholics, or drug dependents. We do know that lots of their mothers and fathers died within the following 5 years of the war, and that there were many women who lost a fiancé or a husband during the war. The term 'spinster' became very common in everyday life. In the 10 years following the end of war, another estimated 60,000 returned soldiers died.

---ooo000ooo---

Ross St. Claire in his book "Our Gift to the Empire" wrote:-

Life after the war was not kind to many ex-54th Battalion men despite them deserving better. How many died early due to their war service is speculative, but it must have been considerable. Of approximately 270,000 AIF survivors of the war only 149,000 were still alive in 1939. The figures speak for themselves. Tens of thousands of ex-AIF men died in their 30's, 40's and 50's. One only has to glean the death notices pages in returned servicemen publications, such as Reveille, to appreciate how many died young. The combined effects of poor diet, prolonged exposure to mud, cold and rain, wounds, gas and physical and mental illness ended many of their lives prematurely.

In March 1936 an article appeared in Reveille, titled

"THE TOLL OF WAR. OUR DEPLETING RANKS":

The first page I look for in "Reveille" - and the one I believe most Diggers turn to - "The Last Post." It is appalling the way our ranks have been depleted. In the February issue alone are given the names of 120 war veterans who have passed on, and doubly tragic is the fact that the average age of death is about 40.

In the last year I have lost good cobbers in their early 40's, and I venture to say that that, too, has been the experience of many of my fellow-Diggers. It's getting that way now that many survivors of the AIF, though only in early middle age, will consider ourselves extremely blessed by the gods if they live to a ripe age.

The author of the article, from which the above was taken, was John Martin, a former member of the 30th Battalion. He went on to detail the reasons for the high mortality rate of ex-AIF men. Their diet was poor, eaten at irregular intervals, under poor conditions and lacked the vegetables and vitamins of a normal diet. Worse, the combination of being wet and cold for days, even weeks on end contributed to rheumatic and arthritic complaints later.

Finally, Martin wrote *"War undoubtedly took its toll from our mental and spiritual qualities."* Sensitive men, those he called *'gentle spirits'*, were affected greatly by the strain and horrors of trench warfare. Suicide accounted for many. But even the hardiest of souls suffered:

Three years in the great holocaust of France - stench! mud! vermin! blood! filth! - dying men, ghastly wounds ... Three years of ceaseless thunder of the guns, the days and nights of peril, the thousand and one nerve shattering incidents and episodes that sapped and undermined the most hardened and prosaic natures.

In his article Martin did not even mention the effects of gas. Even if not afflicted immediately, many Australians suffered from the latent effects of mustard gas. It permeated their clothes and seeped into the ground, only to be released when the sun came out, and it was in the water-filled shell holes from which many a parched man drank. Many men were affected by lung disease after the war. One example was Corporal Harry Arnold. He had left Australia a fit and healthy twenty-one-year-old in 1915, but died after struggling with war-induced illness in 1930 aged thirty-five.

---ooo000ooo---

While discussing the contents of this book with some friends, one of them wrote to me to advise me of the effects of mustard gas on returned soldiers. My friend was born in 1939, and remembers going fishing with his grandmother's sister's husband, in the 1950s at Ulmarra near Grafton. If the returned soldier hooked a large fish, my friend had to pull it in for him, even though my friend was at the time only 10 or 11, because the mustard gas lung damage would cause the returned soldier to cough non-stop if he over-exerted himself.

Confining "Deaths" And "Injuries" To An Event.

It appears from reviewing the Battles Nomenclature Committee report (which will be examined in detail in Chapter 11), and the assembled list of battles from Captain E.A. James' work, that the British Army classified only those who were actually killed between the 'chronological limits' i.e. the dates of their officially pronounced battle or event. Therefore, anyone who died later, from wounds inflicted during the course of those defining dates, were classified as having died of wounds, and their deaths were not included in the official record of deaths during the defined date(s) of the battle or event.

How Was The Service Of A Soldier Recognized Following The End Of The War?

The most obvious way of recording the service of a deceased soldier, is the erection of a headstone in a war cemetery. As part of the process of defining the work to be done by The Commonwealth War Graves Commission, the CWGC only commemorates those who have died during the designated war years, ***while in Commonwealth military service***, or ***of causes attributable to service.***

The armistice was signed on 11 November, 1918, resulting in a suspension of hostilities that is generally credited as being the end of the war. The Treaty of Versailles was still to be negotiated, and it was not signed until 28 June, 1919. This is the date which appears on many memorials around Australia to signify the end of the war, as opposed to the suspension of hostilities.

The Commonwealth War Graves Commission continued to look at the deaths of soldiers which were attributable to service, as being up until 31 August, 1921, almost three years after the suspension of hostilities. In the years since, we have had many soldiers being awarded a headstone due to them being considered to have died from illness sustained during the period up to 11 November, 1918.

So it is clearly a case of service being recognized *if the death of a soldier is attributable to his service*. While numbers prepared for a Daily Report are appropriate on the day for assessment of what is necessary for operational reasons, in the historical analysis, this appears to have been overlooked in properly defining the numbers involved at Fromelles. On the basis of the points raised by Mr. David Noonan referred to earlier, it is likely to be the case for every other action, or Battle for the rest of World War One.

Those Who Were 'Killed In Action' On 21st July 1916

Along with other attacking battalions, the 31st Battalion crossed No-man's Land at 6.00 pm on the 19th July, and captured the German front line in their sector. They then moved a further 350 yards behind looking for the second line of trenches. Despite erroneous Allied Intelligence stating there was a second line, they found only a small hollow in the ground and the 31st had to fall back to the German front line trenches. The 31st held the German front line, until 5.45am on 20th July when, under heavy counter attack, they had to cross No-man's Land back to their own trenches.

By 8.00am on 20 July, the 31st Battalion had more or less finished its fighting at Fromelles.

From 8.12am on 20 July the 31st's Commanding Officer, Lt Col. Toll, commenced arrangements for extraction of the wounded and then the Battalion was withdrawn from the front line to billets at Fleurbaix. The attack was a disastrous introduction to battle for the 31st having suffered close to 600 casualties, over half its strength. Although it still spent periods in the front line, the 31st could play no role in major offensives for the remainder of the year.

Due to the fighting on 19 July, and the retrieval of soldiers on 20 July, the 31st Battalion did not get to carry out its daily reporting on 20 July. Their leader and Company Commander, Captain Charles Mills, had also been wounded and captured. Shaken by such devastating losses and now without their commander at the helm to guide them, it's no wonder that record-keeping became disorganized and chaotic in the aftermath.

But the dead and wounded still had to be retrieved as much as they could be, and account had to be taken of the numbers who had been killed. In the case of the 31st Battalion the numbers were blurred. But why?

When one reads the Australian Roll of Honour pages in Robin Corfield's book 'Don't Forget Me Cobber', you will come to a section on page 457, ***'Noted as KIA 21 July, see page 24'***. There are 40 soldiers' names listed, where it seems, they have been undisturbed by historians as though they were unassailable. There are often references made to the 40 soldiers of 31st Battalion who were (as you will read erroneously) listed as Killed in Action on 21 July, 1916.

But the details of those names are not totally correct. While the names correspond with records of the Court of Enquiry's pronouncements as to the soldiers being Killed in Action on 21 July, 1916, some had died before that date, and some should have been pronounced as having been killed before that date. And three others should have been added to the list for the sake of completeness.

By reference to the Nominal Roll for the battalion, the list as featured in Corfield's summary, and the individual soldier's files from the National Archives of Australia, it becomes appearent that there should be 43 names, not 40, whose information should be looked at in detail.

Some were 'Missing', and many had died before 21 July, some being on 19 July. For the purposes of comparison, a chart showing the various soldiers said to have been killed on 21 July follows:

Soldier	Nominal Roll	Court of Enquiry finding
1033 Alder, Pte Robert	Died 19/20-07-1916	KIA 21-07-1916
570 Anderson, Pte Leonard William	Missing 21-07-1916	KIA 21-07-1916
719 Bean, Pte Reginald Alexander	Missing 21-07-1916	KIA 21-07-1916
1136 Benson (Harrison) Pte Charles Frederick	Died 21-07-1916	KIA 21-07-1916
817 Beseler, Pte George Henry	Died 19/20-07-1916	KIA 21-07-1916
605 Bills, Pte Thomas Henry	Missing 21-07-1916	KIA 21007-1916 (Despite I.D. collected by Germans)

589 Bloomfield, Pte Bertram Benjamin	Died 19/20-07-1916	KIA 21-07-1916
582 Bray, L/Cpl Frank Davies	Died 19-07-1916	KIA 21-07-1916
829 Brown, Pte Leslie Charles	Died 19/20-070-1916	KIA 21-07-1916
585 Brumby, Pte William Christopher	Died 19/20-07-1916	KIA 21-07-1916
620 Cartwright, Pte Arthur Leslie	Died 19/20-07-1916	KIA 21-07-1916
32784 Caughey, Pte Rodger	Missing 19/20-07-1916	KIA 21-07-1916
1004 Chadwick, Pte Fred	Died 19/20-07-1916	KIA 21-07-1916
623 Cullen, Pte William	Missing 21-07-1916	KIA 21-07-1916
629 Denton, Pte Reginald George	Missing 21-07-1916	KIA 21-07-1916
641 Dunn, Pte Leslie Clark	Died 21-07-19	KIA 21-07-19 (Despite I.D. collected by Germans)
2130 Farrell, Pte James Stephen	Missing 2-07-1916	KIA 21-07-1916
795 Francis, Sgt Arthur Ernest	Died 19-07-1916	KIA 21-07-1916
702 Hale, Pte Norman Arthur (I.D. collected by Germans)	Missing 21-07-1916	KIA 21-07-1916 (Despite I.D. collected by Germans
706 Hibbert, Pte Albert George	Died 19/20-07-1916	KIA 21-07-1916
693 Sgt Leslie Ralph Hopkins	Missing 19/20-07-1916	KIA 21-07-1916
2144 Jamieson, Pte William Alexander	Missing 21-07-1916	KIA 21-07-1916 (Despite I.D. collected by Germans)
715 Johnson, L/Cpl Ralph	Died 19/20-07-1916	KIA 21-07-191 6 (Despite I.D. collected by Germans
4231 Jolly, Pte Wesley Clarence	Died 19/20-07-1916	KIA 21-07-1916
900 King, Pte Frederick John	Died 19-07-1916	KIA 21-07-1916
988 King, Pte Henry	Died 19/20-07-1916	KIA 21-07-1916

1000 Law, Pte Edward Manning	Died 19/20-07-1916	KIA 21-07-1916
250 Mackenzie, Sgt Alan	Missing 21-07-1916	POW DOW 21-07-1916
732 Marcus, Pte Reuben	Died 21-07-1916	KIA 21-07-1916
136 Oliver, Pte Stanley	Died 19/20-07-1916	KIA 21-07-1916
739 Peak, Pte Sydney George Barker	Missing 21-07-1916	KIA 21-07-1916
1114 Pearson, Pte George Charles	Missing 21-07-1916	KIA 21-07-1916
3891 Robertson, Pte Robert	Missing 21-07-1916	KIA 21-07-1916
1877 Rowlands, Pte Wesley Richard	Missing 21-07-1916	KIA 21-07-1916
929 Simmons, Pte William Thomas	Died 19/20-07-1916	KIA 21-07-1916
944 Stiles, Pte Albert	Died 19/20-07-1916	KIA 21-07-1916
2491 Taylor, Pte William	Died 19/20-07-1916	KIA 21-07-1916
956 Terrens, Albert James	Missing 21-07-1916	POW DOW 21-07-1916
803 Thompson, Cpl Samuel	Died 19-07-1916	KIA 21-07-1916
2436 Warby, Pte Charles Francis	Missing 21-07-1916	KIA 21-07-1916
983 Willis, Pte Henry Victor	Died 19-07-1916	KIA 21-07-1916 (Despite I.D. collected by Germans)
799 Wilson, Pte Alexander	Died 19-07-1916	KIA 21-07-1916
794 Woodman, CSM Harold George	Died 19-07-1916	KIA 21-07-1916

Ten (10) were noted on their files as being 'Missing', while 15 were noted on their files as having 'died 19/20-07-1916'. The rest were not accounted for until 21 July, on which date they were said to have Died according to the Nominal Roll.

The 21st July was the first opportunity after the battle for Company Commanders to compile their losses, and anyone who was unaccounted for at that moment in time was nominally recorded as missing or regarded to be dead ***'as at'*** that date, rather than ***'prior to'***.

So, the numbers were tallied ***'as at'*** 21 July, and the deaths, instead of being recorded as having occurred on either 19 or 20 July, to show the dates on which the various lost soldiers were killed, they were instead interpreted subsequently and erroneously, as having 41 of that battalion being 'killed in action' on 21 July, 1916. Two more would be pronounced as having been prisoners of war who died of their wounds by that date.

But they were not, nor could they have been, killed on the day of roll call for the battalion, and their date of death should never have been construed in that manner. Being dead before the numbers were tallied could only have been interpreted to show that they were killed in the days prior, of either 19 or 20 July.

The figures for each battalion on 21 July only serve to show that there has clearly been an error with the 31st Bn. Suddenly there are supposedly forty three (43) 31st Bn men who were *'killed in action'* on that day.

German Documentation

In 2009, two hundred and fifty Australian soldiers were recovered from mass graves discovered at Pheasant Wood, just outside Fromelles. In the years since then, one hundred and eighty have been positively identified (up until 24 April, 2024).

These men were killed during the fighting, but after the battle their bodies lay behind German lines. Before being collected and buried at the Pheasant Wood mass graves, the German commander ordered that any identification discs and personal property was to be removed from their bodies, and bagged.

From the gathered information the Germans compiled lists of the dead and forwarded copies as well as the property back to the Allied command via the International Red Cross (IRC) in 1916. These lists became described as the German Death Lists. Documents were recorded in many Red Cross Missing and Wounded case files but also appear in the military service files now housed at the National Archives of Australia.

The German documentation appears in a number of different types, comprising death lists, card vouchers, official notification forms and property returns. They clearly state that the men named in them died on 19 or 20 July.

By the time a Court of Enquiry would consider their situation the following year, in August 1917, the Germans had passed on details of fourteen soldiers who had been part of 31st Battalion, seven of whose names appear on the chart above, and all of those seven soldiers have since been formally identified and now rest at Fromelles (Pheasant Wood) Military Cemetery. They had their identification material collected and returned, so they would have died on 19 or 20 July, but incredibly, the Court of Enquiry listed them all as being Killed in Action on 21 July.

One of the seven soldiers is 715 L/Cpl Ralph Johnson. Until his identification occurred, he was listed as having been buried at Pont-Du-Hem Military Cemetery. Ironically, German documents suggested he was more likely buried at Pheasant Wood, which ultimately proved to be the case.

One other soldier, 585 Private William Christopher Brumby was also on the list of those listed as killed on 21 July, but not on the German Death List. He is also been identified via DNA sampling, and is now buried at Fromelles (Pheasant Wood) Military Cemetery, Fromelles.

Challenging The Official Record Of The Court Of Enquiry

During 2016, Tim Lycett[114], reviewed the records of the 31st Battalion soldiers who were said to have been killed on 21 July, 1916. He prepared a submission to the Office of Australian War Graves, and to Commonwealth War Graves Commission via email on 11 January, 2017, seeking to correct the error of the 21 July attributions for the various soldiers. His research and eventual submission resulted in a telephone call about twelve months later advising that the headstones would be changed to read 19-20 July, 1916. Tim has never received a formal written response to his submission, but it is apparent that his submission had its intended effect. The headstones of the soldiers have all been changed to have their date of death stated as 19-20 July, 1916.

A summary of his Submission follows.

31st Battalion War Diary (AWM4 23/48/12)

A document of primary significance worthy of consideration is the battalion war diary, which detailed the day-to-day operations, movements and casualties of the unit. Examining the 31st Battalion diary for July 1916 reveals some very telling facts. In the lead up to the 19th July attack at Fromelles, the diary records the move up towards the front into billets at Fleurbaix.
Here on the 16th, the diary reports that the battalion suffered nine casualties from enemy shelling. On the following day, on 17th July, a total of fourteen men were noted as having been wounded and on the 18th, the day before the assault, a further twelve men were reported wounded. During the period of the battle over the next two days a brief mention is made of some casualties but it wasn't until the 21st - the day after the battle - that the diary tallied up the casualty figures.

The real challenge which existed at the time, was for Headquarters Staff to quickly and accurately compile casualty numbers, and to be able to keep fighting the day to day war. Frankly, they did the best they could. There was probably no way of knowing with 100% accuracy who was missing, who were killed, wounded, or prisoners. That takes time. But what appears obvious is how the method of recording has led to a conspicuous error in the listing of the dates of death for the 43 soldiers listed in the chart above, all from the same unit. An error which until recently had been left to become official record.

The entry, reproduced below, is particularly revealing in two ways. Firstly it confirms that the necessary summary of operations reports, including casualty return, were compiled on the 21st July and could explain why this day may have been chosen as a nominal date for those whose fate was not yet known.

[114] Former Policeman, and Crime Scene Officer, and with Sandra Playle, author of Fromelles, The Final Chapters

The second point of interest isn't so much what appears in the entry but rather what it does not say. Had forty three men of the battalion been 'killed in action' on that day (and presumably many others also wounded), casualty figures of that magnitude would certainly be documented as would a report describing the action. And yet there is no mention of even a single casualty occurring or a report resulting from any operations on the 21st July.

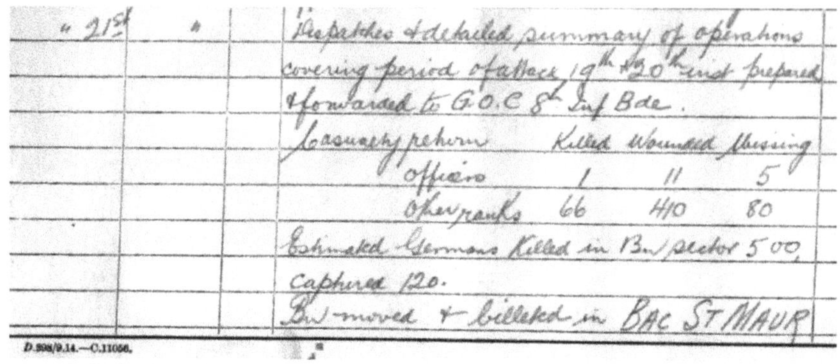
" 21st " | Despatches & detailed summary of operations covering period of attack 19th & 20th inst prepared & forwarded to G.O.C 8th Inf Bde.
Casualty return — Killed Wounded Missing
Officers 1 11 5
Other ranks 66 410 80
Estimated Germans Killed in Bn/sector 500, Captured 120.
Bn moved & billeted in BAC ST MAUR
D.598/9.14.—C.11056.

[Picture 20: War Diary, 31st Battalion – 21st July 1916]

Beyond mention in the war diary, the only casualties that can be located on the 21st July are two men who were presumably from a single platoon left behind to hold Junction Post in the third line defences whilst the remainder of the battalion withdrew to Bac St Maur.

Without question, the war diary entries were of great significance and strongly supported the contention that all the soldiers in question were actually killed during the fighting on 19th-20th July.

The Court Of Enquiry, 1st August 1917 (AWM4 23/48/25)

The Court of Enquiry referred to earlier was held more than a year after the Battle. On the 1st August, 1917, whilst billeted at Racquinghem France, 31st Battalion Headquarters held an enquiry to determine the fate of a number of men still listed as 'missing' after the Battle at Fromelles. It is interesting to note that the war diary for this date states, *"Court of enquiry assembled to enquire into the fate of certain officers and o/ranks reported missing in action in the Operation of Fleurbaix"*, better known now as the Battle of Fromelles that occurred on 19-20 July, 1916.

Even more significant is point (5) in the overall finding of the report two pages later which states, *"That the Officers and Other Ranks whose names appear on the attached list may be presumed to have been 'Killed in Action' on the 19th or 20th July, 1916."*

The list of names referred to as appended to the report includes thirteen of the men found among those still officially listed as killed on 21 July and coincidentally, all of them from 'C' Company. Rather ambiguously, the list still maintains that all but one of the thirteen were killed on 21 July despite the overall findings of the enquiry. The only one it confirms to be killed on the 20 July is 739 Private Sydney Peak who just happens to be the only one of the group whose body was recovered post-war from an isolated battlefield grave and buried at Ration Farm Military Cemetery. The other twelve all fall into the category of missing men whose names appear on memorials.

As alluded to by the enquiry, a further examination of the 43 men supposedly killed on 21 July reveals that a significant proportion of those who were missing and subsequently memorialised originated from 'C' Company. What then becomes apparent is that headquarters of 'C' Company confusingly made the decision to list its missing *as at* the date they became aware of that fact; i.e. the day they compiled their casualty list caused by the action of the previous two days.

Although it ambiguously confirms they were lost in the action at Fromelles, it reinstates the date first recorded in the hours after the encounter, despite whatever new information about each man had since become available. What was originally stated as being '*as at*' was then made the actual date and became official in the 'Report of Death' form, inserted into the individual soldier's service file and conveyed to the man's next-of-kin. From there it was used to compile death certificates, the Commonwealth War Graves register and the Australian Roll of Honour, thus irrevocably embedding the initial error.

31st Battalion Nominal Roll - Deceased (AWM9 23/7)

The 31st Battalion's Nominal Roll is located in the Australian War Memorial archives and although undated, there are clues to when it was compiled from battalion records. The report is a comprehensive list of members of the 31st Battalion who died throughout the war, whether killed in action, died of wounds, succumbed to illness or other reason. It appears in alphabetical order and has a date range of 1916-1918, covering the entire period of time the battalion was on the Western Front. The only anomaly to this is the final few names that are out of order and almost certainly supplementary to the list, added just prior to its release. The three soldiers named in the addendum are dated to have died during February 1919, and are the only ones listed for that year. Therefore it's probable that the document was prepared during February-March 1919.

Of the 43 soldiers listed as killed in action on 21 July, this battalion roll records that 22 of them actually died on 19-20 July. But significantly, 13 of the 22 are for men already in known graves, buried in surrounding cemeteries immediately after the Fromelles engagement.

The other 9 of the 22 that are dated 19-20 July in this report, are to be found spread among the men originally named on memorials.

Other Sources

A number of other sources adding to the evidence questioning the actual date of death can be found within the men's service files.

Scattered throughout the official documents were numerous notations referring to the soldier being killed on 19 or 20 July. These have largely been crossed out at some later date and altered to read the 21st. Nevertheless, it again suggests there must have been some original official information stating the men died on the earlier dates.

Most illuminating are the service files for the eight soldiers buried at Rue Petillion Military Cemetery, more commonly known during the war as Eaton Hall Cemetery. These men, Cartwright, Chadwick, Hopkins, Jolly, King, Simmons, Stiles and Thompson all have a notation on their 'B103 Casualty Form – Active Service' record with details from a burial report compiled by the Graves Registration Unit (GRU). All of these except Chadwick state that the man died on

19/20th July but have had it crossed out. The notation in Chadwick's file is slightly different, stating he was actually buried on 20th July and this remains unaltered.

Interestingly, the entries made in Stiles' B103 Casualty Form clearly state that he both died and was buried on 20th July. The former of these was later altered to read the 21st, but the latter maintains that he was buried the previous day. Additionally, there is a faded note explaining that the original date has been amended to correlate with the 'B2090A – Report of Death' but on closer examination even this form is ambiguous. The B2090A also originally claimed Stiles was killed on 20th July and has been altered to read the 21st , but confusingly his date of burial states the opposite. Where it was initially recorded he was buried on the 21st July, this has been altered to read the 20th July.

So for both Chadwick and Stiles we find undisputed evidence of them having been buried on 20th July even though their official dates of death are the following day. The only possible way this could occur is to either ridiculously assume they'd been buried alive or accept that their official date of death is inaccurate.

Breaking Down the Numbers

The 31st Battalion soldiers reportedly killed on 21st July 1916 can be broken down into two distinct categories.

The **first** is those fifteen (15) men whose bodies were recovered at the time or very soon after the event of their deaths in 1916. These men were subsequently buried in cemeteries and have known graves with named headstones. This group consists of fifteen men whose graves are recorded at four surrounding Commonwealth War Graves Commission (CWGC) Cemeteries: Anzac Cemetery Sailly-Sur-La-Lys, Pont-Du-Hem Military Cemetery, Ration Farm Cemetery and Rue-Petillon Military Cemetery.

Their names are R. Alder, G.H. Beseler, A.E. Francis, W. Taylor, A. Wilson, R. Johnson, S.G.B. Peak, A.L. Cartwright, F. Chadwick, L.R. Hopkins, W.C. Jolly, F.J. King, W.T. Simmons, A. Stiles, and S. Thompson.

The **second** category belongs to twenty-five (25) men whose remains were initially not recovered and had their names recorded on a memorial to the missing. Because the official date provided for their deaths fell one day outside the dates of the Fromelles engagement, they subsequently had their names inscribed more than one hundred kilometres away on the Villers-Bretonneux Memorial, and not on the local VC Corner Memorial, where those men actually recorded as *missing during the battle* were memorialised.

Due to the ongoing identification work undertaken by the Army's Unrecovered War Casualties Unit, and the Fromelles Association of Australia, these numbers have now become 24 soldiers found or identified, with 16 still missing.

Due to more recent discoveries, this second category can now be further sub-divided.

To date, nine (9) of these reinterred in 2010 have been identified among those listed as killed 'on the 21st July'. They are T.H. Bills, W.C. Brumby, W. Cullen, L.C. Dunn, N.A. Hale, W.A. Jamieson, R. Johnson, H.V. Willis, and H.G. Woodman.

For five (5) of these men, the CWGC has acknowledged that they actually died during the battle and has amended their dates of death accordingly. Only one (1) man from this newly identified sub-group still has the 21st July 1916 on his headstone: 794 CSM Harold George Woodman.

For the remaining nineteen memorialised men, the CWGC has also recently acknowledged that they were actually casualties of the Fromelles battle and removed their names from the Villers-Bretonneux Memorial, agreeing to add them to the VC Corner Memorial on an Addenda Panel.

Due to the discovery of documented witness accounts of their deaths, six of these still-memorialised men have had their dates of death slightly adjusted by the CWGC to be 19-21st July 1916, which although still not completely accurate, does more closely reflect the facts. Those soldiers are L.W. Anderson, R.A. Bean, R. Caughey, R.G. Denton, L.C. Dunn, and J.S. Farrell.

Thirty four of the forty three listed men have some reference in official document sources including German death notifications, eyewitness accounts, 31st Battalion Nominal Roll, service documents or any combination of these, stating their deaths occurred during the Fromelles engagement on 19-20th July 1916. Only six remain with no documented proof stating they died on either of those dates.

Given the fact that there appears no comment in the battalion war diary regarding being involved in any further fighting on the 21st July or having sustained any casualties whatsoever, it is only reasonable to conclude that these men all died on either the 19th or 20th July, 1916 and not the 21st as is officially recorded.

The Lasting Response by CWGC

It should now be evident that 43 men of the 31st Battalion were killed in action or died because of their wounds at Fromelles on the 19-20 July, 1916 but their date of death was wrongly attributed as 21st July. This has the immediate effect of increasing the numbers of those killed during, or because of, the Battle of Fromelles i.e. on either 19 or 20 July, 1916.

Thankfully, because of the representations made by Tim Lycett and others to review the position in relation to 40 soldiers who were originally recorded as having been KIA on 21 July, 1916, the Commonwealth War Graves Commission have advised that all the 31st Battalion casualties whose deaths had been recorded as 21st July will be changed to 19-20 July 1916.

While this still leaves the erroneous reference to 21 July it does include the correct timeframe for the deaths of all involved, being 19 or 20 July, and therefore part of the correct overall Battle tally.

Logic has prevailed.

Compare The Common Law 'Year And A Day' Rule.

The common law has a different view of dealing with cause and effect. In the common law, we were previously able to deal with deaths which occurred within a year and a day of the incident, as being attributable to the incident giving rise to the injury itself. If this principle was adopted in the military records of those who died due to wounds incurred in Fromelles alone, the figures will be higher than the hitherto 1917 we have been given without demur from all authorities from that battle.

The Year and a Day Rule is a common law rule, applied in criminal law. It has been removed from our criminal law in each State or Territory, with New South Wales doing so in 1990. While it was in place during World War 1, it does not have a particular place in military situations. I am referring to it in this work, *only by way of comparison*, to show the absurdity of the process of defining when a person, for example a soldier, was killed if it is not conveniently occurring between the dates of the actual defined 'battle'. It is another example of 'cause and effect' alluded to in the introduction at the start of this chapter.

Take for example, an injury to a person, which occurs during the course of an armed robbery. Assume that the accused or the co-accused is armed with a firearm and during the robbery the firearm is discharged deliberately, without care as to the consequences i.e. recklessly. The person injured is rendered unconscious, and is hospitalized with life support, or ongoing intensive care, but does not recover, and dies one year and one day after the infliction of the injury. (A comparable situation might be someone in the trenches, suffering the effects of gassing, who dies after 'a long illness'.)

The person who inflicted the injury, or all those involved in the armed robbery would (at the latest until 1990 in New South Wales) have been charged with murder. This is because if the persons involved in the robbery, if aware that firearms were to be used, could be convicted of murder.

The proofs of the offence of murder are as follows:

- ✓ The victim was a human being;
- ✓ The victim died;
- ✓ The act or omission of the accused caused the death of the victim; and
- ✓ The act of the accused was without lawful cause or excuse; and

The accused:

- ➢ intended to kill or do grievous bodily harm to some person; or
- ➢ foresaw that it was probable that the death of a person would result; or
- ➢ intended to commit a felony punishable by penal servitude for life or 25 years.

So, if the alleged incident occurred in, say, 1989 and the accused was only just now arrested, the Crown would still have to prove beyond reasonable doubt that the victim died within a year and a day of the injury being inflicted.

The point of that law was that if the victim survived for one full year after the incident that the cause of death must have been from some other source. However, now in our advanced and technological age where science and medicine have progressed to levels unthought of earlier, the need for this rule was revised.

---ooo000ooo---

In upcoming Chapter 12, we will learn of the work undertaken by the Battles Nomenclature Committee, convened to determine the formal classification of all actions in which Imperial units were engaged throughout the Great War. In their Report, they considered:

'That, owing to the varying length of the preparatory action, and the difficulty of fixing its actual commencement, this period hould not be included, but that "zero day" should be taken as the first day in all cases.'

'That, on the other hand, the period of consolidation after an attack should, as a general rule, be included.'

It is this author's belief that the preparatory work of Fromelles should also be included in dealing with recording its extent. If we employ the reasoning of the accounting profession toward a project, we look at the activity leading up to, during, and conclusion of the project under consideration. That has become known in accountancy as the 'Total Cost of Ownership'. In this author's, and also that of Robin Corfield, the total cost of the Battle of Fromelles commenced on 14 July, when German forces attacked and killed many Australian soldiers, because they were there, in preparation for the upcoming attack originally scheduled for 17 July, but postponed to 19 July.

For a moment, recall the guidelines for the CWGC, which only commemorates those who have died during the designated war years, *while in Commonwealth military service* or [subsequently] *of causes attributable to service*. There is therefore a qualification for, and a limitation on, just how far the connection with service and injury will go before a soldier is able to be honoured with a CWGC endorsement and burial place.

In comparison to a soldier wounded in the course of battle, the victim of the armed robbery in the example above, would have been initially classified as being 'wounded in action', and later dying from wounds. It is appropriate to consider those who were wounded in battle, and died because of their wounds, to have died because of the battle, just as much as those who fell on the battlefield. A simple case of 'cause and effect'. And similarly, it is also appropriate to consider those who were wounded in battle, were taken prisoner, and died of their wounds, as also having died because of the battle, just as much as those who fell on the battlefield. Again, 'cause and effect.'

By now, it will have become obvious that following this course of not just appropriate thinking, but simple logic, the numbers who were killed *at or because of* Fromelles need to be acknowledged and corrected. There is now no doubt that the 'official' tally of deaths was plainly and simply wrong. It was incorrectly confined.

Determining the Correct Tally of those who died because of the Battle of Fromelles

In the Battle of Fromelles there were soldiers killed in action. There were soldiers wounded in action who died from their wounds. There were soldiers who were taken prisoner. And of those prisoners, some had already been wounded in action, only to die because of those wounds.

They were all ***casualties***. They had all been taken out of the fighting and were no longer effective in terms of battle. Thus the term 'casualties' does not automatically mean deaths, but it does include the number of deaths from that battle. It's a simple example of *'cause and effect'*.

We are routinely informed of the number of ***deaths*** incurred in a Battle, not the full extent of the casualties which accumulated because of the battle.

The cause of a ***death*** is the soldier being killed during battle, or dying from being wounded in the course of the battle taking place, subsequently dying of wounds incurred in the battle, or dying of wounds while a prisoner of war, i.e. *from causes attributable to service*.

So there is a distinct difference between a statement of the 'casualties' compared to a statement of the number of 'deaths'.

Out of the casualties, there were those who were wounded in the course of the Battle, who were taken back from the front line, to a regimental aid post, to a casualty clearing station, or to a hospital. Some of them would return to duty days, or weeks later. Some never did. Again, it's a simple example of *'cause and effect'*.

48 THE WESTERN FRONT [19th-21st July, 1916

TABLE II.—Casualties sustained by the 5th Division at Fromelles between noon on 19th July and 8 p.m. of 21st July :—

	Killed in action	Died of wounds	Prisoners of war	Wounded in action	Total
8th Inf. Bde	477	51	166	1,086	1,780
14th "	438	58	283	948	1,727
15th "	726	98	5	904	1,733
Other units	60	9	16	208	293
Totals	1,701	216	470	3,146	5,533[15]

We have routinely been informed that the total casualties at the Battle of Fromelles was 5,533. This tally comes from the official history publication *'Australian Medical Service Vol. II Western Front Chapter 3, pp35-48 – The Battle of Fromelles'*. It states numbers for the period from noon on 19 July to 8pm on 21 July. How reliable is this tally?

We have routinely been informed that the number of deaths incurred in the Battle of Fromelles was 1,917. This is a simple aggregation of the numbers of 1,701 and 216 in the above tally.

How many of those who were Wounded in Action died from their wounds?

Again, how reliable is this figure? Who were those soldiers?

You will see from the material below that: -

(1) it has been possible to identify all those who were killed in action with no ambiguity.
(2) It has been possible to identify all those who were wounded in action, and died from their wounds.
(3) It has been possible to identify all those taken prisoner of war, out of which we can identify those who died from the wounds they sustained.

The real issue is when we commence our enquiry, and when we conclude the Battle Orders for the Battle of Fromelles are no longer applicable.

The Range of Dates for Determining Casualties.

In upcoming Chapter 12, we will learn of the work undertaken by the Battles Nomenclature Committee, convened to determine the formal classification of all actions in which Imperial units were engaged throughout the Great War. In their Report, they considered:

> *'That, owing to the varying length of the preparatory action, and the difficulty of fixing its actual commencement, this period should not be included, but that "zero day" should be taken as the first day in all cases.'*
>
> *'That, on the other hand, the period of consolidation after an attack should, as a general rule, be included.'*

Adopting that reasoning, most would consider the action on 19 and 20 July alone as the means of defining the Battle at Fromelles. But that would ignore the raids and the bombardments which occurred prior to 19 July, when the soldiers were preparing for the Battle, and which were ancillary to the main battle.

Taking a more modern, but slightly different approach, if we employ the reasoning of the accounting profession toward a project, we look at the activity leading up to, during, and the conclusion of the project under consideration. That has become known in accountancy as the 'Total Cost of Ownership'. It is therefore this author's belief that the preparatory work of Fromelles should also be included in dealing with recording its extent.

We have already noted [in Chapter 5] that by 5 July, 1916 General Haig requested the First, the Second, and the Third Armies to pin down the Germans, by continuing their programme on raids. On 5 July, after the discovery that the 13th Jager Battalion had been sent south from the front of the II Anzac Corps, General Haig reiterated the appeal.

On 7 July, General Godley issued to II Anzac Corps an order that 'Raids must therefore take place immediately, and must be on a larger scale than has hitherto been attempted – about 200 or more men or a Company.'

On 8 July, General Monro had directed Lt. Gen. Haking to draw up plans involving the First and Second Armies.

On 9 July, Haking outlined a scheme for an attack by three Divisions against Aubers Ridge.

Between 10 & 11 July, the Australian 8th Brigade had moved into a position in front of Bois Grenier on the left of the allied front. The 57th and 58th Battalions of the 15th Brigade moved into a position to the right, opposite the eastern face of the Sugar-loaf.

The 14th Brigade filled the centre of those positions with 55th and 56th Battalions.

What followed were a number of re-formulations of the battle plan, with less artillery than originally planned, but for the entire 5th Division to be involved. The artillery bombardment was to commence on 14 July, and to last about three days.

From the time of the arrival of the entirety of 5th Division, the preparations had crystallized into readiness for Battle.

As we've already noted [in Chapter 5], a bomb fight occurred in the trenches of the 58th Battalion during the night of 14/15 July. It was originally preparation for the commencement of the Battle (which was scheduled to commence on 17 July, but postponed until 19 July).

Casualties were already being sustained at Fromelles.

Then the Battle Orders were issued on 15 July, and after two postponements, the Battle of Fromelles began in earnest on 19 July, 1916 with face to face combat.

Varying units continued to hold the front line until 31 July, after which the basis of the activity being undertaken at Fromelles had changed, and the Battle Orders of 15 July had ceased to be the overriding directive for units at Fromelles.

We could assert that the Battle of Fromelles began on 5 July, but that would be 'in spirit' only. It also pre-dates the gathering together of the units which went on to fight at Fromelles. It is clear that the units to be involved had come together by 14 July, and that a set of actions had been determined, the only un-fixed date being the step-off of soldiers in face-to-face battle. Once the planning has been examined, it is obvious that the real start date for the Battle of Fromelles is earlier than on 19-20 July. It is realistic to have 14 July as the date by which planning had turned to readiness.

In determining the date by which the Battle of Fromelles is no longer in process, we need to consider when the actions being taken pursuant to the Battle Order ceased, and by which date the units involved were re-assigned. By examining the various Brigade and Unit Diaries, that date appears to be 31 July, 1916.

How many Died because of the Battle of Fromelles?

What has until now been the most helpful account of those who were killed, and the units they belonged to, could be found in Robin S Corfield's *'Don't Forget Me Cobber, The Battle of Fromelles'*.

Corfield had access to Red Cross records to compile the chart he had put together for his book. It was an ambitious exercise, but more have needed to be added.

Corfield's work included those who were involved in artillery barrages, and raids and who were killed or wounded from 14 July to 18 July, as part of the tally of those involved in the Battle at Fromelles. His tally also includes those who were supposedly killed on 21 July, 1916.

Starting with data from Commonwealth War Graves Commission, I have separately examined records of deaths and burial. The battalions in which those soldiers served were checked against the battalions involved at Fromelles.

There were approximately 200 more Australian and 270 British found by reference to the CWGC records, but on further examination not all those Australians would be found to have died because of the Battle of Fromelles. Some sustained wounds after 31 July, and died soon after.

I have separately searched individual soldier files through National Archives of Australia, via their website at https://www.naa.gov.au to confirm the details of individual soldier outcomes.

Where of assistance, information has also been checked on the website of The AIF Project, at https://www.aif.adfa.edu.au.

A search was also undertaken of prisoner of war lists supplied by the Germans, so that all those captured can be properly recorded in the list of '*casualties*'.

The number of soldiers 'killed in action' and those who 'died of wounds' after 21 July, have had their individual and Unit records reviewed to ensure their cause of death was related to their actual service under the Battle Orders for the Battle of Fromelles. A number have been removed from the first edition of this book as a consequence. There were found to be no more killed in action beyond 26 July, but progressively more who died of wounds.

After July, 1916, the various units were in consolidation activities, or moving to other positions. Their circumstances had changed.

The numbers go on to include those who died of wounds from the Battle, even up until 1944. There are possibly more who died after lengthy and life-consuming illnesses after 1916, and if they become apparent, these numbers will be updated.

I have now added all who became prisoners of war, including those who died while they were prisoners.

Looking at individual soldier files and Unit Diaries, there are some who were killed in action even on 26 July, 1916, as part of the holding of the line.

Those who have been found to have been killed in 'Pozieres', or whose places of burial are on the Somme, have been removed from the Corfield table where that is clearly in error, based on reviewing their service records. Similarly, those who were said to be killed at 'Pozieres', but are buried at cemeteries around Fromelles, are appropriately recorded as being part of the Fromelles numbers, again based on reviewing their service records.

The names of 39 soldiers in the 31st Battalion who were originally listed as having been killed on 21 July 1916, thanks to a Submission made by Tim Lycett to Office of Australian War Gravers, are now officially recorded as having been killed on 19-20 July.

Robin Corfield's book was originally published in 2000 and had a further edition published in 2009. It had to be updated on account of the soldiers who have been reclaimed from the German burial pits at Pheasant Wood outside of Fromelles in 2009 and 2010. Consequently, details of any incorrectly recorded previously (e.g. Herbert 'Nutsy' Bolt as being DOW, and Private Raymond Pflaum previously recorded as POW DOW), are now to be correctly recorded as KIA.

The names of the soldiers who now lie in the Fromelles (Pheasant Wood) Military Cemetery have been updated to reflect their final places of burial, instead of having been recorded on the Memorial Wall at VC Corner Cemetery, or the Villers Bretonneux War Memorial. Changes will take place into the future as more soldiers' names are announced that now lie at Pheasant Wood.

For a moment, recall the guidelines for the CWGC, which only commemorates those who have died during the designated war years, *while in Commonwealth military service* or [subsequently] *of causes attributable to service*. There is therefore a qualification for, and a limitation on, just how far the connection with service and injury will go before a soldier is able to be honoured with a CWGC endorsement and burial place.

By now, it will have become obvious that following this course of not just appropriate thinking, but simple logic, the numbers who were killed *at or because of* Fromelles need to be acknowledged and corrected. There is now no doubt that the 'official' tally of deaths was understated.

Listing the Names of all those who Died because of the Battle of Fromelles

A listing of all those who died by reason of being killed in action, died of wounds, or died of wounds while a Prisoner of War, are set out in **APPENDIX I** of this book, as are those who were captured and remained Prisoners of War. It also records six who died while they were prisoners of war, but for other reasons, such as influenza.

Those who were supposedly 'Killed in Action' on 21 July, are now officially reclassified as having been killed on 19-20 July, but for this study they are still shown separately.

The sheer numbers tell their own story. The following analysis reveals extent of those killed in action, dying of their wounds, or dying of their wounds while prisoner of war. i.e. ***from causes attributable to their service*** in the Battle of Fromelles.

Date	KIA	DOW	POW	POW – DOW	POW - Died
1916.07.14 – Australian	5	1			
1916.07.14 – British	6				
1916.07.15 - Australian	52	2	3		
1916.07.15 – British	3				
1916.07.16 – Australian	8	2			
1916.07.16 – British	7				
1916.07.17- Australian	12	3			
1916.07.17 – British	7				
1916.07.18 – Australian	8	6			
1916.07.18 – British	13				
1916.07.19 – Australian	1070	5	27	2	
1916.07.19 – British	461	12	37	1	
1916.07.19-20 - Australian	147				
1916.07.19-21 - Australian	39				
1916.07.20 – Australian	474	28	406	4	
1916.07.20 – British	35	19	6		
1916.07.21 – Australian	4	31		5	
1916.07.21 – British	1	8			
1916.07.22 – Australian		26		4	
1916.07.22 – British		12			
1916.07.23 – Australian	4	14		3	
1916.07.23 – British		4			

Date	KIA	DOW	POW	POW – DOW	POW - Died
1916.07.24 – Australian 1916.07.24 – British		11 4	1	4	
1916.07.25 – Australian 1916.07.25 – British	1	16 10		3	
1916.07.26 – Australian 1916.07.26 – British	3	7 1			
1916.07.27 – Australian 1916.07.27 – British		5 4		1	
1916.07.28 – Australian 1916.07.28 – British		10 3		1	
1916.07.29 – Australian 1916.07.29 – British		3 5			
1916.07.30 – Australian 1916.07.30 – British		4 5			
1916.07.31 – Australian 1916.07.31 – British		8 2			
1916.08 – Australian 1916.08 – British		57 84		5	
1916.09 – Australian 1916.09 – British		5 70		3	
1916.10 – Australian 1916.10 – British		4 35		2	1
1916.11 - Australian		2			1

Date	KIA	DOW	POW	POW – DOW	POW - Died
1916.12 – Australian 1916.12 – British		2 1			
1917 - Australian		1			
1918 – Australian 1918 - British				 1	4
1920 – Australian		1			
1944 – Australian		1			
Total Australian	**1827**	**253**	**437**	**37**	**6**
Total British	**533**	**279**	**43**	**2**	

As you would have seen from the dates at the beginning of this chart, and for the reasons I have already stated, I have commenced the collation of numbers on 14 July, 1916, and not 19 July, 1916 as set out in the table in the official history publication ***Australian Medial Service Vol II Western Front Ch3 pp35-48 Trench Warfare – The Battle of Fromelles***. That table is from 19 July to 8pm on 21 July. It does not deal with the entirety of the Battle, and it does not tabulate those who later died of wounds.

In the period 14-18 July, which I have determined is the correct range of dates for the Battle of Fromelles, there were 85 KIA, + 14 DOW, + 3 POW - 102 more casualties who should be accounted for. The tally of casualties which includes all these groups must therefore be increased by 103.

This study confirms the following losses:-

14 – 18 July, 1916 – the leadup to the Battle

KIA / DOW / POW	Australian	102
	British	36

19 - 21 July, 1916 – the duration of the Battle

KIA / DOW / POW / POW DOW	Australian	2242
	British	571

From 22 July, 1916 – subsequent to the Battle:

KIA / DOW / POW / POW DOW / POW-Died	Australian	216
	British	250

There was a combined total of Australian deaths and prisoners taken of 2,560 due to the Battle of Fromelles.

Of these, there were 1,827 killed in action.

There were 253 who died of wounds.

There were 480 taken prisoner, of which 37 died of wounds while they were prisoners, and another 6 died from causes not related to the Battle of Fromelles.

By reference to the various descriptors for fatalities, I assert that the number of **Australian** soldiers who died in battle or from their wounds tallies up at **2,117**.

This means that out of the [now revised] Australian casualties of 5,636 (which includes all prisoners), some 3076 would go on to be dealt with medically, or go on to fight elsewhere.

The tally of 'casualties' which has been attributed to the **British** soldiers is 1,547.

By reference to the same descriptors for fatalities, the number of British soldiers who died in battle or from their wounds tallies up at **814**.

Out of the British casualties, 734 would go on to be dealt with medically, or go on to fight elsewhere.

The combined total of Australian and British deaths is **2,931.**

It is beyond this work to seek to document the number of German soldiers and the categories in which they have fallen, so the work from Robin Corfield dealing with the German numbers, has been reproduced without alteration.

---ooo000ooo---

{{Note: German newspaper reports shortly after the Battle makes mention of 481 Australian POWs captured. This research has located 480. If there are additional POWs, then any additional POWs confirmed will need to be added to these numbers. Author, January, 2026}}

{{Note further: British Prisoners of War have not yet been critically confirmed. The General Staff of 61st Division reported following the Battle that there were about 80 taken prisoner.

To date, information in relation to 45 British POWs has been located, and these are included in the above figures. All such prisoners were part of the 2/7 Royal Warwickshire Regiment, and the 182nd Machine Gun Company, which were the only units to have broken through the German lines near the Wick Salient. So, if there are more British POWs, then the final tally will also need to be increased. Author, January, 2026}}

---ooo000ooo---

It is now time for the correct the 'Official' figures. The 'Official' numbers for deaths of **Australian** soldiers have long been stated as 1917. That did not include those who were said to have been 'Killed in Action' on 21 July, 1916, let alone those who died of wounds later. Using the test of the Commonwealth War Graves Commission, the number of **Australian** soldiers who who died **while in Commonwealth military service**, or **of causes attributable to service,** because of the Battle of Fromelles, is **2117.**

It must also be stated that the number of **British** soldiers who died **while in Commonwealth military service, or of causes attributable to service**, because of the Battle of Fromelles is **814.**

The total Australian and British soldiers who died while in Commonwealth Military service, or of causes attributable to service ***because of*** the Battle at Fromelles is therefore **2931** combined. And that is **2931** men, who because of the High Command's wish not to frighten the Australian and British populace, were said to have died in ***some important raids***, not a ***Battle,*** and thereby were not even given the dignity of a place name inscription in a war memorial.

--ooo000ooo---

A POSTSCRIPT:

I would at this point like very much to acknowledge the significant input received in the preparation of this Chapter, from Tim Lycett, in the provision of material relating to soldiers listed as having originally been killed on 21 July, 1916. Thanks Tim.

---ooo000ooo---

10. What's in a Name - A Raid? An Attack, A Feint? Or A Battle?

A rose by any other name would smell as sweet.

What matters is what something is, not what it is called.[115]

---ooo000ooo---

If The Numbers Are Wrong, Then What Else Is?

At the beginning of this book, it was stated that the action that took place at Fromelles on 19/20 July, 1916 was initially described (by British High Command) as a ***raid***. It was a description which I have stated was manifestly inappropriate.

Following on from the explanations given for various terms in Chapter 1, we will look at some terminology in more detail. We will look at the definitions of the plain English terms of '**action**', of '**attack**', of '**campaign**', and of '**offensive**'. All of these terms are used in connection with various '**operations**' which unfolded during the course of fighting on the Western Front in 1914 to 1918. Their application to what happened at Fromelles in 1916 will be critically examined.

Operation

An '**operation**' is a highly organized activity that involves many people doing different things.[116]

It is a very broad expression. We will be looking at 'operations' in the military setting of this work, which will show that a military operation is a coordinated military action taken by a state or non-state actor to resolve a situation. Military operations are complex and can involve many people doing different things. They can take place on the battlefront, in the atmosphere, and across the oceans. The term of 'Operation' will therefore embrace all of the terms in the paragraph above.

Raids often take place within the time and space of a battle, and a raid is therefore likely to be classified as lower in terms of significance than a 'battle'. While having an objective, raids are often not necessarily decisive in the course of an overall battle.

---

[115] Generally credited to William Shakespeare, in Romeo & Juliet, 1600.

[116] See: *Collins COBUILD Advanced Learner's Dictionary. Copyright © HarperCollins Publishers*

A raid is not normally intended to capture and hold terrain, but instead finish with the raiding force quickly retreating to a previous defended position prior to enemy forces being able to respond in a co-ordinated manner or formulate a counter-attack.

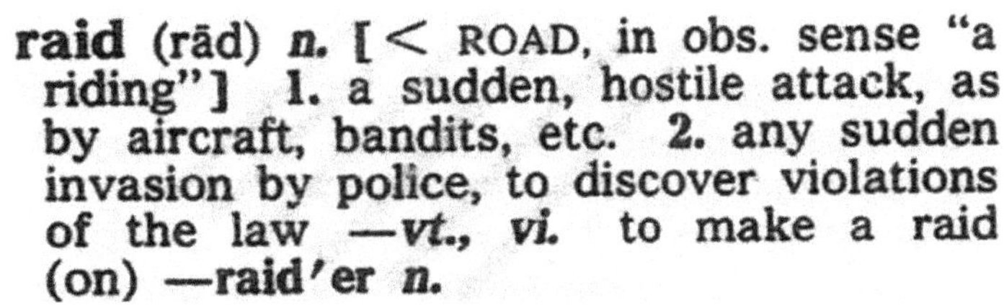

raid (rād) *n.* [< ROAD, in obs. sense "a riding"] **1.** a sudden, hostile attack, as by aircraft, bandits, etc. **2.** any sudden invasion by police, to discover violations of the law —*vt., vi.* to make a raid (on) —**raid′er** *n.*

[Picture 20: Collins Dictionary]

An example of a 'raid', is that of Capt. Wilder Neligan's raid by the 9th Bn on 1-2 July, 1916. 148 men compared to nearly 20,000.

Actions

This term can be applied to all forms of military combat. As can be seen from the definition below, the word 'action' in the ordinary sense is much generalized, and can be applied to a number of outcomes, including military combat (see below).

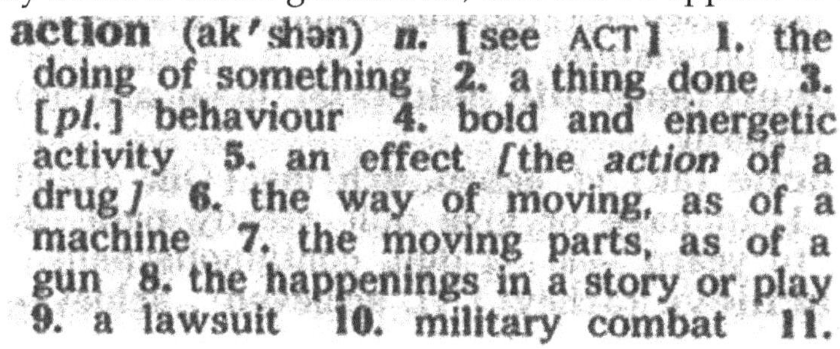

action (ak′shən) *n.* [see ACT] **1.** the doing of something **2.** a thing done **3.** [*pl.*] behaviour **4.** bold and energetic activity **5.** an effect [the *action* of a drug] **6.** the way of moving, as of a machine **7.** the moving parts, as of a gun **8.** the happenings in a story or play **9.** a lawsuit **10.** military combat **11.**

[Picture 21: Collins Dictionary]

Sometimes the terminology used is that of '**engagements**'. The English language is a creature of many words having similar or the same meanings. These are just a few.

Attack

Attacks are ordinarily defined in the following manner

An outsider would see very little distinction between the terms 'action', and 'attack'. But either term ('action' or 'attack') would have the result of downplaying the significance of the Battle and the events at Fromelles.

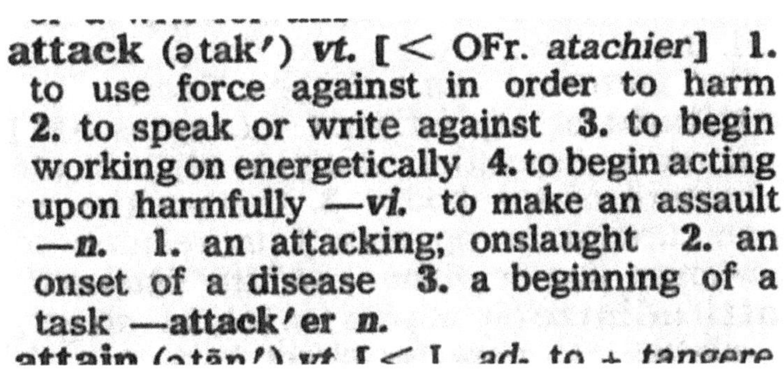

attack (ətak′) *vt.* [< OFr. *atachier*] **1.** to use force against in order to harm **2.** to speak or write against **3.** to begin working on energetically **4.** to begin acting upon harmfully —*vi.* to make an assault —*n.* **1.** an attacking; onslaught **2.** an onset of a disease **3.** a beginning of a task —**attack′er** *n.*

[Picture 22: Collins Dictionary]

In the next chapter, we will see how the Battles Nomenclature Committee applied a very restricted meaning to the term 'Attack".

Battle

What should have become apparent up to now, is the manner in which the word 'battle' as opposed to the use of the more formal descriptor 'Battle' is used by various commentators. The two are invariably used interchangeably, but as you will know from Chapter 1, a 'Battle' is a 'battle' that is given a certain status. Battles are significant events, often deserving of memorialising. They merit inscription on memorials where memorials honour service by reference to individual Battles. And they may or may not be granted the further military status of being granted Battle Honours.

As pointed out in Chapter 1, for military purposes the term 'battle' seems to require a higher level of definition than it does in ordinary usage.

Recall the ordinary definition of 'battle' below: -

In general, a '*battle*' during the 20th century was, and continues to be, defined by the combat between opposing forces representing major components of total forces committed to a military *campaign*, used to achieve specific military *objectives*. Where the duration of the battle is longer than a week, it is often for reasons of staff operational planning called an ***operation***. Battles can be planned, encountered, or forced by one force on the other when the latter is unable to withdraw from combat.

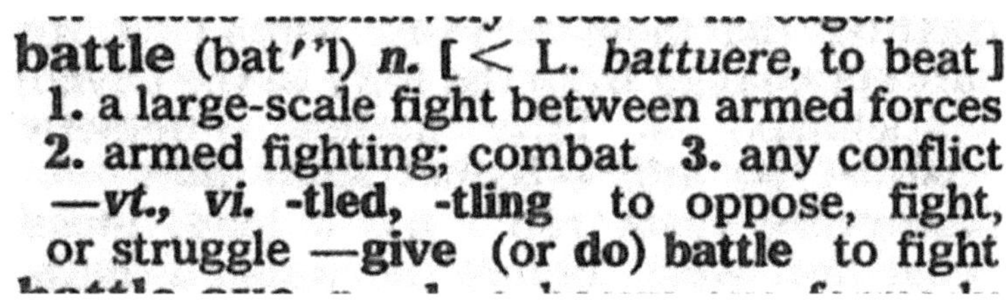

battle (bat′'l) *n.* [< L. *battuere,* to beat] **1.** a large-scale fight between armed forces **2.** armed fighting; combat **3.** any conflict —*vt., vi.* -tled, -tling to oppose, fight, or struggle —give (or do) battle to fight

[Picture 23: Collins Dictionary]

A *Battle* always has as its purpose the reaching of a mission goal by use of military force. A victory in battle is achieved when one of the opposing sides forces the other to abandon its mission, or to surrender its forces, or routs the other, i.e., forces it to retreat or renders it militarily ineffective for further combat operations.

It thus appears that from a military standpoint, it was not simply enough to have a fight with your opponent.

But recording the progress of war is a process of ensuring that fighting is classed into levels of more and more importance or significance. The fighting at Fromelles was, despite its obvious intensity of effort, never classed as big enough or significant enough to warrant the descriptor of "Battle".

Campaign

When the various actions, raids, and battles along the Somme River were spoken of collectively, they came to be known as The Somme Offensive, or 'the Battle of The Somme'. They were each a part of a campaign, a series of military operations with a particular objective.

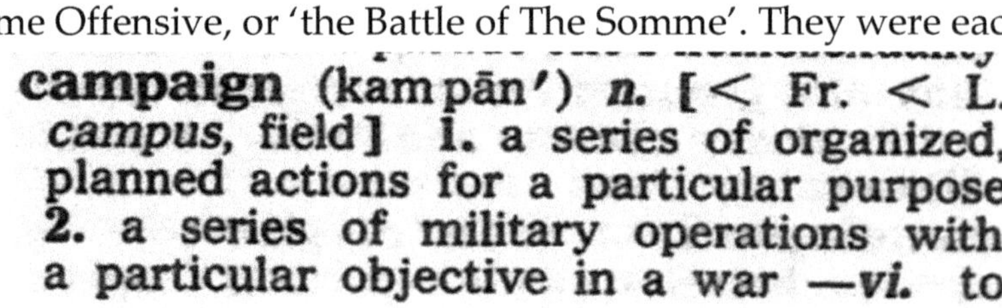

campaign (kam pān′) *n.* [< Fr. < L. *campus,* field] **1.** a series of organized, planned actions for a particular purpose **2.** a series of military operations with a particular objective in a war —*vi.* to

[Picture 24: Collins Dictionary]

There were, of course, many more battles and raids to be fought under the heading of *The Somme*, up until the Battle of Ancre and the Capture of Beaumont Hamel on 18 November,1916, but our study is confined with those up until the Battle at Fromelles.

Offensive

The Battle of the Somme also became collectively described as the Somme Offensive, as defined in the second section of the caption below.

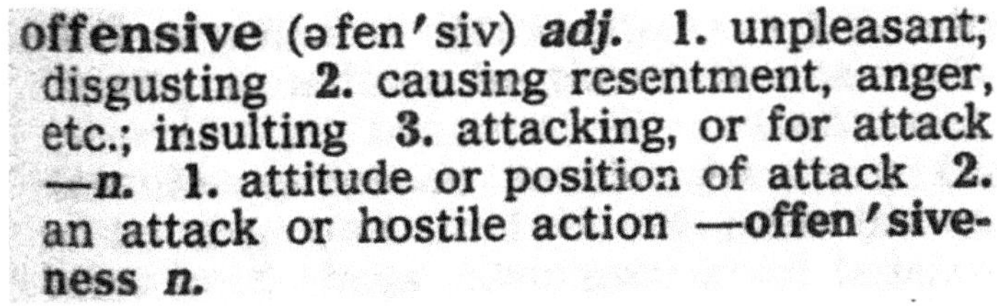

offensive (əfen′siv) ***adj.*** **1.** unpleasant; disgusting **2.** causing resentment, anger, etc.; insulting **3.** attacking, or for attack —***n.*** **1.** attitude or position of attack **2.** an attack or hostile action —**offen′siveness** ***n.***

[Picture 25: Collins Dictionary]

You can easily become distracted by arguing a case of semantics here. One battle, a group of battles, a campaign, an offensive. The lingering impression when referring to 'the Battle of the Somme', is to think isthe singular, and not to see a number of separate battles, each of which are significant in their own right.

For all intents and purposes, when we read of a 'Campaign', it is more or less interchangeable with the term 'Offensive' throughout this book, unless the context requires other applications of their meaning to be applied.

A Comparison With Preceding Actions Near Fromelles.

In Chapter 2 we looked at various actions which occurred near Fromelles prior to July, 1916.

If we look again at the actions which occurred near Fromelles (even on the same ground as in 1916), we can consider the use of the terms which have been defined, and consider those in relation to the scale of what took place in each action, but more significantly, the relative scale of what took place at Fromelles on 19/20 July, 1916.

18 December, 1914 – The First Action At Fromelles.

It was in October 1914 that the British found themselves in the Fromelles and Aubers region.[117] Having departed the now-stabilising Aisne and Marne battlefields, marching by night and resting by day the British were seeking to arrest further German advances, on the Somme, in Artois, and then in French and Belgian Flanders. It was the period known as the 'Race to the Sea'.

On 20 October, elements of the British 19th Brigade, alongside troops from General Louis Conneau's Mounted Division, some on bicycles, patrolled the ridge between the villages of Aubers and Radinghem, looking to push forward.

[117] Barton p.24.

On 21 October German troops in field grey appeared *en masse*. Their approach was announced by gunfire, the heavies that many had yet encountered. Both Briton and Frenchman were forced into withdrawal.

On the morning of 23 October, a defensive line was established between the settlement of La Boutillerie and a small area encompassing a tiny hamlet on the flat plain below Fromelles village known as Rouges Bancs. For the next two weeks, a sequence of hostile assaults was beaten off, enemy fortunes often being hampered by heavy mists, typical of the area. The enemy was digging in.

On 15 November, 1914, the 2nd Scots Guards took over the Cordonnerie Farm sub-sector and on the 27th a small party led by Lieutenant Sir Edward Hulse crept through unharvested turnips to make the first British ***raid*** in the neighbourhood.

The lines were now firmly fixed and the Germans were already installing a designed defensive infrastructure. The great siege had begun.

The December 1914 action at Fromelles was the very first set-piece attack in the sector.[118] The battleground selected lay between the Sailly-sur-la-Lys-Fromelles road (the Rue Delvas, later christened the *Tommy-Strasse* by the Germans) and a track south-east of La Cordonnerie Farm. It was exactly the area that would be assaulted by the British 8th Division on 9 May the following year, and again by the 5th Australian Division in July 1916, over a year and a half hence.

Two battalions were to challenge the German defenders. The orders given to the 2nd Scots Guards and the 2nd Border Regiment (Borders) on 18 December 1914 were chillingly similar to those which would be issued to the Australians in July 1916 – and so was the way the action unfolded. Their attack was also timed for 6.00pm, *but being European mid-winter, this made it nocturnal.*

The troops, carrying spades and sandbags to consolidate the expected gains, were required to assemble in No-man's land and upon the whistle rush and hold the German front trench.

Should it be unoccupied they were to push on to the second line.

The Scots Guards assumed responsibility for the right-hand sector facing Rouges Bancs, which includes the area today occupied by the Australian Memorial Park, while to their left the Borders attacked positions immediately west of Delangre Farm.

The action was unsuccessful. The cross fire from well-placed German machine guns played a big part, accounting for very heavy casualties. Despite the menace which came from the use of machine guns, future allied leaders, in particular General Haig, were not inclined to regard them with appropriate significance at the time.

The battalion's casualties were approximately 180 killed and wounded – 50 percent of the force. The Borders were left with 128 casualties, several of whom still lay in No-man's land. For *voluntarily bringing in wounded 'by daylight on 19 December, 1914 under heavy rifle and machine-gun*

---

[118] Barton p.32.

fire,' Privates Abraham Acton and James Smith, both of the 2nd Border Regiment, were awarded the Victoria Cross – the origin of the place name 'VC Corner'.

On the morning of 20 December, soldiers from both sides by an almost tacit understanding as to no firing, arranged to bring in bodies of soldiers who had been left in No-man's land up until that time.

How is The Action Of 18 December, 1914 Classified in The Battle Accounts?

The Report titled: "Army, The Official Names of the Battles and Other Engagements Fought by The Military Forces of the British Empire During The Great War, 1914 – 1919, and the Third Afghan War, 1919 [the BNC Report] contains **no reference** to any action taking place on 18 December, 1914.

The Attack on Wytschaete, on 14 December 1914 by II Corps, 3rd Division is listed.

The Defence of Givenchy, 1914, 20-21st December by Indian Corps: 1st Lahore, and Meerut Divisions is listed.

But there is no reference to any Battle, Action, or Miscellaneous Incidents on 18 December, 1914. As we will learn in the following Chapter, if it had been included in the BNC Report, it would have been classified as an 'Attack', which was for their purposes an unsuccessful Battle.

Compounding this injustice, in "A Record of the Battles and Engagements of the British Armies in France and Flanders, 1914-1918", by Captain E.A. James, there is no reference to a battle of any description taking place on 18 December, 1914.

Was this an oversight? Or was it a case of just not thinking it was of any significance?

10– 13 March, 1915: The Battle of Neuve Chappelle

Between 10 and 13 March, 1915, this battle was fought by Indian and United Kingdom forces under the command of Field Marshal Sir John French in what was described as the Artois region of France. It was not on the Somme. It was a British offensive some 8 kilometres by road from Fromelles.

The battle was intended to cause a rupture in the German lines, which would then be exploited with a rush to the Aubers Ridge and possibly Lille. It was an action involving 4 Divisions (approximately 40,000 men). After using aerial photographs to identify the ground fortifications, the battle followed. It was the first deliberately planned British offensive, and tactical surprise, and a break-through was achieved after preparation for the attack involved great attention to detail.

Unfortunately, after a set-piece attack, unexpected delays slowed the tempo of operations. Command was undermined by communications failures; and infantry/artillery cooperation broke down. The Germans had time to receive new reinforcements and dig new communication lines.

Although aerial photography had been useful, it was unable to efficiently identify the enemy's strong defensive points. Added to this, poor communications also meant that the British commanders had been unable to keep in touch with each other. The battle became uncoordinated, and this in turn led to the disruption of supply lines.

A German counter-offensive failed, but forced the British to use most of their artillery ammunition and the offensive was postponed on 13 March, and abandoned two days later. Small numbers of German troops in strong-points and isolated trenches, had been able to maintain a volume of small-arms fire sufficient to stop the advance of far greater numbers of attackers.

The Battle showed that a breakthrough in trench defences was possible if the attack was carefully planned, but after the initial shock to the Germans, their defence recovered as the British were beset by delays, loss of communication and consequent disorganization.

Haig had made his intent plain to his subordinates, but they had not grasped the "spirit" of the plan and had failed to press on when initial objectives had been captured, although one of the subordinate commanders later claimed that pressing on would have backfired due to the lack of ammunition.

Cassar called the battle a British tactical success[119] but that the strategic intentions had not been met. Joseph Joffre, the French Commander, praised the results of the first day, then dismissed the significance of the attack "*Mais ce fut un succes sans lendemain*" (But it was a success which led to nothing).

Four divisions took part during the battle, and suffered 7,000 British and 4,200 Indian casualties.

One would have expected the lessons learnt in that battle would not be repeated in the next ones. But were they?

How is The Action Of 10-13 March, 1915 Classified in The Battle Accounts?

In the BNC Report, this action is referenced as the Battle of Neuve Chapelle.

Note that this Battle occurred in a location which was geographically proximate to Fromelles.

9 May, 1915: The Battle of Aubers Ridge

This was fought on 9 May, 1915, just two months later. It was part of the Second Battle of Artois. Again, it was not on the Somme. And it was fought, again, by units from the United Kingdom, and the Indian Empire, under the command of Sir John French.

It was intended to exploit the German diversion of troops to the Eastern front. The objective was Aubers Ridge, in the flat and poorly drained terrain, an area of slightly higher ground 2-3

[119] Cassar, G. (2004) Kitchenerr's War: British Strategy from 1914 to 1916. Washington: Brassy's ISBN 1-57488-708-4. P. 166

kilometres wide marked by the villages of Aubers, Fromelles and Le Maisnil. The same area as had been attacked two months earlier.

The Germans had learnt in the intervening two months, that one breastwork was insufficient to stop an attack. Fortifications were quickly augmented. Barbed wire entanglement was doubled and trebled. The depth of breastworks was increased. Machine guns were sited in emplacements at ground level set to sweep No-man's land from flanking positions. Attacking soldiers would be cut off at the knees. And a second breastwork was begun as general strengthening, about 200 yards behind the front line.

Intelligence about the newly strengthened German positions was not available, or not given sufficient attention. No surprise was achieved. The British bombardment was wholly insufficient to break the German wire and breastwork defences, or to destroy or suppress the front-line machine guns.

British artillery equipment and ammunition were in poor condition: the first through over-use, the second through faulty manufacture. Poor communications meant that it soon became impossible to tell where British troops were; and given the state of artillery and ammunition, accurate close-support artillery fire was impossible.

This battle was an unmitigated disaster for the British army. No ground was won and no tactical advantage gained. It is doubtful if it had the slightest positive effect on assisting the main French attack 15 miles to the south. The battle was renewed slightly to the south, from 15 May as the Battle of Festubert.

The British and Indian forces had suffered casualties and losses of 11,161. The Germans had lost 902.

How is The Action Of 9 May, 1915 Classified in The Battle Accounts?

In the BNC Report, this action is referenced as Attack at Fromelles as part of Battle of Aubers Ridge.

Note that this Battle occurred in a location which was geographically proximate to Fromelles.

15 – 25 May, 1915: The Battle of Festubert

This was fought between 15 & 25 May 1915. The offensive formed part of a series of attacks by the French Tenth Army and the British First Army in the Second Battle of Artois (3 May – 18 June 1915). Again, not on the Somme.

After the failure of the attempted breakthrough at Aubers Ridge (9 May 1915) tactics of a short hurricane bombardment and an infantry advance with unlimited objectives, were replaced by the French practice of slow and deliberate artillery-fire intended to prepare the way for an infantry attack.

The objectives were 1,000 yards (910m) forward, rather than the 3,000 yards (2,700m) depth of advance intended at Aubers Ridge. The battle was the first British attempt at ***attrition.***

The attack was made by the British First Army under Sir Douglas Haig against a German salient between Neuve Chapelle to the north and the village of Festubert to the south. The assault was planned along a 3-mile (4.8 km) front and would initially be made mainly by Indian troops. This would be the first British army night attack of the war.

The battle was preceded by a 60-hour bombardment by 433 artillery pieces that fired about 100,000 shells. This bombardment failed to significantly damage the front line defences of the German 6th Army but the initial advance made some progress in good weather conditions. The attack was renewed on 16 May. By 19 May the 2nd Division and 7th Division had to be withdrawn due to heavy losses. On 18 May, the 1st Canadian Division, assisted by the 51st (Highland) Division, attacked but made little progress in the face of German artillery fire. The British forces dug in at the new front line in heavy rain. The Germans brought up reinforcements and reinforced their defences. From 20–25 May the attack was resumed and Festubert was captured.

The British lost 16,648 casualties from 15/16–25 May. The German defenders had approximately 5,000 casualties, including 800 men taken prisoner. The offensive had resulted in a 3-kilometre (1.9 mile) advance.

How is The Action Of 9 May, 1915 Classified In The Battle Accounts?

In the BNC Report, this action is referenced as the Battle of Festubert.

Note that this Battle occurred in a location which was geographically proximate to Fromelles.

---ooo000ooo---

Before the Battle of Fromelles in July, 1916, just getting to this point in the war had already left a trail of poor decisions, poor planning, and inflated expectations. In the previous two years, there had already been various unsuccessful 'battles' with disastrous outcomes.

19/20 July, 1916 – The Attack At Fromelles

In Chapter 6 we read what took place at Fromelles on 19/20 July, 1916. It is not intended to reproduce that chapter here. But comparatively, the poor planning and outcomes made in the preceding battles were simply repeated at Fromelles on 19/20 July, 1916.

Following the Battle, it is now mid-summer in France in 1916.

Australia, have or will soon have lost 22296 soldiers either killed leading up to or during the battle. Another 289 will die because of their wounds sustained in the Battle, in the months and years to come.

The British similarly have lost 563 soldiers either killed leading up to or during the battle. Another 250 will die because of their wounds sustained in the Battle, in the months to come.

Therefore, 2,964 soldiers will die because of the incompetence of the commanders who were responsible for planning and executing this military 'attack' from the start.

How is The Action Of 19/20 July, 1916 Classified in The Battle Accounts?

In the BNC Report, this action is referenced as 'Battle of Bazentin Ridge, with subsidiary Attack at Fromelles.

How Did The British High Command Report The Fighting At Fromelles?

The initial British High Command Statement following the Battle at Fromelles was:

> *Yesterday evening, south of Armentieres, we carried out some important* ***raids*** *on a front of two miles, in which Australian troops took part. About 140 German prisoners were captured.* (July 20, 2.50 pm, 1916.)

Twenty four battalions and numerous supporting units approximating 20,000 soldiers had been engaged in a battle seeking to capture and hold German trenches. The reference to what had just taken place was an insult to the 5533 Australian casualties, of whom 2151 were killed in action or would die of their wounds. It was just as much an insult to the 1547 British casualties, of whom 813 were killed in action or would die of their wounds.

It can be speculated that by announcing that there were 'some important raids' that there were a number of separate actions (i.e. 'small' raids) when in fact, it was one combined action with nearly 20 000 troops involved and was unquestionably a Battle.

It was a deliberate lie to cloud not just what had happened, but the true classification of what had happened.

The Sheer Numbers Do Not Appear To Matter.

While the British High Command remained steadfast in classifying Fromelles as a raid at this time, it is worth looking separately at the statistics alone, in order to understand the magnitude of just what they did classify as a raid:

- Soldiers engaged in the action at Fromelles: Nearly 20,000.
- Across a frontage of: 4000 yards
- Field Guns to be used: 296
- Shells for Field Guns: 200,000
- Heavy Guns to be used: 78
- Shells for Heavy Guns: 15,000
- Machine Gun groups: 6
- Australians Killed in action: 1815

- from wounds sustained: 299
- or while as prisoner, from wounds sustained: 36
- British Killed in action: 533
- from wounds sustained: 279
- or while as prisoner, from wounds sustained: 2

All of which make the descriptor of a 'raid' ridiculous.

Being careless was one thing, but when it came to the Allied leadership, being callous about the battle in its aftermath was something even more unsettling. And callous it was!

No doubt there will be more to come elsewhere. So just move on. There is a war to be won after all.

After their involvement in the Battle of Fromelles, the various units of the 5th Division were so decimated that they could not be used as an effective Division for some time. Reinforcements were received, and in October / November 1916, the various Battalions of the 5th Australian Division were once more placed into front line service, and again suffered casualties, but this time they were actually *on* The Somme. They were thus moved to the geographic location to which their actions in July were attributed, but from which were geographically separate.

As already noted in Chapter 9, using reference to simple definitions, and also military definitions, The Somme was in reality, a Campaign, an operation made up of many battles. Whether called a Campaign or an Offensive, it commenced 1 July, 1916 and ended 18 November, 1916.

Not long after being moved to The Somme, the 5th Division would be sent elsewhere to serve on the Western Front. The work of the 5th Division and the British 61st Division would continue in France and Belgium until the end of the war. Nowhere else would the places where they served, be capable of any ambiguity as to location.

How Did The Germans See The Fighting at Fromelles?

While the British High Command was not prepared to describe the events at Fromelles as a 'battle', the same cannot be said about our German opponents.

Some of the most comprehensive pieces on Fromelles were written by official war correspondent Wilhelm Scheuermann.[120] On 3 August, the *North German Gazette* published a long piece entitled 'The Battle of Fromelles'.

In it we can see the conversion of Hauptmann Fritz Lubcke's interrogations into unalloyed propaganda. Naturally, Scheuermann concentrated his attentions on the Diggers; it was after

[120] Barton p.305

all, the first significant encounter with a substantial number of prisoners who had come halfway around the world to fight the German nation. The following items are significant:

- *The English troops who took part in this action were Australians. ... They had reckoned on certain victory and had counted on the superiority of their artillery of which they had been told so much.*
- *The officers made a good impression and showed themselves superior in education and bearing to the average English officer. The bravery of the men and their stamina is undoubted.*
- *The English enterprise against Fromelles has caused headshaking everywhere amongst ourselves as well as in neutral countries, and perhaps even among the French and the English themselves, for no one could imagine what the intentions of the English really were. Had it been merely a demonstration against the enormously strong German defences of Lille it would appear to have been undertaken with too large or too small a force so that the whole affair would have been a mystery had we not captured the English Operation Order. From this it appears that the English were of the opinion that we were taking the troops from this part of the front for employment on the Somme. They wanted to prevent this, to bind down our forces and, if possible, cause us to drain our reserves on the Somme.*
- *This plan absolutely came to nothing, for the Bavarian Division which was attacked gained the ascendancy by themselves, without bringing up a single gun from any other position, and without need, even for a moment, of the reserves.*

---ooo000ooo---

Of course, it can be said that the Germans had a vested interest to present a biased account of Fromelles, but it also appears that the Germans were quite capable of defining the fighting at Fromelles for what it was: a Battle. They did not see it as a *raid,* based simply on the level of fighting they were faced with. They saw it as a battle. But it became apparent to them that the purpose of the fighting was ultimately not what it appeared to be. And when all was said and done, it became a failed battle.

The Modern Australian View on The Battle of Fromelles

The Australian War Memorial's description of the fighting at Fromelles on 19 and 20 July, 1916, is stated in the opening paragraph of their website, under the heading of 'the Battle of Fromelles' as follows:

> *The battle of Fromelles on 19 July 1916 was a bloody initiation for Australian soldiers to warfare on the Western Front. Soldiers of the newly arrived 5th Australian Division, together with the British 61st Division, were ordered to attack strongly fortified German front line positions near the Aubers Ridge in French Flanders.*

The attack was intended as a feint to hold German reserves from moving south to the Somme where a large Allied offensive had begun on 1 July. The feint was a disastrous failure. Australian and British soldiers assaulted over open ground in broad daylight and under direct observation and heavy fire from the German lines.

Over 5 500 Australians became casualties. Almost 2 000 of them were killed in action or died of wounds and some 400 were captured. This is believed to be the greatest loss by a single division in 24 hours during the entire First World War. Some consider Fromelles the most tragic event in Australia's history.

A Raid, An Attack, Or A Battle, As Part of A Campaign?

Any student of English language will be able to distinguish that the amount of preparations for, the amount of munitions used in, and the sheer number of soldiers engaged in, and the objectives to be sought, styled Fromelles not as a raid. It was a battle pure and simple. That battle was an adjunct to an overall Campaign which was the Battle of The Somme. The historical classification of Fromelles as a raid up to this point[121], is due more to High Command embarrassment and a lie than any objective assessment of its components. It was nonetheless a lie.

Describing Fromelles as a raid also came about partly because the High Command was predisposed to calling all battles 'raids', until events dictated otherwise, and partly as a result of the embarrassment of the High Command in seeing yet another of their planned actions fall to pieces in front of them.

Some have commented that calling Fromelles a raid was in some way an attempt to deflect what had happened. A cover up. After all, Pozieres was about to begin in less than a week. But that does not stand scrutiny, and is really only a convenient explanation if you seek to reflect on the events as a collective exercise. The High Command's announcement on 20 July, and the reports from German sources on 20 and 21 July were made before the action at Pozieres even commenced.

Even the highest in command, Field Marshal, General Sir Douglas Haig himself had problems describing fighting with appropriate terminology. Every engagement he refers to appears to be described as a 'raid'. His approach was that of a ***'thruster'***, one who pushed and strained to break through the enemy lines, using the thinking of a cavalry rider charging into the opposition.

As recorded in General Sir Douglas Haig's Somme Dispatch of 23 December, 1916, five months later, he wrote:

---

[121] Up to this point it was called a 'Raid', but in the fullness of time, the Battles Nomenclature Committee would style Fromelles as a '*Subsidiary Attack*' to the Battle of Bazentin Ridge fought between 14 & 17 July, 1916. A change of terminology.

> *37. ….. Some idea of the thoroughness with which their duties were performed can be gathered from the fact that in the period of four and a half months from the 1st July some 360* ***raids*** *were carried out, in the course of which the enemy suffered many casualties and some hundreds of prisoners were taken by us.* ***The largest of these operations was undertaken on the 19th July in the neighbourhood of Armentieres [i.e. at Fromelles].*** *Our troops penetrated deeply into the enemy's defences, doing much damage to his works and inflicting severe losses upon him. [Emphasis added]*

The Field Marshal was noted to paint an optimistic picture of the impending success of *all* of his individual actions. He was always about to *break through, by wearing down his opponents*. What is interesting in this Dispatch, is that he elevated Fromelles (although not specifically naming it he singled out the action on 19th July) while other actions incurred much heavier total casualties.

This Dispatch was written at a time when British politicians were grouping against Haig, and had sought his return to England to give a proper account of the battle strategies that had taken place. Haig refused to leave the battle fields, but wrote his Dispatch instead.

According to Tim Travers[122]: - the major reason for Haig continuing the Battle of the Somme seems to have been his own view, derived from his Staff College lectures, that the General who held on the longest would win. Sooner or later, he thought, the enemy would break. Haig had not meant to fight a lengthy campaign at the Somme, but by August he had committed himself, and by now [in October] continuous pressure must be applied so that enemy morale might be kept at the state in which it might collapse.

But as has already been pointed out, The Somme was in reality, a Campaign, or an operation made up of many battles. So while ever the supreme commander is neglecting to acknowledge the fighting at Fromelles as a Battle in itself, those lower in the ranks would not dare to elevate it to that classification.

While we look at Fromelles in the context of the definitions of a raid or a battle, it was described with regularity as a battle in reports, newspapers, and letters from 1916 onwards. Sometimes it was called a 'stunt', which was a soldier's slang reference to a battle of no real consequence, or a poorly planned one, at the time. The reference to it being a 'raid' was seen as inappropriate by those who served, and they continued to call it a Battle from that time on. Its military historical classification will be dealt with in the coming chapter, as part of the work of the Battles Nomenclature Committee.

In 1916, every soldier involved in the fighting at Fromelles knew that what they were taking part in was more than a raid. It was a Battle. And it was very much life and death.

[122] Tim Travers: The Killing Ground p.185

While not the largest single loss of soldiers in a single day, Fromelles was notable as the worst number of casualties in a single day for Australia, ever. And while what happened at Fromelles may have been an adjunct to the Offensive which has been historically described as The Battle of The Somme, the fighting at Fromelles was then, and always will be properly described as a Battle itself.

One has to wonder then, just what 2151 Australian, and 813 British soldiers died for? While the High Command had access to soldiers, to ordnance, and to all facets of warfare, it appears that whoever in the High Command issued the statement on 20 July, 1916 did not have access to a dictionary.

---ooo000ooo---

While it can be, and has been argued that the staggering loss of soldiers' lives in battle is a criminal activity, we also need to remember that it is in the nature of war to have victories, defeats, and futile battles in between.

The actions of Captain Neligan referred to earlier in Chapter 4 is an excellent example of what was a ***raid***. The action that took place at Fromelles is an example of a ***battle***.

We will learn in the following Chapter, that although the AWM description refers to 'the battle of Fromelles', and public sentiment for what took place at Fromelles sees it as such, the action which took place there is officially classified as an 'Attack', a classification of lesser significance.

Fromelles may have been a futile battle, and at the same time, success in a strategic respect, but it should not continue to be demoted to a 'raid' when it was a Battle. Letting it continue to be classified as something lesser than it was will only continue to be disgraceful.

As for the British High Command's statement:

1 person saying the wrong thing is saying the wrong thing.

1000 people saying the wrong thing are still saying the wrong thing.

1,000,000 people saying the wrong thing are still saying the wrong thing.

\- Geoffrey Benn

---ooo000ooo---

11. Peace and Memorials

The Armistice, Peace, and the Recording of History.

On 11 November, 1918, at 11.00am, after battles fought over nearly 4 years, from 1914 to 1918, an Armistice - *"a temporary stopping of warfare by mutual agreement; truce"*[123] -was declared on what has ever since become known as Armistice Day. It has become a day of remembrance, shared around the world.

It [an armistice] is not necessarily the end of a war, since it may constitute only a cessation of hostilities while an attempt is made to negotiate a lasting peace [124]

The Great War for all intents and purposes ended on 11 November, 1918 as hostilities did not reignite.

Peace returned to Europe. Guns ceased firing. Rifles were put down. Church bells began to peal again, and families tried to find lost family members, and what was left of their property across the battlefields and villages of Western Europe.

But the clearing work, the reopening of roads and bridges, the rebuilding of the towns, villages, and hamlets smashed to pieces across the former battlefields, and supervising the transition from armies on a war stance to one of being decommissioned would not finish for many years to come.

The Imperial (later the Commonwealth) War Graves Commission -- which itself was first created during the course of the Great War -- was called upon to undertake a task which had never before been carried out in this manner. The scale, and associated high number of casualties of the war required an entirely new attitude towards the commemoration of war dead. Previous to World War I, individual commemoration of war dead was often on an ad hoc basis and was almost exclusively limited to commissioned officers. However, the war required mobilisation of a significant percentage of the population, either as volunteers or through conscription. An expectation had consequently arisen that individual soldiers would expect to be commemorated, even if they were low-ranking members of the military. That expectation continues to this day.

A committee under Frederic Kenyon, Director of the British Museum, presented a report to the IWGC in November 1918 detailing how it envisioned the development of the cemeteries. Two key elements of this report were that bodies should ***not*** be repatriated and that uniform memorials should be used to avoid class distinctions. Beyond the logistical nightmare of returning home so many corpses, it was felt that repatriation would conflict with the feeling of brotherhood that had developed between serving ranks.

The Paris Peace Conference

[123] Collins Dictionary definition.

[124] See, for example the reference to 'Armistice' searchable on Wikipedia.org.au

The Paris Peace Conference commenced soon after. The work of the Peace Conference is detailed in the extensive work of *Margaret Macmillan: Peacemakers*, and is recommended reading for those interested in the aftermath of World War 1. It is beyond the ambit of this work to examine the Peace Conference in any detail.

But for anyone familiar with Fromelles, it may come as a surprise to find that General Haig's emissary to the conference was none other than Lt Gen Haking, the same person who had presided over several of the war's most bitter actions, in which tens of thousands of British and Empire soldiers were killed or wounded, with little apparent point and precious little gain. The nightmare that was the attack at Aubers on 9 May, 1915; the near destruction of the 21st and 24th Divisions, rushed too late into the Battle of Loos on 25/26 September 1915; the slaughter of the Royal Sussex battalions at the Boar's Head at Richebourg on 30 June, 1916; and most disastrously for Haking's reputation, the appalling attack that was Fromelles on 19/20 July, 1916.

Haking conducted a successful attack at La Bassee in June 1918. Simpson writes that his "performance continued to be far more convincing than earlier in the war". On 17 October 1918 his Corps entered the line as part of Birdwood's Fifth Army. Simpson further wrote that "his promise before the First World War was never realised during it, but he undoubtedly showed far more skill with the soldiers of 1918 than with those under his command in 1915–16".

Reports in the press following the start of the peace conference reported, for example: -

27 Nov 1918: London, Monday.

Lively scenes mark the daily armistice conferences at Spa, where meetings are held in the labor union's premises.

Herr Erzberger and Winterbeldt frequently effect great irritation, striking the green table with their fists, whereupon General Mudant, who is representing Marshal Foch, more coldly exacts a pound of flesh.

General Haking, who commanded the Australians in the battle of Fromelles, represents Field-Marshall Sir Douglas Haig.

So, to this author at least, ironically the same person who was described as 'the Butcher of Fromelles' after the disaster of 1915, and again in 1916, would become an emissary for peace.

The work of the Paris Peace Conference would lead to the signing of a number of treaties, collectively referred to as The Treaty of Versailles. The features of the Treaty of Versailles were:

- The Treaty was just one of the peace treaties at the end of WW1.
- It ended the state of war between Germany and the Allied Powers.
- Other Central Powers were dealt with in separate treaties.
- Although the Armistice, signed on 11 November, 1918, ended the actual fighting, it took six months of negotiations at the Paris Peace Conference to conclude the peace treaty.
- The concluded Treaty of Versailles was signed on 28 June 1919.

Australian Divisional Battlefield Monuments

Before The BNC Acted, The Fromelles Soldiers Did Some Cataloguing of Their Own.

While the official arrangements were being undertaken for classification of individual actions, all of the Australian Divisions, - and in relation to Fromelles - the British 61st Division were undertaking that task for their own historical purposes, i.e., they did not wait for an 'official' cataloguing of the work they carried out, or a listing of the places at which they had fought.

Following the cessation of wartime standing, Sir Joseph John Talbot Hobbs, commanding the Australian Corps, had become deeply involved in the erection of memorials to the Australian divisions, having been appointed to select sites, prepare designs and arrange for construction before he relinquished command of the Australian Corps in May 1919.

Four of the five Divisional Memorial designs were his. Hobbs chose Polygon Wood for the memorial to the 5th Division (which he had commanded), and Villers-Bretonneux for the Australian National Memorial.[125]

This means that the information for the divisional memorial plaques was finalised, presumably in consultation with the divisional commanders, by May 1919.

Significantly, these names for the memorial plaques were settled before the Battlefield Nomenclature Committee was formed and before the Australian Official History was published, and reflected the divisions' views of significant battles.

1st Australian Division

[Pictures 26 & 27 supplied by Jim Munro, FFFAIF.]

[125] Biography - Sir Joseph John Talbot Hobbs - Australian Dictionary of Biography (anu.edu.au). Prior to the war, Talbot Hobbs had been President of the West Australian Institute of Architects, and in his parallel military career had been the staff officer for army engineering services in 1906-12, Colonel 22nd Infantry Brigade and aide-de-camp to the Governor-General. On the outbreak of war in 1914, Hobbs was given command of the artillery of the AIF 1st Division, and at the end of 1916 he was given command of the 5th AIF Division, and after the Armistice succeeded Monash in command of the Australian Corps.

The memorial commemorates the officers, non-commissioned officers and men of the First Australian Division who fought in France and Belgium in 1916, 1917 and 1918. The commemorative tablet on the memorial [below right] lists the major battles for the 1st Australian Division.

Battle Names on the 1st Division Memorial, Pozieres, France.

POZIERES – MOUQUET FARM – LE BARQUE – THILLOY – BOURSIES – DEMICOURT – HERMIES – LAGNICOURT – BULLECOURT – 3 RD BATTLE OF YPRES – MENIN ROAD – BROODSEINDE RIDGE – PASSCHENDAELE – BATTLE OF THE LYS – 2 ND BATTLE OF THE SOMME – LIHONS – CHUIGNOLLES – HINDENBURG LINE.

2nd Australian Division

Positioned at Mont St Quentin, where the Australian 2nd Division captured one of the most formidable defensive positions on the Western Front, the 2nd Division's memorial was unconventional. Instead of an obelisk such as at the other four AIF divisional memorials, the original memorial which was unveiled in 1925 was a statue of an Australian soldier bayoneting a German eagle sprawled at his feet. However, this statue was removed and destroyed by German soldiers in 1940 during World War II, leaving only the stone plinth. A replacement statue, consisting of an Australian soldier standing in full kit was installed in 1971.

[Pictures 28 & 29, supplied by Jim Munro, FFFAIF.]

Battle Names on the 2nd Division Memorial, Mont St Quentin, France.

POZIERES – MOUQUET FARM – FLERS – MALT TRENCH – LAGNICOURT – BULLECOURT –MENIN ROAD – BROODSEINDE RIDGE – PASSCHENDAELE – VILLE-SUR-ANCRE – MORLANCOURT – HAMEL – VILLERS-BRETONNEUX – HERLEVILLE – HERBECOURT – BIACHES – MONT ST QUENTIN – BEAUREVOIR LINE – MONTBREHAIN.

3rd Australian Division

The memorial to the 3rd Australian Division, stands at Sailly-le-Sec on the road between the Somme towns of Corbie and Bray.

On 2 February 1916, after the AIF had withdrawn from Gallipoli and was being reorganised in Egypt, the Australian Government agreed to form a third Division. Assembly began in March

1916 and the Division began to arrive in England in July, 1916. After intensive training it crossed to France in December, 1916 and thereafter fought in many of the major actions of the war. The 3rd Australian Division was initially under the command of then Major General John Monash, who went on to command the Australian Corps and who is generally regarded as amongst the finest military commanders of the war. He was succeeded on 31 May 1918 by Major General John Gellibrand.

[Pictures 30 & 31, supplied by Jim Munro, FFFAIF.]

Battle Names on the 3rd Division Memorial, Sailly le Sec, France.

MESSINES 1917 — THE WINDMILL — 3RD BATTLE OF YPRES —BROODSEINDE RIDGE — PASSCHENDAELE — MORLANCOURT — TREUX — HAMEL — 8 th AUGUST — PROYART — SUZANNE — BRAY-SUR-SOMME — CURLU — CLERY-SUR-SOMME — BOUCHAVESNES — ROISEL — HINDENBURG LINE.

4th Australian Division

For the men of the Fourth Australian Division the war ended on a hill looking south towards the village of Bellenglise, where they built their memorial in France.

This is an isolated place, little frequented by Australian battlefield visitors. The division considered other locations for the memorial, such as Dernancourt, where in April 1918 they had stopped the advance of a greatly superior German force; or Pozières, where in August 1916 they suffered heavy casualties.

But Bellenglise was chosen because, according to the division's own statement, it was 'the culminating point of the Fourth Australian Division's work in the war ... [and] also the scene of an extremely successful operation carried out during the period 18th-24th September 1918'.

[Pictures 32 & 33, supplied by Jim Munro, FFFAIF.]

Battle Names on the 4th Division Memorial, Bellenglise, France

As the plaque on the memorial records, the Fourth Division took part in most of the great battles fought by the Australian Imperial Force on the Western Front.

SOMME — POZIERES — BULLECOURT — MESSINES — YPRES — MENIN ROAD — POLYGON WOOD — PASSCHENDAELE — ARRAS — ANCRE — VILLERS-BRETONNEUX — HAMEL — AMIENS — ALBERT — HINDENBURG LINE — EPEHY.

5th Australian Division

In 1919, the soldiers of the 5th Division AIF nominated the Battles they were involved in for inscription on their own memorial. That memorial stands at Polygon Wood, near Zonnebeke, Belgium. The 5th Division erected its memorial there, not because it was toward the end of their involvement in WW1, but because it was the location of their greatest success.[126]

The memorial stands on top of the buttes, which have given their name to the cemetery below. The original contained a number of German shelters which would later be used by the British as a Brigade Headquarters.

The dedicatory plaque on the front face of the memorial has only a few words in English and French.

"To the Officers, Non–Commissioned Officers and Men of the 5th Australian Division who fought in France and Belgium, 1916, 1917, 1918."

[126] A reference regarding the 5th Division Memorial has been found in the CWGC archive files, from Major Ingpen, in which he states "that whilst travelling back to Ypres at the beginning of October 1919, he was surprised to see the new memorial in place." Although a more precise date is unknown it appears to have been erected/completed in September 1919. Also see the following photos of plaques from the AWM Collection donated by J T Hobbs dated October 1920.

Beneath the dedication are listed seventeen (17) battle locations in France and Belgium. The memorial reflects the Division Commander, General Talbot Hobbs' view of the most important engagements of the Division.

Fromelles has premacy as it was the first of the actions in which the Fifth Division were engaged on the Western Front. The last is Joncourt, after which, the 5th Division were not involved in any further fighting.

[Picture 34, supplied by Jim Munro, FFFAIF.]

The memorial was unveiled by the Prince of Wales on the 10th May 1932, but resources suggest that the obelisk was already in place by 1921, making it the first memorial in Zonnebeke, Belgium.

Buried within the Cemetery is Lt Colonel Alan Scott of the 56th Australian Infantry who was killed just to the left of the butte whilst handing over to Lt Colonel Dudley Turnbull of the 20th Manchester Regiment. The same bullet killing both colonels.

Battle Names on the 5th Division Memorial, Polygon Wood, Belgium

FROMELLES – BAPAUME – BEAUMETZ – DOIGNES – LOUVERVAL – BULLECOURT – 3RD BATTLE OF YPRES – POLYGON WOOD – BROODSEINDE – BATTLE OF AMIENS – VILLERS-BRETONNEUX – MORLANCOURT – 8TH AUGUST – VAUVILLERS – PERONNE – HINDENBURG LINE – JONCOURT.

[Picture 35: 5th Division Memorial, Polygon Wood. Photo by: Geoffrey Benn 20.07.2016.]

VC Corner Cemetery, Fromelles

Commemorating The Missing From Fromelles.

This cemetery came into existence solely as a result of the fighting that took place at Fromelles on 19/20 July, 1916. After the end of World War 1, the battlefields were combed over by groups appointed by the Graves Registration Unit, of the

Imperial War Graves Commission. All the places at which soldiers had been known to have fought were walked over and sites were searched for makeshift graves of soldiers so that all could be consolidated into more lasting war cemeteries.

In the case of Fromelles, the land on which the 250 soldiers were buried at Pheasant Wood was not examined, as soldiers had not fought on that land. Even if it had been, the land would most likely have returned to its flattened nature, and would not have revealed itself as a place of burial.

[Picture 36. Original source unknown.]

In what was No-man's land during the Battle of Fromelles, the remains of 410 discrete soldiers were found but were unable to be identified. Their bodies had long decomposed, leaving only their skeletal remains, and the remnants of any uniform or soldier's equipment that lay beside them. Those remains would be buried separately, but in two pits at VC Corner Cemetery.

VC Corner is the area in which during December, 1914, British soldiers fought so valiantly, that Victoria Crosses were awarded to Privates James McKenzie, Abraham Acton, and James Alexander Smith for their valour. It was effectively the site at which fighting ceased, only for it to feature again in 1915 and 1916.

But there were many other soldiers who were not found, identified, or accounted for. The VC Corner Cemetery came into existence as a place at which all those who could not be found or identified from the Battle of Fromelles on 19/20 July, 1916 could be memorialised.

Some soldiers had been officially recorded as being killed after 19/20 July, so their names were inscribed on the Walls of Remembrance at Villers-Bretonneux War Memorial, despite having served in the Battle of Fromelles from which they went missing. Some of them have since been identified as some of the soldiers recovered from Pheasant Wood in 2009 and reburied in 2019. Others may yet be found to be.

Although the memorial wall refers to the names of 1299 soldiers, the number who should have been immortalized there totals 1356 when all those who are memorialized at Pheasant Wood, at Villers-Bretonneux, and at VC Corner are combined.

It is the only Commonwealth War Graves Cemetery at which only Australian soldiers are named.

61st United Kingdom Division

The UK 61st Division has also recorded its memories nearby, at Laventie. They erected their own memorial, without reference to any place of battle. It was erected by the British in the Town Square of Laventie in 1937, *"to commemorate fallen comrades".*

In a truly egalitarian fashion, no mention was made of officers, non-commissioned or the men, just to 'fallen comrades'.

The 61st Division was a second-line Territorial formation raised in Northampton in January 1915 as a reserve for the first-line battalions of the 48th (South Midland) Division. The men who volunteered for service with the division when it was first formed hailed mostly from Birmingham and the towns and villages of the Malverns, the Cotswolds and the Chilterns. The units they joined included newly formed battalions of the Royal Berkshire, Royal Warwickshire, and Gloucestershire regiments, as well as the Oxford and Buckinghamshire Light Infantry.

[Picture 37: 61st Division Memorial, Laventie, France]

[Geoffrey Benn, 20.07.2024.]

Laventie was the location where the 61st Division began its training on the Western Front, which would involve Fromelles, in 1916. It was thus located near its first official engagement, in much the same way as the Fifth Division had nominated Fromelles as the first of its notable engagements.

In the first few hours of the Battle of Fromelles, over 1,500 men of the 61st Division had been killed or wounded. Over 80 members of the Division who were killed or mortally wounded during the Battle of Fromelles are buried in the Laventie Military Cemetery, at La Gorgue, and 395 of whom are remembered on the walls of the Dump Cemetery and Loos Memorial at Loos-en-Gohelle 24 kilometres to the south.

The Division later erected another memorial in Holy Trinity Church in Stratford-upon-Avon in Warwickshire, the logic for which is that it sits centrally in the areas from which the soldiers of the 61st (2nd South Midland) Division were recruited.

There is also a memorial, a joint one with the 74th (Yeomanry) Division, at St Floris, France.

---ooo000ooo---

12. Cataloguing the War, Classifying the Actions. - The Battles Nomenclature Committee

History will be kind to me, because I intend to write it.

- *Winston Churchill.*

---ooo000ooo---

If Fromelles is part of The Somme, we may just as well say Kalgoorlie is part of Bondi.

- *Geoffrey Benn*

---ooo000ooo---

By 1919, troops in a multitude of fighting forces were being de-mobbed, sent home, and decommissioned. In Britain, the job of the Minister for Munitions, Winston Churchill, finished, and he became the Secretary of State for War and Air with responsibility for demobilisation of Britain's armed forces. More than three million men were processed back into civilian life in the first six months.

A similar process commenced for the Australian Imperial Force under the leadership of General Sir John Monash, now Director General of Repatriation and Demobilisation, supported by Brigadier General CH Jess CMG, DSO, who in July 1919 had become Commandant at AIF Headquarters, London, and officiated as General Officer Commanding, AIF, relieving General Sir William Birdwood, for several months until January 1920. He additionally acted as Director General of Repatriation and Demobilisation, as vice for Monash when required.

1919: The Battles Nomenclature Committee

As a consequence of the end of the war, in August, 1919 Britain's Army Council appointed the ***Battles Nomenclature Committee*** with the following Terms Of Reference: -

(a) To tabulate the actions fought in this war.

(b) To classify these actions with a definite system of nomenclature which will denote their relative importance (e.g. the more important actions might be classed as battles, and the lesser ones as combats, &c.)

(c) To define the geographical and chronological limits of each action. [Report p.3]

The Report, when it was signed off on 9 July, 1920 was given the grandiloquent preamble and conclusion of: -

THE OFFICIAL NAMES OF THE BATTLES AND OTHER ENGAGEMENTS

FOUGHT BY THE MILITARY FORCES OF THE BRITISH EMPIRE

DURING THE GREAT WAR, 1914-1919, and THE THIRD AFGHAN WAR, 1919.

REPORT OF THE BATTLES NOMENCLATURE COMMITTEE,

AS APPROVED BY THE ARMY COUNCIL.

It is commonly abbreviated to simply "the **BNC Report**", which we will do as well.

The committee's report in due course would (rightly or wrongly) define the names, the dates, and the boundaries of important engagements for each theatre of war.

Its work would go on to assist with the defining of Battle Honours by a separate organisation, the ***Battle Honours Committee***, which will be commented upon in coming chapters. But it is with specific reference to Fromelles, that the work of the BNC is considered to be left wanting.

Other than the input of the various Australian members of the Committee, little is recorded as to the interest taken in the ruminations of the committee, in particular with reference to Australian forces involved in actions across the war.

Constitution of The Battles Nomenclature Committee

The BNC first met on 18 August 1919, with Major General Sir John Headlam as President, and five permanent members. One was appointed by the General Staff and of the remainder, one each from Australia, Canada, New Zealand and South Africa. Of the three Australians who served, one was born in England, one in India and one in Queensland. All were highly decorated.

Permanent Members from Australia were: -

- Brigadier General C.H. Jess CMG, DSO, 3rd Division, AIF from 18 to 20 August.
- Lt. Colonel J.L. Whitham, CMG, DSPO, 12 & 52 Battalions, AIF from 19 August to 7 October, 1919, and from 31 November, 1919 to 16 January, 1920.
- Colonel R. Dowse, DSO, Qtr Master 12 Brigade, 4 Div., AIF, from 7 October 1919 to 2 January, 1920.
- Lt. Colonel G.C. Somerville, CMG, DSO, Army HQ AIF, from 26 to 30 January, 1920.
- Lt. Colonel E.C. Plant, DSO, OBE, HQ 1 Div., AIF, from 9 February to 9 July, 1920.

For France and Flanders there were two British members, appointed also by the General Staff. Secretaries came from the (British) Historical Section of the Committee of Imperial Defence which provided information plus advice on the principles and practice of nomenclature and classification.

Temporary Appointees, appointed by the General Staff, for France and Flanders were: -

- Lt. Colonel B.W.B. Bowdler, DMG, DSO, Royal Engineers; and
- Lt. Colonel W.G.S. Dobbie, DSO, Royal Engineers.

Circumstances necessitated several changes to the membership of the BNC Committee, with service from appointees of the General Staff, Australian, British, Canadian, New Zealand, and South African forces.

It was noted that:

"The temporary members mentioned above were appointed to assist the Committee in the consideration of particular campaigns, being specially selected for their local knowledge. For the other theatres of war temporary members were not appointed, but assistance was obtained as required from officers who had served in the campaign under consideration."

---ooo000ooo---

Extracts from the general text of the Committee's Report are reproduced in the following paragraphs. Many of the comments made by the Committee are deserving of support, but many also are cause for criticism in view of the history of the events of the Great War, and what we have come to know in the years since.

Procedure [Report p.4]

The first session was devoted to a general discussion of the scope of the work, an examination of the correspondence on the subject, and a discussion of the principles to be followed. The various campaigns were then studied in detail with the aid of both documentary and personal evidence.

Worthy of note, at page [5] of the Report, is the explanation of the use of: -

Personal Evidence

– To supplement the documentary evidence, the Committee have taken full advantage of the permission to call upon individual officers with special knowledge for their assistance. In most cases these officers attended meetings of the Committee and took part in the discussions, and several were good enough to describe the course of the particular campaigns in lectures. At an early meeting the Committee decided that the evidence of these officers should not be recorded. In coming to this decision the Committee were prompted by two main considerations. In the first place, they felt that they would be able to obtain a better idea of the actual circumstances by informal discussion than by formal evidence. In the second, they considered that only in such a way could they obtain the benefit of hearing the views of officers of high ranks without abrogating their own responsibility for their recommendations.

Personal evidence supplemented documentary evidence. Various letters are enclosed with the Committee's historical records within the National Archives of the United Kingdom.

Evidence was taken informally but not recorded. Thus, the actual basis on which individual actions would be assessed, remains unknown.

Definition of Limits: both geographical and chronological.

As regards *chronological* limits it was decided -

That no attempt should be made to define the time more closely than by the "day" (midnight to midnight).

That, owing to the varying length of the preparatory action, and the difficulty of fixing its actual commencement, this period should not be included, but that "zero day" should be taken as the first day in all cases.

That, on the other hand, the period of consolidation after an attack should, as a general rule, be included. [Emphasis added.]

Those who were supposedly 'Killed in Action' on 21 July, 1916 and referred to in Chapter 9 should therefore have always been included in the details of Battle at Fromelles.

As regards *geographical* limits the Committee decided *to denote the boundaries of battle areas by easily recognisable geographical features which could be followed on such generally accessible maps as those of the 1/100,000 series. In doing so they have been influenced by the conviction that, for the purposes of this Committee, a meticulous accuracy was of less importance than that their recommendations should be generally intelligible without the use of special maps.*

Their stated general principle was:

*'...to include in a battle area only what might be fairly regarded as the **actual "battlefield"**.'*

All actions which the BNC decided took place outside the geographical limits laid down as above were to be regarded as separate engagements, defined as '**subsidiary actions**', and as such: -

',,,should be considered on their merits as separate engagements. In cases, however, where such engagements were definitely undertaken as contributing to the battle this has been indicated.' [127]

Fromelles was fought as part of the Third Phase of the war, and falls within the sequence of fighting tabulated as 'III - The Allied Offensive, 1916'. This also became the case in the work of Captain E.A. James in his text 'A Record of the Battles and Engagements of the British Armies in France and Flanders, 1914-1918', which was published later, in 1924. Sadly, in both the BNC Report and Captain James' work, the focus of the works is "The Battles of The Somme", and take no account of French Flanders as a separate geographical area.

Having any battle fought out during that particular time period tended to label all as being part of "The Somme" when with Fromelles, this was a geographical impossibility. Yet the absurdity continued. Fromelles was well outside the actual battlefields of the Somme. It was 80 kilometres away from Bazentin Ridge to which it has been attached as a "subsidiary attack". It cannot in any sense be logistically subsidiary to any action that distant, let alone one which had finished before Fromelles had even begun.

And the BNC was at pains to justify the application of varying criteria to different theatres of war, distinguishing engagements in France and Belgium from other localities.

Classification of Engagements.

5. It is in devising a system of classification which would be generally applicable that the Committee have been confronted with the most difficult problem. This is due, of course, to the very different conditions prevailing in the different theatres of war. If a scale suitable for some of the minor theatres had been strictly

[127] Report p.6.

applied to the Western Front, it would have involved an innumerable list of "raids" and similar minor operations. If the process had been reversed, it would have meant omitting all mention of any but one or two actions in wars of some years' duration.

The Committee came early to the conclusion that any attempt to classify the engagements in all the theatres of war on hard and fast rules based on the numbers engaged, the casualties suffered, or any similar foundation must work unfairly in practice. The principles they have followed in their classification are explained in the following paragraphs.

Sub-Division of Battles.

*6. The extension of the period covered by the so-called battles in Europe from days to weeks, and even to months, has introduced an entirely new factor, and the term "battle" in the ordinary pre-war application of that word is obviously no longer sufficient. From the dispatches it is plain that the Commander-in-Chief's conception of such operations as the so called "Battle of the Somme" (1916), each delivered with the full force available, after the necessary pause for preparation. The Committee have endeavoured to present them in this form by resolving them into their successive stages: - first the main attack – then the succeeding general attacks marking the opening of new phases. Each of these constitutes in itself a "Battle" of the first magnitude, and the whole operation thus comes to form a "Group of Battles". * (*Compare the French term "Battaille d'ensemble.") This is, after all, only a parallel to the changes on organization. The expansion in the size of the field armies has necessitated their subdivision, not only into Armies, but into "Groups of Armies." We spoke of the "Battles of the Somme." This principle has accordingly been followed by the Committee in all such cases.*

Even with this arrangement, many of the individual battles will still be on so grand a scale that the Committee have seriously considered the possibility of further subdivision by breaking up the front of the great simultaneous attacks and denominating the struggle for each distinctive feature a separate battle. This would no doubt appeal to many units – from corps to battalions – to whom a battle fought on a wide front often means only the fighting for some definite locality. The Committee have the greatest sympathy with such feeling, and have always been anxious to give due regard to them, but it is their duty to consider the commanders who planned as well as ***[P7]*** *the units who executed, and any such subdivision of the front is liable to give a false impression as regards the unity of the plan. It would be misleading if such attacks as those on the 1st July 1916, and the 31st July 1917, were represented as a series of disjointed actions. Moreover, to name even all the villages which were scenes of desperate conflicts would carry such subdivision beyond all reason. Thus at once would arise the risk of injustice to those units who fought just as gallantly for less prominent localities, or for some nameless trench.*

After some discourse on the perception of whether battles accrue to one place or to more, to one unit or more, the committee then stated:

The Committee, therefore, decided that battles fought under a ***single plan*** *on* ***a continuous front*** *should not be broken up, but should each constitute a separate battle.*

At the same time, they should have mentioned some of the more important tactical incidents of the battles when this could be done without risk of obscuring the greater issues involved. It must not, however, be inferred that the fighting at the places was necessarily more severe than on other parts of the front, or that the number of such incidents recorded is any criterion of the greatness of the battle itself.

What was meant by 'a Single Plan on a Continuous Front'?

This part of the Committee's explanation is a master statement of confusion. For example, how can it be said that an action fought at a distance of 80 kilometres from others is fought on a 'continuous front'?

It should be borne in mind that Fromelles was the 'single plan' of the First and Second Armies to the north and south of Fleurbaix/Fromelles. On the Somme there were the Fourth Army and the Reserve Army (later to become the Fifth) Army.

Between the First and Second Armies, and the Fourth and Reserve Armies was the Third Army

There are twelve (12) catalogued Battle Names recorded by BNC. Are we to accept that the 'single plan' is the fighting of all the various actions listed under "The Allied Offensive, 1916"? If so, we might just as easily refer to 'The Great War', and include everything that happened anywhere.

But clearly the action at Fromelles was not fought on the same continuous front as the other battles which made up the Campaign on The Somme. It was fought in the French Flanders region to supposedly provide a distraction to prevent German soldiers being moved to the Somme.

And it is further confusing as to how separate battles 'should not be broken up', when there are twelve catalogued Battle Names listed in the report. Battles occurring within the designated twelve battles did take place, and in some cases, led later to the Battle Honours Committee having to reflect that. But for battles, we have been left with the confusion arising from the BNC Report.

The justification for the various actions being described as 'battles' is misconceived. It has resulted in a process of rationing of recognitions, not rationalisation of the actions fought. The process of classification should not result in the categorisation of a little 'battle', or a medium-sized 'battle', or a large-sized 'battle', or an extra-large-sized 'battle', or a huge-sized 'battle'. It should have simply acknowledged just that a particular action was in fact a battle. And Fromelles was indeed a battle.

Terminology.

When reading this material from the BNC Report, you should recall the ordinary definition of various words which have been set out already in Chapters 1 and 10.

7. As regards the terms to be employed in describing the different engagements, the Committee have had great difficulty in arriving at any system which would adequately cover the very wide range required.

*The only names in the military terminology which convey any real indication of proportion are "battle," "action," and "affair," and these are obviously inadequate to convey all the graduations. As explained above, the Committee decided to supplement them by adopting the plural - "Battles" - for those prolonged operations which were such a feature of the later stages of the war on the Western Front. Even with this addition a gap was still left, for there were no means of differentiating between battles so different in size as, say, Messines and Ctesiphon. The Committee carefully considered the possibility of recommending the use of some such terms as "super-battle" or "battle-royal" * (*Compare the old German term of "Kaiserschlact.") to denote those battles which did not come within the category of "groups of battles," but were yet vastly greater in scale than any single "battle" of the past. It was felt, however, that the objections to such compounds outweighed any technical advantages that might be gained by their use.*

A great deal of time was devoted to terminology. The words "battle", "action", "affair", "capture", "occupation", "attack", "defence" etc. were analysed and the decision taken that:

In the use of the above terms, and also of what may be called descriptive terms, such as "Capture," "occupation," "attack," "defence," &c., the Committee have endeavoured to maintain a certain consistency. ***Thus, the rank of "battle" has been confined as a general rule to engagements of primary importance fought out between forces not smaller than the corps.*** *[Emphasis added]*

A "**Corps**" was not defined. But by reference to the United Kingdom's National Archives, contained within the search result for "British Army Operations in the First World War" the reader will readily find that:

- a Corps consisted of "two or more divisions".

Similarly, a dictionary search for 'Corps'[128] will result in the definition of:

- An operational formation, sometimes known as a **field corps**, which consists of two or more divisions, such as the *Corps d'armee*, later known as *1 Corps* ("First Corps") or Napoleon's *Grande Armee*. It would consist of between 20,000 and 40,000 soldiers and be typically led *by a Lieutenant General* (which was the very case with Lt Gen Haking being in charge at Fromelles).

As we have read in Chapter 6, the action at Fromelles was fought out by XI Corps, of two Divisions. The only battle / engagement / action involving 2 assaulting divisions elsewhere than on The Somme in the entirety of 1916 was at Fromelles.

This being so, it would have been appropriate to consider the action which took place at Fromelles as a **'battle'**.

They may have more readily classified Fromelles as a "Battle" had Fromelles been a tactical success on the ground.

The Committee went on to say: -

---

128 https://www.macquariedictionary.com.au/features/word/search/?word=Corps&search_word_type=Dictionary; https://en.wikipedia.org/wiki/Corps & https://en.oxforddictionaries.com/definition/corps

*The title "**action**" has been employed for the next class, the limit in this case being taken as the division; lesser engagements have been styled "**affairs**."*

The term "combat" has been avoided as having reverted to its original use of a hand-to-hand fight, between individuals, and so become obsolete in the sense in which it was employed in the last century. "Capture" has been reserved for operations where the primary object was the capture of a definite locality, and where this was only accomplished by actual fighting of some importance; where this latter condition was absent "occupation" has been used, and the same distinction has been drawn between "passage" and "crossing."

And then the Committee defined an "***attack***":

*Again, the term "**attack" has been confined to unsuccessful offensive operations**, and **"defence" to successful defences of definite localities**, the term "German (or other) attack" being used in cases of simple resistance to an attack. In some cases, the Committee have made use of the expedient of combination to give recognition to the cumulative effect of a series of small collisions where – as often in bush fighting – a number of these have taken the place of pitched battle.*

The BNC Got it Wrong.

This is where and how the BNC plainly and simply got it wrong in relation to Fromelles.

It was seen at the time as being an attack that was an 'unsuccessful offensive operation', even though it would properly be described as a battle, having been fought by a Corps.

The BNC's errors in relation to the proper classification of the Battle of Fromelles also came about because the BNC relied on input from Australian representatives (none of whom had been involved at Fromelles), and the two Temporary Appointees for France & Flanders, who likewise were of little assistance.

This lack of information on which a proper determination would have taken place effectively continued until CEW Bean published the Official History Vol III, chapters XII and XIII, France 1916, in 1929.

And then, at the conclusion of these deliberations, the BNC arrogated to itself a licence to be as perverse and flexible as it wished, reaching this conclusion: -

These, and other rules which have been formulated from time to time by the Committee, have, however, been intended only as general guides in applying the terms available, and they have not hesitated to depart from them where such a course appeared to be justified by the circumstances. The proportion of the troops engaged to the whole force in the field has been taken into account; moral and political effect has been considered; and the prescriptive right acquired by some engagements which have seized the popular imagination to higher titles than are strictly warranted has been recognized.

Those *general guides* have had the perverse effect that they have in the years since, been held to be quite inflexible. Fromelles was not seen as one of those engagements at that time which had seized the popular imagination to higher titles, at least not by the BNC. But by the misinterpretation of their own terms of reference, the BNC's work would continue to deny proper recognition of the actions of those of the Australian 5th and the British 61st Divisions at

Fromelles as a Battle.

Then followed on pages 7 and 8, the arguments for not using hyphenated names, concluding that year names (such as 'Aisne 1918') were to be preferred rather than 'first battle', 'second battle' and so on.

It is clearly apparent that the BNC would only award a '**battle**' recognition to a successful action. This will be seen as inconsistent below, when Fromelles is compared to other unsuccessful actions which have been classified as Battles.

Fromelles was seen as a tactical failure, given that no ground was won. But it was in some respects, a strategic success, despite the toxic criticism by Lt Gen Haking of the soldiers who fought, because there is clearly evidence that various German units in the French Flanders region did not move for some weeks following the battle at Fromelles. The strategic purpose of the feint had been successful. Soldiers were not moved to The Somme.

The Conclusion of The BNC Report

The Report concludes:

9. A tabulated list of the engagements fought during the War, classified and with their geographical and chronological limits defined as directed, is attached.

As a general rule no reference has been made to the work of the Navy or Air Force, or to Allied troops fighting in line with our own. In some few cases, however, where the other forces have been small, and where the co-operation has been very close, their engagements have been included.

It adds that it has not always followed its own rules and principles, justifying this on the grounds that the application of hard and fast rules would in many cases be unjust and that their recommendations do substantial justice:

*By an arrangement in several columns, by the use of different term and of varieties of type, the Committee have endeavoured to indicate the relative importance of events, in accordance with the principles explained in the preceding paragraphs. It is fully recognised that many cases may be found, which are not entirely in accord with these principles. [Emphasis added]. The Committee believe, however, that any attempt to apply hard and fast rules to such a **[P9]** complex subject would result in much greater inconsistencies, and would, moreover, in many cases be unjust. The members of the Committee have themselves had very varied experiences during the war; and they have had access to the official documents as well as the assistance of a large number of officers specially qualified to speak on the subject; they have carefully considered each case in all its aspects, and they believe that their recommendations do substantial justice.*

JOHN HEADLAM, Major General, President

H. FITZ M. STACKE, Captain, Secretary

9th July 1920.

There then followed the BNC's **TABULATED LIST OF ENGAGEMENTS** in which Fromelles comes in for classification grouped under the heading of **FRANCE AND FLANDERS** at page

16 (see image below), as follows:

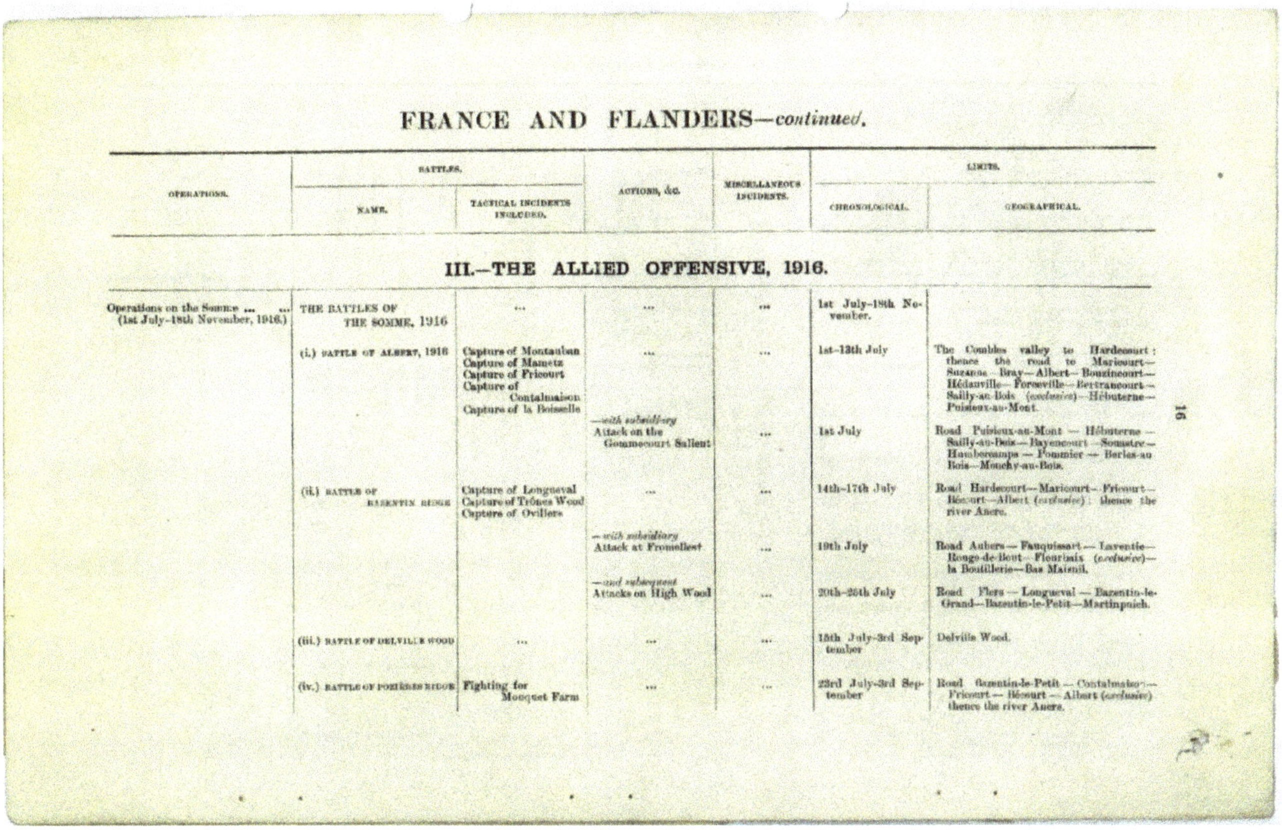

FRANCE AND FLANDERS—*continued*.

III.—THE ALLIED OFFENSIVE, 1916.

OPERATIONS.	BATTLES. NAME.	BATTLES. TACTICAL INCIDENTS INCLUDED.	ACTIONS, &c.	MISCELLANEOUS INCIDENTS.	LIMITS. CHRONOLOGICAL.	LIMITS. GEOGRAPHICAL.
Operations on the Somme ... ... (1st July–18th November, 1916.)	THE BATTLES OF THE SOMME, 1916	...	...	...	1st July–18th November.	
	(i.) BATTLE OF ALBERT, 1916	Capture of Montauban Capture of Mametz Capture of Fricourt Capture of Contalmaison Capture of la Boisselle	...	...	1st–13th July	The Combles valley to Hardecourt: thence the road to Maricourt—Suzanne—Bray—Albert—Bouzincourt—Hédauville—Forceville—Bertrancourt—Sailly-au-Bois (*exclusive*)—Hébuterne—Puisieux-au-Mont.
			—*with subsidiary* Attack on the Gommecourt Salient	...	1st July	Road Puisieux-au-Mont — Hébuterne — Sailly-au-Bois—Bayencourt—Souastre—Humbercamps — Pommier — Berles-au-Bois—Monchy-au-Bois.
	(ii.) BATTLE OF BAZENTIN RIDGE	Capture of Longueval Capture of Trônes Wood Capture of Ovillers	...	...	14th–17th July	Road Hardecourt—Maricourt—Fricourt—Bécourt—Albert (*exclusive*): thence the river Ancre.
			—*with subsidiary* Attack at Fromelles†	...	19th July	Road Aubers—Fauquissart—Laventie—Rouge-de-Bout—Fleurbaix (*exclusive*)—la Boutillerie—Bas Maisnil.
			—*and subsequent* Attacks on High Wood	...	20th–25th July	Road Flers — Longueval — Bazentin-le-Grand—Bazentin-le-Petit—Martinpuich.
	(iii.) BATTLE OF DELVILLE WOOD	...	...	...	15th July–3rd September	Delville Wood.
	(iv.) BATTLE OF POZIÈRES RIDGE	Fighting for Mouquet Farm	...	...	23rd July–3rd September	Road Bazentin-le-Petit — Contalmaison—Fricourt — Bécourt — Albert (*exclusive*) thence the river Ancre.

16

[Picture 38: Page 16 BNC Report]

Column 1 is headed "Operations on the Somme".

Column 2 contains the name of the relevant BATTLE. In this instance Battle of Bazentin Ridge.

Column 3 contains the Tactical Incidents Included in said BATTLE: Capture of Longueval, Capture of Trones Wood, Capture of Ovillers.

Column 4 contains details of relevant ACTIONS: **with subsidiary attack at Fromelles (On the Aubers Ridge),**

Column 5 lists Miscellaneous Incidents. None are listed in this instance.

Column 6 lists the Chronological Limits of the event so described: 14-17 July 1916 for Longueval, Trones Wood, and Ovillers.

19th July, 1916 for Fromelles.

Column 7 lists the Geographical Limits of the event so described.

For Longueval, Trones Wood, and Ovillers: Road Hardecourt -- Maricourt – Fricourt – Becourt – Albert – (exclusive): thence the river Ancre.

For Fromelles this is: Road Aubers – Fauquissart – Laventie - Rouge-de-Bout – Fleurbaix (exclusive) - la Boutellerie - Bas Maisnil.

Earlier in the list under the heading "II - TRENCH WARFARE 1914-1916" on page 15 of the BNC Report, Fromelles is mentioned as a Tactical Incident as 'Attack at Fromelles in the Battle of Aubers Ridge 9 May 1915. Named geographical locations are the same as above. Also listed under the same Trench Warfare heading is the Battle of Loos 25 September 1915, fought in this general area with Bas Maisnil again mentioned by name. Consistent with the geographical location of events which took place, there is no mention among these listings as being within a distant geographical area such as The Somme. Nor was there any need to refer to them as being anything other than Battles where they took place, and not as having some inexplicable connection with another geographical region altogether.

The word 'definite' in the BNC's second Term of Reference is rendered devoid of meaning when departure from it is judged as essential, simply to avoid being unjust. In the event, a substantial and continuing injustice has resulted by the BNC connecting Fromelles with The Somme.

What is 'definite' is that the 'actual battlefield' is geographically far removed from all the other nominated 'battlefields' within the area of the 'Allied Offensive 1916' on the Somme. The boundaries at Fromelles were correctly named but the justification for its juxtaposition 'on the Somme', unlike the two British Battles there in May and September 1915 (which remained in situ) appears from the BNC reasoning to its being 'definitely' [that flexible word again] undertaken as contributing to the battle!

But, if Fromelles is part of the Somme, we may just as well say Kalgoorlie is part of Bondi.

It can only be for reasons of chronological sequence recording that Fromelles is connected to Bazentin Ridge. But this overlooks that the planning for Fromelles was exclusive of that for Bazentin Ridge, as was the geographical location. It becomes plain to conclude that the BNC simply associated Fromelles with Bazentin Ridge because Bazentin Ridge had just concluded. They must have concluded that Fromelles had some sort of supporting role.

But no reason is given and at the very least it seems incongruous that the Battle of Fromelles, 19 July, could have made any contribution to or have had any influence whatever upon the Battle of Bazentin Ridge which commenced 14 July and *ended* on 17 July, just two whole days before the Battle of Fromelles began. To equate 'justice' with that sort of 'flexible definiteness' is beyond bizarre.

Another justification for transplanting Fromelles from far away French Flanders down on to the Somme was that although it was not classified as a battle, it came within their definition of a 'battle' (i.e. The Somme) fought under a single plan on a continuous front. Therefore, it (The Somme) should not be broken up.

One will note when looking at the location of the various geographical locations named in the BNC Report for 1916, that they are within the geographical area of The Somme, *except for Fromelles*. The BNC listing points this out with a footnote drawing the reader's attention to the fact that Fromelles is "On the Aubers Ridge".

There was, and readers will have no difficulty in recognising Fromelles as, a battle. But the 'single plan' concept strains credulity. One wonders what kind of competent High Command could plan to mastermind the debacle at Fromelles, which occurred two days later than scheduled, and yet classify it as an action subsidiary to a battle which had already ended two days earlier.

Moreover, the 'continuous front' was separated by great distance and two separate Army Commands. This significant discontinuity between the Somme front [of some 13 miles] and the Fromelles front [of some 2 miles] was about 40 miles at the nearest point.

On its face, the Report, by its own definition, inferentially identified Fromelles as a battle.

And despite being suppressed in historical significance by BNC as an 'attack', the Official History of the AIF in the Great War has 2 Volumes devoted to The Battle of Fromelles. It is not an action which is easily diminished.

Fromelles deserves to be remembered separately from The Somme, and as such by inscription in its own right. Not as part of a larger offensive or campaign.

Comparison With Other Battles

Some comparison of the various actions at Fromelles is called for in order to be able to attempt to understand the basis for which the BNC decided as it did, and the term of recognition given for the various actions which took place in the same of overlapping locations. Even after doing so, no realistically logical reason can be identified.

Term of Reference	**First Battle of Fromelles**	**Battle of Neuve Chapelle**	**Battle of Aubers Ridge**	**Battle of Festubert**	**Battle of Fromelles**
Date(s) of Action	18 Dec, 1914	10-13 Mar. 1915	9 May, 1915	15 - 25 May 1915.	19 – 20 July, 1916
Military / Geographical Connection	Between La Boutillerie and Rouges Bancs below Fromelles village.	Artois Region. 8 kilometres from Fromelles.	Artois Region. 8 kilometres from Fromelles. Same areas as attacked 2 months previously known as Neuve Chapelle.	Artois Region. Slightly to the south of Battle of Aubers Ridge.	The Somme. The same location as preceding Battle of Neuve Chappelle, Aubers Ridge, Battle of Festubert.
Participants	2nd Scots Guards and the	4 Divisions (40,000 men)	British First Army	British First Army with Indian &	British 61st Division, and Australian 5th

	2nd Border Regiment			Canadian support.	Division. 10,000 – 15,000 British forces. 30,000 German forces.
Status	To hold the German front trench and should it be unoccupied to push on to the second line. Action unsuccessful.	Initial success. Abandoned after third day.	No ground was won and no tactical advantage gained.	The first British attempt at ***attrition.*** After winning some ground, action ceased.	Designed as a feint to prevent German troop movements south to The Somme. A tactical failure, but with the result that other German forces delayed movement for some weeks. Ultimately successful.
Casualties	Approx. 180 of the 2nd Scots Guard and 128 from the 2nd Border Regiment.	7,000 British and 4,200 Indian	British and Indian forces - 11,161. Germans - 902.	British - 16,648 casualties. German - c. 5,000 casualties	British casualties – 1547. Australian casualties – 5533 German casualties – 1,600 – 2,000
BNC Treatment	Not listed, thus not considered. Geographically proximate to Fromelles.	Battle (of Neuve Chapelle). Geographically proximate to Fromelles.	Attack at Fromelles as part of Battle of Aubers Ridge. Geographically proximate to Fromelles.	Attack at Rue du Bois as part of Battle of Aubers Ridge. Geographically proximate to Fromelles.	Subsidiary 'Attack' as part of Battle of Bazentin Ridge. 80 km away from Fromelles.

The Erroneous Connection With Bazentin Ridge

It will be noted by any student of geography that the various sites of actions, ***except for Fromelles***, are in the Somme region, some 80 kilometres away from Fromelles. And in addition, the actions reported under this heading took place between 14 July and 25 July.

The Battle of Bazentin Ridge occurred 14 - 17 July 1916. Fromelles was neither in the Somme, nor could it either conceptually, or factually, be the subject of a battle at Bazentin Ridge. The Battle of Bazentin Ridge had concluded before Fromelles even had a shot fired in anger.

Grouping Fromelles 1916 under the Battle of Bazentin Ridge is clearly an anomaly, simply because The Battle of Fromelles was clearly not part of the Battle of Bazentin Ridge.

A modern motor car can take approximately 80 to 90 minutes to travel from Fromelles to Bazentin (see map below). A soldier in 1916 would be likely to take 3 or 4 days marching in good conditions.

The 'attack' at Fromelles was intended as a feint to hold German reserves from moving south to the Somme. It was also an action to seize and hold ground, so not a raid.

The records show that the original plan for the attack at Fromelles was for Lt Gen Richard Haking's IX Corps to attack with three Divisions. This was subsequently pared back to two Divisions, the Australian 5th, and the British 61st.

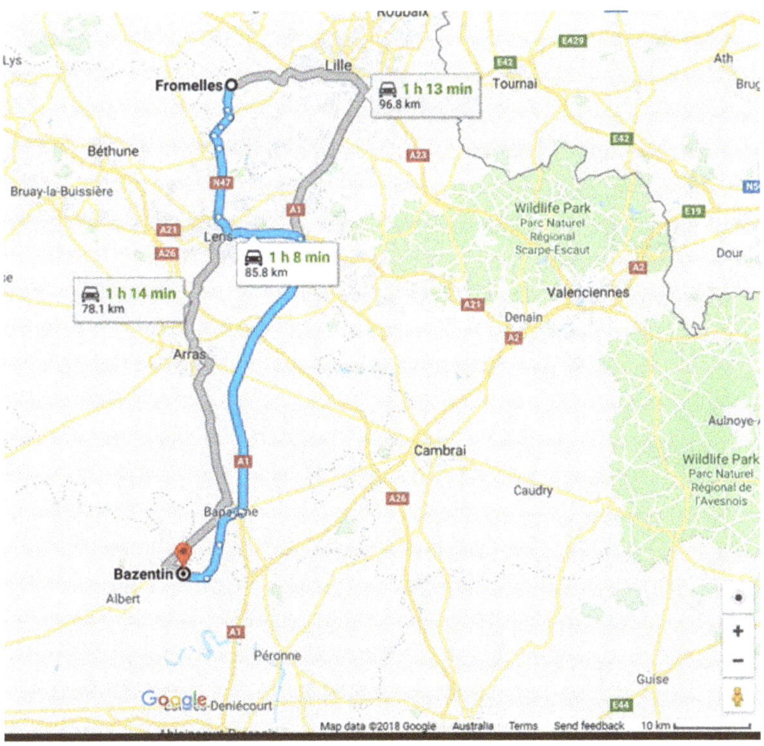

[Picture 39 Map 9: Fromelles to Bazentin. Google.com.]

However, this still constituted a Corps attack. See the widely recognized definition of a Corps being, "a military unit of ground combat forces consisting of two or more divisions and other troops".

It will also be noted by anyone who reads the BNC's Report that "the rank of 'battle' has been "confined as a general rule to engagements of primary importance fought out between forces <u>not smaller than</u> the corps".

The operation was conducted against the 6th Bavarian Reserve Division, supported by two flanking divisions of the German 6th Army. Two Allied Divisions under the direction of a Corps attacked a German Division supported by two flanking Divisions. This also fits the criteria of a 'battle'.

So, from a purely 'technical' point of view, the attack at Fromelles was a 'battle'.

In order to be an effective diversion, it would obviously have had to be some distance away from the Somme.

But it is generally understood that a feint would normally happen ***before*** or ***during the attack it is aimed to support***.

Surely for Fromelles to be an 'effective diversion' it should have occurred before, or coincided with, the Battle of Bazentin Ridge?

However, Fromelles occurred on 19 July, two days after the Battle of Bazentin had finished.

While it can readily be accepted that Fromelles was a feint in support of the overall campaign which became known as the 1916 Battle of the Somme - within the context of the 'The Somme' being the British Expeditionary Force's five British Armies' main effort on the Western Front in 1916 - it does not make sense that the official record should state that, 'the Attack at Fromelles' (with a footnote indicating that Fromelles is on Aubers Ridge) is listed as an Action as part of the Battle of Bazentin Ridge.

This linkage is simply not credible. It would be more credible if Fromelles was listed as an action as part of the Battle of Delville Wood (15 July – 3 September 1916) as it coincided with it, or the Battle of Pozières Ridge (23 July – 7 August) which it preceded. But again, they are on the Somme, and miles away from Fromelles. It also needs to be remembered that it was not until October 1916 that the various Battalions of the 5th Australian Division which fought at Fromelles, saw front line service and suffered casualties on The Somme.

Compare The Treatment of The 'Attack' At Fromelles 1916 With Three Other 'Attacks'.

The BNC classified three other 'Attacks' which were ultimately awarded Battle Honours: -

- BELLEWAARDE 16 June, 1915;
- BELLEWAARDE again on 25/26 Sep 1915; and
- BULLECOURT (11 Apr 1917).

All three attacks ended in stalemate. The second attack at Bellewaarde has striking similarities with the attack at Fromelles. *"The attack on the 25th September on the trenches held by the Germans in the vicinity of Hooge and Bellewaarde Lake was made with the object of distracting attention from a "full-dress" attempt to break through at Loos, away to the southward, and to contain the enemy's reserves. Like so many similar attacks, it entailed heavy losses to the attacking infantry."*[129] Does this sound familiar? You can readily see that it has a striking similarity to Fromelles.

The second attack at Bellewaarde was recognised with a Battle Honour. Don't be distracted by the fact that there were three engagements at Bellewaarde that are covered by the one Battle Honour:

[129] See: http://nlwmemorial.tripod.com/nlwmemorial/pages/pagesmisc/bellewaarde2ndwkelly.htm

- Battle of Bellewaarde 24/25 May 1915 (fought by a Corps of 4 Infantry and 2 Cavalry Divisions);
- the First Attack at Bellewaarde 16 June 1915 (fought by 1 Infantry Division, and a bloody, and fruitless action); and
- the Second Attack at Bellewaarde 25/26 September 1915 (fought by a Corps and 2 Infantry Divisions).

The Battle Honours Committee's rule of thumb was to economise on names and award the same name Battle Honour when fought in similar areas. So these are three different Honours which share the same name.

The similarities between the Second Attack at Bellewaarde and Fromelles are striking. We can surmise that because the Battle Honour BELLEWAARDE was awarded for the first 'battle' in May, it was easy for the BHC to extend it to also cover the attacks (even though they ended in stalemates) made in June and September. This assumption is supported by the BULLECOURT Battle Honours. The Battle of Bullecourt 3-17 May 1917 (fought by 2 Corps and 6 Infantry Divisions, and a success) likely attracted the Battle Honour BULLECOURT, which was subsequently extended to cover the earlier First Attack at Bullecourt 19 April 1917 (fought by a Corps and 2 Infantry Divisions, but which ended as a stalemate). Nonetheless, there is at least one other precedent here. And a potential example of double standards, or more likely showing that other factors came into play to marginalise Fromelles.

Compare Fromelles With Unsuccessful 'Battles' Of The Third Ypres Campaign 1917

The Third Battle of Ypres was the major British offensive in Flanders in 1917. It was planned to break through the strongly fortified and in-depth German defences enclosing the Ypres salient, a protruding bulge in the British front line, with the intention of sweeping through to the German submarine bases on the Belgian coast. The battle comprised of a series of limited and costly offensives, often undertaken in the most difficult of waterlogged conditions - a consequence of frequent periods of rain and the destruction of the Flanders' lowlands drainage systems by intense artillery bombardment. As the opportunity for breakthrough receded, Haig still saw virtue in maintaining the offensives, hoping in the process to drain German manpower through attrition.

The main battles associated with Third Battle of Ypres were:

- Pilckem, 31 July to 2 August
- Langemarck, 16-18 August
- Menin Road, 20-25 September
- Polygon Wood, 26 September to 3 October
- Broodseinde, 4 October
- Poelcapelle, 9 October
- Passchendaele (First Battle), 12 October
- Passchendaele (Second Battle), 26 October to 10 November.

Australian Divisions participated in the battles of Menin Road, Polygon Wood, Broodseinde, Poelcapelle and the First Battle of Passchendaele.

In eight weeks of fighting Australian forces incurred 38,000 casualties. The combined total of British and Dominion casualties has been estimated at 310,000. Estimated German losses were slightly lower. No breakthrough was achieved. The costly offensives, ending with the capture of Passchendaele village, merely widened the Ypres salient by a few kilometres.[130]

In his book *Redcoats to Cams: A History of Australian Infantry 1788-2001*, Mr Ian Kuring states that the offensive known as the Third Battle of Ypres was a series of eleven attacks, of which the Australian Infantry spearheaded five – Menin Road Ridge, Polygon Wood, Broodseinde, Poelcappelle and the First Battle of Passchendaele.

The first three of these were decisive victories, *Poelcappelle was an unsuccessful attack, and the Australians kicked off the First Battle of Passchendaele on 12 Oct which became bogged down due to bad weather and was suspended the following day.*[131] The Australians were relieved by the Canadians, who would eventually capture Passchendeale in the Second Battle some two weeks later (- pages 68/69).

The AWM website has the following entry for the Battle of Poelcappelle:

The battle of Poelcappelle was launched on 9 October with the ridge upon which Passchendaele stood as its objective. Like earlier battles in the Ypres offensive, the aim of the Poelcappelle attack was to secure a series of objectives in turn, protected by a heavy artillery barrage; the troops involved would be drawn from the 49th and 66th British, and 2nd Australian Divisions. Rain, however, had begun to deluge an already poorly drained battlefield, and adequate numbers of guns were unable to be brought within range. The infantry's advance also wallowed in the mud. The Australians were able to secure some of their objectives for a short time, but, with little artillery support and both flanks open, they were forced to withdraw [emphasis added]. *The 2nd Australian Division sustained 1,250 casualties in the battle.*[132]

Note that although the Battle of Poelcappelle was unsuccessful it was the recipient of a Battle Honour. And the unsuccessful First Battle of Passchendaele, which was stopped after 24 hours, was also recognised with a Battle Honour. At Poelcappelle the 2nd Australian Division sustained c. 1,250 casualties in on 12 Oct 1917. At the First Battle of Passchendaele on 12 Oct 1917 the 4th Australian Division suffered c. 1,000 casualties and the 3rd Australian Division c. 3,200 casualties. Compare these to the 5533 casualties the 5th Australian Division suffered at Fromelles on 19 July 1916, which is not recognised as a battle or accorded a Battle Honour. It

---

[130] See https://www.awm.gov.au/military-event/E104/

[131] Ypres - The Third Battle, 1917. The Battle in Brief. The weather again broke and the constant rain turned the battlefield into a quagmire so that further attacks on 9 October at Poelcappelle and on 12 October at Passchendaele failed with heavy loss. http://www.awmlondon.gov.au/battles/ypres

[132] *See:* https://www.awm.gov.au/military-event/E123/

seems that by 1917 those actions which were recognised as Battles, were correspondingly able to be given the further recognition of a Battle Honour.

And further, that after considering the treatment of Aubers Ridge and the treatment of Poelcappelle, it seems that we can only attribute the lack of recognition of Fromelles as a Battle, to a form of 'military cringe' associated with Fromelles having been *only a 'feint'* and therefore not worthy of Battle recognition, let alone the subject of a Battle Honour. This aspect of 'military cringe' has arguably been perpetuated in the AWM website description of the Battle of Fromelles. It is worth repeating the quotation earlier, as follows:

Fromelles was the first major battle fought by Australian troops on the Western Front. Directed against a strong German position known as the Sugar-loaf salient, the attack was intended primarily as a feint to draw German troops away from the Somme offensive then being pursued further to the south. A seven-hour preparatory bombardment deprived the attack of any hope of surprise, and ultimately proved ineffective in subduing the well-entrenched defenders. When the troops of the 5th Australian and 61st British Divisions attacked at 6 pm on 19 July 1916, they suffered heavily at the hands of German machine-gunners. Small parts of the German trenches were captured by the 8th and 14th Australian Brigades, but, devoid of flanking support and subjected to fierce counter-attacks, they were forced to withdraw. By 8am on 20 July 1916, the battle was over. The 5th Australian Division suffered 5,533 casualties, rendering it incapable of offensive action for many months; the 61st British Division suffered 1,547. The German casualties were little more than 1,000. The attack was a complete failure as the Germans realised within a few hours it was merely a feint. It therefore had no impact whatsoever upon the progress of the Somme offensive. [133]

Both at Poelcappelle and Fromelles the respective Australian forces involved captured some tactical objectives but ultimately were forced to withdraw. Yet Poelcappelle has recognition as a Battle and has been accorded a Battle Honour, whereas Fromelles has neither!

The 'Battle' Of Hamel.

By way of comparison, The BNC Report lists the 'Capture of Hamel' under General Monash in 1918 as an 'action'. This is logical as it only involved one Division's worth of infantry in the attack which lasted 93 minutes. However, this has gone down in popular culture as 'The Battle of Hamel'. The AWM shows it as a 'battle'.[134] And, subsequently, owing to the success of the action of Hamel, books and articles also call it a 'battle'.[135] This of course also applies to Fromelles, with many books/articles also referring to it as a 'battle'.

---ooo000ooo---

[133] See: https://www.awm.gov.au/collection/E159/

[134] See: https://www.awm.gov.au/collection/E103/.

[135] For example: Laffin, J. The Battle of Hamel : The Australians' Finest Victory 1999.

Andrews, E.M.; Jordan, B.G. (1991). "Hamel: Winning A Battle". Journal of the Australian War Memorial.

1921: Australian Response to The BNC Report

The BNC Report was signed off by the Committee, on 9 July 1920. It was approved by the Army Council, and then submitted to King George V for approval. It was then presented to the British Parliament by Command of His Majesty, in 1921, and first appears to have been published in 1921.

Ten months after the BNC Report had been delivered by its committee, on 18 May, 1921, Winston Churchill, Colonial Secretary wrote to the Governor-General of Australia to transmit copies of Parliamentary Paper Cmd 1138, containing the *Official Names of the Battles and other engagements fought by the Military Forces of the British Empire during the Great War, 1914-1919, and the Third Afghan War, 1919*, being the Report of the Battles Nomenclature Committee as approved by the Army Council. (NAA File 478/2/600)

But the Australians saw the report as being inadequate in addressing the actions of Australian soldiers. Following protocol, on 2 September, 1921, Defence Department Secretary Thomas Trumble wrote to the Prime Minister's Department requesting that a Statement prepared by the Inspector General and the Chief of the General Staff of the AMF on the BNC Report be forwarded via the usual official channel for the consideration of the Army Council.

The authors of the Statement were Lt Gen Chauvel, Minister G.F. Pearce, and Major General Brudenell White, Chief of the General Staff. The Statement focussed on the operations in which the AMF were involved in August and September, 1918 in which they took a leading part. The Statement noted that although the Report had been approved and presented to Parliament and had been issued as a Command Paper, that it would not be too late to alter certain decisions and to adjust apparent inconsistencies if the matter is re-presented at once.

Significantly, the Statement pointed out that various operations should be classified as Battles, judged on the bases laid down by the Nomenclature Committee. But unlike Fromelles which has been analysed elsewhere, this Statement sought to deal with Lihons, Chuignes, Peronne, and the Outpost Line on the Hindenburg Line. No less than seven (7) operations in Palestine were raised for re-consideration.

The AMF were left wondering how so many operations could fail to be properly classified using the very criteria set out by the BNC in its explanation as to how it came to the making of the findings in its Report, just as this author has done regarding the appropriate classification of Fromelles. (See NAA File 478/2/600)

In a response which was deferential to the workings of the BNC, on 2 December, 1921, Churchill wrote to the Governor-General in Australia, to acknowledge receipt of dispatch No. 316 of 19

September, 1921. Churchill wrote that '*The Army Council had intimated that full consideration will be given to the representations contained in the statement prepared by the Inspector General and the Chief of the General Staff of the AMF*'.

In defence of the BNC Committee, Churchill also wrote that '*they (the Army Council) point out that the classification adopted by the Committee was the result of a careful examination of the evidence, both documentary and oral, and received the approval of the Commanders in Chief concerned, and that in these circumstances it is doubtful whether it will be found possible to make any substation alteration in the Report*.' (NAA File 478/2/600)

1922: Australian Response to The BNC Report

In 1922, a further print of the BNC Report was issued, and on a copy held at the National Archives, UK, is endorsed in handwriting:

"Amended in accordance with recommendations of the Battle Honours Committee".

On 24 January, 1922, Winston Churchill, writing from the Colonial Office, London, responded to the dispatch from Australia of 2nd December, 1921. He was writing to quell Australian scrutiny of the BNC's work. His letter to the Governor-General, which was referred to the Secretary of Defence, justified the BNC's conclusions, saying: '*The Army Council are of the opinion that there are the gravest objections to reviewing the classification of engagements put forward by the Committee and officially approved.*'

The letter went on to state that '*The conclusions of the Committee were only reached after a most careful survey of the documentary records and after hearing the evidence of a large number of distinguished officers intimately connected with the operations under review, and all controversial points received the fullest discussion. The classification finally adopted was approved by the Commonwealth representatives on the Committee and has been accepted both in the United Kingdom and in the Dominions as doing substantial justice to all the forces concerned.*'

If Australia was to seek to review the Committee's deliberations, then all the Dominions would have to be asked to do so as well. The ramifications of acquiescing to Australia were obvious. It would be better if it did not happen. One justification was that it would be extremely difficult to devise satisfactory machinery for dealing with such criticisms. The original Committee were by then dispersed throughout the Empire and it would be practically impossible to re-assemble them. (NAA File 478/2/600)

It appears that any adjustment of the records would be just too difficult.

On 29 March, 1922, Thomas Trumble, Secretary to the Department of Defence wrote in a Minute paper of his concerns in relation to the input of Australian Officers in the deliberations of the BNC. He called on the addressees of the Minute Paper to advise whether they considered it desirable that any further representations be made to the Imperial Authorities in the matter.

He went on to write: *A perusal of the Report discloses that the Australian representatives on the Committee from time to time were officers who served on the Western Front. It is assumed however that these officers were supplied with all the necessary data to enable them to watch the interests of Australian Formations which served in other theatres of war as well as those which served in France and Flanders.* (NAA File 478/2/600)

But were they? Clearly the Secretary was not satisfied that there had been sufficient input to the machinations of the committee on behalf of Australian Forces, and that appropriate Battle Honours might not be attainable as a consequence.

Was There Any Realistic Alternative?

In the modern era, any review of activities by officials has to deal with a process of administrative review. The "decision" which was made is thus the centre of the review process. Was it taken on the correct basis, with all appropriate information available, or were elements ignored which should have been taken into account?

The 'decision' of BNC to attach Fromelles to the Battle of Bazentin Ridge and The Somme, when it was an action outside the geographic construct of the Somme, is an error in itself. But it was forced on the BNC by the High Command which connected all the actions on that timeframe as being part of 'the Somme'. Again, geographically wrong. These actions were part of an offensive, but not in the same region.

The 'decision' of BNC to categorise it as part of the battle of Bazentin Ridge, is likewise, in error, if not perverse. No units involved in Fromelles were involved at Bazentin Ridge. And Bazentin Ridge had finished 2 days before Fromelles was even commenced.

The alternative readily available by following its Terms of Reference would have been for the BNC to tabulate Fromelles as it was: a Battle in French Flanders. It did not need to be referenced to the military plans that were made that would unfold on The Somme. Another entry in the BNC Report under the heading of French Flanders would have been enough.

Having done that, it would have been soldierly to recognise that a Corps was involved in the battle, and to denote the conflicting result of a tactical failure, but strategic success.

The third of the terms of reference was something completely ignored in that Fromelles was not in fact defined by geographic limits, but by some strategic decision-making elsewhere.

The Committee could not find a fresh page for this "attack" in a geographical location far removed from the Somme, so it was labelled a 'subsidiary attack' to Bazentin Ridge. On the BNC's reasoning, events in Kalgoorlie can therefore be closely related to events in Bondi.

The Australian 5th Division and the British 61st Division, and ourselves are 100 years later, dealing with the unintended consequences of this 'flexibility' and resulting incongruity.

---ooo000ooo---

Despite the scale of loss in July, 1916, the Battles Nomenclature Committee had continued to dignify the Fromelles attacks with the title 'action'. The term 'battle' would not appear in official publications for many years to come.

But the action of May 1915 ***was*** classed as a 'battle' so, although on 19 July 1916 the Diggers had suffered from its operational legacy, there was a bizarre ultimate 'benefit': the ground was to be carefully swept by the British simply because the earlier encounter was on the official list as the Battle of Aubers Ridge[136], which had occurred one year prior.

The troops carrying out the sweep of the area were not looking in the area where the burial pits were located for the 250 soldiers buried by the Germans in 1916. And nor would they detect any indications of burial pits while doing so. That would only come about more than 90 years later due to the efforts of Lambis Englezos and others, in 2007 and 2008.

Some Comments on The BNC's Findings

It can be readily concluded from the evidence presented:

- Fromelles was not intended to be an action on The Somme, but an action undertaken in another region, to draw attention from The Somme, as part of an overall Campaign.
- The action at Fromelles was undertaken by a Corps, and that by BNC's own criteria, it is was a 'Battle'. The BNC defined "battles" as "engagements of primary importance fought out between forces not smaller than the corps".
- The decision which was made by the BNC was a failure of consideration of the significance of the action at Fromelles.
- That the timing of the action made it attractive to assert that it was subsidiary to Bazentin Ridge, but this is in fact an absurdity when considering the first bullet point above. It was listed in a chronological sequence by those charged with the Terms of Reference, but otherwise has no relativity with The Somme or Bazentin Ridge. It should have been listed as an action separate to The Somme. After all, The Somme was only one of many campaigns being undertaken in France at the same time.
- The Battle of Bazentin Ridge has a date range of 14-17 July 1916. The Attack at Fromelles occurred on 19/20 July 1916. Logic suggests that for the Attack at Fromelles to be a

[136] Barton p.30

subsidiary attack it should at least occur as a precursor to, or at the same time as the main attack.

- That the action may have been a tactical failure, and accordingly be tagged as an 'attack', but strategically it was a success, and thus can justifiably be recorded as a battle.
- A "battle" recognition was not awarded where the engagement was not considered a successful outcome. Fromelles met the requirements of being described as a battle, but because it was not considered to be a 'successful' outcome, it was relegated to being classed as an 'Attack', and failed to achieve recognition because it was a tactical failure. But in contrast, Hamel was an Action, and was able to achieve Battle Honours due to its success on 4 July, 1918.
- The BNC consistently made inconsistent interpretations, failed to properly follow their own Terms of Reference, made and broke its own rules as it went along and was as 'flexible' or as 'inflexible' as it felt it needed to be.
- The evidence presents a compelling case, and effectively proves that a Battle was fought at Fromelles in July, 1916. It is commonly recognised as such by the populace at large, politicians, governments, and the families of all those who fought there.
- Our politicians have expressed publicly that Fromelles was a Battle, which we accept, but none have come to have that sentiment converted into a political act, by taking steps to have it formally confirmed. Inscription on memorials thus remains illusory.
- Any reference to the soldiers dying at Fromelles being part of The Somme is a geographical absurdity, given the correctness of the geographical descriptions for the earlier battles fought in the same location.
- That Fromelles deserves to be honoured by inscription wherever Australian War Memorials seek to do so, by reference to a place of battle. The fact that it is as yet not recognised in official memorials by reference to battles, or inscription, is and remains an affront to those who served, and the families who saw them serve.

---ooo000ooo---

13. The Awarding of Battle Honours

Recognition:

Babies cry for it;

Men die for it.

---ooo000ooo---

In Chapter 1, we looked at the differences between the terms of 'raids', of 'battle', and of 'battle honours'. As pointed out earlier in the course of this book, the awarding of *Battle Honours* is not the central purpose of this work; it is instead the due recognition of the action at Fromelles as a Battle. But it would be remiss not to look at the work of, and the process for awarding of Battle Honours, as Fromelles is directly affected by that process.

In the previous chapter, we saw how the Battles Nomenclature Committee (the BNC) was tasked with identifying and categorising the various engagements undertaken in France and Flanders – and the Great War as a whole - between 1914 and 1918. That work would have the follow-on effect that various engagements would be able to be awarded battle honours. The further work of determining which engagements deserved the recognition of a battle honour would be undertaken by another body altogether: ***The Battle Honours Committee*** ('the BHC').

It is again worth pointing out that unless you had a successful engagement, you could not *ordinarily* hope for it to be the subject of a Battle Honour. But it is possible, and has occurred on only two occasions.

So it was again, when the BHC adopted the findings of the BNC and then proceeded to determine the recognitions which would be awarded for the various engagements.

This Chapter will be focused on the history of, and the process of awarding of Battle Honours generally, and will look at some of the results of the BHC.

What Is A "Battle Honour"?

As noted in Chapter 1, a battle honour is an official public recognition of a battle, action, or engagement by the inclusion on a unit's Colours, Standard, or Guidons. Australia's oldest battle honour is Suakin, 1885. This was granted to the 1st, 2nd, and 3rd Australian Infantry Regiments by King Edward VII in 1907 to acknowledge the contribution of the New South Wales Infantry to the war in the Sudan.[137]

In dealing with the classification of Battles, and ensuing Battle Honours, the following points will emerge: -

[137] See: Australian War Memorial Website: https://www.awm.gov.au/articles/encyclopedia/battle_honours

1. Battle Honours are usually presented in the form of a name of a country, region, or city where the unit's distinguished act took place, usually together with the year when it occurred.

2. Not every battle fought will automatically result in the granting of a Battle Honour.

   Conversely, a regiment or a battalion might obtain more than one Battle Honour over the course of a larger operation.

3. Nor does a Unit have to successfully defeat their adversary to earn a Battle Honour. But there are only two examples where this has been the case, as we will read later.

[Picture 40: Regimental Colour, 4th Battalion, AIF. AWM REL39960.001]

AUSTRALIAN WAR MEMORIAL REL39960.001

Battle Honours Of The British and Imperial Armies

From their institution until the end of the Second World War, awards were made by, or in consultation with, the British government. For example, there are two hundred and sixty-three officially-named battle honours commemorating battles and campaigns fought before The Great War. At first, awards were random, irregular and highly selective. It was not until 1965 that the last awards for the period prior to 1914 were confirmed by a South African Battle Honours Committee.[138] Since 1945, the individual countries of the former British Empire have awarded battle honours to their forces independently.

Origins[139]

The first battle honour was the motto *Virtutis Namurcensis Præmium* (Reward for valour at Namur), ordered by King William III to be emblazoned on the colour of the 18th Regiment of Foot, later the Royal Irish Regiment, for their part in the Siege of Namur in 1695.

The first battle honour to be awarded in the British Army was awarded to the 15th Hussars in 1768 for the Battle of Emsdorf fought in 1760.

The first battle honour displayed on the colours in the modern manner was awarded in 1784 when four infantry regiments that took part in the defence of Gibraltar of 1779–83 were ordered to display the word Gibraltar on a scroll on their Second (now Regimental) Colour. Although

---

[138] Rodger p.9.

[139] See: https://en.wikipedia.org/wiki/Battle_honour : Battle Honours of the British and Imperial Armies; reproduced under Creative Commons Licence.

this award was made promptly after the event, this is not always the case: the oldest battle honours, Tangier 1662–1680 and Tangier 1680, were awarded in 1909, over 220 years after the temporary but tumultuous occupation of that port.

The majority of battle honours are displayed simply as the name of the award inscribed on a decorative scroll. A number of honours, particularly those which were considered to be of particular significance, were awarded with a badge that in some way represented the engagement or theatre. Early examples were the castle and key for Gibraltar, mentioned above, and the sphinx for Egypt. Tiger and elephant badges were frequently awarded for engagements on the Indian sub-continent.

Development and Formalization

The procedure for awarding battle honours was originally extremely arbitrary. For example, the victories of the Duke of Wellington in the Napoleonic Wars were copiously honoured, but those of the Duke of Marlborough in the War of Spanish Succession were entirely ignored. By the mid-19th Century, honours were being awarded for contemporary actions that were little more than skirmishes compared with the great European battles of the 18th Century.

Until 1832, battle honours were awarded to a specific unit and, if it was disbanded, the honour was lost. After this date, honours were awarded to the parent regiment of the battalion whose actions led to the award. During the Second Boer War, however, some honours were awarded to the Militia battalions of infantry regiments in their own right.

Also in 1832, the motto *Ubique* ("Everywhere") was awarded by King William IV to the Royal Artillery and the Royal Engineers in recognition of their universal service. It was stipulated that this was considered to be a battle honour, substituting for all other prior and future distinctions. As such, it did not appear on the badges or appointments of the Territorial, Militia or Volunteer regiments of these Corps. Despite it being the sole battle honour of the Royal Artillery, a number of Artillery officers who served in the First China War were awarded the personal honour *China*. This honour was displayed on their appointments and was not extended to other ranks or to the regiment as a whole. It was a unique award which was allowed to lapse in time, and the procedure was never repeated.

In 1844, the display of honours and badges on the infantry's colours was standardized. Embellishments that had previously been borne on either, or both, of a battalion's colours were, after this, only permitted to appear on the Second Colour, which was renamed the Regimental Colour. The First Colour was renamed the Royal Colour and was to be free of decoration other than for the Royal Crown and the regimental number.

In 1886 when the awards for the Nile Campaign of 1884-85 were being considered, it was ruled that only regiments south of Assouan on or before 7 March 1885, in the expedition to relieve General Gordon, would qualify for such awards.[140] This was the first time that the 'battlefield' was defined by what would become known later as 'geographical and chronological limits'.

---

[140] Rodger p.15

During the Boer War many more corps, therefore, became eligible for campaign honours than had been the case in any previous war: including the Regular Army, Yeomanry, Militia and Volunteers, a total of 196 British regiments were awarded South Africa with appropriate year dates between 1899 and 1902. The award was also made to a further 22 Canadian, 37 Australian, 23 New Zealand and 12 South African regiments. As pointed out above, the awards made to the Militia battalions lapsed when the Militia was disbanded.

[Picture 41: Regimental Colour of the 18th Regiment of Foot showing the earliest battle honour (for Namur) and the badges later awarded for *Egypt* and *China*.[141]]

It was at this time that the rule was instituted that, for a cavalry regiment or infantry battalion to be eligible for an award, *the Headquarters and fifty per cent or more of its strength* must have been present. Exception was made for the Yeomanry regiments, which had contributed company-sized contingents to the Imperial Yeomanry, but not their individual regimental headquarters. These regiments were considered to be eligible if parties of 20 or more had been present.

In 1917, in recognition of their sacrifice in the Great War, the battalions of the Territorial Force were permitted to carry the honours of their parent regular battalions on their colours, badges and other appointments, a practice that had previously been forbidden.

What Is The Significance of Battle Honours?

Battle Honours indicate in outline the fighting history of a unit and appear not only on the unit colours but in the Forces List.

For example, it would be illogical to grant 'Gallipoli' Honours to say, the 60th Battalion, AIF when no such unit was present on the Peninsula. A Battle Honour is granted to a unit not to individuals. The honour is the individual's by virtue of his having served in the unit. (See for

[141] Source: Richard Cannon (1848). Historical Record of the Eighteenth, or the Royal Irish Regiment of Foot (https://archive.org/stream/recordofeighteen00cannniala#page/n41). London: Parker, Furnival, & Parker. P.1. Author: Madeley.

example AWM File 72/1/195, where Administrative Staff, Army Headquarters wrote on 22 October, 1924 to 3rd District Base, Melbourne to confirm the significance of Battle Honours.)

It remains a tradition that whenever military personnel meet a colour or guidon, it must be saluted. This is not only because it is an object which represents the authority of the Crown, but also because the colours contain a regiment's battle honours, and thus represent the regiment's history and its dead.

For example, the 7th Battalion, Royal Australian Regiment on 8 February, 2019, undertook a ceremony whereby the Battalion's Colours were 'uncased', and returned to their Cabinet. It was described as 'a unique opportunity to parade the Colours through the Battalion and emphasises the start of the New Year.'

Saluting a colour or guidon is thus a pivotal act in retaining an awareness of regimental history and traditions, key in the functioning of the regimental system. It remains common for army instructors to ensure that their recruits have memorised and are able to recite all of their regiment's battle honours. Such methods are meant to bring the new soldier into the regimental ethos and sub-culture by means of imprinting shared history.

The French Also Have A History of Battle Honours.

In France, Battle Honours, known officially as *inscriptions des noms de batailles au drapeau*, were created during the Wars of the French Revolution. The first honour such awarded is Valmy, 1792, the last one to date is Koweït, 1991. The honours are sewn in gold letters on the regiment's flag or standard under the form <battle name>, <date> (such as Austerlitz, 1805).

Depicted below is the Flag of 1st Regiment of Marines (1 RIMA). It illustrates that the Regiment has been awarded fifteen battle honours.

[Picture 42: Flag of 1 RIMA, France.]

In some cases, the honour refers to a whole conflict and take the form <Geographical designation>, <date> (such as AFN, 1952-1962) or simply <war>, <date> (such as Grande Guerre, 1914-1918). Some military schools and academies have their motto sewn on their flags as a battle honour, such as École Polytechnique's Pour la Patrie, les Sciences et la Gloire (for Fatherland, Sciences and Glory) in addition to the "true" battle honour Paris, 1814.

During the 19th century, honours were limited to eight on a regiment's flag, leading to a need to erase some honours in order to make room for newer ones. This limitation was upped to twelve honours after the Second World War.

Some units, such as the 2 RIMA, have been awarded more than twelve honours, but their flag displays only the twelve considered the most important in the history of the regiment.

Other units, such as the 1 RIMA, have been granted a dispensation allowing them to emblazon all their battle honours, regardless of their number.

In addition to battle honours, military units can be awarded collective military decorations and collective mentions in dispatches, leading eventually to the fourragère award.

Additional Recognition in The Form of Battle Clasps.

Individuals are also entitled to be awarded medals and battle clasps for their own recognition of achievement.

In tandem with the process involved in the selection of Battle Honours, is the process involved in determining the appropriate 'Battle Clasps'.

Battle Clasps had become a well-established addition to War Medals to identify which battle/battles in a conflict an Officer or Soldier had fought in.

For example: Peninsular War 1807 to 1814, there were 21 Clasps; Crimean War 1853 to 1856, there were 4 Clasps; Second Boer War 1899-1902, there were 26 Clasps.

Battle Clasps to individual Officer and Soldier Medals essentially corresponded to Unit Battle Honours – thought there are as usual a few anomalies.

Now consider The Great War. There are a total of 255 Battle Honours. Eighty-six (86) of these are 'theatre' battle honours, leaving 169 'campaign' and 'battle' Battle Honours. Based on the history of Battle Clasps this equated potentially to 169 Battle Clasps.

It can readily be understood that the cost of producing potentially literally millions of battle clasps is what would lead to the decision in 1923 *that on account of present financial stringency, the Army Council have decided not to proceed with the proposal to issue Battle Clasps for the Great War.*

In this chapter, we will also look at how the process of deciding on Battle Clasps affected the soldiers who served from Australia. In looking at the determination of battle honours and clasps between the world wars, we will follow a chronological approach.

--- 1919 ---

Moving quickly following the Armistice in November, 1918, the first meeting of the Battle Clasps Committee (whose creation was at the behest of Winston Churchill after some representations) was held at 27 Pilgrim Street, London on 24 March, 1919. The Committee would from the start include a representative of the Admiralty, of the Air Force, and of India, and should also have a representative of each Dominion, namely, Australia, Canada, New

Zealand and South Africa. It was felt 'That for such a victorious war it would be unforgivable not to award Battle Clasps'.

The questions to be resolved were to become: -

- What were the defined limits of any Battles;
- Who had taken part in those, and
- Who would be entitled to receive a Battle Clasp.

At the second meeting of the Battle Clasps Committee, on 25 March, 1919, it was made known that Sir Douglas Haig would like every man who took part in raids preliminary to, or otherwise connect with, a decisive engagement, to receive a clasp for that engagement. For example, that all the people who obeyed his orders in any operation connected with the Somme should receive a clasp for The Somme.

Ahead of the Battle Clasps Committee's meeting to be held on 15 April, 1919, a number of Memoranda were prepared by the officer in Charge of Military Branch, Historical Section of the Committee of Imperial Defence.

One such memorandum focussed on the battle of Neuve Chapelle, and the factors warranting a Battle Clasp being awarded for the action from 10 to 13 March, 1915.

Festubert, being an offensive launched by the First Army on 9 May, 1915, and 18-26th May, 1915 against German positions at Fromelles on the Aubers Ridge was also scrutinized in the same memorandum.

A separate memorandum was prepared in relation to The Somme. While it was analysed to the point of showing that there were multiple actions within the time limits of The Somme Campaign itself, there was no mention of the activities undertaken at Fromelles on 19 – 20 July, 1916. None. On a reading of the memorandum, it would appear to an outside observer that nothing happened there at all.

On 7 May, 1919, The Battle Clasps Committee sent a copy of its second report to Major General Howse, Headquarters AIF, London. All operations catalogued under 'The Somme' would receive the same Battle Clasp. The proposal to limit The SOMME to one clasp was carried by one vote of the Committee.

In a meeting of the Battle Clasps Committee, on 18 July, 1919, Lt Gen H.G. Chauvel was present to discuss awards of Clasps for Palestine. From the minutes of that committee, it is clear that Chauvel sought more recognition for the Australian and New Zealand soldiers' service in Palestine, and that he argued for it forcefully.

Recommendations were made for a clasp for each of Egypt, Western Front; Suez Canal; Romani; Sinai; Jerusalem; Syria; Palestine; and Damascus. (NAA File 478/2/600)

--- 1920 ---

As we saw in the previous chapter, the BNC first met on 18 August 1919. From the Minutes of the Meeting of the Battle Clasps Committee on 20 June, 1920, the Chairman told the Committee that the first thing he wished to do was to co-ordinate their suggestions for Clasps with the suggestions of the BNC, and that, unless there was any particular reason they should not, this Committee wished to come into line with the BNC Committee. (NAA File 478/2/600)

Therefore, the BNC determinations would be critical to assessing the recognition given to soldiers for their efforts throughout the Great War. They only had to wait for the following month, as the Report of the Battles Nomenclature Committee, as chaired by Major General Sir John Headlam KBE, DB, DSO, as President of the Committee, was delivered on 9 July, 1920.

The final report of the Committee on Battle Clasps would essentially follow the naming process adopted in the BNC Report. Consequently, there would be no Battle Clasp awarded for operations at Fromelles in December, 1914 (which did not appear in the BNC's list of operations etc.), or for July 1916 (which was listed in the BNC Report as being under The Somme).

There would however, be a Battle Clasp for Neuve Chapelle from 10 – 13 March, 1915, and for Festubert for 15-25 May, 1915. All operations catalogued under 'The Somme' would receive a Battle Clasp, as previously determined by a one vote majority. In the event that this was not found to be appropriate, then Clasps should be awarded for Fricourt, Bazentin, Delville Wood, Pozieres, Guillemont, Courcellette-Flers, Thiepval-Combles, and Beaumont Hamel. Once again, Fromelles was not recongised for either a Battle recognition, or as a consequence, for a Battle Clasp.

In eleven months after its work began, by 9 July 1920, the Report of the BNC was ready, and was published the following year without amendments. Author Alexander Rodger expressed the view that the BNC Report of 1921 is:

> ***"...rightly accepted as a classic piece of staff work."***[142]

At the macro level, as Rodger commented, the Report has been universally recognised as a very good document. And it can be readily acknowledged that it is not perfect. For example, it does not cover some of the micro actions that took on National importance in Dominions such as Australia, Canada, Newfoundland and New Zealand.

The basic structure of the BNC Report was to be followed by others subsequently. There is therefore, no surprise at all that the Battle Honours Committee ('the BHC') would also follow the classifications of the various battles, actions, and other engagements, by the BNC when determining how Battle Honours would be awarded. But it was in the Battle Honours Committee's use of the BNC Report where problems arose.

--ooo000ooo---

On 23 July, 1920, the Second Report of the Committee on Battle Clasps was issued. It was recorded in the Report that Battle Clasps were intended to reward risk or responsibility. Nine members met to formulate a proposal for the issue of such clasps. One member – Rear-Admiral

[142] Page 77, Rodger: Battle Honours of the British Empire and Commonwealth Land Forces 1662-1991.

Rudolph W. Bentinck - did not concur with the proposals, and outlined the history of awards since the Battle of Waterloo.

--- 1921 ---

On 22 February, 1921, The War Office, London, wrote to the Officer Commanding, AIF Headquarters in London, to advise that the question of supply of Battle Clasps had not been definitely settled, and that, for financial reasons, 'no decision will probably be reached in the near future'. (NAA File 478/2/600)

A copy of the Final Report Battle Clasps Committee was posted from London to Department of Defence, Melbourne, on 11th August.

Subsequently, after four months, on 18 December, 1921, AIF Headquarters London, wrote to Department of Defence, Melbourne. No further developments regarding the issue of Battle Clasps had transpired, the Imperial Authorities concerned intimating that the existing financial position 'renders the provision of clasps very unlikely for some time'. The whole issue hinged on cost. (NAA File 478/2/600)

---ooo000ooo---

The **Battle Honours Committee** was convened by the end of 1921. Its Terms of Reference were (as stated in its Report of 11 April, 1922):

To consider and report on the selection of Battle Honours to be borne on the Colours of Regiments and Corps for the Great War, with special reference to: -

(a) Whether Honours to be emblazoned on the Colours and to be shown in the Army List should differ in any way.

(b) Whether in the selection and allotment of Battle Honours the guiding principle should remain that Headquarters and at least 50 per cent, of the effective strength of the unit in a theatre of operations should have been present at the engagement for which the Honour is awarded.

(c) Whether the Honours for Territorial units and Special Reserve (Reconstituted Militia) units should be the same as those that may be authorized to be borne on the Colours of their respective regular units and *vice versa*.

(d) What action, if any, should be taken regarding the Honours to be borne on the Colours of the New Army battalions which have now disappeared?

(e) What should be the maximum number of Battle Honours to be borne on the Colours that may be allotted to a single Regiment or Corps for the Great War.?

--- 1922 ---

The work of the Battle Honours Committee was now in progress. On 3 January, 1922, The Battle Honours Committee met for the second time. The Terms of Reference were discussed by the various Committee members under the Chairmanship of General Sir Archibald Murray. A number of matters were provisionally agreed upon. Following that meeting, the BHC sent correspondence to as many unit headquarters as it was able to do so, in which it asked the recipient to give its opinion on the Terms of Reference, and the issues provisionally agreed upon. In the covering letter to each of the units, it was pointed out that one of the provisionally agreed upon points from the Committee's discussions, Item B (ii) was:

Whether Regular Battalions should only be awarded the honours earned by Regular Battalions, Territorial Battalions, the honours earned by Territorial Battalions and Service Battalions the honours earned by Service Battalions.

It was stated in the covering letter that:

"It is felt that if "B" (ii) should be accepted, the list of Battles would be too long, and if a selection should be made under (e) of the Terms of Reference, the Regular, Territorial, and Service Battalions of the same Regiment would elect very different Battles, and so defeat the laudable object of welding them together with the same Battle Honours."

What had become apparent was that the sheer scale of the Great War led to a previously unheard-of number of honours being awarded and it was simply impractical to emblazon every one of them on the Regimental Colours. The Gloucestershire Regiment, for example, was awarded over 80 honours.

The issue of Battle Honours was very much alive after the BNC delivered its report for the approval of the King, and before the Battle Honours Committee had been able to deliver a suitable determination. The following extract from Hansard for the House of Lords, commencing at p.126, on **14 February, 1922** deals with the issue as recorded at that time[143]:

'BATTLE HONOURS'.

Lord Nunburnholme rose to ask His Majesty's Government the reason why battle honours have not been conferred on county regiments for the Great War. The noble Lord said:

My Lords, in rising to ask the Question which stands in my name I may, perhaps, say a few words upon a subject in which a great many people in this country are interested, and that is the subject of battle honours and clasps for engagements in the late Great War. It is now some three years and three months since the war terminated, and many committees who have erected county war memorials would like to have the battle honours of the county regiments inscribed thereon. We are all proud of the engagements in which the various arms of His Majesty's Service took part in the war. At the present time there is a great deal of talk of doing away with many units of His Majesty's Forces, and it will be a pity if [127]these units are abolished without the ordinary recognition they have always received after a great war in the shape of honours for the engagements in which they have fought.

This is, of course, a very difficult subject for the various Committees which have been sitting at the War Office to deal with. I trust I am voicing your Lordships' opinion when I say that I hope the time has now arrived when we should have some decision on the matter. I do not speak on behalf of those who served in the Royal Navy, but I notice a letter from one in The Times to-day. He says that he has received his war

[143] HL Deb 14 February 1922 vol 49 cc126 [126]

medals, is very thankful to have lived long enough to get them, and hopes that some day he may receive a clasp. If the Government can give any satisfactory information on this Question I am sure we shall be greatly obliged.

In reply, The Chancellor of The Duchy Of Lancaster (**Viscount Peel**):

My Lords, I quite understand that there is much public interest in many quarters on the points raised by the noble Lord as to the question of the battle honours of different units. May I point out that it is a very complicated subject. Before you can decide what battle honours can be borne on the colours of the different regiments you have to decide what a battle is, and that has been a matter of very great difficulty. [Emphasis added.] The old definition of a battle will not do, in view of the prolonged engagements, lasting over several weeks, with which we are now familiar. A Committee called the Battle Nomenclature Committee was appointed, which reported on what were officially the actions of the war and what were the geographical and chronological limits of those actions and their relative importance. This Report was published last year.

Having defined what the battles were, the next point was to set up another Committee to ascertain what units had taken part in them. That again was a matter of considerable difficulty. Let me give your Lordships the terms of reference to this Committee. They had to consider and report on the selection of battle honours to be borne on the colours of regiments and corps for the Great War with special reference to five separate points. I will not trouble you with all the five points, but I will quote two in order to show the difficulty of the problem. One point in the [128] selection and allotment of battle honours was whether the guiding principle remained, that headquarters and at least 50 per cent of the effective strength of the unit should have been present in engagements for which the honour is awarded. It was rather unfortunate that certain units had to be disbanded without knowing the battle honours to which they were entitled, but a much more difficult question is raised in connection with the temporary units which have been disbanded in large numbers. Another question the Committee have to consider is what action, if any, should be taken regarding the honours to be borne on the colours of the New Army battalions which have now disappeared. It will be some time before the Committee, which was set up in December last, will be able to complete its Report.

It may interest your Lordships to know how long it was before decisions were finally reached in the bestowal of honours in previous wars. The periods vary. The battle honours for the South African war, which ended in May, 1902, were awarded in December, 1904, some two and a half years afterwards. Those of the Crimean war were awarded in October, 1855, whilst the war was still in progress. The battle honours of the Peninsula war, which ended in 1814, were awarded in 1818, or four years after the termination of the war. It is interesting to observe that the battle honours for the Marlborough Campaign in Flanders were not finally settled until the year 1882, a very considerable time after the termination of that campaign. I do not say that such a long period is to elapse now, but I feel sure the noble Lord will agree that some time must pass before this question of battle honours is finally decided. I agree with him that it should be settled as soon as possible.

---ooo000ooo---

Despite the Battle Honours Committee having already commenced its work, Secretary Winston Churchill from the Colonial Office, London, caused a letter to be sent to 'The Official Secretary of the Commonwealth of Australia' on **25 February, 1922**, to request that he inform the High Commissioner that it has been decided by the Army Council to refer the question of selection of Battle Honours for the Great War to a Committee under the Chairmanship of General Sir A.J. Murray, G.C.M.G., K.C.B., D.V.O., D.S.O.

The terms of reference for the committee were outlined, and would become incorporated into Army Orders or Military Orders. He advised that the Committee were to meet for the 'first' (sic) time on 28 February, 1922.

But the Committee had already met before this. By this time, the Dominions were being advised that they may wish to appoint members to serve on the Committee. (See NAA File 421/5/2)

The Official Secretary responded, two days later on **27 February**, advising that Colonel P.N. Buckley R.A.E., Military Adviser will represent the Commonwealth. (NAA File 421/5/2)

By **3 March, 1922**, the Battle Honours Committee had minuted its 9th Meeting. Apart from decisions as to limiting the number of Battle Honours to be claimed, it was suggested that the various units select their honours from the List of Honours to be shown in *the Army List*. (NAA File 421/5/2)

In a Minute Paper issued **8 March, 1922**, headed 'Preparation of Rolls in Connection with Battle Clasps' from the Department of Defence, early arrangements were being proposed for a committee to review with Base Records to determine the best method of compiling Unit Chronological Diaries. (NAA File 478/2/600)

On **23 March, 1922**, Inspector-General Lt Gen Chauvel urged the Secretary Department of Defence to circulate a copy of the Final Report of The Battle Clasps Committee. After all: '*it is hardly conceivable that clasps for this War will never be issued and the longer the compiling of rolls is delayed the more difficult it will be.*' (NAA File 478/2/600)

In what must be a record for any Committee involved in Army affairs to make a decision of any magnitude, the Battle Honours Committee signed off on its Report in response to the Army Order which gave rise to its work, on **11th April, 1922**, just four months after its work began. The formal Report of the Battle Honours Committee, under the Chairmanship of General Sir Archibald J. Murray was delivered. The Secretary was Captain C.N. Jervelund. The report was over 100 pages in length inclusive of Appendixes. Although this was its Report, the final report would not be completed until 1925.

Just three months after its second meeting, in its Report the Committed stated that it had: - 'consulted 137 Officers, including all the Commanding Officers of battalions serving at Home, on the question of the award of battle honours for the Great War. The 125 replies received show *a great divergence of opinion* on the subject.'

Recommendation 3.(iv) was that:

Regiments of Cavalry and Yeomanry, Battalions of Infantry, Regular, Militia (Imperial Reserve) and Territorial, shall have emblazoned on their Standards and Colours not more than twenty four Honours of which ***not more than ten*** *shall be "Great War" Honours, to embrace the whole history of the Regiment* *concerned, the date on which it was raised from the date on which it was raised to the end of the Great War, such Honours to be selected by Regimental Committees from the List of Honours to be shown in the Army List. In order to weld further the Regular, Militia and Territorial Battalions of the same Regiment together, the Honours emblazoned on the Colours shall be the same for all units comprising the Regiment concerned. The honours to be emblazoned on the Colours to be shown in the Army List in thicker type.*

And further, in Recommendation 3.(v):

That the guiding principle in the selection and allotment of Battle Honours shall be that Headquarters and at least 50% of the effective strength of a unit in a Theatre of War, (exclusive of drafts which, although in a Theatre of War, had not actually joined the unit), must have been present at the engagement for which the honour is claimed.

In a letter of **23 June, 1922**, from Colonial Office Secretary Winston Churchill, Downing Street, London, to Lord Forster, Governor-General, Australia, a copy of the Battle Honours Committee Report was provided. The letter stated:

I understand that the recommendations contained in the Report have now received approval and that the necessary instructions will be issued in due course; but, in the meantime, the Army Council request that the Report should be regarded as confidential. (NAA File 1926/525.)

On **28 July, 1922**, Secretary Trumble, of the Department of Defence wrote to the Prime Minister's Department, advising of matters arising from the Report of the Committee on Battle Clasps. The issues raised were subsequently the subject of a letter from the Governor-General to Downing Street, London, on 23 August, 1922. (NAA File 478/2/600)

At a meeting of the Battle Honours Committee on 18 August, 1922, it was decided that units shall be allowed to select up to a maximum of 10 for emblazoning on their respective Regimental Colours. Territorial Units would not be treated as having formed an integral part of units which had served in battle. (War Office File 32/5921)

Then **Army Council Instruction No. 458 of 1922** was published on 8 September, 1922 to state (pertinently) that: -

2. **If Regiments or Corps wish to select a Battle Honour from the column under the heading of "Actions etc." in the Battles Nomenclature Report, they may do so under the following conditions**: -
    (a) That a separate application is made for such a claim, giving full reasons for wishing to depart from the normal procedure.
    (b) That such claims will be a case of *substitution*, i.e. that any honour selected from this column will necessitate the abandonment of a claim to the same honour under the headings of columns 1 and 2.
    (c) That such claim so made will be considered *on its merits*.
3. If Regiments or Corps wish to claim an honour for an engagement in the Great War not mentioned in the Battles Nomenclature Report, they must make a separate application for it giving full reasons why it is desired. Such claims will be considered and a decision given.

4. A Battle Honour must only be claimed from one column of the Battles Nomenclature Report, e.g. if an honour is claimed from the column under the heading of "Tactical Incidents Included" the same honour cannot be claimed under the heading of the other columns.

Thus, it was ordered, in September 1922, that regiments should select up to 10 honours to be emblazoned on their Regimental Colours along with previous awards, up to a total of 24.

---ooo000ooo---

The appropriate recognition to be given to soldiers as a result of the Committee on Battle Clasps ignited a series of concerns by the Australian Prime Minister. At the instance of the Prime Minister, the Governor-General, Lord Forster wrote to the Secretary of State for the Colonies on **23 August, 1922**, and advised that with regard to France and Belgium, with the exception of the following actions the battles in which the Australian Division in France were engaged are covered by Battle Clasps. The **exceptions** were: -

(1) **Fromelles, July 1916**;

(2) The recapture of Villers Bretonneux, 25th April, 1918;

(3) The successful actions at Ville-sur-Ancre, May 1918 by the 2nd Australian Division; at Morlancourt, May 1918 by the 3rd Australian Division; at Morlancourt, June 1918 by the 2nd Australian Division, at Morlancourt, July 1918 by the 5th Australian Division.

(4) The successful attack of the 2nd Australian Division near Monument Wood (Villers Bretonneux) July 18th, 1918.

(5) The successful battle of Hamel, 4th July, 1918.

(6) The capture of Chuignes, 21st – 23rd August, 1918.

As regards the action at Fromelles, it was advised that:

This attack of the 5th Australian Division was the first offensive undertaken by the Australian troops in France. Apart from the fact that the 5th Division suffered very heavy losses, this action, it is urged, deserves recognition because it was successful in preventing the despatch of several German divisions to the Somme area at a critical time. <u>Though Fromelles is not geographically in the Somme area, it is considered that as the action was mounted in connection with operations on the Somme</u>, [emphasis added] *it would be fitting that the Battle Clasp "The Somme 1st July – 30th November, 1916" should be awarded to those units which took part in this action.*

Fromelles had been overlooked completely prior to this representation. This would give those who served at Fromelles at least some connection to the Somme, but it was not the ideal solution.

The other actions pointed out in the correspondence had their own individual analysis included in the correspondence.

(NAA File 478/2/600)

---ooo000ooo---

The **Army Council Instruction No. 458 of 1922** was drafted so that it in turn became **Army Order 338** on **4 September 1922**, in which we see that the **Army Battle Honours Committee** published the conditions of award of battle honour. These were contained in six short concise sub paragraphs:

1. *His Majesty the King has been graciously pleased to approve the award of battle honours to regiments and corps under the following conditions: -*
   - *(a) Regiments and corps will have awarded to them, and recorded … the honours due to them for taking part in battles enumerated in … the tabulated list of engagements given* ***in the report of the Battles Nomenclature Committee.***
   - *(b) Following the honours previously earned and at the head of the List of Honours granted … will be placed 'The Great War' and the number of battalions of the regiment taking part.*
   - *(c) Within the meaning of regiments and corps will be included Cavalry and Yeomanry regiments. An infantry regiment or corps will include the Regular, Militia (or Special Reserve), Territorial and Service Battalions … There will be only one Honours List for a regiment or corps.*
   - *(d) Regiments of Cavalry and Yeomanry, battalions of Infantry, Regular, Militia (or Special Reserve) and Territorial, will have emblazoned on their Standards, Guidons and Colours not more than 24 honours of which* ***not more than 10 will be 'Great War' honours*** *… such honours to be selected by Regimental Committees … The honours emblazoned on the Colours will be the same for all units comprising the regiment concerned, and will be shown in the Army List in thicker type.*
   - *(e) The guiding principle in the selection and allotment of battle honours will be that* ***Headquarters and at least 50 per cent of the effective strength of a unit … must have been present at the engagement for which the honour is claimed.***
   - *(f) Regimental Committees under the chairmanship of their regimental colonels … will be set up to select the particular honours for Regimental Colours.*

---ooo000ooo---

Thus, it was ordered, in September 1922, that regiments should select up to 10 honours to be emblazoned on their Regimental Colours along with previous awards, up to a total of 24.

Understandably, it led to a storm of protest, since many regiments would have had to remove previous honours. The order was, therefore amended the following December to allow each infantry regiment to select up to 10 honours to be emblazoned on its King's Colour, honours from other conflicts continuing to be displayed on the Regimental Colour. (See (British) Army Order 470/1922.)

---ooo000ooo---

Army Council Instruction No. 458 of 1922 was then published on **8 September, 1922** to state that: -

- Should Regiments or Corps wish to select a Battle Honour form the heading of 'Actions' set out in the BNC report, the conditions were set out under which they could do so.

- Most significantly, it was stated that such claims would be made *in substitution* for any which would have been listed under the Battles headings.
- But each claim would be considered ***on its merits***.

---ooo000ooo---

Any claim for a Battle Honour for Fromelles would therefore be seen as being *in substitution* for something else if it was made. Would anybody therefore be prepared to contest a claim for an Action in substitution for a Battle Honour on that basis?

Fromelles' was listed under the heading of 'Actions, &c.'. It had been attached in the BNC Report to Bazentin Ridge as a subsidiary Attack. It was not classified as a Battle, let alone a successful battle.

In the months following, units were advised of the Army Council's decisions:

That Regiments and Corps shall have awarded to them, and recorded in the Army List, in addition to those already shown, the Honours due to them for taking part in the Battles enumerated in Columns 1 and 2 of the Tabulated List of engagements given in the report of the Battles Nomenclature Committee (A.2325).

The limitation of not more than 24 Honours, of which not more than 10 shall be "Great War" Honours was reiterated. At the same time, the units were informed that: -

*I am able to state that should you wish to select "**Tactical Incident**" or "**Action**" contained in Columns 3 or 4 of the Battles Nomenclature Committee Report you may do so under the following conditions –*

(a) That a separate application is made for such a "Tactical Incident" or "Action", giving full reasons for wishing to depart from the normal procedure.

(b) That it is to be understood that such action will be a case of substitution i.e. that any "Tactical Incident" or "Action" selected under Columns 3 or 4 will necessitate the abandonment of a claim to the same Operation or Battle under Columns 1 and 2.

(c) That such claim so made will be considered on its merits by the Army Council and a decision given by them.

I am to request that you will convene a Committee of your Regiment or Corps, to which a representative of every Battalion whether still existing or not must be invited, to prepare lists of Battle Honours as claimed by the Regiment or Corps as a whole, for the purpose of printing in the Army List, on the attached proforma, and a further list showing the Honours selected by your Committee for emblazoning on the Colours. These lists to be forwarded in triplicate to this Department.

---ooo000ooo---

On **5 October, 1922**, by letter from Secretary Winston Churchill, Colonial Office, Downing Street, London, Lord Forster, Governor-General Australia, was advised of the making of **Army Order 338 of 1922** and **Army Council Instruction 458 of 1922**, which was now being communicated to the Dominions. The subject of the letter sent confidentially, referred to on 23 June, 1922, now became official. (NAA file 1926/525)

The Intercession of Major General Sir John Headlam.

Then there was an interesting development in the United Kingdom, which does not appear to have been advised to the Dominions, and which this author has only found revealed after 70 years of being classified.

On 10 October, 1922, Major General John Headlam, President of the BNC, wrote to the Secretary of the War Office, referring to Army Order 338. His comments are worth reading here so that we can put them specifically in context in relation to Australian Units. He wrote to address what he saw was a misapprehension of the BNC Committee intentions, which, if uncorrected,

> ***'.... must lead to difficulties when the Regimental Committees get to work'***

and to put on record that a claim for an "Action" could not in fact be in substitution for the honour of a "Battle".

He wrote that:

> ***But, as a matter of fact, there can be no question of the 'same honour' between the two columns, for the engagements recorded in the Actions column are entirely separate and distinct from those in the Battles column, and in no case do they bear the same name.***
>
> ***It is true that in some cases actions are aligned to battles by such phrases as "With subsequent", "With subsidiary", etc., but this was done only to bring out the strategical connection, and was never intended to imply that these actions formed any part tactically of the battles to which they are so linked. An examination of the limits of time and space assigned to each action will make it clear that they were absolutely distinct engagements.***

There was therefore the distinct possibility that the British 61st Division, and the Australian 5th Division could claim 'Fromelles' as an honour under the "Actions" heading. But sadly, it appears that this did not take place. The letter of Major General Headlam does not appear to have been passed on to the Australian authorities.

Although Headlam offered a method by which this misconception could be cured in Army Orders, it was not taken up by the Army Council; instead it was retained in correspondence and not given any action in response.

The possible consideration of *ACTIONS* being given Honours, in addition to *BATTLES* as opposed to being in substitution for such, was therefore not taken up. It was never allowed to be.

---ooo000ooo---

Two months later, on **7 December, 1922**, the BHC questioned whether ***unsuccessful engagements*** should be awarded as Battle Honours. Fromelles was an example of an unsuccessful engagement.

The Committee noted and accepted that in the past, campaigns that were ***unsuccessful engagements have never been awarded as Battle Honours*** [emphasis added]. Except for two: -

(i) Monte Video 1807; and

(ii) Mangalore 1783.

The Committee diplomatically noted that all engagements successful or not ultimately led to the result of the victory. Ultimately, the Committee sought guidance on the issue, and questioned if, during the administration of Battle Honours, any glaring case came to light where a Regiment claiming an Honour did not maintain the high traditions of the Service, that a special report on the case can be submitted by them to the Army Council. By 27 December, 1922, the Army Council had decided the BNC decisions should be followed.

A Battle Honour for Fromelles appeared to be doomed.

(War Office File 32/5921)

On **16 December, 1922**, The British War Office issued a further '**Army Council Instruction' number 619 of 1922**, in that: -

The following amendment will be made to A.C.I. 458 of 1922: -

For Para 6, substitute: -

6. Regimental Committees, to which a representative of every battalion, whether still existing or not, must be invited, will be convened by the Colonels of Regiments and Corps, to prepare lists of Battle Honours as claimed by the Regiments or Corps as a whole, for the purpose of printing in the Army List, and further lists showing the "Great War" Honours selected for emblazoning on the King's Colours, Standards and Guidons.

On the same day, in Army Order 470 of 1922, it was proclaimed (among other statements) that:

The Honours emblazoned on the King's and Regimental Colours will be same for all units comprising the regiment concerned and will be shown in the Army List in thicker type.

This was aimed at the British Regiments and Corps which consisted of multiple Battalions. During peacetime a British Infantry Regiment usually consisted of two or three battalions. In wartime the Regiment would recruit and raise more battalions. If the Regiment had a large recruiting base (City/Country) they raised many wartime battalions. For example, The Northumberland Fusiliers swelled to 52 battalions, which were scattered and fought on the Western Front, Italy, Macedonia, Gallipoli and Egypt.

If a Regiment had a small recruiting base they raised only a few battalions. For example, The Hertfordshire Regiment raised 4 battalions for the Great War, all of which only fought on the Western Front.

The Northumberland Fusiliers' 52 battalions were scattered across many theatres of war in WW1. Its separate battalions collectively earned over 70 battle honours which were all consolidated and contribute to The Northumberland Fusiliers' list of Battle Honours. From this list 10 were chosen to display on The Northumberland Fusiliers' King's and Regimental Colours.

In other words, the individual battalions that earned the various battle honours did not have separate Colours to show the separate Battle Honours that each had respectively earned. Hence the 'Order' that the Honours emblazoned on the King's and Regimental Colours will be same for all units [battalions] comprising the regiment concerned.

The Hertfordshire Regiment's 4 battalions earned a total of 26 Battle Honours, 10 of which were chosen for display on their Colours.

---ooo000ooo---

Four years after fighting had ceased on the Great War, Secretary Devonshire in Downing Street wrote on **19 December, 1922** to the Governor-General in Australia advising that no decision had yet been taken by His Majesty's Government on the report of the Committee on Battle Clasps. The Governor-General subsequently referred the correspondence to the Prime Minister's Department.

---ooo000ooo---

Army Order 470 of 1922 was subsequently published on **22 December, 1922**.

It dealt with obviating the necessity of removing honours at present emblazoned on the colours of regiments and corps. In addition to those already emblazoned *a maximum of 10* may be added in addition to those already carried.

Regimental Committees were to be set up to select the honours, and select those for emblazoning.

---ooo000ooo---

By the end of 1922 The Battle Honours Committee had used the findings of the BNC to determine eligibility for Battle Honours. Each British or Dominion unit (infantry battalion, cavalry regiment, and later, armoured regiment) was permitted to apply for battle honours from the list of more than 180 on the BNC list, of which ***no more than ten could be displayed***.

Divisions or Brigades were 'formations', and not 'units'. Accordingly, only units, being infantry, cavalry or armoured units could be eligible to make claims.

The selections required approval by the British War Office before being finally approved by the King. There are a number of units that were awarded over 50 battle honours from the Great War, but only display ten.

Almost every Australian Infantry Battalion was awarded over 20 Great War Battle Honours, but only display ten.

--- 1923 ---

Further Intercession by Major General Sir John Headlam.

Again, on **20 January, 1923**, Major General Sir John Headlam wrote to General Sir Archibald Murray, of the Battles Honours Committee. Headlam was not satisfied that the manner in which recognition for Battle Honours being implemented was appropriate. All recognitions for Honours appeared to be determined on the findings of the BNC in determining what the relative ranking of Battles, Actions, and Tactical Incidents among others. His letter stated: -

> ***You will of course remember that there was no mention of Honours in our Terms of Reference. We were only asked to name, define, classify, and tabulate the engagements fought in the War, and we therefore looked upon our task as primarily historical. Fortunately the idea that our Report might eventually be utilised in the award of Honours was not entirely absent from our minds, but even so, the problem has proved a complicated one.***

More significantly, and putting it plainly:

> ***... contrary to general practice in the past, engagements should not be debarred from consideration for Honours on account only of want of success.***

What then of Actions which were grouped under Battles? Headlam had some clear comments to be made on that as well:

> ***The engagements recorded in the "Actions" Column differ absolutely from those Tactical Incidents just considered, in that they are in all cases engagements separate and distinct from those in the Battles Column. It is true that in some cases they are linked to Battles by such phrases as "With subsequent" ... "With subsidiary" ... and so forth. But these connecting phrases were introduced only to bring out the <u>strategical</u> connection and were never intended to imply that they formed any part <u>tactically</u> of the Battles to which they are so linked. <u>Each of the engagements in the Actions Column must therefore be considered on its own merits as regards its suitability as an Honour.</u>***

(War Office File 32/5921)

---ooo000ooo---

On **19 February, 1923**, the UK War Office circulated **Army Council Letter 3/Gen. No. 209 of 1923** to be passed to the various Dominions as well as the units in the United Kingdom, regarding the selection of Battle Honours. That letter was received in Australia in due course, and a copy was sent to the Governor-General, on 9 March, 1923. The War Office advice contained the direction that:

The only Battle Honours which can be selected by Regiments and Corps from the Tabulated List of Engagements, as shown in the Report of the [BNC], are those contained in the attached list marked "B".

The List marked "B" was prefaced to be:

(In addition to any that may be granted for the special reasons referred to in para: 3 of Army Council Instruction No. 458 of 1922.)

List B contained reference to Somme 1916, under which were recorded:

Albert,	Bazentin,	Delville Wood,	Pozieres,
Guillemont,	Ginchy,	Flers-Courcelette,	Morval,
Thiepval,	Transloy,	Ancre Heights, and	Ancre 1916.

Several names of particular significance to Australia were not included in the initial list of battle honours. As a result of a submission by the Governor-General to the British War Office, some modifications were later made to the list to satisfy Australian requirements. Amendments included the addition of Mont St Quentin, Hamel, Péronne, and Chuignes in France, and a dispensation to use Sari Bair–Lone Pine for the fighting on Gallipoli. (NAA File 1926/525 & War ffice File 32/5921)

Fromelles of course was nowhere to be seen on the list as the BNC had already relegated it to being an Attack, and a subsidiary Attack at that. And the Honours Committee was not about to offer up an unsuccessful engagement.

---ooo000ooo---

The Canadians were perplexed. So much so that on **27 March, 1923** they wrote to the War Office at Ministerial level, to ask what had happened.

The revised List of Battles forwarded to Canada, and to Australia, omitted all operations which were classified by the BNC as "Actions". The only exception appeared to be Villers Bretonneux.

---ooo000ooo---

Under **Army Council Instruction 458 of 1922** a unit was permitted to make special application to have an "Action" placed upon its Colours. The War office letter of 19 February, 1923 stated however, that the only Battle Honours which could be selected from the list of engagements as shown in the BNC Report are those which were contained in the list attached to the War Office letter. Was it to be understood that the permission contained in the ACI referred to previously was no longer applicable?

Furthermore, in the List of Battles attached to the War Office letter, "Tactical Incidents" were also omitted. Under ACI 458 of 1922, such tactical incidents might be claimed by a unit as a Battle Honour provided the same honour was not also claimed under the major heading of the battle or operation of which it formed part. As tactical incidents did not appear in the revised list, was it to be understood that no choice could now be made between the tactical incidents shown in the BNC Report and the major battle headings?

At this time, it is interesting to note the proceedings underway in relation to those actions which were fought in the area of Fromelles in 1915, in which British and Indian forces were involved.

On **13 June, 1923**, Captain Jervelund of the BHC sought clarification from the Historical Section of the War Office in relation to two actions in 1915, whose actions were listed under 'Aubers' and 'Rue Du Bois'. Should they not both be grouped together under one Battle Honour? The significance of these actions has been revealed previously: they were fought in essentially the same area as that of Fromelles.

The answer would come eleven months later, on 10 May, 1924. **Military Order 167 of 1924** was published, stating that the two actions should receive the one Battle Honour of 'AUBERS'. (See entry for that date in due course.)

---ooo000ooo---

The question of the award of Battle Clasps was finally resolved on **2 July, 1923**. Lord Devonshire from the Colonial Office, London, wrote to the Governor-General of Australia, to advise that in view of the present financial stringency, the Army Council had decided not to proceed with the proposal to issue battle clasps for the Great War. (NAA File 478/2/600)

---ooo000ooo---

In dealing with the returns from various Units, The War Office on the command of the Army Council advised generally on **25 July, 1923**, that some further explanation was required.

They proceeded to give the example of fighting under the Battle Honour of BAZENTIN. But in closing they further advised that while many Regiments had distinguished themselves in the highest degree of fighting at localities which have not been included in the list previously provided, it had been decided ***not to make exceptions in such cases unless the action of the unit has formed the subject of a Special order of the Day by a British Commander-in-Chief in the Field***. (War Office File 32/5921)

The means of deciding claims for Battle Honours were being restricted even further, without a specific Order or Instruction. (War Office File 32/5921)

---ooo000ooo---

Adopting this reasoning, it would be apparent that the ***Attack at Fromelles*** was capable of being awarded a Battle Honour, even if it was not successful at first instance being classified as a Battle.

Perhaps a claim for a Battle Honour for Fromelles was not doomed after all.

But Fromelles does not have any such Special Order of the Day.

---ooo000ooo---

By 1 August, 1923, Australia had published Australian Military Order 336 of 1923 by command of the Military Board. This adopted the principles in place to date from the British Instructions, and Orders, and would form the basis upon which Australian units would proceed in their own quests for battle honours.

The processes which unfolded in Australia will be the subject of the following Chapter. The rest of this Chapter will look at the way in which Battle Honours were finalised in Great Britain.

---ooo000ooo---

Instructions for submission of claims for Battle Honours had been previously contained in UK **Army Council Instruction No. 458 of 1922**, dated 8.09.1922, published as Appendix III to this Order, but: -

- Not more than 10 Battle Honours may be selected for emblazonment on guidons of Light Horse Regiments, or Colours of battalions.
- Committees of not less than 6 but not more than 10 were to consider such claims.

--- 1924 ---

On **1 February, 1924**, the Secretary of State, Downing Street, London, wrote to advise the Governor-General in Australia of the tabulated list of decisions which the Army Council had reached, concerning the award of Battle Honours to Regiments and Corps.

(NAA 1926/525.)

The first of ten lists of awards for The Great War (related to the British Army) was published in **February 1924** in **Army Order 49**. Curiously it contained three battle honours not to be found in the Report of the BNC: -

- '**Pursuit to Mons**' had been introduced by the Battle Honours Committee as an additional honour to cover operations in Picardy between 17 October and 11 November 1918, and could be claimed only by corps which took part in these operations, but did not qualify for the award of either Selle, Valenciennes or Sambre.
- Although many individual claims for additional battle honours were disallowed, '**Bois des Buttes'** was approved for The Devonshire Regiment for the 2nd Battalion's action on 6 June, 1918, when IX Corps was under the command of the French Fifth Army.
- List No. 2 also promulgated a second additional award – '**Bligny**' – to the King's Shropshire Light Infantry on behalf of the 1/4th Battalion, the King's (Shropshire Light Infantry), which carried out a critical counterattack on 6 June 1918, again under IX Corps.

Although these battle honours (Pursuit to Mons, Bois des Buttes, and Bligny) were not identified as engagements by the BNC, the BHC created them in response to applications that were received from the Regimental Colonels of the respective units involved – which presented convincing cases. This showed that the BHC was not limited to the engagements identified in the BNC report.

---ooo000ooo---

Military Order 167 of 1924 was published on **10 May, 1924**. The order referred to MILITARY ORDER 336 of 1923, in which the Battle of Aubers Ridge (1915) was shown as consisting of (1) Attack at Fromelles; and (2) Attack at Rue du Bois. These Attacks were set out on page 15 of the BNC Report. The order concluded by stating that '**AUBERS**' was selected as the Battle Honour for those operations. Again, note that these two attacks took place in 1915. And that they were both unsuccessful.

The reason for us looking at the AUBERS Battle Honour is that in comparison, the action in July 1916 was also unsuccessful, but as we will read later, no Battle Honour was awarded for Fromelles in 1916.

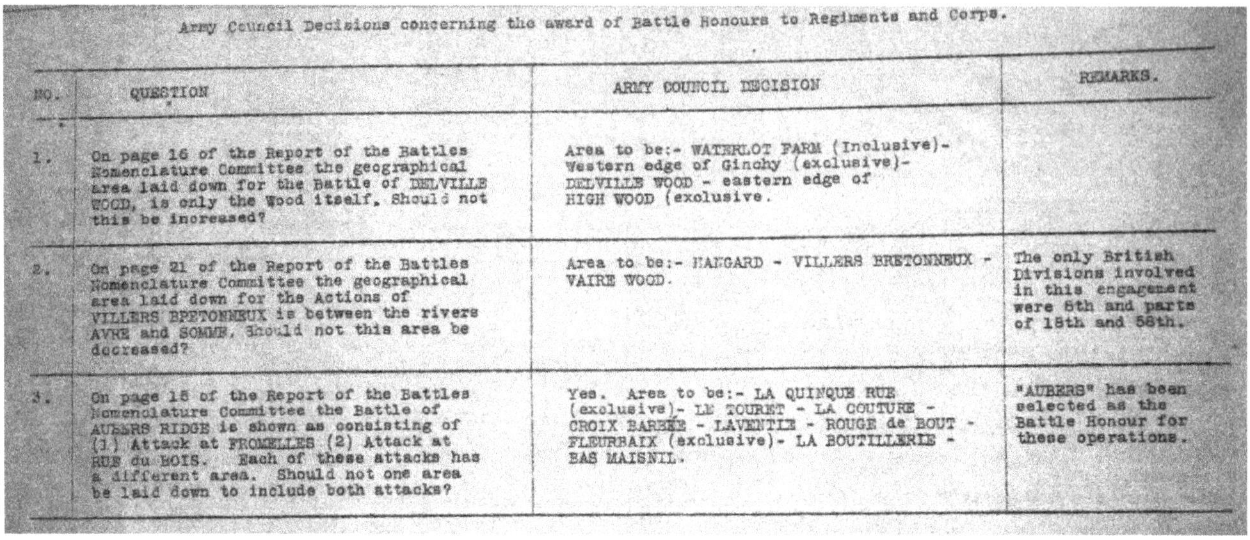

Army Council Decisions concerning the award of Battle Honours to Regiments and Corps.

NO.	QUESTION	ARMY COUNCIL DECISION	REMARKS.
1.	On page 16 of the Report of the Battles Nomenclature Committee the geographical area laid down for the Battle of DELVILLE WOOD, is only the Wood itself. Should not this be increased?	Area to be:- WATERLOT FARM (Inclusive)- Western edge of Ginchy (exclusive)- DELVILLE WOOD - eastern edge of HIGH WOOD (exclusive.	
2.	On page 21 of the Report of the Battles Nomenclature Committee the geographical area laid down for the Actions of VILLERS BRETONNEUX is between the rivers AVRE and SOMME. Should not this area be decreased?	Area to be:- HANGARD - VILLERS BRETONNEUX - VAIRE WOOD.	The only British Divisions involved in this engagement were 8th and parts of 18th and 58th.
3.	On page 18 of the Report of the Battles Nomenclature Committee the Battle of AUBERS RIDGE is shown as consisting of (1) Attack at FROMELLES (2) Attack at RUE du BOIS. Each of these attacks has a different area. Should not one area be laid down to include both attacks?	Yes. Area to be:- LA QUINQUE RUE (exclusive)- LE TOURET - LA COUTURE - CROIX BARBEE - LAVENTIE - ROUGE de BOUT - FLEURBAIX (exclusive)- LA BOUTILLERIE - BAS MAISNIL.	"AUBERS" has been selected as the Battle Honour for these operations.

[Picture 43: Extract from (UK) Military Order 167 issued on 10 May, 1924.]

Military Order 300 of 1924 was published on **26 July, 1924**, (in similar fashion to Army Order 338 of 1922 published on 4 September, 1922) proclaiming that the ONLY Battle Honours which can be selected by Regiments and Corps from the tabulated list of engagements, as shown in Appendix V of Military Order 336/1923, are those contained in Appendix II thereto.

Appendix II proclaimed that 'Somme 1916' would include various battles, but Fromelles was not mentioned. It had not been classified as a Battle by the BNC.

Under the heading "Battle Honours", it is stated thus: -4

3. *The only Battle Honours which can be selected by Regiments and Corps from the tabulated list of engagements, as shown in Appendix V. of Military Order 336/1923, are those contained in Appendix II.*

Appendix II Battle Honours went on to state:

(In addition to any that may be granted for the special reasons referred to in para 3. of Army Council Instruction No. 458/1922, vide Appendix III of MO 336/1923.)

Festubert, 1914	Somme, 1916.
Givenchy, 1914	- Albert
Neuve Chappelle.	- Bazentin *(*Fromelles*)*
Hill 60.	- Delville Wood
Ypres 1915	- Pozieres.

- Gravenstafel.	- Guillemont
- St. Julien.	- Ginchy
- Frezengerg	- Flers-Courcelette
- Bellewaarde	- Morval
Aubers	- Thiepval
Festubert, 1915	- Le Transjoy
Hooge, 1915	- Ancre Heights
Loos.	- Ancre, 1916
Mount Sorrel	Bapaume, 1917.

Fromelles was included in this table only because it had been listed as a subsidiary Attack to the Battle of Bazentin.

---ooo000ooo---

Military Order 480 of 1924 was published on **1 November, 1924**, 'in order to make it quite clear to all concerned, that honours may be claimed by units as under: -

(a) Battle Honours, vide appendix II to Military Orders 300/1924;

(b) Honours under the heading 'Action, &c.', vide para 2 of Army Council Instruction 458 of 1922 (Appendix III of Military Order 336/1923). *This is a case of substitution.*

(c) "Tactical Incidents" vide para 4 of Army Council Instruction 458 of 1922. *This is a case of substitution.*

(d) Special Claims, vide para 3 of Army Council Instruction 458 of 1922.

(e) Theatre of War Honours, vide Military Order 300 of 1924.

Under the heading 'Battle Honours', it was stated that: -

"With reference to Military Order 300/1924, the question has been raised whether the list of Battle Honours given in Appendix II thereto excludes those honours given in Appendix V to Military Order 336/1923 under the column "Tactical Incidents included". This is not so.

In answering the question it was stated that Honours may be claimed as under: -

(a) Special Claims, vide para. 3 of Army Council Instruction 458 of 1922.

(b) Honours under the heading 'Action, &c.', vide para 2 of Army Council Instruction 458 of 1922 (Appendix III of Military Order 336/1923). *This is a case of substitution.*

(c) "Tactical Incidents" vide para 4 of Army Council Instruction 458 of 1922. *This is a case of substitution.*

(d) Special Claims, vide para 3 of Army Council Instruction 458 of 1922.

(b) (e) Theatre of War Honours, vide Military Order 300 of 1924.

Thus Special Claims ***could*** be made. Fromelles could in fact become the subject of Special Claims after all.

--- 1925 ---

UK Army Order 55 of 1925 was issued in **February, 1925**. By this time, the British had felt that they had classified all that were to be classified for Battle Honours, and that their job was now complete, at least in relation to the British units. All previous lists had been cancelled. The only lists now in effect would be those to be published in the forthcoming '**Complete List of Battle Honours Awarded for the Great War**'. The Order stated that:

'No further submissions concerning Great War battle honours will be made'.

(War Office File 32/5921)

The exasperated hope of the Battle Honours Committee or the Army Council was not to be.

Over the years since the Great War, several amendments have since been made to awards published in the Final List."[144]

The **Complete List of Battle Honours Awarded for the Great War** was issued on 27 February, 1925, with the UK Army Orders for 1925. The list applied to United Kingdom units, not Dominion ones, as much was still being done in Australia and the Dominions to finalise their claims for Battle Honours.

---ooo000ooo---

The Complete List included the Battle Honours for the Royal Warwickshire Regiment, for the Gloucestershire Regiment, and for the Oxfordshire and Buckinghamshire Light Infantry. Significantly, each of these units fought in the action in 1915, and would each again fight at Fromelles in 1916.

Reading it, we can see that the Royal Warwickshire Regiment was able to claim Battle Honours for 'NEUVE CHAPELLE', for 'AUBERS' and for 'FESTUBERT', all in 1915.

And that the Gloucestershire Regiment was awarded a Battle Honour for 'AUBERS', again in 1915.

And that the Oxfordshire and Buckinghamshire Light Infantry was awarded a Battle Honour for each of 'AUBERS' and 'FESTUBERT', again each in 1915.

Just as it would be the case for Australian units, neither of these units would be able to claim a Battle Honour for Fromelles in 1916. Any connection with Fromelles in 1916 was to be made via the Battle Honour of 'Somme 1916, '18', which as we have already noted, is a misapplication of geographic and tactical reality.

---

[144] Rodger p.85

---ooo000ooo---

The Battle Honours which were to apply for British units were now finalised, and subject to any future claims for amendment, this saga had come to an end for the soldiers from Britain.

14. Why is there No Battle Honour for Fromelles?

You can claim as many as you like, but you can only display 10.

So choose wisely.

---ooo000ooo---

In Chapter 12, we looked at the formal process of classification of battles and other actions during World War 1, by the Battles Nomenclature Committee (BNC). The conclusion reached by the BNC was that Fromelles was an Attack, subsidiary to the Battle of Bazentin Ridge. But not a Battle.

The work of the BNC would have the follow-on effect that various engagements would be able to be claimed for battle honours by the soldiers who fought in them. The further work of determining which engagements deserved the recognition of a battle honour would be undertaken by another body altogether: ***The Battle Honours Committee*** ('the BHC').

In this Chapter, you will read of the manner in which claims for Battle Honours were dealt with in Australia. You will also read a history of events whereby no Australian unit which fought at Fromelles was permitted to claim 'Fromelles' for a Battle Honour.

While the object of this chapter will be the application of such matters to Fromelles for Australian units, elsewhere in separate actions, the British 61st Division's Commander McKenzie sought to claim 'Fromelles' as a Battle Honour, but was also unsuccessful.

It is again worth pointing out that unless you had a successful engagement, you could not *ordinarily* hope to be the subject of a Battle Honour. But it is possible, and has occurred on at least two occasions. So it was when the BHC adopted the findings of the BNC and then proceeded to determine the recognitions which would be awarded for the various engagements.

---ooo000ooo---

Australian Military Order 336 of 1923 was published on 1 August, 1923, promulgated by command of the Military Board, with the following specific provisions: -

(1) The conditions under which Battle Honours may be awarded to the Light Horse Regiments and Infantry Battalions of the AIF are set out in **Army order 338 of 4 September, 1922**, and **Army Orders 470 dated 16 December, 1922**.

(2) The Battles Nomenclature Committee Report contains a tabulated list of engagements in respect of which Battle Honours may be claimed.

(3) Headquarters and at least 50 percent of a unit must have been present at the engagement.

(4) All Battle Honours granted will be included in the records of the various units, but not more than 10 may be selected for emblazonment on unit Colours or guidons.

(5) Regiments and Battalions are responsible for furnishing and substantiating their claims.

(6) to (9) also published, but are not being dealt with further for present purposes.

Instructions for submission of claims for Battle Honours had been previously contained in UK **Army Council Instruction No. 458 of 1922**, dated 8.09.1922, published as Appendix III to this Order, but: -

- Not more than 10 Battle Honours may be selected for emblazonment on guidons of Light Horse Regiments, or Colours of battalions.
- Committees of not less than 6 but not more than 10 were to consider such claims.

--- 1924 ---

On **1 February, 1924**, the Secretary of State, Downing Street, London, wrote to advise the Governor-General in Australia of the tabulated list of decisions which the Army Council had reached, concerning the award of Battle Honours to Regiments and Corps.

(NAA 1926/525.)

The first of ten lists of awards for The Great War (related to the British Army) was published in **February 1924** in **Army Order 49**. Curiously it contained three battle honours not to be found in the Report of the BNC: -

- '**Pursuit to Mons**' had been introduced by the Battle Honours Committee as an additional honour to cover operations in Picardy between 17 October and 11 November 1918, and could be claimed only by corps which took part in these operations, but did not qualify for the award of either Selle, Valenciennes or Sambre.
- Although many individual claims for additional battle honours were disallowed, '**Bois des Buttes'** was approved for The Devonshire Regiment for the 2nd Battalion's action on 6 June, 1918, when IX Corps was under the command of the French Fifth Army.
- List No. 2 also promulgated a second additional award – '**Bligny**' – to the King's Shropshire Light Infantry on behalf of the 1/4th Battalion, the King's (Shropshire Light Infantry), which carried out a critical counterattack on 6 June 1918, again under IX Corps.

Although these battle honours (Pursuit to Mons, Bois des Buttes, and Bligny) were not identified as engagements by the BNC, the BHC created them in response to applications that were received from the Regimental Colonels of the respective units involved – which presented convincing cases. This showed that the BHC was not limited to the engagements identified in the BNC report.

---ooo000ooo---

Military Order 167 of 1924 was published on **10 May, 1924**. The order referred to MILITARY ORDER 336 of 1923, in which the Battle of Aubers Ridge (1915) was shown as consisting of (1) Attack at Fromelles; and (2) Attack at Rue du Bois. These Attacks were set out on page 15 of the BNC Report. The order concluded by stating that '**AUBERS**' was selected as the Battle Honour for those operations. Again, note that these two attacks took place in 1915. And that they were both unsuccessful.

The reason for us looking at the AUBERS Battle Honour is that in comparison, the action in July 1916 was also unsuccessful, but as we will read later, no Battle Honour was awarded for Fromelles in 1916.

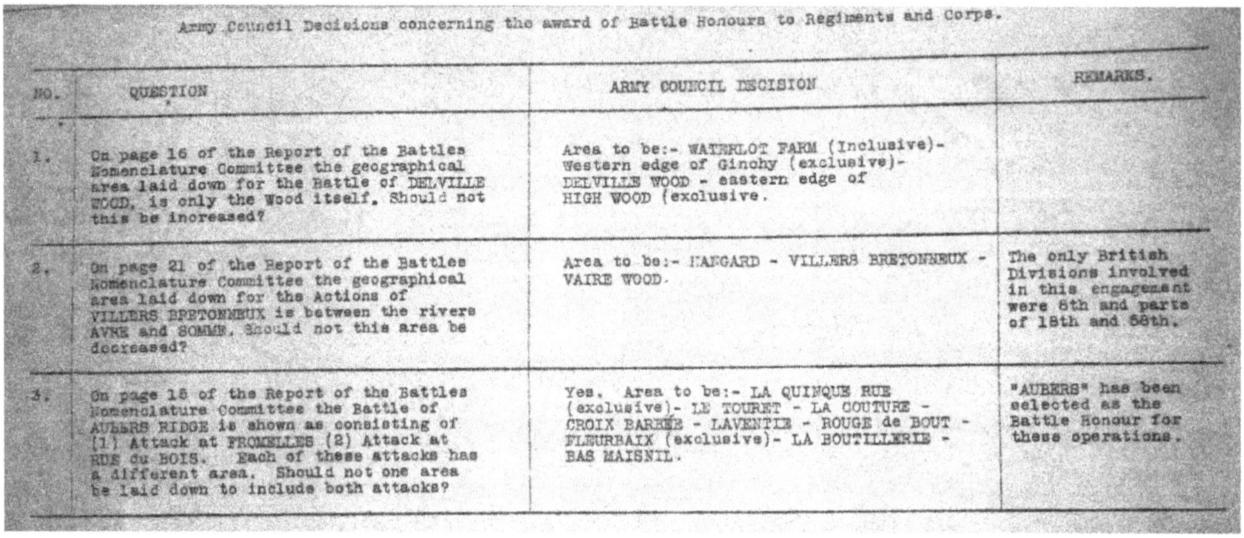

Army Council Decisions concerning the award of Battle Honours to Regiments and Corps.

NO.	QUESTION	ARMY COUNCIL DECISION	REMARKS.
1.	On page 16 of the Report of the Battles Nomenclature Committee the geographical area laid down for the Battle of DELVILLE WOOD, is only the Wood itself. Should not this be increased?	Area to be:- WATERLOT FARM (Inclusive)- Western edge of Ginchy (exclusive)- DELVILLE WOOD - eastern edge of HIGH WOOD (exclusive.	
2.	On page 21 of the Report of the Battles Nomenclature Committee the geographical area laid down for the Actions of VILLERS BRETONNEUX is between the rivers AVRE and SOMME. Should not this area be decreased?	Area to be:- HANGARD - VILLERS BRETONNEUX - VAIRE WOOD.	The only British Divisions involved in this engagement were 8th and parts of 18th and 58th.
3.	On page 15 of the Report of the Battles Nomenclature Committee the Battle of AUBERS RIDGE is shown as consisting of (1) Attack at FROMELLES (2) Attack at RUE du BOIS. Each of these attacks has a different area. Should not one area be laid down to include both attacks?	Yes. Area to be:- LA QUINQUE RUE (exclusive)- LE TOURET - LA COUTURE - CROIX BARBÉE - LAVENTIE - ROUGE de BOUT - FLEURBAIX (exclusive)- LA BOUTILLERIE - BAS MAISNIL.	"AUBERS" has been selected as the Battle Honour for these operations.

[Picture 44: Extract from (UK) Military Order 167 issued on 10 May, 1924.]

Military Order 300 of 1924 was published on **26 July, 1924**, (in similar fashion to Army Order 338 of 1922 published on 4 September, 1922) proclaiming that the ONLY Battle Honours which can be selected by Regiments and Corps from the tabulated list of engagements, as shown in Appendix V of Military Order 336/1923, are those contained in Appendix II thereto.

Appendix II proclaimed that 'Somme 1916' would include various battles, but Fromelles was not mentioned. It had not been classified as a Battle by the BNC.

Under the heading "Battle Honours", it is stated thus: -

3. *The only Battle Honours which can be selected by Regiments and Corps from the tabulated list of engagements, as shown in Appendix V. of Military Order 336/1923, are those contained in Appendix II.*

Appendix II Battle Honours went on to state:

(In addition to any that may be granted for the special reasons referred to in para 3. of Army Council Instruction No. 458/1922, vide Appendix III of MO 336/1923.)

Festubert, 1914	Somme, 1916.
Givenchy, 1914	- Albert
Neuve Chappelle.	- Bazentin *(*Fromelles*)*
Hill 60.	- Delville Wood

Ypres 1915	- Pozieres.
- Gravenstafel.	- Guillemont
- St. Julien.	- Ginchy
- Frezengerg	- Flers-Courcelette
- Bellewaarde	- Morval
Aubers	- Thiepval
Festubert, 1915	- Le Transjoy
Hooge, 1915	- Ancre Heights
Loos.	- Ancre, 1916
Mount Sorrel	Bapaume, 1917.

Fromelles was included in this table only because it had been listed as a subsidiary Attack to the Battle of Bazentin.

---ooo000ooo---

On **28 July, 1924**, Admin. Staff of 1st Cavalry Division wrote to 1st, 6th, 7th, 12th, and 15th Light Horse Regiments, advising that any changes in the Regimental Committees List of Battle Honours necessitated by **Military Order 300/1924** would be discussed in upcoming conference of C.O.'s on 12 August, 1924. (NAA File 472/1/87).

No doubt, similar advices were being forwarded by Infantry Divisions to their component Battalions.

On **13 August, 1924**, the 45th Battalion were requested to re-submit their claim in conformity with MO 300/1924. Of interest was that 'Attack on Fromelles' had been indicated in relation to 30th, 53rd, 54th, 55th, 56th Battalions, despite not being included in the Appendix II to MO 300/1924. The Army personnel in charge of selecting the appropriate Battle Honours were promoting 'Fromelles'.

In a response, on **19 August, 1924**, Brig General Bennett advised that 'in the Battle of Fromelles the 30th Battalion should be inserted – this unit was omitted in error.' It should thus be included along with 53rd, 54th, 55th & 56th Battalions. There would therefore now be **five** battalions which were being advised should make claim to 'Fromelles'.

General Chauvel wrote on **25 August, 1924**, to Lt.-Col. E.M. Williams, from Army Headquarters, Melbourne, advising of the 27 Battle Honours for which he understood the various Cavalry units were variously entitled. (NAA File 472/1/87)

On **9 September, 1924**, 30th Battalion AIF lodged its amended Claims for Battle Honours, with the Secretary, District Battle Honours Committee. The amended claims sought the inclusion of **FROMELLES**, BAPAUME and LAGNICOURT.

In relation to Fromelles, the claim provided the following background:

The Battle of FROMELLES was the first Action of the 30th Battalion participated in. A Company under Major BEARDSMORE, followed the Advance of the 31st Battalion & commenced workon digging the Sap to connect the British Front line to the German. The remainder of the Battalion were shortly after absorbed in support of the Troops fighting in the German Lines & were finally withdrawn to hold our original Line when the Captured Enemy Trenches became untenable. Our Casualties were over 300. (AWM File 72/1/195.)

Military Order 480 of 1924 was published on **1 November, 1924**, 'in order to make it quite clear to all concerned, that honours may be claimed by units as under: -

(a) Battle Honours, vide appendix II to Military Orders 300/1924;

(b) Honours under the heading 'Action, &c.', vide para 2 of Army Council Instruction 458 of 1922 (Appendix III of Military Order 336/1923). *This is a case of substitution.*

(c) "Tactical Incidents" vide para 4 of Army Council Instruction 458 of 1922. *This is a case of substitution.*

(d) Special Claims, vide para 3 of Army Council Instruction 458 of 1922.

(e) Theatre of War Honours, vide Military Order 300 of 1924.

Under the heading 'Battle Honours', it was stated that: -

"With reference to Military Order 300/1924, the question has been raised whether the list of Battle Honours given in Appendix II thereto excludes those honours given in Appendix V to Military Order 336/1923 under the column "Tactical Incidents included". This is not so.

In answering the question it was stated that Honours may be claimed as under: -

(c) Special Claims, vide para. 3 of Army Council Instruction 458 of 1922.

(b) Honours under the heading 'Action, &c.', vide para 2 of Army Council Instruction 458 of 1922 (Appendix III of Military Order 336/1923). *This is a case of substitution.*

(c) "Tactical Incidents" vide para 4 of Army Council Instruction 458 of 1922. *This is a case of substitution.*

(d) Special Claims, vide para 3 of Army Council Instruction 458 of 1922.

(d) (e) Theatre of War Honours, vide Military Order 300 of 1924.

Thus, Special Claims ***could*** be made. Fromelles could in fact become the subject of Special Claims after all.

--- 1925 ---

Australian Army Order 37 of 1925 was published on **24 January, 1925**, advising that approval had been given for the formation of a sub-committee of the Military Board to deal with applications for Battle Honours by Light Horse Regiments and Infantry Battalions of the AIF as they are received from formations (vide para 8 of MILITARY ORDER 336/23)

A.A.O. 37 stated thus: -

Approval has been given for the formation of the following Sub-committee of the Military Board to deal with applications for Battle Honours by Light Horse Regiments and Infantry Battalions of the Australian Imperial Forces as they are received from formations (*vide para. 8 of M.O. 336/23): -*

The President was originally going to be Lt. Col. W.E.H. Cass (who had served at Fromelles), but instead General Brand was appointed to take on that role.

The Members were to be Batt. Col. W.J. Foster, Staff Corps, with Batt. Lieut. Col. J.L. Whitham, Staff Corps. *The Secretary* was to be Mr. V.C. Duffy.

---ooo000ooo---

UK Army Order 55 of 1925 was issued in **February, 1925**. By this time, the British had felt that they had classified all that were to be classified for Battle Honours, and that their job was now complete, at least in relation to the British units. All previous lists had been cancelled. The only lists now in effect would be those to be published in the forthcoming '**Complete List of Battle Honours Awarded for the Great War**'. The Order stated that:

'No further submissions concerning Great War battle honours will be made'.

(War Office File 32/5921)

The exasperated hope of the Battle Honours Committee or the Army Council was not to be.

Over the years since the Great War, several amendments have since been made to awards published in the Final List."[145]

The **Complete List of Battle Honours Awarded for the Great War** was issued on 27 February, 1925, with the UK Army Orders for 1925. The list applied to United Kingdom units, not Dominion ones, as much was still being done in Australia and the Dominions to finalise their claims for Battle Honours.

---ooo000ooo---

The Complete List included the Battle Honours for the Royal Warwickshire Regiment, for the Gloucestershire Regiment, and for the Oxfordshire and Buckinghamshire Light Infantry. Significantly, each of these units fought in the action in 1915, and would each again fight at Fromelles in 1916.

Reading it, we can see that the Royal Warwickshire Regiment was able to claim Battle Honours for 'NEUVE CHAPELLE', for 'AUBERS' and for 'FESTUBERT', all in 1915.

[145] Rodger p.85

And that the Gloucestershire Regiment was awarded a Battle Honour for 'AUBERS', again in 1915.

And that the Oxfordshire and Buckinghamshire Light Infantry was awarded a Battle Honour for each of 'AUBERS' and 'FESTUBERT', again each in 1915.

Just as it would be the case for Australian units, neither of these units would be able to claim a Battle Honour for Fromelles in 1916. Any connection with Fromelles in 1916 was to be made via the Battle Honour of 'Somme 1916, '18', which as we have already noted, is a misapplication of geographic and tactical reality.

On **12 November, 1925**, the **58th Battalion** lodged its ***Special Applications*** for Battle Honours in addition to the circulated list provided previously. First on the list was **Fromelles** for which it advised that approximately 95.6% of the strength engaged including headquarters, was present. It also noted that this was the first battle in which the Australian Imperial Forces were engaged in France and Flanders.

It was sought to have Fromelles, Bapaume and Actions of Villers Brettoneux emblazoned on Battalion Guidons and Colours. (AWM File 72/1/195.)

On **19 December, 1925**, The Secretary of the District Battle Honours Committee wrote to Headquarters 3rd District Base to return all claims to units for the purpose of comparing them with other units of the Brigade to which they respectively belong and, when necessary of making additional claims. The Schedule attached showed the following list of Battle Honours applied for by the Units of the 15th Brigade A.I.F.: -

- 57th Battalion: Defence of Suez Canal, BULLECOURT, Ypres, 1917, POLYGON WOOD, Poelcapelle, First Battle of PASCHENDAELE, First Battle of The Somme, 1918, Avre, Ancre, 1918, AMIENS, ALBERT, occupation of PERONNE (Special Claim), Hindenburg Line, BELLICOURT TUNNEL.
- **58th Battalion**: Defence of Suez Canal, **FROMELLES** (Special Claim), BAPAUME, 1917, BULLECOURT, Ypres, 1917, POLYGON WOOD, Poelcapelle, First Battle of Passchendaele, Second Battle of PASSCHENDAELE, First Battles of The Somme, 1918, Avre, Ancre, 1918, VILLERS BRETONNEUX, AMIENS, Second Battles of The Somme 1918, ALBERT, MONT ST. QUENTIN, Hindenburg Line, BELLICOURT TUNNEL.
- 59th Battalion: SOMME, 1916, 1917, 1918, BAPAUME, 1917, BULLECOURT, YPRES, 1917, POLYGON WOOD, VILLERS-BRETONNEUX, AMIENS, Capture of HAMEL (Special Claim), AVRE, CANAL du NORD, ST. QUENTIN CANAL.
- 60th Battalion: BULLECOURT, Ypres, 1917, Menin Road, POLYGON WOOD, Poelcapelle, First Battle of PASSCHENDAELE, AVRE, AMIENS, Albert 1918, MONT ST. QUENTIN, HINDENBURG LINE, **FROMELLES** (Special Claim) BAPAUME 1917, VILLERS-BRETONNEUX.

On **24 December, 1925**, Army Headquarters, 3rd Military District wrote to 22nd, 24th, 37th, 39th, 57th, 58th, 59th and 60th Battalions with a compiled claim for Battle Honours for each Battalion.

Reference was made to MOs 336/23 – 167/24, 300/24 and 480/24. The Battalions were asked to compare the honours on the list, against those of other units, when making any additional claims. 5th Division were to be entitled to Theatre of War Honours for France & Flanders 1916, and France & Flanders 1918.

--- 1926 ---

A Cable was sent from London to Australia on **2 March, 1926**, at the urging of Australia's Military Representative as follows:

War office very anxious receive claims units Battle Honours without further delay otherwise their trained staff unavailable later for assistance checking and compilation (stop) War Graves Commission also early require lists Unit Battle Honours for inclusion memorial registers already under preparation (stop) Cable approximate date claims may be expected. (NAA File 421/5/2.)

On **16 March, 1926**, Headquarters **57th Battalion**, North Fitzroy wrote to Headquarters 15th Infantry Brigade, advising that they sought the following Battle Honours: **Fromelles**, Morchies, Bullecourt, Polygon Wood, Defence of Amiens, Villers Bretonneux, Morlancourt, Harbonnieres, Peronne, and Hindenburg Line.

On **30 March, 1926**, 60th Battalion lodged its claim for Battle Honours. **Fromelles** was listed as the first of such, and also for emblazoning on Colours and Guidons.

By this date, "Fromelles" had been claimed as a Battle Honour by 30th, 53rd, 54th, 55th, 56th, 57th and 60th Battalions.

There were by now **seven** Battalions which had claimed 'Fromelles' as a Battle Honour.

On **6 April, 1926**, 3rd Division Headquarters were advised that 'From your list of battle Honours of 15th Brigade, you have omitted the ***Special Application*** of 57th Bn: - **Fromelles**; Bapaume; Villers Bretonneux.

On **30 April, 1926**, The Military Board, Army Headquarters, Victoria Barracks issued advice to 3rd District Base, Victoria Barracks, Melbourne, setting out the various Battle Honours for which various Battalions were entitled. '**Fromelles**' was specifically included for 29th, 32nd, 57th, 58th, 59th, and 60th Battalions.

When you add those which had already signified their claim for Fromelles, this made the number of battalions either being suggested they make a claim, or having made a claim, as **eleven (11)** battalions. The only one which was inexplicably not indicated was 31st Battalion.

On **3rd May, 1926**, 1st Cavalry Division, Victoria Barracks, Sydney, wrote to 1st Light Horse, to 7th Light Horse, to 12th Light Horse, advising of the Battle Honours suggested by Army Headquarters, Battle Honours Committee. The list omitted such Tactical Incidents as Walkers Ridge, Lone Pine, Russell's Top, Hill 60 (Anzac), Maghdaba, Beersheeba, Es Salt, Amman, and Abu Telul. It was advised that notwithstanding this omission, ***Army Headquarters propose to strongly urge the War Office to grant these as Battle Honours.*** [emphasis added]

It was also pointed out that some [Cavalry] units [just as did some Infantry units] claimed special honours in accordance with Military Order 336 of 1923. But as there was no *Special Order of the Day by a British Commander in Chief in the Field* issued setting out the merits of such action, Army Headquarters regret that the claims cannot be entertained. (NAA File 472/1/87)

Two days later, on **5 May, 1926**, there was a Memo from the Secretary, District Battle Honours Committee 3rd Military District Headquarters, Melbourne to:

5th, 6th, 7th, 8th, 14th, 21st, 22nd, 23rd, 24th, 29th, 32nd, 37th, 38th, 39th, 46th, 52nd, 57th, 58th, 59th, 60th Battalions, passing on a letter from Army Headquarters Battle Honours Committee attaching a list of honours to which it was thought the various Battalions were entitled.

The list was prepared in accordance with the instructions in Military Order 336 of 1923. A perusal of the list of Honours in Appendices I and II of Military Order 330 of 1924 omitted such AIF Tactical incidents and actions such as The Chessboard, Quinns Post, Lone Pine, Hill 60 Anzac, **Fromelles**, Thilloys, First Attack on Bullecourt, Lagnicourt, Hamel, Mont St. Quentin and Peronne.

Again it was stated that notwithstanding this omission it was proposed to strongly urge the War Office to grant them as Battle Honours.

It was stated that many units submitted claims for special honours in accordance with para. 3 of Appendix III to British Army Order 470 of 1922 re-published in Military Order 336 of 1923. As no special Order of the day by a British Commander in Chief in the Field was issued setting out the merits of such action it is regretted the claims cannot be entertained.

AWM File 72/1/195.

On **11 May, 1926**, the **29th Battalion** advised of 19 Battle Honours being sought, of which **Fromelles** was but one. Ten others were sought for the purpose of emblazoning on unit Colours.

On **11 May, 1926**, the **58th Battalion** advised that it concurred with the list of honours submitted to them for claims (which included **Fromelles** by way of ***Special Claim***).

On **11 May, 1926**, the **59th Battalion** advised that it concurred with the list of honours submitted to them for claims.

Going even further, on **11 May, 1926**, the **60th Battalion** concurred with the recommendations for battle honours made by Army Headquarters Battle Honours Committee, except that:

*"I consider '**FROMELLES**' should be emblazoned on the regimental colours instead of 'ALBERT 1918' for the reason that the Battalion was heavily engaged at Fromelles, about 90% being killed or wounded, while there were very few casualties in the Battalion at Albert, 1918."*

The claim went on to state that the 60th Battalion [in 1918] was within the geographical and chronological limits of the Battle of AVRE and not of ANCRE 1918. (AWM File 72/1/195.)

And on **25 May, 1926**, the **60th Battalion** concurred with the list previously suggested, with the following exceptions:

(1) Desire **Fromelles** to be emblazoned on colors in lieu of Albert, 1918.

(2) Consider that Bn. is entitled to Battle of Avre and not Battle of Ancre 1918.

AWM File 72/1/195.

On **13 July, 1926**, from Headquarters **57th** Battalion, North Fitzroy to the Secretary, 3rd District Base, Battle Honours Committee, in their reply to being asked previously, it was advised that the Battle Honours being sought included '**Fromelles**'.

The 'Omission' of Fromelles From Battle Honours.

In the United Kingdom, the 'Complete List of Battle Honours Awarded for the Great War' had been published on 27 February, 1925. There was pressure being exerted to have the Dominions finalise their claims for Battle Honours. This pressure had come down to those in charge in Australia, such that on **29 July, 1926**, just sixteen days after the 57th Battalion had advised its claims for Battle Honours, Lieutenant General Sir John Monash, Major General Sir Brudenell White, and Lieutenant General H.G. Chauvel met to discuss their recommendations regarding Battle Honours to the Defence Minister. The Defence Minister at the time was Sir (formerly Major General) Neville Howse VC. He had been the Minister since 16 January, 1925. Was there an imperative in their doing so?

Seven days later, on **5 August, 1926**, Lt Gen Chauvel, as Chief of the General Staff wrote to the Secretary, Defence Minister's Department, advising that they had examined Army Order 338 of 1922, Army Council Instruction 458 of 1922, letter from the Army Council 3/Gen. No. 209 of 1923, and the claims of Regiments and Battalions.

Point 2 of the letter noted that claims were to adopt the names as set out in the BNC Report under 'Battles' etc., and that not more than 10 of these might be selected to be emblazoned on standards, Guidons and Colours.

It was noted further that if a Battle Honour was to be selected from 'Actions' then it would be in substitution for a claim to the same honour in columns (1) and (2), and that each claim would be considered on its merits.

It was noted further that Army Council letter 3/Gen. No. 209 gave a list of the only Battle Honours which could be selected by Regiments and Corps out of the 'Tabulated List of Engagements' referred to above i.e. from the BNC Report.

Point 5 of the letter noted that a considerable number of 'Actions, etc.' had been claimed which were included in the original list but not in the list which accompanied the letter from Army Council above. Army Headquarters Committee had already cut out a number of such claims which were considered not appropriate.

Among those which were left, under France and Flanders, were **Fromelles**, Thilloys, Mouquet Farm, Lagnicourt, Hamel, Chuignes, Mont St. Quentin, and Peronne.

Point 6 of the letter was effectively the guillotine for any hope Australian soldiers who served at Fromelles might have had for a Battle Honour for Fromelles. It stated that:

> ***In view of the opinion of the Minister*** [emphasis added] **that no claims should be pressed except in the case of Engagements considered of extreme importance to Australia, we recommend that this list be further curtailed by the omission of the following: -**
>
> (inter alia) **Fromelles**, Mouquet Farm, Thilloys, Lagnicourt.

(NAA File 421/5/2.)

Were The Soldiers of Fromelles Betrayed, Dishonoured, or Sacrificed?

Disregard the contents of Military Order 336 of 1923.

Never mind the claims lodged by the various units, adopting the principles which permitted them to do so.

Never mind the significance of Fromelles as the first Battle to be fought by Australians on the Western Front.

Never mind the 'extreme importance to Australia' of Fromelles.

Never mind that eleven battalions had adopted the same guidelines and reasoning in making or being suggested they make a claim for 'Fromelles'.

Disregard any suggestion of dealing with such claims 'on their merits' before disregarding them.

The Committee of Monash, Brudenell White, and Chauvel were not prepared to adopt a reading of the very same rules which others had done up to this point.

The reasons given for omission do not stand scrutiny against the various bases for being allowed to be considered. The decision for omission is in contradiction to the reasons for Fromelles being allowed to be claimed. Fromelles was culled. The real question is: Why?

Australia's own top brass had dumped Fromelles, not the British, not the soldiers who fought there, and not the committees which were formed to compile the very submissions to name those engagements which deserved Battle Honours. It was done in what appears to be in order to appease the Minister at the time, Sir Neville Howse VC.

It was the top brass themselves, acting 'In view of the opinion of the Minister'. And who in the military structures below them would even dare to question their authority and judgment? Chauvel had led the Light Horse Regiments from the Suez to Damascus. White had engineered the evacuation of Gallipoli. And Monash had formulated the tactics to push the Germans back to the Hindenburg Line, bringing the war to a close at least a year earlier than expected.

Tellingly, none of these committee members had any involvement at Fromelles.

The Secretary of the 3rd Military District Headquarters must also have been shattered, after writing to the various battalions on 5th May, 1926, and advising that: 'Notwithstanding this omission it was proposed to strongly urge the War Office to grant [Fromelles, Mouquet Farm, Thilloys, and Lagnicourt and others] for Battle Honours.'

---ooo000ooo---

But even so, various claims for Battle Honours continued to be discussed. On **11 August, 1926**, claims were submitted by Regimental Committees on the forms prescribed in Army Orders, and had been revised by the Committees in each District, and finally a representative Committee at Army Head-Quarters.

Following that process, the Secretary, Department of Defence, Australia wrote to the Secretary, Prime Minister's Department on **11 August, 1926**, and the contents of the letter give rise to an insight into the decision-making process in Australia.

I am directed to request that the attached claims on behalf of Cavalry and Infantry Units of the Australian Imperial Force for Battle Honours in respect of service in the Great War be forwarded through the usual channels to the Secretary for transmission to the War office.

These claims are made in accordance with the provisions of UK Army Orders No. 338 and 470 of 1922 and Army Council Instructions Nos. 458 and 619 of 1922. It will be observed, however, that they are not completely in accord with the further instructions contained in the letter from the Army Council No. 3/Gen.No./209 of 19/2/23, in that they include the following engagements which were clearly allowed, under certain conditions, by the Army Orders and Army Council Instructions above quoted but which were excluded from the List which accompanied the letter No. 3/Gen.No./209 of 19/2/23.

DARDANELLES

Lone Pine.

FRANCE AND FLANDERS

Hamel

Chuignes.

Mont St. Quentin

Peronne.

EGYPT AND PALESTINE

Magdhaba.

Beersheba.

Es Salt

Amman.

In putting forward this list, it was advised that: -

These names are of particular significance to Australia, and it is felt that the historical and sentimental value of the regimental records of Australian Cavalry and Infantry Units will be considerably depreciated if these names, which are so intimately linked with Australia's war effort are allowed to fall into oblivion. Regimental Committees have been permitted by this Department to apply for them, and in many cases have selected them to be emblazoned on their Colours.

Sadly, **Fromelles** would not be referred to again.

Although Generals Chauvel, Monash and Brudenell White had effectively removed Fromelles from consideration, Chauvel still pressed for consideration of Cavalry units being awarded Battle Honours for JERUSALEM, and for NABLUS. In doing so, he wrote on **12 August, 1926** to Sir George Milne advising that the formation of Regimental Committees took some doing and some delay. He wrote that claims were considered in terms of Army order 338 and 470 of 1922, and Army Council Instruction 458 of 1922. But those claims had not been 'vetted' by the time that Army Council letter 3/Gen. No. 209 of 19 February, 1923 (3 ½ years previously) was promulgated, giving a new list: -

'...shutting out a number of engagements etc., which were of considerable importance to the units concerned and were quite clearly included under the original Army Orders and Instructions. This caused great dissatisfaction and confusion and led to much correspondence between Regimental Committees and District Committees. Eventually, in view of the keen interest displayed and the known importance to the Australian Forces of many of these engagements the Military Board informed the Regimental Committee that, provided their claims were in accordance with the Army Orders above quoted and were passed by the District and Headquarters Committees they would be recommended for acceptance. These claims have now been submitted after careful revision by District and Army Headquarters Committees, through the usual channels to the War Office.' (NAA File 421/5/2)

---ooo000ooo---

By **12 August, 1926**, the Secretary to the Military Board at Victoria Barracks Melbourne wrote to the Senior Military Representative, Australia House, London, forwarding a communication addressed to the Prime Minister's Department regarding claims for Battle Honours on behalf of Cavalry and Infantry Units of the A.I.F.

Also attached were copies of a minute by Generals Chauvel, Monash, and Brudenell White regarding claims which were being made for certain Honours. Fromelles was not one of them. Fromelles had already been 'omitted' and abandoned. (NAA File 421/5/2)

On **17 August, 1926**, after being processed at the Prime Minister's Department, the content of the letter was sent in substantially the same format, when the Governor-General wrote to the Secretary of State for Dominion Affairs, respectfully putting the request to the United Kingdom authorities.

The letter referred to particular Honours being sought, and advising they were sought after long and deliberate consideration. Those for France and Flanders included Hamel, Chuignes, Mont St. Quentin, and Peronne.

In asserting claims for these Honours, the reasoning would just as easily and just as readily have applied in the case of a claim for Fromelles: -

These names are of particular significance to Australia, and it is felt that the historical and sentimental value of the Regimental records of Australian cavalry and infantry units will be considerably depreciated if these names, which are so intimately linked with the war effort of this country, are allowed to fall into oblivion.

Fromelles had been abandoned by those who had the power, the authority, and the influence, to pursue it for Battle Honours, despite it being the subject of Special Claims in accordance with the various Orders and guidelines. It could have been the third unsuccessful engagement to have earned a Battle Honour, but it was omitted, and thereby culled. Nearly 20,000 soldiers in Australia's 5th Division, and the UK 61st Division were left without appropriate recognition for their service.

Nobody on the Army Council was prepared to adopt the reasoning of Major General Headlam of the BNC. And Australia's military leaders do not appear to have been made aware of it. If Headlam's opinion on the matter had made its way into the determination process, the result could have been very different.

By **29 November, 1926**, the names of the various Battle Honours had in effect been finalised. Australia's representatives in Australia House, London, wrote to the Dominions Office, Downing Street, to record their agreement. In finalizing, it was stated that the Regiments and Battalions of the Australian Military Forces desired to have their approved Honours emblazoned on their Colours in time for the arrival in Australia of Their Royal Highnesses the Duke and Duchess of York in the coming year.

By the end of 1926, the various Battle Honours which were able to be claimed for the Western Front were:

- Somme 1916
- Pozières
- Bapaume 1917
- Arras 1917
- Bullecourt
- Messines 1917
- Ypres 1917
- Menin Road
- Polygon Wood
- Broodseinde
- Poelcappelle
- Passchendaele
- Villers-Bretonneux
- Lys
- Hazebrouck
- Kemmel
- Hamel
- Marne 1918
- Tardenois
- Amiens
- Albert 1918 (Chuignes)
- Mont St Quentin
- Hindenburg Line
- Epehy

- Somme 1918
- Arras 1918
- Avre
- Ancre 1918
- St Quentin Canal
- Beaurevoir
- France and Flanders 1916–1918

The Secretary of the Department of Defence wrote on **22 December, 1926** to the Acting Secretary for the Prime Minister's Department, Melbourne:

The claims have been revised in accordance with the decisions of the Army Council and are forwarded herewith for favour of transmission to the Secretary of State in order that the sanction of His Majesty the King may be obtained.

It is suggested that the Secretary of State be asked to convey to the Army Council the Commonwealth Government's great appreciation of the very favourable consideration given to the special claims.

I shall be glad if the sanction of His Majesty may be communicated by cablegram as it is desired to promulgate the Battle Honours and proceed with the work of emblazoning the Honours on the Regimental Colours at the earliest possible date.

NAA File 1926/525.

On **27 December, 1926**, the Governor-General forwarded a letter to the Secretary of State for Dominion Affairs, in terms as sought in the Department of Defence letter of 22 December, 1926.

1927 And Thereafter

On **9 March 1927**, **Australian Army Order AO 112** was published in order to announce 'the award of Battle Honours for the Great War to Cavalry and Infantry units of the Australian Imperial Force'. [146]

The Preamble to the Orders states:

1. *His Majesty the King has been graciously pleased to approve of the award of Battle Honours for the Great War to Cavalry and Infantry Units of the Australian Imperial Force, as shown here under.*

Ultimately, Fromelles was not awarded recognition as either a Battle, an Action, a Tactical Incident, or the beneficiary of a Special Claim for any unit which served there. Any such units were to be satisfied with 'Somme 1916', even though that award could realistically only relate to action which took place in November of that year.

The Battalions of the Australian 5th Divn arrived on the Somme in October 1916. Although they did not fight in one of the listed battles for October and November, they were nonetheless entitled to the Battle Honour 'Somme 16' because they were actually there.

[146] Rodger p.91.

The Battles of the Somme 'officially' concluded on 18 November, 16. The Australian 3rd Division crossed the English Channel on 21 November 16. While 3rd Div Battalions served on the Somme in December 1916, they did not get the Battle Honour 'Somme 16' because they arrived too late. But they did arrive just in time to qualify for 'France and Flanders 16'.

After listing the various Regiments and Battalions and the Awards to which they would be entitled, the Order continued: -

2. *The above Battle Honours have been allotted to the units of the Australian Military Forces which have inherited and how hold the titles, designations and traditions of the corresponding units of the Australian Imperial Force.*

3. *These Battle Honours will be published in the Army List of the A.M.F. under the names of the units concerned, and such units will henceforth be entitled to emblazon on their Guidons or Regimental Colours the ten selected honours which are shown in thick black type above.*

4. *The order in which the Honours are promulgated bears no relation to the chronological sequence of the A.I.F. battles themselves, but corresponds to the order of precedence, approved by His Majesty the King for recording and for emblazoning honours on the Guidons and Colours of the military units of the Empire.*

One might be left thinking that the final position regarding Battle Honours was that their selection ended in 1927. But that is not the case at all. **Australian Army Order No. 117 of 1936** was published in order to extend the awards of 'ANZAC', 'Gallipoli 1915', 'Egypt 1915-1917', and 'Palestine 1917-1918' to include a further seven light horse regiments '*in recognition of the numbers of officers and men supplied by the .... Regiments and by their regimental areas to Light Horse units of the Australian Imperial Forces.*'[147]

---ooo000ooo---

1928 saw another case of recognition for service finalised. It was far less glorious than what had been originally planned, and was for what was originally referred to as The Gallipoli Star.

The Gallipoli Star was a military decoration proposed by the British Empire never awarded but later privately struck. Originally to have been called the ANZAC Star, it was proposed by Lieutenant General Birdwood for Australian and New Zealand troops who served at Gallipoli. Despite approval by George V, protest at a medal that could not be awarded to other Commonwealth troops who took part in the operation (the majority deployed) was raised by British parliamentarians and media. In 1918, after ribbons, but not medals, had been shipped to the two nations, the award was cancelled and instead eligible soldiers were awarded the 1914 – 1915 Star.[148]

The 1914 Star was authorised under **Special Army Order no. 350** in November 1917 and by an **Admiralty Fleet Order** in 1918, for award to officers and men of the British and Indian Expeditionary Forces who served in France or Belgium between 5 August and midnight of 22–

[147] See Rodger p.92.

[148] See: https://en.wikipedia.org/wiki/Gallipoli_Star_(Commonwealth)

23 November 1914. The former date is the day after Britain's declaration of war against the Central Powers, and the closing date marks the end of the First Battle of Ypres.

Altogether 378,000 1914 Stars were awarded.[149]: The creation of the 1914 Star for the original British Expeditionary Force, who served in France and Flanders up to November 1914, led to demands from the Australian and other dominion governments for medallic recognition for their original contingents who had fought at Gallipoli and elsewhere. This led to a proposed Gallipoli Star, for award to New Zealanders and Australians of the Australian and New Zealand Army Corps (ANZAC), but not to the British forces serving alongside them. Following protests by British parliamentarians and news media, the Gallipoli Star was never awarded and the 1914-15 Star, with its wider eligibility, instituted instead.

The 1914–15 Star was instituted in December 1918 and was awarded to officers and men of British and Imperial forces who served against the Central European Powers in any theatre of the Great War between 5 August 1914 and 31 December 1915, provided they had not already received the 1914 Star: https://en.wikipedia.org/wiki/1914–15_Star.

An interpretation of this could be that while the British created the '1914 Star' for the original BEF contingent, the British quashed any similar initiative by 'colonial' Dominions. The 1914/15 Star being created as the 'compromise'. But there were still 'special' British (and Indian) members who got the '1914 Star'.

Yet another example of how soldiers were not honoured, despite having had a champion, in this case that of Lieutenant General Birdwood.

---ooo000ooo---

The final awards, to date, were made in 1958 when the Royal Tank Regiment, now eligible as a formed part of the Royal Armoured Corps, received its belated battle honours for the Great War.[150]

The procedures after the Second World War were similar to those following the First. The re-formed Battles Nomenclature Committee made every effort to avoid using names that had been used by the Great War committee but, if this was not practicable, the awards were differentiated by year date (e.g. Baghdad, Baghdad 1941). If two separate engagements took place at the same location in the same year, they were differenced by Roman numerals (e.g. Cassino I; Cassino II). In several instances, the Battles Nomenclature Committee felt it was desirable to indicate the aim or nature of the operation, particularly when two separate operations took place at the same location (e.g. Capture of Tobruk; Defence of Tobruk).

---

[149] See: https://en.wikipedia.org/wiki/1914_Star

[150] Rodger p.86

An Example of the Display of Battle Honours: -

The 54th Battalion's Battle Honours.

Taking the **54th Battalion** as an example, the Battle Honours awarded to them, and entitled to be emblazoned on the regimental colours as pictured on the first page of this chapter (from left to right in order of precedence as approved by His Majesty the King) are:

Somme 1916, 18	Bullecourt
Ypres, 1917	Passchendaele
Villers Bretonneux	Amiens
Mont St. Quentin	Hindenburg Line
France and Flanders 1916-18	Egypt 1916

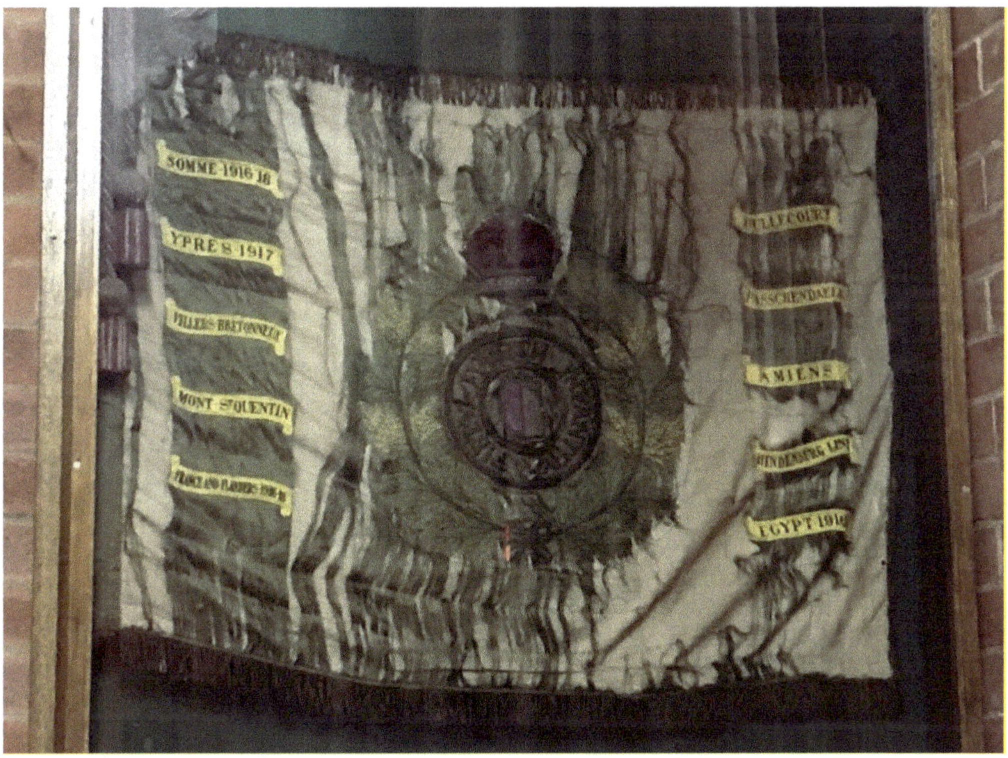

[Picture: 45: Battalion Flag and Battle Honours, All Saints Cathedral, Bathurst NSW.

[Photo: Lee Ford.]

Other battle honours (not for emblazoning) *but awarded* included:

Menin Road	Polygon Wood

Poelcappelle	Ancre 1918
Albert 1918	St. Quentin Canal

The Battle Honour of Somme 1916 supposedly provided recognition for service for the 54th Battalion, and all those who fought at Fromelles, but it was in all reality only for their time from 21 October, 1916 near Flers. The Battle Honour 'Somme 1916, 18' supposedly enabled them to cover both 1916 conflicts, although not giving pre-eminent recognition to Fromelles.

The soldiers of the 54th and those other five battalions knew this was not appropriate, but without a champion to pursue their recognition, the true status of Fromelles was lost.

---ooo000ooo---

Some Exceptions Regarding Awards of Battle Honours

In the France and Flanders Theatre, there are a number of 'substitute' Battle Honour names – created along National lines – in order to honour the claims made by the various participants. This is not an exhaustive list, but these are the most readily apparent.

- **'Albert 1916'** and **'Albert (Beaumont Hamel) 1916'** – for Newfoundland. A 'substitution' by bracketed inclusion of a place name not identified in the BNC Report.

Battle honours of the Great War were almost invariably only awarded for engagements specifically named by the Battles Nomenclature Committee. A particularly poignant exception to this rule is that of the Royal Newfoundland Regiment.

In short the Battle of the Somme was the Royal Newfoundland Regiment's first major engagement, and during an assault that lasted approximately 30 minutes the Regiment was all but wiped out. The Regiment applied for the honour 'Beaumont Hamel' in memory of the first day of the Battle of the Somme.

The corps submission for awards had been made promptly, but the delay was caused by the insistence of the 'Newfoundlanders' to claim 'Beaumont Hamel' for the 1st Battalion on 1 July 1916, when it suffered seven hundred and ten casualties.

The Royal Newfoundland Regiment, arguably one of the most distinguished corps in military history, was otherwise absent from this series of awards.

The regiment, raised as The Newfoundland Regiment in 1915, was re-designated the Royal Newfoundland Regiment in 1918 and disbanded in 1919.

From a 'Battle Honours' perspective what is of interest is the fact that instead of just accepting the listed Battle Honour of 'Albert 1916' the Newfoundlanders lobbied for the Battle Honour 'Beaumont Hamel'.

The BHC at first refused the request on the grounds that it was not listed in the *Report of the Battles Nomenclature Committee*. The awards of Bois des Buttes and Bligny did not support this view.

But the Newfoundlanders were determined that the Regiment's conduct on 1 July 1916 should be recognized with a Battle Honour.

There matters remained, with much correspondence between the Governor and the then Colonial Office, until a compromise was proposed that the claim should be allowed in the form Albert (Beaumont Hamel) 1916, provided that it was specifically approved by the King.

Accordingly, a royal submission was prepared and presented to George V, who, it is recorded, did not hesitate to grant this specially worded Battle Honour.

The full battle honours of The Royal Newfoundland Regiment were finally published separately in 1925 and are perpetuated today by The Royal Newfoundland Regiment, raised 1949, a regiment of the Royal Canadian Infantry Corps."[151]

The moral of this story is that lobbying can work. But you also need to have notable people on your side. In this case the Governor of Newfoundland.

- **Bullecourt – for Australia.**

The BNC identifies the First Attack on Bullecourt 11 April 1917 as a flanking attack with operations fought during the Battle of Arras (9 April – 4 May 1917), coinciding with the First Battle of Scarpe (9 - 14 April 1917). The Battle of Bullecourt 3 - 17 May 1917 was classified as a 'Battle'. The Australian 4th Division fought at the First Attack on Bullecourt 11 April 1917 while the Australian 1st, 2nd, and 5th Divisions fought at the Battle of Bullecourt 3 - 17 May 1917.

The Battalions of the 4th Division could have selected Arras 1917, or Scarpe 1917 as their Battle Honour. Instead, they selected the Action 'First Attack on Bullecourt' and, together with the Battalions of the 1st, 2nd and 5th Divisions have the Battle Honour '**Bullecourt**'.

- **'St. Quentin**' and **St Quentin 1918** – for British (Royal Tank Regiment).
- '**Bapaume 1918'** and '**Mont St. Quentin'** – for Australia. A full substitution of an 'action' for a 'battle' as identified in the BNC Report.

The BNC identifies the First Battle of Bapaume on 24/25 March 1918 and the Second Battle of Bapaume from 31 August to 3 September 1918. The BHC created the Battle Honour '**Bapaume 1918'** which covers both these Battles. This followed their 'policy' of creating one Battle Honour for multiple battles that occurred in the same location in the same year. The AIF did not have any troops that fought at the First Battle of Bapaume. However, during the Second Battle of Bapaume the Australian 2nd Division, supported by the 3rd and 5th Divisions, captured Mont St. Quentin. The 'Capture of Mont St. Quentin' is listed in the BNC Report as a 'Tactical Incident' that occurred during the Second Battle of Bapaume. To better reflect the AIF's involvement the decision was made to substitute 'Mont St. Quentin' for 'Bapaume 1918'. Consequently, Battalions of the three Australian Divisions involved have the Battle Honour '**Mont St. Quentin**'.

---

[151] See: Rodger, pages 84/5.

- **'Amiens**' and **Amiens 1918** – for Australia.
- '**Albert 1918'** and '**Albert 1918 (Chuignes)'** – for Australia. A 'substitution' by bracketed inclusion of another action identified in the BNC Report.

The BNC identifies the Battle of Albert 21-23 August 1918, with the 'Tactical Incident' of the Capture of Chuignes. The BHC created the Battle Honour 'Albert 18'. However, in recognition of the five Australian Infantry Battalions involved in the capture of Chuignes, the substituted Battle Honour '**Albert 1918 (Chuignes)'** was created.

- **'Sambre'** and 'Sambre **(Le Quesnoy)'**– for New Zealand. A 'substitution' by bracketed inclusion of another action identified in the BNC Report.

The BNC identifies the Battle of Sambre 4 November 1918 with the 'Tactical Incident' of the Passage of the Sambre, and the Capture of Le Quesnoy. The BHC created the Battle Honour 'Sambre'. The New Zealand Division was involved in this Battle. 11 New Zealand Regiments have the Battle Honour '**Sambre'**. However, in recognition of the NZ Wellington Regiment's involvement in the capture of Le Quesnoy, the substituted Battle Honour '**Sambre (Le Quesnoy)**' was created for them.

Finding A Champion.

In response to a request from Australia for a Battle Honour for the Capture of Hamel 4 July 1918, the British initially proposed to group it - geographically - with the Actions of Villers Bretonneux 24/25 April 1918, and have one Battle Honour with two different qualifying dates (i.e. Villers Bretonneux 24/25 April and 4 July 1918). However, Australia persisted with the request for a separate Battle Honour. The British acquiesced and the Battle Honour 'Hamel' was created. When one looks deeper into this it quickly becomes apparent that the Battle Honour 'Hamel' was championed by General Monash. Without the intention of casting aspersions, Monash was far from being a modest man. He had superb qualities for any large job which might crop up: absolute self-confidence, skill in the manipulation of men to his forceful will, a magnificently developed administrative competence and a powerful intellect[152].He proved to be an astute leader. Nonetheless he did look after the soldiers, if on occasions driving them hard. Without Monash's support and drive the Battle Honour 'Hamel' would likely not have been created.

A similar case can be made for General Chauvel championing a few of the Battle Honours against the Turks being modified from 'British Gaza' to 'ANZAC Gaza-Beersheba' and 'Sari Bair' to 'Sari Bair-Loan Pine'; and the bracketed versions of 'Jordan (Es Salt)' and 'Jordan (Amman)'. There simply was no equivalent Australian or British champion for Fromelles – and the forgotten front in general.

Perhaps if Chauvel or Monash or Brudenell White had been involved in the Battle at Fromelles, there would have been a special claim made unhesitatingly.

---ooo000ooo---

[152] See: http://adb.anu.edu.au/biography/monash-sir-john-7618

A recent example of being able to look back and give due recognition to the actions of Australian soldiers was revealed on 13 May, 2018. In a press release titled:

GALLANTRY AWARD FOR SERVICE DURING THE BATTLES OF CORAL AND BALMORAL.[153]

We were informed that: - The extraordinary gallantry displayed by members of the 1st Australia Task Force (Forward) and associated units deployed to Area of Operations SURFERS during the Battles of Fire Support Bases Coral and Balmoral is being recognised by the Unit Citation for Gallantry today in Canberra.

Minister for Defence Personnel, the Hon Darren Chester MP, said the Battles of Fire Support Bases Coral and Balmoral were among the largest and most protracted battles fought by Australians in the Vietnam War.

"The men who fought at Fire Support Bases Coral and Balmoral displayed collective gallantry which is worthy of the Unit Citation for Gallantry," Mr Chester said.

"Throughout the 26 days of the battle a force comprising of up to 3,000 Australians prevailed against a well prepared and numerically superior enemy.

Twenty-six Diggers died and 99 were wounded, the worst casualties for any Australian battles in the Vietnam War.

"This year sees the 50th Anniversary of the Battles of Fire Support Bases Coral and Balmoral and it is timely that we are able to recognise the courageous service displayed by the members of the units involved."

The Unit Citation for Gallantry was created in 1991, along with the Meritorious Unit Citation. The Unit Citation for Gallantry recognises extraordinary gallantry in action.

Although a Unit Citation and a Battle Honour are different, they are arguably similar in nature. Whereas a Unit Citation is a device worn by all the members of the Unit, a Battle Honour is an award of a right by a government or sovereign to a military unit to emblazon the name of a battle or operation on its flags ("colours"), uniforms or other accessories where ornamentation is possible.

However, both recognise conspicuous service by the Unit as a whole.

Summing Up:

- Australia's Battle Honours were governed by the determinations of the Battle Honours Committee, and the Army Council of the United Kingdom. We were therefore restricted by what was seen as appropriate by those agencies when it came to the determination of Battle Honours from World War One.

[153] See: - https://www.minister.defence.gov.au/minister/dchester/media-releases/gallantry-award-service-during-battles-coral-and-balmoral

- Fromelles had not been classified as a Battle in its own right, so it was going to be difficult to claim it for a Battle Honour, given that only two previous unsuccessful Battles had been able to be awarded Battle Honours.
- Fromelles could have been made the subject of a Battle Honour for an Action, but the comments of Major General Headlam of the Battles Nomenclature Committee were never passed on to the Australian Authorities. So that course was never taken up. If it was, then perhaps a champion for the cause would have emerged, but that did not occur.
- British and Indian soldiers who fought at Fromelles (under the Attack at Fromelles and the Attack at Rue du Bois in 1915, were able to claim the Battle Honour of AUBERS for those actions, despite those actions also being unsuccessful.
- Australian and British soldiers who fought at Fromelles in 1916 should have been able to claim a Battle Honour for FROMELLES 1916, but they were denied that possibility by none other than Monash, Chauvel, and Brudenell White when they 'omitted' Fromelles from the list of awards being sought.
- In doing so, Australia's Commanders in Chief themselves overruled the selection of eleven (and prob ably all twelve) battalions which all sought or were recommended to claim 'Fromelles' as a Battle Honour, and all of which had adopted the guidelines inherent in Military Orders and Instructions. Those battalions were in the process dishonoured.
- The Australian Government's recent award of Unit Citations for Gallantry for the fighting at Fire Supports Bases Coral and Balmoral during the Vietnam War shows that retrospective review of past military conflicts and battles is possible.

---ooo000ooo---

Sydney amateur historian Neville Kidd, author of a self-published work dealing with Major Harrison of the 54th Battalion, has previously commented that:

*The award of a Battle Honour is to the Battalion, and not for the individual men. The choice offered was **'a Hobson's Choice'** since the Battle of Fromelles was never awarded a Battle Honour and so could not be selected. Whatever the process was for the selection of Battle Honours, it operated to ensure the exclusion of FROMELLES for nomination.*

A '**Hobson's choice**'[154] is a free choice in which only one thing is offered. Because a person may refuse to accept what is offered, the two options are taking it or taking nothing. In other words, one may "take it or leave it." The phrase is said to have originated with Thomas Hobson (1544–1631), a livery stable owner in Cambridge, England, who offered customers the choice of either taking the horse in his stall nearest the door or taking none at all.

[154] See: https://www.Wikipedia.org.

The term "Hobson's choice" is often misused to mean a false illusion of choice, but it is not a choice between two equivalent options, which is a Morton's fork; nor is it a choice between two undesirable options, which is a dilemma.

A Hobson's choice is therefore a choice between something, or nothing.

And the soldiers at Fromelles got a nothing.

---ooo000ooo---

The Australian soldiers who fought at Fromelles in 1916 were forever denied the recognition they so justly deserved.

They would for the rest of their lives carry their own Battle Honour, their 'Badge of Honour' if you like, that of being known as ***'The 19th of July Men'***, because when all was said and done, for their service at Fromelles, they were given no other recognition at all.

They had to create their own.

---ooo000ooo---

15. The Inexcusable Absence of the Victoria Cross for Fromelles

Battle Honours collectively go to the battalions which fought.

Honours in the form of mentions in dispatches, or medals, go to the individual soldiers who fought.

The most significant awards to individual soldiers were as follows: -

Mentioned in Dispatches.

Mentioned in Dispatches (MID) is a public commendation for distinguished service or bravery in the armed forces. It's one of the oldest forms of recognition for gallantry in the Imperial military.

What it means is that your armed forces forebear did something that had warranted their name to be included in an official account written by a superior officer, which had then been sent on to the War Office.

Military Medal

A military medal is a British award for bravery in battle on land, given to non-commissioned personnel for bravery in battle or for service in a campaign.

It was established in 1916 and discontinued in 1993.

Military Cross

The Military Cross (MC) was established by King George V on December 28, 1914, initially intended for commissioned officers of the British Army holding the rank of Captain or below, as well as Warrant Officers, to recognize "distinguished and meritorious services" during active operations against the enemy on land; essentially created to award bravery in combat to lower-ranking officers during World War I.

Distinguished Service Order

The Distinguished Service Order (DSO) was established by Queen Victoria on September 6, 1886, to recognize acts of meritorious or distinguished service by military officers during wartime, typically in actual combat against the enemy; essentially, it was intended to reward acts of gallantry that were not considered worthy of a Victoria Cross but still deserving of recognition.

Distinguished Conduct Medal

The Distinguished Conduct Medal (DCM) was established in 1854 during the Crimean War, by Queen Victoria to recognize gallantry in the British Army. It was the second highest award for gallantry for non-commissioned officers and other ranks, after the Victoria Cross.

The DCM was awarded to soldiers, non-commissioned officers, and warrant officers.

The DCM was discontinued in 1993 and replaced by the Conspicuous Gallantry Cross.

Victoria Cross

In the case of honours, there is no higher recognition for Australian and British soldiers than the Victoria Cross.

The Victoria Cross, awarded ***for acts of extreme bravery in the face of the enemy***, was established by Queen Victoria in 1856 to recognize gallantry displayed by British soldiers during the Crimean War (1854-1856), making it retrospective to cover actions from that conflict. It was designed to be awarded to all ranks of the armed forces, regardless of their position, marking a significant change from previous military awards.

---ooo000ooo---

Following the Battle of Fromelles, and up until 3 September, 1916, a number of steps were taken to record and nominate soldiers for various awards for the gallantry of the individual soldiers.

Company Commanders made recommendations for awards, to Commanding Officers.

Commanding Officers gave their inputs to Brigade Commanders.

Brigade Commanders made recommendations to Division Command, and finally, Division Command made recommendations to the Corps Commander.

The first distribution of medals for actions at Fromelles occurred on 3 September, 1916. Sir Henry Plumer GOC 2nd Army had consented to present the ribbons. After a special church parade at the Divisional Bombing School, the first medals from the deeds of Australians at Fromelles were distributed.

Lt. W. Smith of 53rd Battalion wrote in Reveille, July 1936 of his recollection of witnessing that distribution of medals:

> *About a month later it was my good fortune to stand in the front rank of a hollow square – representative of every unit in the division – to witness the presentation of some fifty decorations by General Plumer. In his brief speech to the troops General Plumer said:*
>
> *'That although the attack was not widely known to the public, and when the history of the war came to be written, it would probably be described as a failure. Yet, the number of decorations presented to 5th Division was proof of the importance of which the General Staff attached to it, and its appreciation of the heroism displayed.'*

On 12 October, 1916, another presentation of medals from Fromelles was made by General Sir William Birdwood. According to 5444 Pte E C Schaeche of 59th Battalion: -

> *Medals were bestowed on chaps for conspicuous gallantry on the 19th July stunt.*

Robin Corfield has tallied the medals awarded as a result of the Battle of Fromelles as DSO – 9, MC – 48, DCM – 45, MM – 63, Foreign Decorations – 14, MID – 38, making an overall total of 217.[155]

The Missing Victoria Crosses

Prior to Fromelles, there had been 9 awards of the Victoria Cross for actions at Gallipoli.

Prior to Fromelles, there had been 1 award of the Victoria Cross on the Western Front. It was amply illustrated by the following article which appeared in the Sunday Mail, Brisbane, on 17 March, 1940:

HEROES OF THE SOMME. V.C'S FOR VALOUR[156]

The fighting on the Somme in July, 1916, is associated with some of the most glorious pages in the history of the A.I.F.

Actually, the first Australian V.C. on the French front was won by young W. Jackson, of the 17th Battalion, during a raid near Armentieres — he was then only 19years old. His performance was of the time-honoured type — the rescue of a comrade under fire. Jackson joined up in the backcountry of New South Wales at the age of 17, and his youthful gallantry in the early days of 1916 was something his fellows marvelled at.

A party of Australian raiders organised under the command of Captain Heritage, of the 19th Battalion - he was killed later near Pozieres — suddenly found itself held up by a deep ditch behind which a German post had established itself in comparative force.

However, by skillful leadership the garrison was overcome and the survivors were sent back to the Australian lines. Many of our men had been wounded in the fray, and after escorting some prisoners through No-man's land, young Jackson picked up one of the sufferers, and in a hail of fire, managed to remove him to the shelter of his own trenches.

Nothing daunted he climbed out into the open again and carried another man in without mishap. The fury of the fight had fully developed by now, and the German artillery took a hand, but the call from a stricken sergeant out in the open between the two trenches was too much for Jackson.

He rushed over the parapet for the third time, and picking his man up had almost reached the Australian lines when one of his arms was almost ripped off.

Despite this, he went out to search for more of his wounded comrades, and it was only when he himself collapsed that he was removed to the Regimental Aid post.

Readers are then informed that: -

---

[155] See Corfield p.274

[156] The full page article can be accessed at: - https://trove.nla.gov.au/newspaper/article/98245615?searchTerm=THE%20fighting%20on%20the%20Somme%20in%20July%2C%201916%2C%20is%20associated%20with%20%20%20%20%20%20%20%20%20%20%20&searchLimits=

After weeks of comparative quiet, the Australian Divisions were next massed for the slaughter of Fromelles - or Fleurbaix.

Nothing could have been more heroic than the way the newly-formed Fifth Division undertook that difficult and hopeless operation.

It was one of the bloodiest actions in which Australian troops ever took part. Curiously enough, there were no V.C.'s awarded for Fleurbaix - probably because the individual acts of heroism were too numerous.

The article went on to recount the fighting on the Somme, and the awards of Victoria Crosses to a number of soldiers in the 1st and 2nd Australian Divisions.

---ooo000ooo---

Not one soldier who fought at Fromelles was awarded the ultimate military honour of the award of the Victoria Cross. There were of course many individual recognitions of honourable action by soldiers in all participating Battalions.

So who was recommended to receive a VC, and why were they not successful?

The efforts of at least four soldiers stand out as glaring absences: -

In the 56th Battalion, **2436 Private Thomas Charles Rowley**

In the 57th Battalion, **3101 Sgt Simon Fraser**

In the 58th Battalion, **4621 Private Edgar Williams**.

In the 58th Battalion, **Major Arthur Justin Sanford Hutchinson**.

Despite the actions of those in the 5th Division, and in particular the actions of Sgt. Simon Fraser, and Major Hutchinson, no awards of the Victoria Cross were ever made to a member of the 5th Division, for what took place during the Battle of Fromelles.

---ooo000ooo---

In the 56th Battalion, **2436 Private Thomas Charles Rowley**[157] was the subject of notice by Lt. Col A H Scott when he wrote:

> *... several very brave acts and devotion to duty were noticed in the Bn, the most noticeable being 2436 Private T C Rowley of A Coy'.*

General McCay in his own handwriting put Rowley forward for the Victoria Cross. Statements were made by Capt. Fanning and Capt. Sheen, 2nd Lt R A Reid and CSM S M Dykes, on around 3 August, 1916. All those recommendations are no longer part of Rowley's file.

McCay though enough of the eventual award of the DCM to put this note in Routine Orders 19 August: -

[157] See Corfield, p.285

The GOC Division has great pleasure in promulgating the following communication from General Headquarters. Under authority granted by His Majesty the King, the Commander in Chief awards the Distinguished Conduct Meal to No. 2436 Pte T C Rowley, 56th Bn AIF.

The extraordinary deeds of Pte Rowley were vividly described in his citation: -

At Petillon on 19th/20th July 1916, Private Rowley, during the retirement of the 8th Infantry Brigade, proceeded to the parapet of the enemy's trench and standing up on the parapet in full view, threw grenades which such effect that the enemy had to withdraw from the parapet over which they were firing and so enabled 80 men of the above-mentioned Brigade to reach their lines in safety. The gallantry displayed by Private Rowley also prevented the enemy from gaining the head of the communication trench which had been completed across 'No-man's Land' and so prevented the enemy from cutting off the 14th Brigade's left flank thus preventing numerous casualties. Private Rowley continued throwing grenades for some three hours until his arm was so swollen that he was unable to use it. During this period Private Rowley had to retire for a fresh supply of grenades, returning alone to the enemy's parapet and continued to hold them back. By his action, he proved himself to be a man of great presence of mind and coolness and his action also denotes personal bravery of the very highest order.

---ooo000ooo---

Sometimes the mention of being someone whose actions are appropriate for the award of a Victoria Cross comes from the soldier's own battalion ranks and not the Officers commanding them. In this case, **4621 Private Edgar Williams** of 58th Battalion.[158]

Edgar was born in 1884 and became Junior Teacher at Long Gully, in 1900; assist at Eaglehawk in 1904, and subsequently Head Teacher at Mundoona North, Narrapoort and Ouyen. He was the second Principal or Head Teacher of Ouyen Primary School, where he remained until 4 September, 1915 when he enlisted. He embarked with the 14th Reinforcements to the 5th on the Transport Themistocles in November, 1915. In Egypt he was allotted to the 57th, but was transferred on 5 April, 1916 to the 58th.

He proceeded to France and after a few days' training at Steenbecque, he entered the fighting zone. He was detailed to manage a gas apparatus, but obtained a transfer to stretcher bearer. He refused an offer to enter a school of instruction on 18 July, and came safely through the battle, but it was in rescuing wounded after the battle that he lost his life from a stray bullet. He was awarded the Military Medal for bravery in the field.

In a letter to Mrs. Williams of Ouyen, his widow, Corporal Vernon wrote: -

There were many left out wounded after the attack. Private Williams worked for sixty hours getting in as many as he could. If any man deserved the VC, he did. He had not a thought for himself, so long as there were others who needed attention. His friends tried to persuade him to take some rest, but there were still two men whom he was anxious to bring in, and he refused to

[158] See Corfield p.288-289.

stop until he had got them. As he was carrying one of them, he was hit in the head. He could be seen from our trenches and was still able to walk but apparently blinded.

The children of his old school at Ouyen later obtained an enlarged framed photograph of their Head Teacher to hang in the school room as a memorial. The Education Department tribute of Williams was that he was '*a very good teacher, active, capable and conscientious*.'

---ooo000ooo---

Another stretcher bearer who was Mentioned in Dispatches was **4744 Cpl Walter John Brown**[159], of 58 Battalion, who '*with Cpl W C Davies – since KIA – rescuing wounded … five trips before he himself severely wounded* (Elliott). These were the stretcher bearers that went out after signalling to the Germans their mission and brought back a badly wounded man. They paused to be able to get their man into safety and a German sniper who must have been following their journey fired a single shot, killing the man on the stretcher and severely wounding Brown. The event was witnessed by many and caused great anger among the men.

---ooo000ooo---

In the 57th Battalion, the actions of **3101 Sgt Simon Fraser** have been seen as worthy of a Victoria Cross, especially when comparing the extent of his actions in seeking to recover fallen comrades, over the days from 19 to 23 July. Brigadier General Pompey Elliott wrote of Fraser: -[160]

He was responsible for some very gallant work during the period 19/23 July in leading parties into No-man's land and rescuing many wounded. His example has been of the greatest value to his company.

Estimates vary, but Fraser has been credited with gathering in of up to 60 of his comrades, under cover of darkness, and at great risk to his own safety. It was during one of those recoveries that he heard the plaintive cry of a fellow soldier of 'Don't forget me cobber!' Compare therefore Fraser's actions of the first award of the Victoria Cross, to 'young W. Jackson'.

Sgt Fraser was **Mentioned in Dispatches**.

Given that there had not been a general truce ordered following the Battle, and given that soldiers such as Fraser continued after the battle to search for their fellow diggers under cover of darkness for another three days, it perhaps comes as no surprise that those sorts of actions were not to be rewarded, especially by way of any medals. On 5 October, 1916, 5th Division Routine Orders contained the statement that:

In future the Victoria Cross or other immediate reward will not be given for the rescue of wounded excepting to those whose duty it is to care for such cases. Such attempts, more often than not, result in the death of the would-be rescuer and rescued. Moreover it depletes the fighting strength of units perhaps at the most critical moment.

---

[159] See Corfield p.288

[160] See Corfield p.288

For Fraser being Mentioned in Dispatches may have been enough to satisfy those in command at the 5th Division, but the soldiers who risked their lives were still as likely to be targeted by the Germans as if they were trying to recover their mates during daylight hours.

---ooo000ooo---

Commenting on the lack of the highest award for ***acts of extreme bravery in the face of the enemy,*** the Victoria Cross, and its absence at Fromelles, journalist Andrew Rule wrote of the efforts of Sgt Simon Fraser, in an article published in the Sydney Morning Herald, Sydney on **20 July, 2010** titled:

LOST AND FOUND IN FROMELLES[161]

The pertinent part of the article reads as follows: -

One of the fine physical specimens recruited in 1915 was Simon Fraser, a 38-year-old Western District farmer who had served in his local mounted militia for nine years and had been a rifle club shooter for more. He was more than six foot high — tall for that time — and lean. And he must have been strong: he was mentioned in dispatches for carrying wounded soldiers to safety from the open-air abattoir of no-man's land after the Fromelles debacle.

As he passed one maimed Australian, the man called out "Don't forget me, cobber" and Fraser later returned for him.

The phrase - and Fraser's gallantry - was commemorated in the bronze statue known as "Cobbers", which depicts him carrying a wounded mate on his shoulders. The original was unveiled at Fromelles in 1998; its twin at the Shrine in Melbourne in 2008.

Fraser, promoted to second-lieutenant, was killed at Bullecourt in May, 1917. His body was never found.

The Army sent home his leather case. In with his books, riding boots and pipe was a lock of hair. Sister or sweetheart? Who now knows?

---ooo000ooo---

Now compare the service of one who was recommended for the Victoria Cross, and the unexplained removal of the recommendation by Brigadier General Elliott for the award of a Victoria Cross to **Major Arthur Justin Sanford Hutchinson** which was put forward to General McCay, in Elliott's own handwriting. The citation was written by Elliott: -

> *At Petillon on the 19th/20th July 1916, Major Hutchinson displayed conspicuous and gallant leadership. On the evening of the 19th 20th July, 1916, a message came from 5th Division that the 61st Division on our right would renew the attack at 9.00pm on the Sugar-loaf Salient and notwithstanding that the previous attack by a Battalion had manifestly failed, Major Hutchinson led the two Companies of the 58th Battalion under his command in the most gallant manner under an appalling fire until he fell riddled with machine gun bullets close undfer the*

161 https://www.smh.com.au/national/lost-and-found-in-fromelles-20100719-10i1u.html

German parapet. His life and the lives of his men were gallantly given in the hope of aiding the attack of the 61st Division, which unfortunately was not made.

Records show that enclosure were made in the form of statements by Lt. H J Boyd and 2nd Lt L H Dardier.

Hutchinson received nothing and was not even Mentioned in Dispatches. On his file is written 'No trace of award'.

---ooo000ooo---

What happened to Hutchinson was recounted nearly four years later, by the Editor of The Argus, Melbourne, in a lengthy article on 10 April, 1920, titled: -

FROMELLES, 1916! A GLORIOUS FAILURE. WHAT REALLY HAPPENED[162].

The article went on to detail the work of Major Hutchinson:

The general position of the 5th Divisional front at 7.30 p.m. on the 19th was that the attack was definitely held up from the right brigade sector, and successful on the central and left sectors. The 59th and 60th Battalions had suffered terribly, and in the53rd, 54th, 31st, and 32nd Battalions, the percentage of losses, especially amongst the officers, was very high, and still mounting steadily. The line held was an indifferent one. Consolidation was difficult; the line was not continuous, and later communication along it was irregular and uncertain.

General 'Pompey' Elliott received official news of the failure of the 61st Imperial Division (on the right) at about 7.30 p.m., by which time he was also aware that the 59th and 60th Battalions were badly cut up, and quite unable to advance without assistance. On receipt of information at 7.52 that he could use two companies of the 58th to support his attack, in conjunction with the attack of the 184th Imperial Brigade on the Sugar-loaf, he took immediate steps to make the necessary arrangement. Command of the attack was entrusted to Major Hutchinson.

When General Elliott was notified that the British would be making another attempt at the Sugar-loaf at 9.00pm and wanted assistance from his brigade, he arranged for two companies of the 58th Battalion to make a simultaneous attack. This half-battalion was led by Major Hutchinson.

At 9.00pm Hutchinson 'displayed conspicuous and gallant leadership' in leading his two Companies in the attack on the Sugar-loaf salient under heavy fire. According to Charles Bean: These Australians were very eager. As they waited for their moment to enter the battlefield, they were 'straining like greyhounds on the leash', Bean wrote. They sprang forward 'with splendid dash opposite the Sugar-loaf', inspiring some 59th survivors to accompany them.

The companies pressed on, losing at every step, but undaunted to the end. They reached the remnants of the 59th and 60th Battalions, where they lay grimly waiting in their shallow, improvised positions. They carried them on towards the enemy by the impetus of their own heroic charge. Impeded by broken ground

[162] For full article, see: -
https://trove.nla.gov.au/newspaper/article/1690351?searchTerm=Fromelles%201916!%20A%20glorious%20failure%20%20%20%20%20%20%20%20%20%20%20%20&searchLimits=

and shell holes, the thinning line searched brokenly forward, reeling under the enfilade, enduring everything but the thought of failing.

General Elliott was shocked to find no sign of the British attack that Hutchinson's half-battalion had gone over the top to support. As we know now the supposed British advance did not eventuate, without Elliott knowing it. While Commanding Officer of the Fifth Division, General McCay did not receive a message about the cancellation of the British attack until 8.35pm, Elliot did not receive the message about the cancellation until 9.35pm, well after his men had entered the fray.

The Sugar-loaf's defences were thus enabled to concentrate the whole of their organized machine-gun fire on the one thin Australian line which now endeavoured to penetrate it.

Hutchinson and his men proceeded purposefully across no-man's-land until they, too, were cut down by lethal enfilade fire from the Sugar-loaf machine-guns. Bean described it as 'one of the bravest and most hopeless assaults ever undertaken by the Australian Imperial Force'.

At the enemy wire the fire became hellish, irresistible. Major Hutchinson, his body riddled with bullets perished gloriously, close to the German parapet, and closer to it than anyone had been able to reach before him. It was in vain. The attack melted into nothingness.

Who Was Major Arthur Justin Sanford Hutchinson?[163]

Until now, little has been known about the 21-year-old army major who may well have received one of military's highest distinctions if he had survived what was effectively an impossible mission on the Western Front on this month 100 years ago.

Many Australians, through the film 'Gallipoli', have heard of the charge of the 8th and 10th Light Horse at the Nek on Gallipoli, but few know of the assault by two companies of the 58th Battalion towards the 'Sugar-loaf' at Fromelles in support of the 59th and 60th Battalions. Charles Bean described it as 'one of the bravest and most hopeless assaults ever undertaken by the Australian Imperial Force.'

Hutchinson's father was a clergyman. Hutchinson himself was an apt scholar and sportsman at Launceston Church Grammar. Hutchinson won a university scholarship, but he made the fateful decision to enlist in the army and train at the Royal Military College at Duntroon instead, against the wishes of his parents.

In 1911, applicants for admission to the Military College at Duntroon had to contest a series of competitive examinations. English, history, geography, basic mathematics, and general knowledge were all compulsory. Aspirants also had to select two further subjects to be examined in from four options: mathematics, physics, chemistry, and either French or German. The examinations took place in December 1911. With 103 candidates applying from all around Australia, Tasmanian Justin Hutchinson was fourth on 871.[164]

---

[163] See generally Stefan Petrow: Remembering a Tasmanian Hero: Major Justin Hutchinson ISBN 978 0 646 93516 4

[164] Ross McMullin: Farewell Dear People, p.111

Hutchinson and his confreres began at Duntroon on 7 February 1912 in the second batch of cadets. Their physiques were checked on their first day. He would graduate from Duntroon on 2 November, 1914. Three years after his entry to Duntroon, he was taken from the college to join the 9th Light Horse Regiment and leave home to fight in World War I.

[Picture 46: Major. Arthur Hutchinson. Photo: AWM P11131.001]

He fought in Gallipoli with the 9th Light Horse Regiment, in the machine gun section, and was hampered by illness on several occasions. He experienced a few near misses from bullets - including one that, according to one of his letters – passed through his hair.

Hutchinson would later transfer from the light horse to Brigadier Pompey Elliott's 15th Brigade in 1916, and be appointed to lead two companies. He was promoted to the rank of Major within two and a half years of overseas service.

19 July, 1916: The Battle of Fromelles, and The Attack Led By Hutchinson

Silencing the Sugar-loaf machine-guns had been mainly a task for the artillery; it was the British infantry alongside Pompey Elliott's 15th Brigade who were supposed to occupy the Sugar-loaf and quell any remaining resistance. Neither the artillery — which was inexperienced, inadequate, and inaccurate — nor the British infantry had accomplished these objectives, so the Sugar-loaf machine-guns were able to annihilate the 59th and 60th AIF Battalions.

Moreover, at the point of advance of the 15th Brigade, no-man's-land was more than 365 metres wide compared to 90 metres on the left where the 14th Brigade battalions attacked. As the 59th and 60th Battalions left their positions and proceeded across no-man's-land they were shot to pieces. Hundreds were killed and wounded and those unharmed took cover

At 5.50pm it was the second wave's turn. Major Tom Elliott indicated it was time to go. He ascended a ladder, and over he went. Tom Elliott and his men advanced through high damp grass and occasional ditches. They had to run a daunting gauntlet of enemy projectiles. Tom Elliott kept encouraging men around him until he was hit. It was a bursting shell, very close.[165]

An eye-witness said afterwards that he saw the Major thrown into the air before he fell. Another observer, Private A. Forbes, was looking across no-man's-land as he waited his turn in a following line. Forbes saw Major Elliott fall, and then struggle to his feet, wrestling with his tunic. It seemed like he was burning with pain. 'He had a big gash in his back, as if from high explosive', Forbes noticed. Elliott managed to get the tunic off, and then collapsed. Forbes passed Elliott when he moved across no-man's-land with his wave, and never saw him again. 'Major Elliott was one of the best', he said.[166]

---ooo000ooo---

Pte. 3257 G.H. Smythe, 58th Bn, gave a statement which is now part of the Red Cross file No. 1400406, on 4 August, 1916: "I was with Major Hutchinson as we went across in the charge at Fleurbaix on July 19th. We got across nearly to the first line German trenches and laid down for the next advance. He said 'Don't get too close together boys, spread out in the next run.' He got up and went forward. As we rose, he was hit and fell (in a heap). I was wounded soon after and managed to crawl back to our lines. When I got back, I heard one of the officers ask a wounded man how Major Hutchinson was getting on. The man replied that he was dead. Evidently, he died where I saw him fall."

Pte 3624 S.G. Parker, 58th Bn (patient 1st Australian Auxiliary Hospital, Harefield, England), gave a statement of his own on 12 August, 1916: "Went over the parapet with us and about half way across I saw him, he got up [,] gave an order and fell back dead – about ½ an hour after I saw him still lying there, I knew he was dead … "

After the Battle

The 59th and 60th Battalions had been massacred. Half the 58th under Justin Hutchinson, had been slaughtered. The upshot of this disaster for the 15th Brigade was that its shattered trenches were crammed with dead and dying men. Further casualties were crawling back from no-man's-land. Eye-witnesses struggled for words to describe the distressing and chaotic scenes; some resorted to analogies with butchers' shops.

Pompey Elliott's first definite information that his brigade's attack had failed comprehensively came from Charles Denehy[167]. Having served under Pompey in both the pre-war militia and the 7th AIF Battalion, Major Denehy was one of the brigadier's most trusted officers. Pompey had plucked Denehy from the 57th just before the Fromelles attack to fill the vacancy at the helm of the 58th caused by Jackson's removal. Denehy mentioned Majors McCrae, Tom Elliott, and Hutchinson in his chilling message to Pompey:

---

[165] Ross McMullin: Farewell Dear People, p.152
[166] Ross McMullin: Farewell Dear People, p.153
[167] Ross McMullin: Farewell Dear People, p.153

> *Men of all battalions are coming back from no-man's-land and fully expect that they will gradually drift back to the line. Many men are wounded, many are not. Very many officers are casualties, including Majors McCrae, Elliott and Hutchinson, all of whom are reported dead, and seems impossible to reorganise … Report seems to be unanimous … that not a single man of the 15th Brigade is now arrived in enemy's trench, as enemy's flares are coming from the whole of the front allotted to this Bde. I am now organising the defence of our original trenches.*

The Victoria Cross Recommendation

Describing Hutchinson, Brigadier Elliott, his commanding officer, wrote of his energy, competence and enthusiasm combined with a modest and unassuming manner. Hutchinson had led his men forward against the 'Sugar-loaf' into 'murderous machine gun fire' with 'great dash and resolution'. Indeed, Elliot recommended Hutchinson's conduct as worthy of a Victoria Cross. The Tasmanian clergyman's son had led his men 'in the most gallant manner under an appalling fire until he fell, riddled with machine gun bullets close under the German parapet, while searching on his own for a gap in the wire.'

In his submission, Brigadier Elliott said Major Hutchinson had led his men "with great dash and resolution", advancing at least 200 yards further than the attack point previously reached.

He said that Major Hutchinson had ordered his men to take cover while he undertook reconnaissance. Elliott said despite enduring "hostile enfilading machine gun fire", he "persisted in his gallant attempt to discover a breach in the wire until he was killed."

He described Major Arthur Hutchinson, a Duntroon graduate who led the assault, as a 'boy of the finest type this country produces'.

But his recommendation for this award was denied. The VC recommendation was rejected. Pompey was not surprised. The context was too controversial. Elliott believed there had been a 'shameful betrayal of gallant men' of his brigade who had been 'hopelessly and helplessly massacred'. We know now that the attack should never have been launched, and failed due to lack of communication between the Division Headquarters and the Brigade.

Hutchinson had exhibited the gallantry deserving of an award of the Victoria Cross. Instead, as with so many others, he was awarded the 1914-15 Star, the British War Medal, and the Victory Medal.

Grieving The Loss.

Hutchinson's fiancée, Gladys Forrest, later wrote to the authorities trying to find out what had happened to him:

"Being engaged to Major AJS Hutchinson I am naturally very worried at having received a telegram forwarded on home to Reverend Hutchinson saying 'Officially reported missing twentieth July, Major AJS Hutchinson Ninth Light Horse Regiment' … I am writing to ask if you would be kind enough and can tell me? Is Major Hutchinson now in the 58th in France? Or in Egypt back in the 9th LH? … It would help so much in every way, if you would be able to find out any particulars as to if he is still missing, as the anxiety is so great".

---ooo000ooo---

In 1934, Launceston Grammar School planted an avenue of remembrance, for students who had served and not returned. Margaret Challis (mother of George Challis) planted a tree in tribute to former students, including Justin Hutchinson and Price Morgan. The tree had originated as a seedling from Gallipoli, as had another 97 trees that were planted in tribute to former students. A speaker at the ceremony said that the trees were a splendid 'way of perpetuating the memory' of these students. Regrettably, the avenue of honour has been since demolished.[168]

---ooo000ooo---

On the memorial wall at V.C. Corner, Fromelles, the names of 1,299 Australians are listed. Of those, 595 (46 percent) were from three of the four battalions of the 15th Brigade – the 58th, 59th and 60th – from Victoria, commanded by Brigadier General Harold 'Pompey' Elliott.

Major Arthur Hutchinson's name is remembered on Panel 13 at VC Corner along with 52 other men of the 58th Battalion killed, in the main, in the assault on the 'Sugar-loaf' on 19 July 1916.

His body was never recovered, he has no known grave, and inadequate recognition for his service.

---ooo000ooo---

[168] Ross McMullin: Farewell Dear People, p.229

16. Remembering the Battle of Fromelles - The 19th of July Men

In the aftermath of The Great War, Fromelles did not attain the heady recognition given to Gallipoli, despite it being singularly the largest loss of life in a day, for an Australian action in the war, before or since. But its significance was not lost on the soldiers who served, or the populace generally, from one coast of Australia to the other, as the following material and articles will attest. As the nation continued to stop for remembrance services each year on ANZAC Day, 25th April, those who had served in the Battalions of the 5th Division continued to honour their service and involvement in the Battle of Fromelles on 19th July as well. They became known as *the 19th of July men.*

---ooo000ooo---

Within a year of the end of World War I, Captain Alexander Ellis, late 29th Battalion, 8th Brigade, Fifth Australian Division, had produced his history of the division, the only Australian divisional history to come out of the 'Great War'. Ellis proudly, and perhaps boldly, subtitled his work – '*Being an Authoritative Account of the Division's Doings in Egypt, France and Belgium*'. Marshal Ferdinand Foch, the Frenchman who had been the supreme Allied commander of 1918, was prevailed upon to pen a foreword in French and Ellis ended his book with these stirring words:

The Division's Glorious Dead! … Of all those things of which the 5th Australian Division is proud, it is proudest of its Dead. Them it most reveres. The memory of them it cherishes above all things. May all Australian hearts be softened by their love, animated by their courage, ennobled and uplifted by their example until the end of time.[169]

The Fifth Division's tribute to those who fought with it between 1916 and 1918 in France and Belgium is in Polygon Wood, Zonnebeke, Belgium, near Ieper (Ypres). The way into the memorial lies along a narrow grass way bounded by a low rubble wall. This is Australian land, acquired by the division after the war, and the memorial itself, a stone obelisk, sits on a long, tall bank known as the 'Butte', approached up a steep flight of steps. Beneath the memorial are the 2,103 headstones of Buttes New British Cemetery and the New Zealand Memorial to the missing while beyond stretches Polygon Wood. A glance through the trees behind the memorial reveals that it is close to the northern edge of the wood. The 5th Division's Memorial was commissioned and erected in 1919, the year following the armistice. (See chapter 11 for image.)

[169] Captain Alexander Ellis, The Story of the Fifth Australian Division, London, 1919, p.402. [See also article referred to later, published on 20 March, 1920 Sydney Morning Herald.]

The dedicatory plaque on the front face of the memorial has a few words in English and French. The English simply declares that the memorial is 'To the Officers, Non–Commissioned Officers and Men' who served in the division between 1916 and 1918. The French includes the rather more gracious phrase 'A la Mémoire des ..', 'To the memory of …'. Beneath the dedication the Fifth Division listed 17 battle locations in France and Belgium, places it wished the visitor to realise epitomised for the soldiers of the division their struggle and sacrifice on the Western Front.

Many of these names mean little today but they would have been etched in the minds of Australian families who lost sons, husbands, relatives and friends at these far away places. The division's last action was fought on 1 October, 1918 in France on the Hindenburg Line and that location, along with the nearby village of Joncourt, is recorded on the memorial at the end of the list.

Private Vincent Harvey, 57th Battalion, aged 19, died on 1 October 1918 from gas inhaled during these last attacks of his battalion between 29 September, and 1 October 1918.

Another who died within sight of the end on 1 Oct. 1918 was Private Gordon Edmund Ewart, 58th Battalion (Victoria), aged 20. His father, Alexander Ewart, wrote on his 'Roll of Honour Circular':

Deceased was engaged in following battles – Fleurbaix, Somme, including Morchies, Bapaume, Bullecourt, Ypres, Polygon Wood, Westhoek Ridge, Messines, Villers–Bretonneux, Morlancourt and Bellicourt.[170]

Mr Ewart's chronology might not have been too accurate but his message was clear – his son had fought in virtually every significant action engaged in by the 58th Battalion of the Fifth Division since its first disastrous battle at Fromelles (Fleurbaix) in France on 19 July 1916. And Fromelles, seemingly in the place of honour, is the first battle location on the Fifth Division's memorial. In his listing of the places of battle at which his son had fought, he was particular in distinguishing Fleurbaix (Fromelles) from 'Somme', unlike the officials of the Battles Nomenclature Committee in the years to come.[171]

Soldiers' Reunions:

In the years after the cessation of hostilities, and the end of the war, soldiers who served at Fromelles met on or close to 19th July, the anniversary of the Battle of Fromelles itself, and took the time to remember their service, those who served with them, and what they had come away

---

[170] Private Gordon Edmund Ewart, Honour Roll Circular: www.awm.gov.au/cms_images/awm131/017/017056.pdf

[171] The Australian Remembrance Trail in France and Belgium.

https://anzacportal.dva.gov.au/history/conflicts/australians-western-front-19141918/australian-remembrance-trail/tyne-cot-2

from. They came to be known as the *'19th of July men'* because of their continuing observance of the Battle of Fromelles on its anniversary.

Why did they hold their annual reunions in July? While those who served at Fromelles recognized the community's homage of ANZAC Day, they did not see it being appropriate to simply merge their remembrance with another date of significance, when theirs was another day deserving in its own right.

Given that the survival rate from Fromelles was so horrendous, and for the survivors of that battle to have survived many more battles to be fought later in the war, it is truly remarkable that they were to meet on one particular day, special to them. Why? Was it because for so many of them it was their birth of fire, their first exposure to industrial scale battle warfare? Those who served at Gallipoli would always have ANZAC Day to observe. Yet so many also observed Fromelles' anniversaries on or close to 19th July.

Their announcements of gatherings to commemorate the Battle of Fromelles, and their collective observance of the Battle of Fromelles are set out in Appendix II, due to the sheer number of such events. The pages in the Appendix are taken up with the various units of the Great War issuing notices for reunions, and participating in commemorative services, of which they were justly proud.

The 32nd, 54th and 55th Battalions all held their reunions in July. Those units which held reunions in Sydney can be seen to have done so in July also. Other units held reunions at different times. For example, the 30th Battalion in Newcastle held their reunion in November, on the anniversary of the date on which their unit sailed for the Great War, again not only on ANZAC Day.

In the course of gathering material for this book, I have come across various strains in which parties have commented that the men didn't do much to observe what happened, or that interest in Fromelles died off once the soldiers came home. What I have found, and what you will see, clearly says something very different to those commentators. In fact, the opposite seems to be true as the contents of Appendix II will attest.

One more thing is abundantly clear: those 19th of July men remembered, commemorated, and marched together, on or as close to 19th July as they could do so, right up until the last of those servicemen simply could do so no longer.

In 1966, more than 500 veterans of the 5th Division marked the 50th Anniversary of the Battle of Fromelles with a Parade from the Archibald Fountain to the ANZAC Memorial, Sydney. They joined about 300 people unable to march at the ceremony by the Pool of Reflection. It was the last Parade for the veterans and they carried a request that the Trustees of the ANZAC Memorial always remember their service and the sacrifice oftheir comrades on 19 July 1916. That service continues every year on 19th July.

Their service would not be diminished.

Newspaper Articles

Invariably, correspondence also ensued in the various newspapers dealing with the event by way of letters to the Editor, or by publishing of Articles by various newspapers around Australia. The following articles and extracts are just some of the letters to the Editor, or articles published by Newspapers, and as such were publicly advertised to all who read them across the length and breadth of Australia.

In Chapter 8 we read of the exchanges in the Sydney Morning Herald following the publication of a letter from Lt. Colonel MacLaurin. Among those to respond was one J. Williams, whose letter was published on 30 July, 1919, and among other comments includes the following:

Lieutenant-Colonel MacLaurin gives it as his opinion, that if the 2nd Division had been entrusted with the operation the result would have been entirely different. I beg to differ, and go as far to say that no division on the Western front could have succeeded where the 5th Division failed.

[The 5th Division] took over the line from the 11th English Division, in front of Flers on Oct. 21, 1916, and with the exception of a 10 days' rest in Nov., 1916, was never out of the forward area until Apr. 24, 1917, and was in the line again at Bullecourt in the early days of May, 1917.

Fromelles (or Fleurbaix) was a ghastly failure, in which an Australian division was butchered for no purpose. The infantry paid the price, not so much of their impetuosity, but of the blunder of some "brass hat." The true tale of July 19-20, 1916, has yet to be told, and why so many valuable lives were wasted.

Let the readers of the Lieutenant-Colonel's story not get the idea into their heads that the 5th Division had any disgrace to live down through their conduct in this action. The attacking battalions were set a task made impossible by Fritz knowing the full details of the operation beforehand, a total lack of staff work; the position to be attacked being one of immense strength; and then, if our artillery outnumbered the German, the latter certainly used his with better effect.

The infantry have known for a long time that the easiest way to get the "real" details of a stunt in which they have played the main part is to apply to a member of a unit which occupied a "full-back" position in the game of war.

I am, etc., J. WILLIAMS. July '27.

There was no way that J. Williams was going to let it be said that the performance of the 5th Division had been a failure.

---ooo000ooo---

The [Australian and British] soldiers who served at Fromelles were not diminished by having participated in what could only be called a disaster. Nor would they accept unfounded criticism of their state of preparedness. They learnt to wear their badge with pride, as later achievements would attest.

Writing in **1920**, in The Story of the 2/4th Oxfordshire and Buckinghamshire Light Infantry, Captain G.K. Rose MC, devoted one paragraph to the events of 19 July:

The harassing warfare had a crisis in July. The operations of July 19, which were shared with the 61st Division by the 5th Australian holding trenches further north, were designed as a demonstration to assist our attack upon the Somme and to hold opposite to the XI Corps certain German reserves, which, it was feared, would entrain at Lille and be sent south. That object was achieved, but at the cost of several casualties to the divisions engaged, which were launched in daylight after artillery preparations which results proved to have been inadequate, against a trench-system strongly manned and garrisoned by very numerous machine-guns. The objectives assigned to the 61st Division were not captured, while the Australians further north, after entering the German trenches and taking prisoners, though they held on tenaciously under heavy counter-attacks were eventually forced to withdraw. 'The staff work' said the farewell message from the XI Corps to the 61st Division three months later, 'for these operations was excellent'. Men and officers alike did their utmost to make the attack of July 19 a success, and it behoves all to remember the sacrifice of those who fell with appropriate gratitude. It was probably the last occasion on which large parties of storming infantry were sent forward through 'sally ports'. The Battalion was in reserve for the attack. C Company, which formed a carrying party during the fighting, lost rather heavily, but the rest of the Battalion, though moved hither and thither under heavy shelling, suffered few casualties. When the battle was over, companies relieved part of the line and held the trenches until normal conditions returned. Soon after these events the Battalion was unlucky to be deprived of Colonel Ames, a leader whose energy and common sense could ill be spared. This was the first change which the Bn had in its Commanding officer, and it was much regretted.

This is obviously a somewhat more restrained account than that of J. Williams, but still pointing to the 'success' of the objective.

---ooo000ooo---

In the aftermath of war, books began to be written, in order that history would not forget the part played by the various soldiers, and their units. In the release of the book by Captain A.D. Ellis: The Fifth Division: - Its Great War Record, and as reviewed by F.M. Cutlack under the description of 'A Splendid Story', in the Sydney Morning Herald on 20 March, 1920[172], in an edited version of same, it was written that:

The divisions lighting in France opened with the battle of Fromelles on the evening of July 19, 1916. The attack was finally launched by the Fifth Australian (with all three brigades in the line) and the 61st British Divisions, all new troops, after postponement from the 17th; it was hurried in preparation and was not meant to be more than a feint, or rather a holding action against German troops which were being thinned out for reinforcement against the bigger Allied attacks in the Somme area.

[As we have previously noted, the actual order given to the soldiers was to *capture and hold* the German (second line) trench system.]

The thanks of all the A.I F. will be due to the author for his ample account of this action, which has never hitherto been adequately recorded. He includes, too, an amazing preliminary order from the XI Corps

---

172 See full article: - https://trove.nla.gov.au/newspaper/article/15881522?searchTerm=the%20Fifth%20Division%3A%20-%20its%20great%20war%20record&searchLimits=

commander (General Haking), which should be preserved from oblivion. It was written before British generals knew how to address Australian troops, and the Fifth Division fortunately had not to rely on it for their enthusiasm. It reads: -

"As you know, we were going to have a fight on Monday, but the weather was so thick that our artillery would not see well enough to produce the very accurate shooting we require for the success of our plan. So I had to put it off, and G.H.Q. said do it as soon as you can. (sic). I have fixed Zero for Wednesday, and I know you will do your best for the sake of your lads who are fighting else down south (Signed) R. Haking."

It might have been a notice of postponement of a "come-if-you-can" invitation to lunch.

The Fromelles attack was a disastrous though heroic failure - in itself at any rate, though the holding purpose was served - and it could hardly have been anything else in all the circumstances. But it proved at once the splendid stamina and gallantry of the Fifth Division. The division's total casualties here were 178 officers and 5335 other ranks, a blow to strength from which it took six months to recover. Many promising battalion officers fell.

---ooo000ooo---

In the West Australian, Perth, on 19 July, 1924, an article (edited below) appeared titled:

FLEURBAIX-FROMELLES BATTLE.
FIRST A.I.F. ENGAGEMENT IN FRANCE.
EIGHTH ANNIVERSARY TO-DAY.[173]

Today is the eighth anniversary of the battle of Fleurbaix-Fromelles, the first in which Australian troops took part in France. Writing with regard to this memorable engagement, Mr. E. T. Crain, of Central-avenue, Inglewood, says: -

'It was an attack carried out by the 61st British and the 5th Australian Divisions. The immediate intention was to capture and to hold the front and support lines of the enemy on a front of about 4,000 yards. And was hoped that, if the attack succeeded, it would be the prelude to a further and more important offensive against vital enemy positions on the Aubers Ridge. There are few incidents of the war which deserve a closer attention from the Australian public.

It was an hour of intense silent excitement that preceded the first frontal attack made by Australian troops in France. A crowded front line of trenches, men armed with grenade, rifle and bayonet, machine gunners standing by, orders whispered, while the artillery were pounding the enemy stronghold. At last the signal came, and from our lines leaped battalions of the finest men that ever breathed, only to be met with a withering fire from machine guns and field artillery at point blank range. The waves of men pressed forward steadily but just as steadily the enemy fire grew hotter and hotter, and the enemy front line was

---

173 See full article: -
https://trove.nla.gov.au/newspaper/article/31242896?searchTerm=FIRST%20A.I.F.%20ENGAGEMENT%20IN%20FRANCE.%20EIGHTH%20ANNIVERSARY%20TO-DAY.%20%20%20%20%20%20%20%20%20%20%20%20&searchLimits=

seen to be thickly manned. The losses mounted rapidly as the men pressed forward gallantly in the face of terrific odds. One brigade lost no fewer than 35 out of 39 of the assaulting officers and with them most of the N.C.O.'s.

In the circumstances the survivors could only hold on determinedly to what had been won and await such further action as might be devised to meet the situation.

"On the right sector the English Division met a somewhat similar fate, and the losses were enormous. To even attempt to describe the battle is beyond human power, but the following official casualty list for the period from noon, July 19 to noon, July 20 (24 hours) will suffice to show the intensity of the operation:

8th Infantry Brigade. 54 officers; 1.846 other ranks;

15th Infantry Brigade. 51 officers; 1.725 other ranks;

14th Infantry Brigade. 65 officers; 1652 other ranks:

Div. Engineers. 3 officers; 85 other ranks;

making a total of 173 officers and 5,308 other ranks.

"The infantry battalions that suffered most heavily were the 60th, the 32nd, and the 59th, with 757, 718 and 695 casualties, respectively. The 53rd, 31st and 54th also lost more than half their strength, and it is doubtful that any British division in the whole war suffered equally heavy losses in the same space of time. The casualties in the English brigades were 577, 336, and 400, in the 182nd, 183rd, and 184th brigades respectively.

"It should not be forgotten that while the operation in itself was an unqualified local failure its main purpose in preventing the withdrawal of troops to the Somme was doubtless to some extent achieved.

---ooo000ooo---

Seventeen years after the Battle at Fromelles took place, the reunions of participants did not see a lessening of their pride in their involvement. In a much-abbreviated report in The Advertiser, Adelaide, on 20 July, 1933, it was reported: -

HOLOCAUST OF FROMELLES. AUSTRALIAN DISASTER IN EARLY DAYS IN FRANCE - MANY BLUNDERS.[174]

Members of the 32nd Battalion A.I.F., at their annual reunion on Saturday night, at the Returned Soldiers' Club, will recall with mixed feelings the Battle of Fromelles on July 19, 1916, at which the battalion received its baptism of fire. Few units have had a worse baptism. Few of those who took part in it are left today, but these who are, and other members of the battalion, can look back with pride on the part played by it on that occasion.

---

[174] See full article: https://trove.nla.gov.au/newspaper/article/46988793?searchTerm=AUSTRALIAN%20DISASTER%20%20IN%20EARLY%20DAYS%20IN%20FRANCE%20%20%20%20%20%20%20%20%20%20%20%20&searchLimits=

Fromelles was the first serious engagement fought by Australian troops in France, and it was also the most disastrous. From start to finish it was a series of blunders on the part of the commanders and staffs concerned. Its only redeeming feature being the heroism of the men whose lives were so needlessly sacrificed.

The object of the operation … was to prevent the Germans withdrawing troops from Flanders for use on the Somme … On no account were the troops to go beyond the enemy's support line, which was said to be about 150 yards behind his front trenches.

In the centre of the position to be assaulted was a small salient called the Sugar-loaf, for the capture of which the battalion on the left of the 61st Division was made responsible.

The Germans, from their observation posts on the Aubers-Fromelles ridge, about a mile and a half behind their front trenches, had an excellent view of the British lines, and saw the men forming up. In fact, most of the preparations for the attack during the two or three days preceding it seem to have been made in daylight in full view of the enemy, with the result that he knew it was coming. ... [B]oth divisions attacked gallantly and seized the German front line on either side of the Sugar-loaf. The British troops opposite it however, were unable to force their way through the barbed wire, and were mowed down by machine guns. The two battalions of the 15th Australian Brigade, which attacked immediately north of it, met with the same fate.

It was ascertained afterwards that the heavy guns which should have bombarded the Sugar-loaf before the attack did not do so. ... the battery commander was afraid to fire at it lest his shells should fall short on the British trenches. Consequently, the occupants of the Sugar-loaf post - the strongest of the whole sector — were able to sweep the assaulting troops with rifle and machine gun fire practically unhindered, and without any great risk to themselves.

What were shown as support trenches on the maps issued for the attack turned out to be either ditches, or abandoned trenches overgrown with rank grass and half full of water. The mistake in the selection of the final objective appears to have been due to using maps made from aeroplane photographs taken the previous summer when those trenches were first dug.

In spite of the gallantry of the men both British and Australian, the attack had failed. But with fully trained and experienced infantry and, competent artillery it might quite easily have succeeded, had it been properly organised and conducted.

---ooo000ooo---

In what appears to be more of an editorial piece than a public notice, it was reported on 24 April, 1936 in Daily News Perth:

FIRST INTERSTATE UNIT REUNION[175]

[175] See: - https://trove.nla.gov.au/newspaper/article/84224685?searchTerm=FIRST%20INTERSTATE%20UNIT%20REUNION%20%20%20%20%20%20%20%20%20%20%20%20&searchLimits=

On Sep. 25, 1915, two companies of infantry left Fremantle for Adelaide to join up with South Australian troops, to form the 32nd Battalion, A.I.F.

Leaving Australia in Nov., 1915.the battalion served with the 5th Australian Division on the Suez Canal and in France and Flanders.

That was 21 years ago. A party of those who served in this famous A.I.F. unit will embark, it is anticipated, on Sep. 25, to take part in a re-union, to be held in Adelaide.

This being South Australia's Centenary year, a big re-union is part of the general celebrations, and this is the first occasion on which numbers of ex-servicemen of one unit will travel from several States to gather in one capital city. And not only ex-32nd A.I.F. men will be present, for the band of the 32nd Battalion C.M.F. from Melbourne, with King's and regimental colors and a colour party, will arrive in Adelaide about Sep. 27.

[Although published in April, 1936, the reunion would take place in the coming September.]

---ooo000ooo---

Writing to the Western Mail, Perth, "Toxites" (llth), from Ballidu, on 4 June, 1936 pointed out the larrikins among the soldiers, in his letter titled:

IN THE NURSERY. ALL QUIET AT FLEURBAIX.[176]

Dear 'Non-Com."- In our early days in France when all the boys were keen and eager, it used to give one a thrill of pride to go along the trenches at stand-to in the mornings. We were in the "Nursery" at Fleurbaix and every one would be standing religiously at the firestep, thinking all sorts of things no doubt and giving an occasional eye to the "babblers" cooking the morning meal on charcoal fires in sheltered spots among the collection of sand bag dug outs.

It was generally very peaceful then and we used to listen to the larks rising in no-man's land and hear the black-birds and thrushes starting to pipe in the dusk of those mild May mornings.

One would saunter along the line and chat to old friends and discuss matters like well-sinking or shearing with some of the boys from the outback, until one came to "Old Murphy," who was generally as cantankerous as a camel at that early hour and who used to be finding fault with everything both on our side and on that of our friends the enemy.

Soft words were of no avail - they only started him off on a fresh grouse. And then the order to stand down would come, and the men would clean their rifles and get ready for breakfast.

But Murphy always laid his rifle against the parapet with the muzzle about 45 degrees elevation and fired a shot in the direction of the enemy's country.

"Why do you do that, Murphy?" I asked him one morning.

---

176 See: - https://trove.nla.gov.au/newspaper/article/37774362?searchTerm=IN%20THE%20NURSERY.%20ALL%20QUIET%20AT%20FLEURBAIX%20%20%20%20%20%20%20%20%20%20%20&searchLimits=

"Sure," says he, "it's the best way to loosen any dirrht that might be sticking in the barrel, be gob, and ye niver know but what I might be hittin' wan av thim Fritzies away ahint there, an him walkin' about unheedin' like."

And all the boys grinned and the sentries began to keep their heads down because a few shots would come spattering over in the ever-increasing light of early morning.

---ooo000ooo---

As for being a 'Nursery', there is an often held perspective that the area Fromelles was in (that of Aubers Ridge, Armentieres) was a quiet section of the Western Front, a 'nursery', where units were sent to adjust and acclimatize. We can appreciate from the material in Chapter 3 that this was not always the case. The language (almost a tactic) used in marginalizing and diminishing service in the area by referencing it as a 'nursery' was arguably a contributing factor to it becoming referred to as the 'Forgotten Front'.

---ooo000ooo---

Twenty years after the Battle of Fromelles, the following article appeared on 20 July, 1936 in The Courier Mail, Brisbane:

FAMOUS BATTLE RECALLED.

SURVIVORS HONOUR THE FALLEN.[177]

Anzac Square again resounded to the marching tread of comrades in arms yesterday, [i.e. 19th July] when 25 of the survivors of the Battle of Fromelles (Fleurbaix) fell-in to do honour to the host of officers and men who lost their lives in that engagement.

Yesterday was the 20th anniversary of the battle, and the survivors commemorating it were members of the 31st (Queensland — Victorian) Battalion — a unit of the 8th Infantry Brigade, 5th Australian Division. Led by the commanding officer of the battalion, Lieutenant-Colonel F. W. Toll. D.S.O., V.D., the men marched into the Anzac Memorial crypt and gathered around the 31st Battalion's memorial tablet to take part in a short service. In addition to the returned soldiers, the service was attended by 30 relatives of the fallen.

The gallant deeds of the battalion in the battle in which, so many Queenslanders were killed, were not recalled without a number of the women present — principally aged mothers — being visibly affected.

Captain (Dr.) Eustace Russell, who was the battalion's original officer, in an address, said that seldom had a finer body of men been formed than the 31st Battalion. By their example of sacrifice, courage, and devotion to duty they had shown what Australians could do in the eyes of the world.

This tribute to those who participated in the Battle of Fromelles was not only deserved, but the event was one which should be commemorated regularly on that day each year. In British history there were many

[177] See: -
https://trove.nla.gov.au/newspaper/article/38486144?searchTerm=FAMOUS%20BATTLE%20RECALLED.%20SURVIVORS%20HONOUR%20THE%20FALLEN%20%20%20%20%20%20%20%20%20%20%20%20&searchLimits=

feats of arms and events creditable to the British soldier, but nothing more creditable than the feat of the 31st Battalion in the Battle of Fromelles.

After Lieutenant-Colonel Toll had placed a wreath at the foot of the battalion's memorial tablet, Sergeant C. McKay recited the well-known lines, "They shall not grow old ..." The Last Post" and "Reveille" were sounded by Sergeant-Major. M. Polley.

---ooo000ooo---

Following on from the announcement made in April, 1936, of the upcoming gathering of Diggers, The News, Adelaide reported on the gathering which took place, on 26 September, 1936 under the heading:

GATHERING OF DIGGERS FOR 32ND BATTALION REUNION.[178]

Long before the Melbourne Express reached Adelaide today former members of the 32nd Battalion gathered on the platform. The express was bringing to Adelaide a band, color party, and rifle team from the 32nd Battalion, A.M.F., and Adelaide men who fought overseas with the unit during the war were eager to extend a warm welcome to the visitors.

There were 60 men in today's party, but others are coming from all parts of Australia, and it is expected that 400 will attend the reunion of the battalion at the R.S.S.I.L.A. Club on Monday.

The battalion band, which has won a high reputation in the other States, made a gay show when it formed up outside the railway station before marching to the R.S.S.I.L.A. Club for breakfast. Green collar bands and pipings and gold aiglettes all added a dash of color to the assembly.

Members of the 32nd are proud of the battalion flag. It carries the maximum number of battle honors -- 10 for the Great War and one for South Africa. A member said today that a selection committee had been necessary to choose honors for the flag. If a maximum had not been fixed, the flag could have been covered.

Bapaume. 1917: Ypres, 1917; Ancre, 1918; Albert, 1918; Somme, 1916-18; Bullecourt, Polygon Wood, Amiens, Hindenburg Line, St. Quentin Canal are the honors the flag carries, and each of them recalls grim exploits.

On Monday a party of 20 West Australian members will arrive in the Duntroon.

The Commanding Officers and Fromelles.

Clearly in the press, the various units which participated at Fromelles continued to publicly announce their recollection of the Battle. The soldiers and their commanders had no hesitation in recording their collective pride at being part of this action.

So it therefore comes as something of a surprise to reveal just how those in overall charge at higher levels of command recorded their reflections on the events at Fromelles.

[178] See: - https://trove.nla.gov.au/newspaper/article/132027840?searchTerm=GATHERING%20OF%20DIGGERS%20FOR%2032ND%20BATTALION%20REUNION%20%20%20%20%20%20%20%20%20%20%20%20&searchLimits=

In 1939, in his autobiography published by General Walter Riddell Birdwood titled *'Khaki and Gown'*, there is no mention by him of Fromelles, although he had commanded the ANZAC Corps. In his book on Birdwood, titled 'The Soul of Anzac', John Dermot Millar comments that:

The striking feature of Birdwood's autobiography, Khaki and Gown, is that he devotes only 30 pages (out of some four hundred) to his period of service on the Western Front as commander of the 1 ANZAC Corps. What attention he does give it, deals mainly with personal matters. Although he gives fifty pages to the period he spent at Gallipoli, offering by no means a comprehensive history of the Dardanelles campaign, it contains far more detail than he gave the Western Front period of service. As Birdwood in explanation disingenuously stated, a full treatment was hardly necessary because the Australian official historian, Charles Bean, had produced an 'incomparable account' of all the operations concerned. Of course, Bean had done the same for Gallipoli. (p.105)

There is a good reason for Birdwood's lack of mention of Fromelles. In February 1916 the Australian and New Zealand contingents, back in Egypt, underwent reorganisation to incorporate the new units and reinforcements that had accumulated during 1915: the Australian and New Zealand Army Corps was replaced by two corps, I ANZAC Corps and II ANZAC Corps, with Godley (British Army but commanding New Zealand forces) in command of I ANZAC Corps and Birdwood in command of II ANZAC Corps.

I ANZAC Corps initially comprised of the three 'veteran' ANZAC divisions – the Australian 1st and 2nd Divisions and the newly formed New Zealand Division.

II ANZAC Corps initially comprised of the two 'new' ANZAC divisions – the Australian 4th and 5th Divisions – and the ANZAC Mounted Division. With the move from Egypt to France the ANZAC Mounted Division remained in the Middle East and the Australian 3rd Division forming in England was earmarked to eventually join the II ANZAC Division.

When I ANZAC Corps became the first to depart for France in early 1916, Birdwood, as senior corps commander, took over command of I ANZAC Corps.

II ANZAC Corps followed and arrived in France, under Godley, in June 1916.

As Godley was also the commander of the New Zealand Expeditionary Force, it was decided that the New Zealand Division would be more appropriately located under his command so it swapped with the Australian 4th Division.

Thus at the start of the Somme Offensive on 1 July 1916 Birdwood was in command of the I ANZAC Corps comprising of the Australian 1st, 2nd and 4th Divisions and Godley was in command of the II ANZAC Corps comprising of the Australian 5th Division and the New Zealand Division.

Therefore, it should come as no surprise that there is no mention by Birdwood of Fromelles in his biography as he was not in command of II ANZAC Corps when the Australian 5th Division fought at Fromelles. Godley was in command.

Godley went on leave for a short time during this period of transition for his corps. In July, he returned to duty and II ANZAC Corps took over the section of the front line previously occupied by the I ANZAC Corps, near Armentieres. Later that same month, the 5th Division participated in the Battle of Fromelles in support of the neighbouring British XI Corps.

The Legacy of Those in Command.

Of all the Generals involved it is the controversy surrounding the commander of the Australian 5th Division that is the most interesting. Wikipedia notes the following regarding Lt. General Sir James Whiteside McCay[179]:

McCay is a controversial figure in Australian history. Disagreement about his significance and reputation began before his death and continues to the present. General Sir Brudenell White considered McCay to be "one of the greatest soldiers that ever served Australia, greater even than Monash." McCay's achievements included the creation of the Military Board and the Australian Army Intelligence Corps, and the development of the Staff Corps, "laying the foundations on which the Australian Army was built." In writing Volume III of the Official History of Australia in the War of 1914–1918, which covered 1916, official historian Charles Bean exonerated McCay of blame for Fromelles. This led to a public debate in the pages of The Bulletin in 1929 between critics of McCay and his defenders, led by Elliott. A revival of interest in Australian military history and the rediscovery of graves at Fromelles in the 21st Century led to a number of books being written about the battle, which tended to be critical of McCay.

What remains of most interest when looking at the contribution of General McCay is that McCay became ill in 1930 with cancer. In his last months he destroyed all his papers. Nobody thus was able to know what he did or did not record about Fromelles.

It would be convenient to have someone to blame, if that is appropriate. Many Australians have done just that, in laying all the blame for the planning and disaster at the feet of Lt.-General Haking. But it is realistically not appropriate to do so.

Haig started the ball rolling, not Haking. Monro and Plumer decided the attack area and therefore the Sugar-loaf, not Haking. The attack changed from gaining Aubers Ridge, to an artillery attack, to being carried out by three divisions instead of two, and then back to two. That the newly arrived II ANZAC should take part instead of the seasoned I ANZAC. That inexperienced artillery should be used. That the operation was on and off. And the disaster that befell the 58th Battalion was caused by errors in the 5th Division HQ, not by Haking.

Certainly there were a number of factors at play. The dithering of General Haig at GHQ and his Somme myopia, the half-hearted support from Army Commanders Monro and Plumer, Corps Commander Haking's delusional optimism, and McCay's amateur naivety, each cumulatively played their part. None have come to be recorded as a champion of the Battle of Fromelles. In the final analysis, not one individual General can be blamed for what happened at Fromelles.

And if the end, the result of the Battle of Fromelles was in fact preventing troops being moved south to the Somme, then the numbers killed at Pozieres in the following week may be less than

---

[179] Extract from full article: Wikipedia contributors, "James Whiteside McCay," Wikipedia, The Free Encyclopedia, https://en.wikipedia.org/w/index.php?title=James_Whiteside_McCay&oldid=882390849 (accessed February 26, 2019).

would have been the case otherwise. If this is the case, then perhaps we should in fact be giving Lt Gen Haking some credit where none has been given previously.

---ooo000ooo---

In the military magazine "Contact", published on **20 July, 2016**, there was reference to the Battle of Fromelles. It was reported that "Around 80 members of the Australian Defence Force honoured the service of those who fought on the Western Front during the First World War overnight with commemorative services to mark the 100th anniversary of the Battle of Fromelles on 19 July."

It may be said that these soldiers were merely carrying out an official function, but on meeting with them at Fromelles at the time, and talking with them, it was immediately apparent they were expressing their pride in attending the commemoration, and honouring those who had gone before them. I noticed that not all their eyes were dry.

---ooo000ooo---

17. Remembering The Battle of Fromelles by the Public at Large

What happened was not a battle: it was slaughter so one-sided it is sickening to imagine. A tragedy, certainly. A nightmarish mistake by culpably callous commanders, probably. But no battle.[180]

---ooo000ooo---

In the years following the end of the Great War, it was not only the soldiers who were concerned to ensure that the proper recall of events at Fromelles would take place. The public at large were regularly invited by newspapers and other means, to remember the occurrence of, and the significance of the events and the Battle at Fromelles.

Eight months after the cessation of war, on 17 July, 1919, in this letter to Sydney Morning Herald[181], which was also published in the Western Star and Roma Advertiser in Toowoomba Qld, "**FLEURBAIX**" wrote:

Sir,- As an Australian, I would particularly wish to draw attention to the fact that it was on July 19 and 20, 1916, that the Australians first went into action in France, at the village of Fleurbaix. How many brave Australians laid down their lives at that action, one of the most sanguinary fights of all they fought, the "In Memoriam" columns of the "Herald" tell year by year, and many touching tributes from officers commanding various battalions testify.

The 53rd, 54th, and 30th especially are mentioned, though no doubt many others were equally glorious in their achievements. Now, while Gallipoli holds "pride of place" in Australian annals as being the first great achievement of Australian soldiers, surely Fleurbaix must ever rank second! Pozieres was only the completion of Fleurbaix, which was fought to draw off German reinforcements from that quarter.

How many hearts will feel too sad on July 19 and 20 to give themselves entirely up to rejoicing, though yet their thankfulness is most deep that the sacrifices of Fleurbaix were not in vain.

It seems to me that some special tribute-perhaps from the eloquent pen of Mr. C. E. W. Bean, himself an Australian would be a suitable tribute to those brave men who simply, nay, even joyfully, went to death at Fleurbaix on the eve of July 19 (the day we celebrate peace restored) in 1916.

If Australia is to become the great nation we all hope to see it, such deeds as the charge of our men at Fleurbaix must be emphasised and impressed on future generations of Australians as Balaclava, Waterloo, etc., are upon English youth, and their proper place given to them in Australian history.

I am, etc., FLEURBAIX.

The writer's entreaty that Fromelles be given its 'proper place' in Australian history appears not to have been heeded. As we have already seen in Chapter 12, the Battles Nomenclature

---

[180] Andrew Rule, Sydney Morning Herald 20 July, 2010.

[181] See article at: https://trove.nla.gov.au/newspaper/article/15858958?searchTerm=The%2053rd%2C%2054th%2C%20and%2030th%20especially%20are%20mentioned&searchLimits=

Committee were not of the same opinion two years later. Battle status was not granted to Fromelles.

Nearly four years after the Battle of Fromelles took place, the Editor of The Argus, Melbourne, wrote a lengthy article on Fromelles, on 10 April, 1920, which has been set out in the previous chapter, but the introduction is reproduced here: -

FROMELLES, 1916! A GLORIOUS FAILURE. WHAT REALLY HAPPENED[182].

"The English attack in the region of Fromelles was carried out, as we have ascertained, by two strong divisions. The brave Bavarian Division, against whose front the attack was made, counted on the ground in front of them more than 2000 enemy corpses. We have brought in so far 481 prisoners, including 10 officers, together with 16 machine-guns." — German communique, July 21, 1916.

"Yesterday evening, south of Armentieres, we carried out some important raids on a front of two miles, in which Australian troops took part. About 140 German prisoners were captured." — British communique, July 20, 1916.

In this form readers of "The Argus" received their first information of the great action by Australian troops on the Western front, which has since come to be known as "the attack at Armentieres," or "the fight at Fleurbaix," but was really "the battle of Fromelles." For a long time, the secrecy of war kept a veil drawn over the detail of this sad page in the history of the Australian Imperial Force, but closely censored letters from members of the 8th, 14th, and15th Infantry Brigades soon began to indicate that something more serious had happened than was at first suggested.

Since then, more or less accurate accounts of the battle have been in circulation, but the most interesting features had necessarily to be withheld until after the armistice. The first consecutive record, prepared from official documents and elaborated by commanding officers who actually took part in the engagement, has been prepared by Captain A. D. Ellis, M.C., of the 29th Battalion, and it is included in his "Story of the Fifth Division" (London: Hodder and Stoughton).

In this the movements of the various units from the day the intention to attack at Fromelles was first announced until the work of clearing up the battlefield had been completed are carefully followed with the aid of diagrams, and even the most unmilitary reader will find much to stir him in this account of "our most glorious failure."

The article went on to detail the events leading up to the battle, and accounts of the battle itself, some aspects of which were more grandiose than realistic, but much of which we have featured previously. It also contained a comment on the work of 21 year old Tasmanian, Major Justin Hutchinson of the 58th Battalion (set out at length in the previous chapter), who many commentators believe should have been awarded the Victoria Cross for his actions.

[182] For full article, see: - https://trove.nla.gov.au/newspaper/article/1690351?searchTerm=Fromelles%201916!%20A%20glorious%20failure%20%20%20%20%20%20%20%20%20%20%20&searchLimits=

---ooo000ooo---

Then, on 20 July, 1920, four years after the Battle, it was written in The Sydney Morning Herald, Sydney that[183]: -

Today is the fourth anniversary of the Battle of Fromelles. No name on the long scroll of distinguished actions in which the A.I.F. participated on the Western front recalls more bravery, more glorious endeavor to defeat the German army, than does this tragic phase of the Australians' share in the war.

After outlining a summary of the unfolding of the Battle, and its ultimate end, the author went on to write: -

Thus, the Fifth Division earned the right to take its place alongside the older divisions as a fighting unit, but it was a tragic start. Pre-eminently the operation was used by the British staff as a threat against the vital German defences behind this portion of the front and to weaken the rigid enemy resistance at Pozieres by the removal of their troops to the threatened area in the Armentieres-Fleurbaix sector. Tributes to the gallantry of the Fifth Divisional troops were paid by Sir Douglas Haig, Generals Plumer, Godley, and Haking - gallantry that will live in the annals of history.

---ooo000ooo---

The following article was published on 22 July, 1920, in The Advocate, Melbourne. While not stating such in the newspaper, the article appears to be a reproduction of a public address given by Fr. Kennedy, the Chaplain of the 14th Brigade. Sadly for Fr. Kennedy, his comments are not all supported by a full appreciation of the events as they unfolded.

BATTLE OF FROMELLES, "A HOPELESS BUNGLE."[184]

With regard to the battle of Fromelles, I was chaplain of the 14th Brigade on that occasion. The Battalion to which I was attached had acquitted itself most gloriously all through. Of military strategy I only know what the tyro could see for himself. But I have sufficient intelligence and sufficient powers of memory to realise that that affair was a hopeless bungle from start to finish. General Elliott explained in Bendigo that sometimes it is necessary to sacrifice brave soldiers in order to permit of further successes of the army in general. What right had England, or the War Office, or the High Command to sacrifice a whole division of gallant Australians, and spare English brigades and battalions who were resting away at the back? (Applause.)

There are differences of opinion regarding the casualties and likewise with regard to the English troops associated with us in the affair. The 61st British Division was associated with us. Certain objectives were allotted to us, and also to the 61st. The 5th Division AIF, despite a tremendous barrage of artillery - hell let loose - never failed or quaked. They hopped over like heroes, and stopped not until they gained the

---

183183 See: - https://trove.nla.gov.au/newspaper/article/15898436?searchTerm=Today%20is%20the%20fourth%20anniversary%20of%20the%20Battle%20of%20Fromelles%20%20%20%20%20%20%20%20%20%20%20&searchLimits=

184 For full article, see: - https://trove.nla.gov.au/newspaper/article/171049345?searchTerm=%27%27With%20regard%20to%20the%20battle%20of%20Fromelles%2C%20I%20was%20chaplain%20of%20the%2014th%20Brigade%20%20%20%20%20%20%20%20%20%20%20%20&searchLimits=

second line of the enemy trenches. (applause) The 61st Division was allotted the Sugar-loaf as an objective, but only as a section got into the enemy trenches, and were compelled to retire. Whether they failed in their objective through cowardice or through inefficiency, does not matter very much to us, but it did matter to the shattered remnants of the 5th Division, which later came up to the cross roads to bury their dead.

TERRIBLE HOLOCAUST

The English papers said the Australians gained the enemy trenches for a short time, and succeeded in bringing back some prisoners and valuable information. Not for four or five months afterwards did Australia know the toll that the 5th Division had paid, and it was not realised or understood that the tremendous holocaust of men was used and wasted for no good purpose what-ever. (Applause)

FAILED TO DO THEIR JOB

Not only at Fromelles was it my misfortune, and the misfortune of the troops with whom I served, to be associated with English soldiers who failed to do their job, but right through our career, with very few exceptions, our experience was the same. (Applause) This is not a mere statement from me. I have been associated with tens of thousands of Australians, and I say that during the years I served as chaplain I have not heard more than five Australian soldiers speak other than in terms of contempt and disgust of the English troops with whom they served. (Applause)

As stated at the beginning of this extract, Fr. Kennedy's comments on what happened to the UK 61st Division did not extend to advising his audience of their numbers being less than the Australians, or the barbed wire in front of them not being shelled and broken, or that the distance to their objective was much longer than some of the Australian units. Sadly, in the absence of critical examination of the role played by, and the manner in which the 61st Division were to carry out their assignment, it became all too easy to portray the British as 'having failed to do their job'. The reality is that the 61st Division suffered severely and faced the same problems across a deeper front than the Australians.

Fr. Kennedy's portrayal became yet another spreading of mythology, which is not deserving of repetition other than to dispel it. The rumour had crossed the battlefields before the truth could catch up with it.

The article went on to inform the audience and the reader of the manner in which Australian soldiers were regarded by the English and by the Europeans.

AUSTRALIAN SOLDIERS EULOGISED

Fr. Kennedy read the following extract from an article written by Mr. Grover: -"All the same, the Australian soldier is not a popular soldier with the English soldier. He is not understood: he is a foreigner. The Tommy officer does not understand the Digger, and the Digger does not understand him, which is not to be wondered at!" Fr. Kennedy then continued: - "The article refers to the high esteem and intense admiration in which the Australian soldier is held in France.

DIGGERS IN ENGLAND

An officer from Horseferry-road, in speaking to me on one occasion, referred to the lack of decent hostels in London where our men could board and be properly housed according to their limited means. The same officer told me that a particular department dealing with the issue of invitations to Australians into the homes of respectable families in London ceased to recommend the soldiers after a time. Oftentimes, when the visit took place, the Australians were made to feel that they were colonials, and of an inferior caste to the English. There were instances where the diggers were asked to take refreshment with Susan Jane and Jemima in the kitchen. (Laughter.)

---ooo000ooo---

On 31 July.1920 A Memorial Service, held on the preceding Sunday night, was the subject of a report in The Methodist Sydney, under the heading of:

MEMORIAL SERVICES. FLEURBAIX AND POZIERES.[185]

Senior-Chaplain Green conducted a memorial service in the Paddington Methodist Church on Sunday night in connection with the battles of Fleurbaix and Pozieres, and referred to the danger of forgetting what we owed to those who fought and died.

If we would do honour to ourselves, we must not forget to bring stories of remembrance out of the dark days of war, and get them up in our life and history in Australia for ever.

---ooo000ooo---

In the Newcastle Morning Herald and Miner's Advocate, NSW on 19 July, 1933 the following article appeared:

FLEUBAIX ANNIVERSARY. AUSTRALIANS' HEAVY CASUALTIES.[186]

Today, July 19, is the 17th anniversary of the severe fighting at Fleurbaix, in France, during the Great War. The Fifth Division of the Australian Infantry Forces went into action at the village of Fleurbaix, on July 19, 1916, and was engaged for two days, during which about 8000 casualties were recorded.

The object of the attack was to draw off the overwhelming reserves of the enemy, threatening to smash the lines in the southern area in that theatre of the war, where other Australians were fighting for existence.

The 'southern area' was a reference to The Somme. Given that this was the first engagement of Australian troops on the Western Front, this passage can only be seen as over-stating the true situation, even jingoistically. While British and French forces had been fighting 'in the southern

---

[185] See: - https://trove.nla.gov.au/newspaper/article/155273522?searchTerm=Senior-Chaplain%20Green%20conducted%20a%20memorial%20service%20in%20the%20Paddington%20Methodist%20Church%20%20%20%20%20%20%20%20%20%20%20%20%20&searchLimits=

[186] See: - https://trove.nla.gov.au/newspaper/article/135094678?searchTerm=Today%2C%20July%2019%2C%20is%20the%2017th%20anniversary%20of%20the%20severe%20fighting%20at%20Fleurbaix%20%20%20%20%20%20%20%20%20%20%20&searchLimits=

area', it would not be until four days later that Australian troops would fight on The Somme at Pozieres, and not as stated in the passage here.

The Fifth Division went over into the awful No-man's Land, and into the enemy breastworks, with bayonet and bomb, and destroyed all who resisted their advance.

Again, the statement that the soldiers 'destroyed all who resisted their advance' is not a true reflection of what took place. Certainly, many were killed who resisted, but many were taken as prisoners of war.

The advance stopped the attacks of the enemy, compelled the German command to draw off reserves, and caused considerable losses to the enemy forces.

Again, somewhat over-inflated, but with some foundation in terms of its strategic effect.

---ooo000ooo---

Twenty-three years after the events at Fromelles, on 19 July, 1939 (just two months before the start of World War II), a commentary on the battle, comparing it to other operations in the area, was reported in the Goulburn Evening Penny Post:

FROMELLES - AUSTRALIANS IN FRANCE[187]

Today, July 19, is the 22nd [sic] anniversary of the battle of Fromelles, when for the first time in the history of the Great War an Australian division went over the top in France.

The assault at Fromelles, opposite the township of Lille, and immediately south of Armentieres, was for the purpose of intercepting the picked 6th Bavarian Reserve Division which was being transferred from Belgium to the Somme. In anticipation of the Pozieres stunt which was to be launched a few days later.

The 8th Brigade, which comprised the 30th, held the left flank, joined by the 14th on the right, with the 15th further south, and the latter in turn linked up with the 61st London Household Division[188], a detachment which was inspired by Lord Kitchener shortly before his death. Each division covered a mile of front, the enemy line being at varying distances of from 70 to 250 yards removed from the Australian and British division line.

Each man, according to Brigade Order No. 23, was to carry 150 rounds of ammunition, two bombs and two sand-bags, and a pick and shovel between each three men. It was a sector that claimed the graves of many members of British Battalions prior to the arrival of the Australians in France. On May 9, [1915], an assault by a Tommy Brigade carried out an attack at Aubers Ridge in the same neighbourhood, and again on Sep. 25 [1915].

The total losses amounting to 192 officers and 4490 men. The article's author summarized the Battle itself as "A STRANGE RUSE".

---

187 See: - https://trove.nla.gov.au/newspaper/article/103163955?searchTerm=The%20assault%20at%20Fromelles%2C%20opposite%20the%20township%20of%20Lille%20%20%20%20%20%20%20%20%20%20%20&searchLimits=

188 It was actually the 61st (2nd South Midland) Division.

---ooo000ooo---

On 17 March, 1940 an article appeared in the Sunday Mail, Brisbane Qld titled:

HEROES OF THE SOMME. V.C.'S - FOR VALOUR[189]

The article recounted the fighting on the Somme, and the awards of Victoria Crosses to a number of soldiers in the 1st and 2nd Australian Divisions. Despite the actions of those in the 5th Division, and in particular the actions of Sgt. Simon Fraser, and Major Hutchinson, no awards of the Victoria Cross were ever made to a member of the 5th Division, for what took place during the Battle of Fromelles.

[This issue was discussed in more detail in Chapter 15.]

Lambis Englezos' Search for The Missing, and The Re-Burial Of 250 Soldiers At Fromelles

The location of the missing soldiers at Fromelles became the focus of the now historically famous and long-running investigation by Lambis Englezos, Ward Selby and John Fielding. Their investigation resulted in presentations being made to an Expert Panel in 2005 advocating for a search of the site suspected of being a burial site, next to the Pheasant Wood. The panel was insufficiently convinced and unmoved until in 2006, the German Burial Order of 1916 was found in archives in Munich, Germany.

The Australian Army then established a process by which the excavation of the site at Pheasant Wood could be undertaken, leading to confirmation of the existence of soldiers' remains, and their eventual recovery.

2007 saw an exploratory non-invasive survey undertaken at Pheasant Wood, led by Tony Pollard of Glasgow University Archaeology and Research Department. In that investigation two medallions were found at Pheasant Wood, which were of Australian origin, in a location at which Australian soldiers had not reached in battle. There must be soldiers buried at Pheasant Wood after all.

In 2008, a preliminary and limited excavation was able to be undertaken. Human remains were found. This led to the approval of the excavation of the remains of soldiers now confirmed as being buried next to Pheasant Wood in 1916.

2009 saw the unearthing of the burial pits at Pheasant Wood, and the beginning of an archaeological excavation in which 250 soldiers' remains were located. The process resulted in

[189] The full-page article can be accessed at: - https://trove.nla.gov.au/newspaper/article/98245615?searchTerm=THE%20fighting%20on%20the%20Somme%20in%20July%2C%201916%2C%20is%20associated%20with%20%20%20%20%20%20%20%20%20%20%20%20&searchLimits=

detailed scientific analysis of the remains found, and a multitude of small items found with them.

One of the archaeological team was Louise Loe of Oxford Archaeology, who led the excavation of the clay pit graves and has been serving as a subject matter expert on the data analysis team, formed specifically for the identification process. In correspondence with the author, she wrote that: -

The identification process was of the utmost scientific rigour from beginning to end. Subject matter experts systematically analysed all the pieces of available evidence, including all available military documentation, information from families, family trees, the results of the DNA analysis, items found with the soldiers, the results of the scientific analysis of the bodies and photographs of the soldiers as well. On the basis of this, recommendations were made to the Joint Identification Board for their consideration and the first identifications of the soldiers were made.

Since then, the data analysis team has met annually up until 2019 (with the exception of 2015) and has continued to recommend identifications to the Joint Identification Board . Unfortunately, the board could not meet [in 2020] because of the [Covid-19] pandemic.

In 2010 the new Pheasant Wood Military Cemetery at Fromelles was dedicated with all the remains unearthed in 2009, reburied in their own dedicated plots. The dedication ceremony took place on 19 July, 2010. The remains of the final soldier to be re-buried were interred at the ceremony, 94 years after the battle which claimed his life.

The identities of 96 soldiers had been determined before that dedication took place. It was the result of a long and detailed process of scientific analysis culminating in the first Joint Identification Board at Australia House, London, in the Spring of 2010. Names could now be inscribed in headstones bearing the identity of soldiers lost since 1916.

Dr Loe was shown photographs of some of the soldiers identified, after the soldiers had been reburied, but before the new cemetery at Fromelles was dedicated. Dr. Loe wrote: -

'The first time that we saw the faces of the names we had matched with individuals who we had recovered was incredibly poignant.'

One of those soldiers whose photograph was shown to Dr Loe, was Robert Courtney Green.

Robert had left Nottinghamshire, England for a better life in Australia, only to encounter the ravages of drought on a farm outside Fremantle. He enlisted and would serve in the 32nd Battalion. He was hit with shrapnel in the chest and bled to death in the German 2nd line trench.

---ooo000ooo---

On 16 July, 2010, Lucy Carne wrote in The Advertiser, Adelaide SA, of the work done by Dr. Louise Loe, of Oxford Archaeology, in her article:

HEROES OF FROMELLES SET FOR GOODBYE[190]

After almost a century of lying forgotten in an unmarked mass grave, this soldier was handed back his identity and could now be remembered. It was the moment a gaping wound in Australia's history finally healed.

Robert Courtney Green's photograph was featured in the article. The article went on ..

On July 19, 1916, the young man bled to death in a muddy field in Fromelles, north France, in what was one of Australia's most futile World War I battles. In just 24 hours, 5533 soldiers were killed, wounded or taken prisoner. The Germans dumped[191] 250 Australian and British soldiers in an unmarked mass grave in a nearby wood. There they lay forgotten, deemed officially missing, until now. On Monday, 94 years after they were callously[192] buried, the soldiers will officially be laid to rest. For Dr Loe, of Oxford Archaeology, who led the excavation of the clay pit grave, the most moving moment of the painstaking dig was the first time a photo of a soldier was matched to his remains.

[Picture 47: Robert Courtney Green. AWM P10624.001]

Dr. Loe was sitting in Australia House, London, serving on the data analysis team for the Joint Identification Board at the time.

"I was standing there wondering `what were they thinking when that photo was taken?' she said. "It was very powerful and seeing everything, after all this work, come together, uniting a body number to a name, to a person, it was amazing. It was the most emotional project and the one I have felt most connected to."

At the time of this article (16 July, 2010) DNA tests from living relatives have identified 96 Australian soldiers by name, with 14 South Australians, 55 from New South Wales, 18 Victorians, eight West

---

190 https://www.adelaidenow.com.au/news/south-australia/heroes-of-fromelles-set-for-goodbye/news-story/3d5715d92b5cd9209baf0c22f8156387

191 There remains some controversy as to this claim. Remains appear to have been wrapped in the gas capes of German soldiers before they were laid out in two rows in the mass graves, according to Professor Tony Pollard of the initial archaeological survey team in 2007-2008.

192 Again, a comment that may or may not stand scrutiny given the need to have the bodies buried on account of the health requirements prevailing at the time.

Australians and one from Queensland. A further 109 soldiers have been identified as Australian, but their names are unknown. Forty-two soldiers remain totally unidentified. DNA testing will continue until 2014 in the hope all will be accounted for.

"Even if some of them have not been named, 250 soldiers who fought and died in that battle and who have been lost for all these years will now finally be remembered," Dr Loe said.

---ooo000ooo---

On 20 July, 2010, an article by Andrew Rule was published in The Sydney Morning Herald, under the heading: -

LOST AND FOUND IN FROMELLES[193]

From it, we can begin to understand how the public at large must have felt in learning of the extent of the Battle of Fromelles, and the steps taken in recovering the missing.

IT'S hard to say why a war fought so long ago still has such a potent effect after so many years. But in Fromelles these last few days you see proof of it and it brings a lump to the throat.

Here and at all the other battlefields in Flanders and the Somme, the Great War still echoes. Not just for relatives and descendants of soldiers who served but for others, too.

The 250 "lost diggers" re-interred in the new Pheasant Wood cemetery here were found — and 96 of them so far identified — because of Lambis Englezos, a Greek-born Australian with no connection to the 1914-1918 conflict except a feeling he cannot explain or ignore.

There were a lot of people and a little pomp, though not much pageantry: it was, after all, the dedication of a war cemetery, not a coronation or a royal wedding. But there was a prince, great grandson of the man on the throne when hundreds of thousands of his subjects died on the western front.

Every year, farmers — like the Demaissets on whose land the burial pits were uncovered — plough up countless bullets, shrapnel, buttons and the like, an annual crop that means the past can never be forgotten. The war is part of life here, as it is on every other Western Front battlefield, places whose names are chiselled on monuments and in memories around the English-speaking world: Bullecourt, Pozieres, Ypres, Passchendaele, Menin Road, Polygon Wood, Messines, Mont St Quentin and the rest.

From now on, Fromelles will be on that heartbreaking list. In a war littered with fatal mistakes, what happened near here might have been the worst of all.

WE call it the Battle of Fromelles but it wasn't really — neither a battle nor Fromelles.

What happened was not a battle: it was slaughter so one-sided it is sickening to imagine. A tragedy, certainly. A nightmarish mistake by culpably callous commanders, probably. But no battle.

Battles are fought between forces that both have some chance of victory. That was never the case on the night of July 19, 1916, on the flat killing fields north-west of the town, which sits on a rise that gave the Germans an unbeatable advantage.

---

[193] https://www.smh.com.au/national/lost-and-found-in-fromelles-20100719-10i1u.html

Waves of brave, frightened men — including thousands of young Australians — were thrown against defences and weapons that pulped them like meat forced through a mincer.

Nearly 2000 Australians were killed that night. It was a gut-churning example of what happens when heavy artillery and machine guns on high ground are turned on young men sent by old ones still practising tactics learned on the hunting field and parade ground decades before.

Many bodies were blown to pieces. Many more could not be retrieved — because of an Australian commander's refusal to agree to a truce — and were eventually disposed of by the Germans in the days and weeks afterwards. Wounded men died awful deaths.

GEORGE CHALLIS – JUST ONE OF THOSE KILLED AT FROMELLES.

To give a human face to the soldiers who were killed at Fromelles, Andrew Rule in the previous article, recounted details of a Northcote plumber called Bill Barry who was to become a prisoner of war, and who survived to write a first-hand account of what it was like to "die" on the field. Knocked unconscious by a shell blast that blew holes in his legs, he was tossed on a heap of bodies by a German burial party.

Barry woke to find himself surrounded by dead men "on the edge of a hole about forty feet long ... and into this hole the dead were being thrown without any fuss or respect."

He would survive against all odds. Robbed and beaten at the front, he was sent to a German hospital where he was treated well and had his leg amputated. He was then released to England, where he would take tea at Windsor Castle before returning to Northcote with his extraordinary 65,000 word diary.

When Lambis Englezos read Barry's vivid account of the burial pit it confirmed his belief that at least 200 Australian dead were unaccounted for — and that all or most would be in the mass grave Barry had seen at what the Germans called Pheasant Wood, behind their lines in a field next to Fromelles village.

[Picture 48: George Challis, courtesy of Carlton (Australian Rules) Football Club, Victoria.]

Bill Barry had seen other things, too. Four days before the suicidal attack of July 19, he wrote, "I passed a chap carrying something wrapped up in a blanket and when he was asked what he had he answered with a gloomy look on his face 'This is all that we could find of George Challis, the Carlton champion footballer.' He had been blown to pieces."

GEORGE Challis's name won't mean much except to a few ardent Carlton fans. But to capture what it was like in the Australian winter of 1916, imagine a hero of last year's premiership team being killed on a foreign battlefield nine months after the Grand Final.

Think of a Jimmy Bartel or a Tom Harley, brave players showing promise of a long and useful life off the field. That's the sort of loss Challis was.

One of eight children of a Tasmanian butcher, young Challis was gifted. As a boy he had practised kicking his football through a barn half-door, honing the accuracy that made him one of the island's best players when he went to Launceston High School. A keen student, he went to university at a time when few did, and became a teacher before joining the Prime Minister's department in Melbourne.

He played for Tasmania and for Carlton and seemed assured of a bright future after that. But he could not ignore the call of "duty" after the Anzacs' Gallipoli campaign in 1915.

He was 24 years and five months old when he signed up, just before the 1915 football season ended. By the following July he was but a few bloody body parts in a blanket. Imagine 2000 others killed the same week then add 3500 wounded and you get an idea of what happened on the fields below this town.

There's a postscript to the story of George Challis. Here was one of Australia's best footballers, renowned for speed, reflexes and intelligence. But the army knocked him back the first time he tried to enlist — because one of his toes was bent.

Before 1916 only perfect physical specimens were good enough to be turned into carrion for King and country. After Fromelles, the requirements for trench fodder became less demanding.

"By 1918," says the historian and author Les Carlyon, "they would have taken a budgerigar.

George Challis, of 58 Battalion, is buried in Rue-Petillon Military Cemetery. Plot I, Row K, Grave No. 109, France, and is listed in Panel 165 of the Roll of Honour, Australian War Memorial, Canberra.

---ooo000ooo---

What can we learn from what the public remembered?

The returned veterans and the general public never forgot Fromelles.

The Australian Army website defines a "Battle Honour" as "the public recognition and commemoration of an Australian achievement on the battlefield by a unit or formation of the Australian Army"[194]

If 'public recognition' is the standard by which Battle Honours are measured, then the recognition that the Battle of Fromelles received from the public since before the Great War had even ended, and through the decades of post-war reunions and remembrances, the time when

[194] https://www.army.gov.au/our-history/battle-honours.

the last veteran of Fromelles died of old age, and its enduring present day consciousness, stands testament to the epic standing this Battle has in Australian history.

It is a Battle without even 'Battle' recognition, let alone a Battle Honour.

---ooo000ooo---

18. Treatment of Fromelles by Governments and Politicians

In the aftermath of the Great War, we could expect to see our service men and women given due respect for the service which they gave. That service translated into the construction of official places of honour, and reflection, and places at which the people of Australia could gather to learn of what had been endured by those who served. The process of doing so was still a State by State process, and each had their own method to translate that honour and reflection into monuments which have stood ever since.

What is interesting when looking at the different State monuments in the various capital cities, is just how that process resulted in two distinctly different approaches.

At the beginning of this book, it was pointed out that "Fromelles" does not appear inscribed on memorials in Australia, where those memorials honour the service of our soldiers *according to the place at which they fought*. That is a very different matter to having a place included in relation to a Battle Honour, which is then the subject of an inscription.

Ironically, as also stated on page 1 of this book, and which will be set out in this chapter, the only place at which you might see the inscription of 'Fromelles' within the geographic outline of Australia, is on the memorial within the grounds of the French Embassy in Canberra, which is in terms of international law, a part of France, and not a part of Australia.

We will therefore look at the methods used for honouring the service men and women at various memorials of significance, around Australia.

Brisbane, Queensland: The Shrine of Remembrance.

Across the street from Central Railway Station, and sitting in ANZAC Square between Ann Street and Adelaide Street in Brisbane, Queensland, The Shrine of Remembrance is the Queensland war memorial dedicated to the Australian and New Zealand Army Corps.

Funds were raised by public subscription for a memorial to fallen soldiers in World War I and in 1928 a competition was held for its design. The design winners proposed a Greek Revival structure. The design contains 18 columns in a circle, which symbolize 1918, the year when hostilities ceased.

The Shrine took two years to build and was dedicated on Armistice Day 11 November 1930 by Governor John Goodwin with a dedication plaque. There is also a World War I memorial sculpture on the Shrine of Memories external wall.

Inside the Memorial's circular walls are listed in a circle above the head of the observer, the names of various places of honour, only one of which did not receive a Battle Honour, being places at which Queensland soldiers were involved.

Fromelles is not listed, and nor is The Somme despite Queensland units – such as 31st Battalion - having been involved at Fromelles and The Somme.

[Picture 49: The Shrine of Remembrance, Brisbane. Photo by 'Figaro'.[195]]

The places of which are listed are

VILLERS BRETONNEUX, MONT ST QUENTIN, HINDENBURG LINE,

ANZAC, COCOS ISLAND, **ROMANI, JERUSALEM, DAMASCUS,**

POZIERES, BULLECOURT, MESSINES, YPRES, and AMIENS.

[Items in bold text have been awarded Battle Honours.][196]

Melbourne, Victoria: The Shrine of Remembrance.

Sitting alongside St. Kilda Road, in the grounds of the Kings Domain, The Melbourne War memorial, more commonly known as the Shrine of Remembrance is the largest memorial in all of Victoria and is probably Melbourne's most visited and recognised landmark. It was built to remember the soldiers and others who died during the Great War and subsequently honours all armed peacekeeping missions since.

The design of the Shrine is based on the ancient Mausoleum of Maussollos at Halicarnassus, and the Parthenon in Athens. It is a structure of square plan roofed by a stepped pyramid and entered on the north and south through classical porticos, approached by wide flights of steps

---

[195] Wikipedia contributors, "Shrine of Remembrance, Brisbane,"*Wikipedia, The Free Encyclopedia,* https://en.wikipedia.org/w/index.php?title=Shrine_of_Remembrance,_Brisbane&oldid=845163372(accessed January 5, 2019).

[196] Material sourced from an article in Wikipedia

which rise in stages to the podium on which the Shrine sits. The east and west facing fronts are marked at the corners by four groups of statuary representing Peace, Justice, Patriotism and Sacrifice.

[Picture 50: The Shrine of Remembrance, Melbourne.Photo: Jorge Láscar.197]

Around the outer stone balustrade that marks the Shrine's external boundary are the 16 stone "battle honours" discs. Not all received awards of battle honours, but they commemorate Australia's contributions to the following battles:

LANDING AT ANZAC (GALLIPOLI); SARI BAIR; RUMANI; GAZA-BEERSHEBA; THE NORTH SEA; THE COCOS ISLANDS; **MEGIDDO; DAMASCUS; VILLERS-BRETONNEUX; AMIENS; MONT ST QUENTIN; THE HINDENBURG LINE; YPRES; MESSINES; POZIERES**; and **BULLECOURT**. [Again items in bold text have Battle Honours.]

As in Brisbane, the places listed as being significant honours, do not fully reflect actual places of battle, but Battle Honours awarded to Australian units which served in the Great War. Again, *Fromelles* is not listed, nor is *The Somme*. Yet the soldiers of the Victorian Battalions, including the 31st, the 29th, the 57th, the 58th, the 59th, and the 60th, served at Fromelles. Some such as the 60th under Brigadier Pompey Elliott were decimated there.

197 See: Wikimedia Commons contributors, "File:Shrine of Rememberance (11884180023).jpg," *Wikimedia Commons, the free media repository*, https://commons.wikimedia.org/w/index.php?title=File:Shrine_of_Rememberance_(11884180023).jpg&oldid=281158880 (accessed January 5, 2019).

Fromelles does however, find an honour in one area of the Shrine's grounds. On 19 July 2008, being the 92nd anniversary of the Battle of Fromelles, a replica of the 1998 sculpture by Peter Corlett – "Cobbers" - in the Australian Memorial Park, at Fromelles was unveiled. This depicts Sergeant Simon Fraser, 57th Battalion, (a farmer from Byaduk, Victoria), rescuing a wounded compatriot from No-man's land after the battle.[198]

Perth, Western Australia: State War Memorial

The State War Memorial Cenotaph commemorates Western Australian servicemen and women who have served in all wars and conflicts in which Australia has been involved.

The monument was announced on 2nd July 1927, and was dedicated on Sunday, 24 November, 1929, one year after the intended date of the original dedication.

Its designer was to be Sir Talbot Hobbs, who also designed the 5th Division Memorial at Polygon Wood, in Belgium. The memorial is almost a replica of the Australian Imperial Force Memorials erected in France.

[Picture 51: State War Memorial, Perth. Photo: Geoffrey Benn, 2013.]

The heavy concrete foundations are supplemented by heavy brick walls which enclose an inner chamber or crypt. The walls surrounding the crypt are covered with The Roll of Honour; marble tablets which list under their units the names of more than 7,000 members of the services killed in action or as a result of World War One.

The inscription at the front of the monument reads:

Erected by grateful citizens in remembrance of men of this State who at the call of duty gave their lives for freedom and humanity in the Great War 1914 – 1918 And in remembrance of the men and women of this State who gave their lives in subsequent conflicts. Lest We Forget.

---ooo000ooo---

[198] Edited from an article in Wikipedia.

Hobart, Tasmania: State War Memorial.

The monument was designed by Hobart architectural firm Hutchinson and Walker, after their entry had won a public competition held in 1923 for the structure's design. The design was for an obelisk, which stands at a height to 75 feet (23m). It is an Art Deco reinterpretation of a traditional Egyptian Obelisk.

The Hobart Cenotaph was originally designed as a memorial to the Tasmanians who died serving their country in World War I. However it was later decided that the Cenotaph would be altered to commemorate those who had died in subsequent conflicts as well.

The original inscription reads: "Lest We Forget", and "1914–1919".

Although World War I ended on 11 November 1918, the inscription is dated '1919' in commemoration of the Treaty of Versailles which was signed on 28 June 1919. There are no names recorded upon the Cenotaph itself.

An Anzac Day commemoration was held there in 1925 during construction. During the ceremony, a casket of solid zinc which bore the names of the 522 Tasmanians who were killed in World War I was set into the base of the shaft. It was unveiled on 13 December 1925 and replaced a previous wooden structure.

[Picture 52: State War Memorial, Hobart. Photo: Bjorn Christian Torrissen; http://bjornfree.com]

The first addition was for soldiers who were killed in World War II, and was placed directly below the "1914–1919" inscription, and simply reads:"1939-45". Subsequent inscriptions memorialising Tasmanian war dead in more recent conflicts on the face of the plinth below those of the two world wars are: "The Korean War", "The Malayan Emergency", "The Indonesian Confrontation", "The Vietnam War", and "Peace-Keeping Operations".

On each of the four faces of the monument are the crests of the four service organisations – the Royal Australian Air Force, Royal Australian Navy, Australian Commonwealth Military Forces and the Royal Australian Army Nursing Corps (which was added in 1991). Floodlights on each corner illuminate the shaft of the Obelisk at night.

The site of the Hobart Cenotaph in the Queens Domain, was deliberately chosen for its prominence. Not only is it a location with its own military legacy, having been the site of the Queens Battery from 1838 until 1923, but the site also has excellent views of the city, Mount Wellington and the Derwent River. Following the completion of the Cenotaph, the site was landscaped extensively with a paved avenue lined with Poplar trees, and named "Anzac Parade", leading from the Tasman Highway to the Cenotaph.[199].

Adelaide, South Australia: National War Memorial.

Almost 35,000 South Australians served in the First World War. Of those who served, over 5,000 South Australians died. In response to these deaths, a memorial commemorating the victory and the sacrifice of those who had fought and fallen was approved by Parliament in 1919. The South Australian Government thus became the first in Australia to elect to build a memorial to the soldiers of the First World War.

[Picture 53: National War Memorial, Adelaide. Photo: AWM 131552]

It was decided by parliament that the new memorial should be referred to as the "National War Memorial", even though it was to be a purely South Australian monument, and in spite of the term already being used to describe the memorial to the South African War of 1899–1902. There have been at least two perspectives offered as to why the term 'National' was employed. First,

[199] Information Sourced from: Wikipedia.org.

the name may have been chosen to emphasise the government's intention that the memorial should commemorate all who served during the war, not just those who came from South Australia[200]; and second, that the name may have reflected the perception (still held in spite of federation) that the "province is a nation".

Two competitions took place. One in 1924, the second in 1926.

In the preamble to the original conditions of entry, it was stated that the new memorial was to serve the purpose of:

"...perpetually commemorating the Victory achieved in the Great War, 1914–1918, the Supreme and personal sacrifice of those who participated in that War, and the National effort involved in such activities".

Construction of the memorial began in 1928, and it became the fourth state World War I memorial to be opened when it was unveiled on Anzac Day, 25 April, 1931 before a crowd of almost 75,000 people. The first dawn service would be held on Anzac Day, 1935.

At the insistence of W. F. J. McCann, president of the Returned Soldiers' League, bronze tablets were cast to line the walls of an inner shrine, on which are listed the names of all South Australians who died during the Great War.

The monument is designed to honour both the war dead and all who served in the war — one face being inscribed to those who died in the war, while the other is dedicated to "all who served".

The monument has two sides, referred to by the architects as the *reverse* and the *obverse* of the work, which they likened to the two sides of a coin. These two aspects represent the prologue and the epilogue of war.

On the obverse side is inscribed the words: -

"*To perpetuate the courage, loyalty, and sacrifice of those who served in the Great War 1914–1918*".

To represent the prologue to the war, the obverse of the monument (the side facing North Terrace) features a relief of the Spirit of Duty appearing as a vision before the youth of South Australia, represented in the work by a sculptural group consisting of a girl, a student and a farmer abandoning the "symbols of their craft". The three are depicted in normal dress, as they are not yet soldiers and are currently unprepared for the war that is to come, and they are facing away from the world as they look to the vision before them.

On the reverse side is inscribed the words: -

"*All honour give to those who, nobly striving, nobly fell that we might live*".

On the reverse side of the monument, facing away from the traffic, is a relief carved into the marble representing the epilogue of the war and depicting the Spirit of Compassion as a winged spirit of womanhood bearing aloft a stricken youth. Beneath the figure is situated the Fountain

[200] https://en.wikipedia.org/wiki/National_War_Memorial_(South_Australia) - cite_note-Richardson1998ap1-7

of Compassion, the flow of water representing the "constant flow of memories", while the lion's head from which it emerges, (and which bears the Imperial Crown), is representative of the British Commonwealth of Nations.

Above the two entrances to the inner shrine were to be inscribed the names of the major theaters in which Australians served in the Great War. Originally it was suggested that this was to be EGYPT, GALLIPOLI and PALESTINE on one side, with FRANCE on the other, but in the final work BELGIUM was added to the list.

Although the central square mile of the City of Adelaide is designed to the points of the compass, the monument sits on the corner of North Terrace and Kintore Avenue, adjacent to the grounds of Government House, on the north edge of the city centre. It sits at a 45 degree angle to North Terrace. The placing of the monument to face a north-west direction allows it to be in line with both the Cross of Sacrifice and St. Peter's Cathedral. In addition, the diagonal positioning of the memorial permits the dawn sun to fall on the facade.[201]

Sydney, New South Wales: The Anzac War Memorial

This memorial was built as a memorial to the Australian Imperial Force of World War I.

[Picture 54: ANZAC War Memorial, Sydney. Photo: Adam, J.W.C.[202]]

[201] Information sourced from: Wikipedia.org

[202] Wikimedia Commons contributors, "File:War memorial hyde park Sydney.jpg," Wikimedia Commons, the free media repository,

Fund raising for a memorial began on 25 April 1916, the first anniversary of the Australian and New Zealand Army Corps (ANZAC) landing at Anzac Cove for the Battle of Gallipoli.

The ANZAC Memorial owes its legislated existence to the Anzac Memorial (Building) Act 1923, the Preamble to which nominates that its principal reason for being is *'to serve as a memorial of the achievement of the Australian Imperial Forces'*. The Trustees appointed under the Act from time to time are charged with giving effect to this purpose.

A competition for the design of the memorial was commissioned in July 1929 and a month later the prize-winner was Bruce Dellit. The successful contractors for the building works were Kell & Rigby.

Perhaps it was thought that the memorial would be used for commercial purposes, as it was reported that:

A strong protest against the Anzac Memorial Building being used for club and business purposes was made at a meeting of the Anzac Fellowship of Women, which was held to-day.It was pointed out that all the money subscribed for the erection of the memorial had been collected by women, and it was felt that they should have been consulted in the matter.[203]

In 1931 it was estimated that the building would cost about £70,000/- to build, and was expected to be completed by the end of the year. The Great Depression did not permit the building to be financed as quickly as it was originally hoped, for in 1934 it was felt that owing to a shortage of funds it was possible that the Anzac Memorial would not be opened until 1935. Mr. Robb, the president of the Returned Soldiers' League, was reported to say that about £15,000 was needed and he suggested that the money be publicly subscribed.[204]

But the Anzac Memorial's foundation stones were laid on 19 July 1932, the 16th anniversary of the beginning of the Battle of Fromelles. They are inscribed with the words 'A SOLDIER SET THIS STONE' and 'A CITIZEN SET THIS STONE'. The soldier referred to on the first stone was Sir Phillip Game, the Governor of NSW. Sir Phillip had served as an officer in the Royal Artillery and later the Royal Flying Corps of the British Expeditionary Force on the Western Front during the Great War. The citizen referred to in the second stone was Bertram Stevens, then Premier of New South Wales. The decision to make anonymous those who laid the foundation stones is an important statement about the Memorial's recognition of the community who suffered the impact of the loss inflicted by the Great War and its aftermath rather than a statement about individuals.[205]

The ANZAC Memorial, was completed in 1934, some 18 years after its inception, and was opened on 24 November, 1934 by His Royal Highness Prince Henry, Duke of Gloucester.

---

https://commons.wikimedia.org/w/index.php?title=File:War_memorial_hyde_park_sydney.jpg&oldid=65943757 (accessed January 6, 2019).

[203] See the Newcastle Morning Herald and Miners' Advocate, Newcastle, 25 July, 1929

[204] See: The Northern Star Lismore, 7 February, 1934

[205] From the website of the Anzac Memorial: https://www.anzacmemorial.nsw.gov.au/our-stories/memorial-foundation-stones

It remains the main commemorative military monument of Sydney, and New South Wales. The interior is largely faced in white marble, and features a domed ceiling adorned with 120,000 gold stars – one for each of those men and women from New South Wales who served during World War I. The exterior is adorned with monumental figural reliefs and sculptures by Rayner Hoff.

The main focus of the interior is Rayner Hoff's monumental bronze sculpture *Sacrifice* of a deceased youth, representing a soldier, held aloft on his shield by a caryatid – three female figures, representing his mother, sister and wife.

Four Niches of Remembrance are spaced evenly around the edge of the Hall of Memory. Each of these niches is 26' (8 m) high by 12' (3.7 m) wide. *The Book of the Anzac Memorial* describes the interior of the niches as ***inscribed [with] the names of the more important battles in which the AIF participated*** **[emphasis added].** On the floor of each niche, set jewel-like in the rising sun of the Australian forces, are stones taken from overseas – from Flanders and Gallipoli, from Palestine and New Guinea. The Niches of Remembrance are said to represent each of the *principal theatres of war* in which Australians served during the Great War.

The Gallipoli niche lists the following battles: the battle of **THE LANDING** from 25 April to 2 May 1915; **KRITHIA** fought at Cape Helles on 8 May; **LONE PINE** fought from 6 to 9 August; **SARI BAIR**; HILL 60 21 to 27 August; and THE EVACUATION on the 19th and 20th of December. (Bold text denotes that a Battle Honour was awarded also.)

The Western front niche beneath the heading France and Belgium includes: **THE SOMME 1916**, the campaign fought between July and November; **BULLECOURT**, two battles in April and May 1917; **MESSINES**, the great Australian victory in June 1917; **YPRES**, the battles of the salient from July to November 1917; **SOMME 1918**, from the defence of Villers-Bretonneux to the Amiens offensive; and the **HINDENBURG LINE**, breached by the AIF in September 1918.

The New Guinea and the high seas niche includes COCOS ISLANDS, WEST ATLANTIC, NORTH SEA, OTRANTO BARRAGE, and MALAYSIA.

The Sinai Palestine niche includes **RAMANI, MAGDAHABA, GAZA – BEERSHEBA, MEGIDDO** and **DAMASCUS.**[206]

Canberra, Australian Capital Territory: Australian War Memorial.

The Memorial forms the core of the nation's tribute to the sacrifice and achievement of the more than 102,000 Australian men and women who died serving their country, and to all those who served overseas and at home.

---

206 [Material sourced from Wikipedia.org under the Creative Commons Attribution-ShareAlike License, http://www.anzacmemorial.nsw.gov.au, and other information in the public domain.]

The Inspiration:

After the Gallipoli campaign of 1915, the Australians of the 1st AIF (Australian Imperial Force) and their official war correspondent, Charles Bean, moved on to the major battle theatre of the Western Front in France and Belgium. The Australians' first big battles were at Fromelles and Pozières in July 1916. It was at Pozières that Bean resolved these men and their ordeal should not be forgotten. A month later the idea of a memorial museum for Australian soldiers was born. In 1918 Bean imagined how the Memorial would appear:

> *... on some hill-top – still, beautiful, gleaming white and silent, a building of three parts, a centre and two wings. The centre will hold the great national relics of the AIF. One wing will be a gallery – holding the pictures that our artists painted and drew actually on the scene and amongst the events themselves. The other wing will be a library to contain the written official records of every unit.*

[Picture 55: Australian War Memorial, Canberra. Photo: By "Bidgee"[207]]

Guiding Ideas:

In keeping with the sombre, commemorative tone of the Memorial, Charles Bean was from the start concerned that it should not be seen to be glorifying war or triumphing over the enemy. He urged the Directors and others not to speak about "trophies", preferring the term "relics". He also urged that captions and text should not use derogatory terms for former enemies, such as "Hun" or "Abdul".

---

[207] Wikimedia Commons contributors, "File:Australian War Memorial courtyard.jpg," Wikimedia Commons, the free repository, https://wikimedia.org/w/index.php?title=File:Australian_War_Memorial_courtyard.jpg&oldid=265032110 (accessed January 6, 2019).

In the 1950s Bean drew up a list of exhibition principles, suggesting among other things that the galleries should "avoid glorification of war and boasting of victory" and "perpetuating enmity ... for both moral and national reasons and because those who have fought in wars are generally strongest in their desire to prevent war". In general, he decided, former enemies should be treated as generously as were Australians. The exhibitions also needed to be made relevant and engaging, for example, by presenting relics as objects with their own story rather than as just examples of a type. Bean also thought of the future exhibits should be "so described and displayed as to be understood and interesting 75 years after the events".

Melbourne and Sydney Exhibitions.

After the First World War it took a long time before the Memorial's building in Canberra was constructed. Initially there were delays in arousing public and government enthusiasm. Then the financial crash and the subsequent Great Depression intervened. In the meantime large, long-running exhibitions were held in Melbourne and Sydney. The Australian War Museum opened on Anzac Day 1922 in the Exhibition Building, Melbourne. This exhibition of war relics was enthusiastically received by press and public, and attracted large crowds. The exhibition closed in 1925 and was moved to Sydney, where it remained until 1935.

In 1935 the decision to build the Australian War Memorial was promulgated. The Memorial's design was a compromise between the desire for an impressive monument to the fallen and a budget of only £250,000/-. An architectural competition in 1927 failed to produce a satisfactory single design for the building. A joint design was later accepted and forms the basis of the building we see today, which was completed and opened to the public on Remembrance Day, 11 November, in 1941.

A central Commemorative Area flanked by arched cloisters houses the names of the fallen on the bronze panels of the Roll of Honour. At the head of the Pool of Reflection, beyond the Flame of Remembrance, stands the towering Hall of Memory, with its interior wall and high dome clad in a six-million-piece mosaic and illuminated by striking stained-glass windows. Interred within the Hall of Memory since 1993, is the Unknown Australian Soldier, sprinkled with soil from the Pozieres windmill. Silent sentinels above him are fifteen stained glass windows, each a depiction of a serviceman and nurse of the First World War. At the base of each window is a single word, representing values seen in these men and women, essential not just for victory in battle, but for depth and breadth of character. They are:

RESOURCE,	**CANDOUR,**	**DEVOTION,**	**CURIOSITY,**
INDEPENDENCE,	**COMRADESHIP,**	**ANCESTRY,**	**PATRIOTISM,**
CHIVALRY,	**LOYALTY,**	**COOLNESS,**	**CONTROL,**
AUDACITY,	**ENDURANCE,**	and	**DECISION.**

The Memorial ranks among the world's great national monuments. Sharply etched grandeur and dignity in its stylised Byzantine profile contrast with a distinctively Australian setting among lawns and eucalypts at the head of a wide ceremonial avenue, Anzac Parade. Kangaroos from nearby bushy hills occasionally add to the physical effect.

The Memorial is more than a monument. Inside the sandstone building, with its copper-clad dome, selections from a vast National Collection of war relics, official and private records, art, photographs, film, and sound are employed to relate the story of the Australian nation's experience in world wars, regional conflicts, and international peacekeeping.

The names of battles, campaigns, and theatres in the Memorial's Commemorative Area, relating to the First World War are:

INDIAN OCEAN, **GALLIPOLI**, MESOPOTAMIA, NORTHERN FRANCE,

FLANDERS, and PALESTINE. Only Gallipoli has a separate Battle Honour.

Recited every day at the Last Post Ceremony are said these words:

Here is their spirit, in the heart of the land they loved; and here we guard the record which they themselves made. - Charles Bean, 1948.

The official opening was reported in the Canberra Times the following day, 12 November, 1941. The full report can be found using the search facility on the National Library's search site: Trove. (trove.nla.gov.au)

It was reported: *The impressive service of tribute to the deathless army which placed the seal on Australian nationhood came to its climax when the Governor - General Lord Gowrie closed a switch, and the great bronze doors of the memorial swung silently open. Thus, an ideal conceived on the battlefields of the Old World had been realised and the splendid monument of the Empire's youngest capital became the property of the Australian people.*

Diplomatic representatives of the United States, Canada, China and Japan were prominent among the official guests. Ironically Australia would be at war against Japan very soon after.

"The Australian War Memorial would not only house the relics of war but would keep alive the sacrifices and services which were the cornerstone of a peoples' greatness", said Prime Minister Mr. Curtin.

A Comment On The Different Approaches – A Criticism of The 'Battle Honours' Approach to Inscription.

Compare the approach taken with the various memorials in Perth, in Melbourne, in Brisbane, in Adelaide, in Hobart, in Sydney, and in Canberra.

The memorials in Perth, in Hobart, in Adelaide and in Canberra, do not specify particular battle locations. They thus honour all soldiers in whatever locations they were involved. By being generalist in their attribution, they effectively became inclusive of all those who were involved in all actions, in all Battles, in World War 1 regardless of the issue of Battle Honours.

The memorials in Melbourne, in Sydney and in Brisbane, however, each effectively refer to places at which Battle Honours were awarded. Whilst this may have had the intention of placing significance on those places giving rise to the Battle Honours so honoured, it had the unintended effect that this method memorialises, directly or by implication, every Battle of the AIF on the Western Front except Fromelles.

In previous chapters we saw that Fromelles is in all reality a Battle in every sense of the description, and should be honoured as such, but because it was classified as a subsidiary action to another Battle, it was not then, and so far has not been, separately or properly recognised on any of these memorials. We also saw that Fromelles had no geographical connection with the Somme battles. At one point I suggested that by accepting this approach, you might just as easily say that Kalgoorlie W.A. is part of Bondi N.S.W.

Take for example, the Anzac Memorial in Sydney. Despite the original Trustees' wish to have the niches in the walls of the Memorial *'inscribed [with] the names of the more important battles in which the AIF participated*, Fromelles does not appear. The first major engagement on the Western Front, long remembered as suffering the largest number of casualties in less than 24 hours, is supposedly included by reference to the "SOMME 1916" (embracing Pozieres etc.), the campaign fought between July and November.

As Fromelles was categorised by the BNC as part of The Somme, it was wrongly described geographically as well. It has thus been prevented from being listed as either an 'area of battle' or a Battle, separate from The Somme, and which is a geographically different region in France altogether.

Had the Trustees adopted in the same way as later occurred in Canberra, by referencing France and Belgium or France, Flanders, and Belgium, then that would by its sheer inclusivity have embraced and honoured ALL Battles by the AIF on the Western Front, without exception. The first battle was that of Fromelles, on 19 July, 1916. The last was that of Montbrehain on 5 October, 1918.

---ooo000ooo---

Significantly, on its disbandment on 19 July 1966, the still-surviving 19th of July men, through the 5th Divn. Assn asked the Trustees of the ANZAC Memorial, Sydney, to honour Fromelles by placing a wreath in the France and Belgium Niche Wall of the Hall of Memory annually on the anniversary of the Battle.

Each year thereafter until 2016, a wreath would be placed on a stand in the Niche Wall of the Hall of Memory, by the Trustees of the Anzac Memorial, Sydney.

Then on 19th July, 2016, for the 50th anniversary of the 5th Divn. Assn asking for Fromelles to be commemorated, the Families and Friends of the First AIF (FFFAIF) took the honorable action of instigating a solemn procession commemorating the Battle of Fromelles, which left the Archibald Fountain in Hyde Park, and proceeded to the ANZAC Memorial for a commemorative service.

The place of departure, the Archibald Fountain in Hyde Park commemorates the association of Australia and France in The Great War. The distance of the procession is approximately 400 yards, which equates to the widest part of No-man's land at the Battle of Fromelles.

The solemn procession re-enacts the veterans of the Australian 5th Division marching to the ANZAC Memorial on the 50th anniversary of the Battle to pass to the Trustees the responsibility for remembering their service and sacrifice, and that of their fallen comrades in perpetuity.

That solemn procession and a commemorative service following, have taken place every year since, with the exception of a Covid 19–interrupted commemoration in 2021.

Pozieres veterans, by way of comparison, have made no similar request as they were clearly involved in, and part of the Somme Offensive, between 1 July, and 18 November, 1916. There can be no question that Pozieres veterans are acknowledged and honoured under the listing of 'SOMME 1916' at the Anzac Memorial, Sydney. Regrettably, Fromelles is not.

A Different Approach – By Our French Allies.

It can easily be said that the honouring of Fromelles can be done, and has been done. Our own leaders should have no difficulty in doing so, when our allies have done just that. The following article, worth reading in its entirety was reported in The Canberra Times, on 28 July, 1961:

FRANCE HONOURS OLD BLOOD-TIE[208]

By a Special Reporter.

This afternoon the Prime Minister, Mr. Menzies will unveil a memorial erected by the French Government in the grounds of the French Embassy to honour the memory of Australians who fell on French soil in the 1914-18 War.

At the ceremony as the French Government's special envoy will be General Zinovi Pechkoff, a son of Maxim Gorki, and General Appert, Commanding Officer of the French Pacific Forces.

The memorial, a stone masonry column standing 15 feet high is surmounted by a gilded bronze figure symbolic of "Winged Victory." On the front of the column in French and English is the following inscription in gold lettering:

"To Those Australians Who Fell On French Soil In Memory Of Their Sacrifice For The Freedom Of All Peoples 1914-1918."

On each side of the column are six battle names selected as representative of the many in which the Australians fought in France in the First World War.

They are: Armentieres, ***Fromelles****, Hazebrouck, Pozieres, Bullecourt, Amiens, Villers-Bretonneux, Dernancourt, Mont St. Quentin, Somme, Hargicourt and Bellicourt.*

---

[208] See: - https://trove.nla.gov.au/newspaper/article/103081990?searchTerm=FRANCE%20HONOURS%20OLD%20BLOOD-TIE%20%20%20%20%20%20%20%20%20%20%20%20%20&searchLimits=

The memorial, however, recalls not only the efforts and sacrifices of the Australians in France in the 1914-1918 War but the close and cordial relations which developed between the men of the 1st A.I.F. and the people of the French countryside.

"Nos Australiens" — "Our Australians" the French civilians called — as men of the 14th Battalion marched towards Hebuterne to stop the German thrust towards the important French railway centre of Amiens.

"In cottage after cottage," relates Dr. Bean, official historian of World War I, "the troops found on the walls photographs of the Australians who, in the old Somme days, had been taken into the family circle, and who were still eagerly remembered."

For their part the Australians grew more and more to admire and respect the French people: for the way in which the women, the boys, and the old men kept things going while their men-folk were away fighting the Boche; for the wit and humour of the French girls! and for the convivial atmosphere they found in the village estaminets behind the lines during rest periods. The battle names on the column will remind men of the 1st A.I.F. of many a hard-fought engagement.

At Fromelles on July 19, 1916, for example, the 5th Australian Division, although it had arrived from Egypt only three weeks before, lost 5,533 officers and men in one night. In the same year, from the middle of July to early Sep., 1916, at Pozieres and Mouquet Farm the 1st, 2nd and 4th Divisions sustained a series of attacks and counter attacks by and against Saxons, Bavarians and Rhinelanders, in a series of engagements which cost the Anzac Corps 23,000 officers and men in seven weeks.

Little wonder that when the 1917 battles, Bullecourt, Messines, Ypres, and Polygon Wood were past, and the German offensives of March, 1918, had begun, the Australians were observed at each roadside halt grimly intent as seldom before on cleaning their rifles "Not if they could help it," wrote Dr. Bean, " would the Germans get through to the old folk and children who greeted them with cries of 'Vive l'Australie' and 'les bons Australiens' as they marched forward."

When the German attack had been repulsed the Australian Corps, now under General Sir John Monash, itself turned to the offensive and fighting through Harbonnieres, Lihons, Chipilly, Chuignolles and Clery, shocked the German High Command by its brilliant capture of Mont St. Quentin.

Peronne, Hargicourt, Beaurevoir and Estrees followed and the Hindenburg Line was broached at Bellicourt.

Unveiling of the French Australian memorial this afternoon recalls these and other actions in which Australians fought on French soil and will cement further the bond of friendship between the people of France and Australia begun so well by the fighting "Diggers" of the First A.I.F. in the 1914-1918 War.

The Changed Political and Governmental Attitude to Fromelles.

The different attitudes to Fromelles which had been evident in the building of the various memorials was turned on its head when Lambis Englezos was successful after a long campaign, to convince authorities of the merits of exploring for burial pits next to Pheasant Wood in

Fromelles. The revelation of 250 soldiers' remains brought to the forefront a latent interest by families of the soldiers, in paying homage to these men, and to having them identified.

A new Commonwealth War Graves Cemetery – Pheasant Wood, at Fromelles – was officially dedicated on 19 July 2010. The officials in attendance included His Highness Prince Charles, Prince of Wales and his wife the Duchess of Cornwall, along with Australian Governor-General Quentin Bryce.

There were thousands of family members and other officials present on that hot summer day. There was no ambiguity in the official recognition of Fromelles by this time. People from the United Kingdom also shared in the opening, as was reported on 20 July, 2010 in the online website: This is Nottingham. The following information is edited from a report by Riah Matthews.[209]

SOLDIER FINALLY LAID TO REST IN FROMELLES.

Much of the article was taken up with reporting on the discovery of Robert Courtney Green (whose image is reproduced in Chapter 16), which will be set out in the following chapter, but some of his story was outlined in the following paragraphs:

Robert Courtney Green, who was killed in action in the First World War and dumped in a mass grave, was finally buried yesterday in an emotional ceremony.

Robert was just 30 years old when he was shot in the chest and killed on the night of July 19, 1916. His body was discovered in a mass grave last year. And the brave soldier was finally laid to rest yesterday – exactly 94 years after his death – when he was re-buried with full military honours, along with 249 other British and Australian soldiers whose bodies were discovered in the same grave.

The bright white tombstones line up in a new war cemetery in the French town of Fromelles. It is the first western front war cemetery to be dedicated in 80 years.

Prince Charles and The Duchess of Cornwall joined a walking procession through the village and the last soldier to be found in the mass grave was the last to be buried.

In a speech Prince Charles said: "I am profoundly humbled by the outstanding bravery of these men who fought so valiantly in the indescribable mud and carnage, many thousands of miles from their families and homes."

The Battle of Fromelles is known as the worst 24 hours in Australia's history – more than 5,000 Australians and 1,500 Britons were killed or captured by the Germans. The dead were dumped in communal graves which were left unmarked.

---

[209] Unfortunately, the original link to the article online is no longer available. The text in this chapter is reproduced in part from the original post.

However, the Germans did return Robert Courtney Green's blood-splattered identity disc, which was to be used to identify him nearly 100 years later.

---ooo000ooo---

By 8 April, 2011, the following year, more soldiers had been identified, prompting a media release by **Warren Snowdon MP**, Minister for Defence and Science Personnel and Minister for Veterans Affairs, and Federal Member for Lingiari (Northern Territory):

MORE BATTLE OF FROMELLES DIGGERS IDENTIFIED[210]

A further 14 World War One Australian soldiers who fought at the Battle of Fromelles in France have been identified. Mr. Snowdon said the soldiers were originally from the New South Wales, Victoria, Queensland and South Australia and were among 250 Australian and British WW1 soldiers recovered from Pheasant Wood in France in 2009.

"I am pleased to announce today that a Joint Identification Board held on 4 Apr. 2011, has identified these 14 soldiers by name, Mr. Snowdon said.

"This is very significant. These latest 14 soldiers bring the total number of Australians identified by name at Fromelles to 110

"We are determined to identify as many of these brave Australians as possible. We are encouraged by the success, made possible by the large number of extended family members, both in Australia and Britain, who have provided DNA samples to assist with identification.

The Commonwealth War Graves Commission will now erect new headstones with the identified men's details and they will be dedicated on 19 July this year during the annual commemoration of the Battle of Fromelles.

The Chief of Army, Lieutenant General Ken Gillespie, praised the efforts of the Fromelles Project team, who have made contact with the relatives of the newly identified soldiers this week.

"The additional identifications demonstrate the tenacity and dedication of a wonderful team and also demonstrates how the latest scientific methods and great research can produce outstanding results, Lieutenant General Gillespie said.

"While identification of remains is an extremely complex process, we are hopeful that we will identify more soldiers in coming years. It is important that we are able to identify these soldiers who made the ultimate sacrifice and assist in giving closure to the families.

"The involvement of the families of those diggers that remain unaccounted for has been vital to this process. We currently have almost 3000 family members' details in our records but we still need more. If you think you might be related to a soldier who remains unaccounted for from the Battle of Fromelles, please get in contact with the Army.

---

210 http://www.warrensnowdon.com/2011/04/08/more-battle-of-fromelles-diggers-identified/

[The Army's Fromelles Project team can be contacted by phoning 1800 019 090 or by accessing their website at www.army.gov.au/Fromelles.]

---ooo000ooo---

15 July, 2016 saw the publication, for example, of a news release by **Martin Hamilton-Smith**, South Australian Minister for Veterans Affairs, titled: -

COMMEMORATING SOUTH AUSTRALIA'S WWI SACRIFICE IN FRANCE[211].

He would be in France as part of an official Defence Industries and Trade and Investment visit to France and the United Kingdom, and also able to be present to commemorate the 100th anniversary of South Australia's WWI loss and sacrifice on the French battlefields.

At the Bastille Day Parade on July 14, the colours of the Royal South Australian Regiment will take part in the traditional Parade along the Champs-Elysees. This year, at the invitation of the French Government, the parade will be led by approximately 160 members of the Australian Defence Force, including South Australian members in honour of the contribution of Australian forces during the Battle of the Somme in July 1916.

The Minister will also attend the Last Post Ceremony at the Menin Gate in Ypres on July 18. The following day he will commemorate the 100th anniversary of the Battle of Fromelles by visiting the Fromelles battlefields and attending commemorative services at the Pheasant Wood Military Cemetery and the VC Corner Australian Cemetery and Memorial.

Quotes attributable to Veterans' Affairs Minister Martin Hamilton-Smith were: -

For many Australians, the Western Front really is a sacred place. Being part of these important events marking the centenary of battles fought 100 years ago will be incredibly moving and is a great honour.

The fact that Australia was asked to lead the Bastille Day celebrations on July 14 in this centenary year is also testimony to the great and lasting friendships that exist between our nations dating back to 1916.

To this day Australians are warmly welcomed in those sacred places, where the fiercest battles were fought and where thousands of lives were lost. Here, there are beautifully cared for cemeteries, their keepers dedicated to ensuring those who are buried in foreign lands, alongside their own countrymen, are not forgotten.

This historic period was a momentous time when South Australians from all backgrounds fought and died in the Great War.

---ooo000ooo---

Another parliamentarian who expressed his interest in the Battle of Fromelles and the ripples that the modern-day re-emergence of families' interest in the events at Fromelles, was the Federal Member for the seat of McMillan, based in Victoria, **Mr. Russell Broadbent**. In an

211 https://veteranssa.sa.gov.au/2016/07/commemorating-south-australias-wwi-sacrifice-in-france/

opportunity to speak, and recorded in HANSARD on Thursday, 4 Feb. 2016 at Page 625, he told the house[212]: -

Last Sunday, I was honoured to be present at the opening of a place of reflection, an Anzac memorial put together by the people of Leongatha.

It is a beautiful place — modern in the capacity of today's memorials. It has four parts to it, where all 330 names from the Great War of 1914-18 are now displayed.

There were so many gathered there that day at the place of reflection. They were all the families of those who currently live in Leongatha and district. But this was not particularly about the district; it was specifically about those who went to war from Leongatha. *The Battle of Fromelles* on 19 July 1916 caused the community of Leongatha and district much sadness and grief. I have spoken about the grief of those two world wars and other wars since that continue today, in my opinion. In this respect, perhaps the town was a microcosm for the nation. Over 5,500 Australians became casualties during that bloodthirsty event. On this memorial are the names of nine men who died as a result of the battle and two who died in the lead up to the battle. Harry Cosson was on guard duty when he fell from a moving train on 1 July, 1916, and was killed. John (Jack) Daniell, formerly a junior school teacher at Leongatha Primary School, was killed by stray shell on 15 July.

This was a great event, held and loved by the community. The names of these men are still synonymous with the local area. There were boys named after their uncles—the late Jack Edny and Phil Bellingham who were killed in action on 19 July.

---ooo000ooo---

On 20 July, 2016, former Federal politician, and now Director of the Australian War Memorial in Canberra, **The Hon. Brendan Nelson**, spoke at the National Press Club in Canberra. The speech was titled **'From Fromelles to Pozieres – We Remember'**[213]. It was a speech which explored the development of the Australian character, referencing the actions at Fromelles, and Pozieres in July, 1916.

Mr. Nelson reflected on the paradox that what is most important to us we tend to take for granted, including the political, economic and religious freedoms given us as Australians. Much of what he spoke of has been the subject of different parts of this book. Nelson concluded his speech by reference to the work of Sgt. Simon Fraser, and others in the recovery of wounded soldiers.

Sergeant Alexander Ross of the 57th Battalion reveals the Australian character, volunteering to recover wounded from No-man's land. Captain Hugh Knyvett was with him (Over There With The Australians, published 1918):

212 https://www.openaustralia.org.au/debates/?id=2016-02-04.152.1

213 For full content of speech, see: - https://www.awm.gov.au/commemoration/speeches/Fromelles-Pozieres-we-remember

We found a man on the German barbed wire….so badly wounded that when we tried to pick him up, one by the shoulders and the other by the feet; it almost seemed that we would pull him apart.

Blood was gushing from his mouth, where he had bitten through lips and tongue, so that he might not jeopardize, by groaning, the chances of some other man who was less badly wounded than he.

He begged us to put him out of his misery, but we were determined we would get him his chance, though we did not expect him to live.

But the sergeant threw himself down on the ground and made of his body a human sledge.

Others joined us, and we put the wounded man on his back and dragged them thus across two hundred yards of No-man's land, through the broken barbed wire and shell-torn ground, where every few inches there was a piece of jagged shell, and in and out of the shell-holes.

So anxious were we to get to safety that we did not notice the condition of the man underneath until we got into our trenches; then it was hard to see which was worst wounded of the two.

The sergeant had his hands, face, and body torn to ribbons, and we had never guessed it, for never once did he ask us to "go slow" or "wait a bit."

Such is the stuff that men are made of.

After the bloodbath at Fromelles, Sergeant Simon Fraser spent three backbreaking days bringing in the wounded from No-man's land.

A lone voice pleaded through the fog, "Don't forget me cobber".

He didn't.

We won't.

We never will.

We are Australians.

We did advance, and we will.

We are young, and we are free.

---ooo000ooo---

At the beginning of this Chapter, I expressed the view that what is interesting when looking at the different State monuments in the various capital cities, is just how that process resulted in two distinctly different approaches: -

One process honours locations by name, which then becomes a process of (unintended) exclusion of deserving action by Australian soldiers.

The other process honours all those who served, by not including any particular locations by name at all.

Since the emergence of public awareness of the burial pits in Fromelles, there has been a significant move to honouring the soldiers who fought in the Battle of Fromelles, but as yet, that has not resulted in 'Fromelles' being inscribed on any monuments where service is acknowledged by inscription.

The closest that we have come to that situation is at the ANZAC Memorial in Sydney. The memorial was refurbished and extended in 2018. Despite advocacy to do so, 'Fromelles' was not inscribed in any of the walls of the Memorial. However, the inclusion of a soil sample from Fromelles as one of the 100 Sites of Signficance to NSW does give greater recognition to the Battle of Fromelles.

---ooo000ooo---

19. Families Connecting with Their Ancestors Through Fromelles

This is not ancient history. For some it is their uncle. For some it is their grandfather.

For others, the family never told them of their missing family member.

There remains a real connection between the soldiers who served, and present-day family members.

- Lambis Englezos

---ooo000ooo---

We read in Chapter 17 of a Northcote plumber called Bill Barry who survived the Battle of Fromelles to write a first-hand account of what it was like to "die" on the field. Knocked unconscious by a shell blast that blew holes in his legs, he was tossed on a heap of bodies by a German burial party.

Barry woke to find himself surrounded by dead men "*on the edge of a hole about forty feet long ... and into this hole the dead were being thrown without any fuss or respect.*"

When Lambis Englezos read Barry's vivid account of the burial pit it confirmed his belief that at least 200 Australian dead were unaccounted for — and that all or most would be in the mass grave Barry had seen at what the Germans called Pheasant Wood, behind their lines in a field next to Fromelles village.

In 2009, after opposition from Australian Government and Army, and a series of attempts to have it done, Lambis Englezos was successful in having the suspected mass grave at Pheasant Wood, Fromelles identified and revealed. There were, and there continue to be, remains identified by use of not only DNA, but items found on the bodily remains of soldiers unearthed.

The attention this garnered in Australia, in the United Kingdom, and around the world, brought a series of registrations with the Army's Fromelles Project unit from families of soldiers who had been killed, presumed killed, or missing, since 1916.

Many did not return, but we now know where some of them lie, following the revelation of the burial pits at Pheasant Wood, near Fromelles. For example, the ongoing work of the Australian Army's Fromelles Project team, at Unrecovered War Casualties Unit, and the work of volunteer genealogists working with the Fromelles Association of Australia is continuing to locate relatives able to provide DNA sampling and family histories. In 2018, 9 identified remains were the result of work by the Fromelles Association of Australia working with the Army's Fromelles Project team.

---ooo000ooo---

Recall for a moment, George Challis, who we read about in Chapter 17. He was 24 years and five months old when he signed up, just before the 1915 football season ended. By the following

July he was but a few bloody body parts in a blanket. Imagine 2000 others killed the same week then add 3500 wounded and you get an idea of what happened on the fields below this town.

And that's just Fromelles — worst of a string of murderous battles that would kill more than 61,000 Australians and send home 156,000 more with physical wounds. Of 200,000 others, many had mental scars that would affect families for generations.

---ooo000ooo---

There were 1299 possible soldiers in the list of those killed and unaccounted for at Fromelles, but only 250 bodily remains were found at Pheasant Wood. So, as a consequence, many families of those soldiers would be and still are disappointed, at not having been able to connect with their lost ancestor.

Interest in the project also caused many families to investigate their connection with Fromelles as new inquisitors. For example, more than 3000 DNA samples have been registered with Unrecovered War Casualties Unit, Army, in the hope of connecting with such a lost ancestor.

Many have found details of ancestors who had served and returned. Groups formed on Facebook, such as 'I Honour the Dead of Fromelles', and 'The Fromelles Association of Australia'. They have continued to disclose information relating to the soldiers who fought, and some of whom have been found, at Fromelles. A French citizen, Pierre Seillier, who lives in the area of Fromelles, has and continues to produce Tribute Cards to those who fought and died at Fromelles. His artwork graces the cover of this book.

166 of the 250 discovered bodies have been identified to date. The work of the Fromelles Association of Australia and the Army's Unrecovered War Casualty Unit continues. It should therefore come as no surprise to read of the extent of the reactions, the feelings of elation, and the discovery of a part of the family jigsaw which had been missing for so long. Some of their stories make up this chapter. While their stories are all different, those who shared the experience of finding a lost ancestor all share a sense of connection that only the identifications provided.

BOLT, HERBERT THOMAS 'NUTSY' [With thanks to Josie Shelley.]

Herbert Thomas "Nutsy" Bolt was born 10th September 1893, at Newtown NSW, to Frederick and Mary Bolt. He was the youngest of 7 children, having 4 brothers and 2 sisters. Herbert attended Newtown Superior Public School and when he left school he worked as a brick-maker at Bedford Steam Works, Newtown.

He loved all sports, was a very good Boxer and excelled at Rugby League, having played over 50 games between 1912 and 1915 for the Legendary Newtown Bluebags and also represented NSW twice against Queensland in 1913, playing alongside such greats as Dally Messenger.

Herbert met a young woman named Jennie Margaret Hughes. They married in 1914, when they were both 21 years old. Their baby daughter Mary Monica was born in February, 1915 but Herbert always called her his "Little Mona".

On Thursday 20th June 1916, Herbert set sail from Egypt for France, and disembarked on Friday the 30th June.

On the evening of the 19th July, 'Nutsy' and his mates of the 55th Battalion were in the thick of the battle. As one described it: "The air was thick with bullets, swishing in a criss-cross lattice of death".

When dawn broke on the 20th July 1916, 'Nutsy' was still alive, and had even reached the first of the German Trenches. Sadly, he was killed that morning. His great mate Frank Johnston was with him in the trenches that fateful morning and in his account to the Red Cross, he said:

"He and I were close to one another, when we were attacked by the Germans. He got more than 6 of them with his bayonet and the butt of his rifle, when he got a bullet through the head. He fell instantly, being killed outright. He was as game as any man, and was a well-known Newtown Footballer".

He was just 22 years old and left behind his young wife Jennie and baby daughter, his little Mona, now 17 months of age. Sadly, Little Mona never knew anything of her Father, her mother never discussed him at all, the only knowledge she had was that he fought in WW1 and was a very good Rugby League player who played for the Newtown Bluebags.

So with this very scarce information Josie Shelley, daughter of Mary Monica, searched for more information on her grandfather Herbert 'Nutsy' Bolt, killed in action at Fromelles, on 20th July 1916.

[Picture 56: 'Nutsy' Bolt, Jennie, & Mona. Photo: Josie Shelley, family collection.]

In 2004 Josie and her husband visited the War Cemetery at Villers Bretonneux where Nutsy's name was displayed on the wall, along with thousands of other soldiers with no known grave. "It was such an emotional feeling to see his name there and so sad to think that this dreadful war had deprived us of this wonderful brave young grandfather."

A few years later Lambis Englezos persevered with his quest to find the missing soldiers who he firmly believed were buried at Pheasant Wood, Fromelles, and finally received Government approval to investigate and hopefully recover their bodies.

"I received a letter from the Australian Army Headquarters to inform me that 3009 Corporal Herbert Bolt was included on the working list, however they stressed that it did not confirm he was buried at Fromelles or that his remains might be successfully identified. I remained very hopeful."

Josie and her 5 children registered their interest, but unfortunately, the female line was unsuitable as a testing candidate. Then one of Herbert's brothers' great grandson also registered his interest and gave a DNA sample with the Army's Fromelles Relative's Database. And match it did.

---ooo000ooo---

So, 100 Years On, when you ask this grand-daughter of a soldier who fought at Fromelles what does Fromelles mean to her 100 years after that disastrous Battle, you will find it means many things. Josie Shelley probably best sums up the range of emotions with the following comments: -

Fromelles means so very much to me and it will always have a very special place in my heart. It means so many things, where my grandfather fought and died, where he lay buried for 94 years in a mass grave, before he was discovered exhumed and identified, and where he is now buried in hallowed ground. So many memories and emotions for me.

Pride: My grandfather, Corporal Herbert Thomas Bolt, always known as "Nutsy", was a brick-maker, a well-known football player, and enlisted for service three days before his 22nd birthday. He sailed for Egypt 2 days before Christmas, leaving behind his young wife Jenny and his baby daughter Mary Monica his "Little Mona", just 10 months old. He underwent training in Egypt before going to France. He had already been promoted to Corporal in May, 1916.

Anger: He was just 22 years old, killed in action just 7 weeks before his 23rd birthday.

Sadness: It is so sad he never got to celebrate his family birthdays, Christmas, or his baby's 1st birthday! Sadness that it was thought that his body would never be recovered, only to have his name etched on the wall of the beautiful memorial cemetery at Villers Bretonneux. So sad too that my mother never knew her father and her children never got to meet and know their grandfather. I was fortunate to visit Villers Bretonneux twice, the 2nd time I had my entire family with me and it was so very sad for all of us to see his name on the wall along with so many thousands of other young men and women!

Disbelief: In 2009 I received a phone call from the Army to say Herbert could be one of the soldiers found in the mass grave and they needed DNA to prove it was him. I could not believe after 94 years lying in this mass grave, one of these poor soldiers could perhaps be my grandfather!?!

Gratitude: I will be forever grateful to Lambis Englezos for having the passion, willpower and dedication to pursue his dream of discovering this mass grave in Pheasant Woods. I am very grateful also, to the people of Fromelles for their continued respect and gratitude to our Australian soldiers - they even named their school "Ecole Cobbers" in honour of Australia's battlefield mateship. I have made some wonderful friends in Fromelles, who very kindly visit Nutsy for me. They put flowers on his grave and send us photos; it is so very kind of them to do that for us.

The village of Fromelles will always have bitter sweet emotions for me, bitter that my grandfather was lost there in that dreadful slaughter, and sweet that the Fromelles people have honoured and respected our fallen so very much and continue to do so.

Joy: I cannot tell you the absolute joy I felt when I received the phone call from Major Jason Kerr on the morning of 16th March, 2010, to tell me that my grandfather's body had been absolutely identified and his grave was to be dedicated on the 19th July 2010 in Fromelles.It was a very emotional call for me and I thought a miracle as well. There was much joy and celebration within our family having received such wonderful news, to think Herbert would finally be laid to rest after 94 years buried in the beautiful Australian War Grave Cemetery at Fromelles, in hallowed ground with his own headstone, and with his fellow soldier mates.

Satisfaction: My husband and I, and 9 of our family were present on the 19th July 2010 for the Dedication & Burial Ceremony at the beautiful "Pheasant Wood" Military Cemetery at Fromelles for all those lost soldiers, including my grandfather. What an honour it was to be there for this very emotional event and then to see his video up there on the big screen telling his story was one of such heartfelt sadness, but such satisfaction as well, to know he now had his own grave and headstone buried with his mates in such hallowed ground!

After the ceremonies, some Australian soil, as well as soil from Gallipoli was spread on Nutsy's grave. When the soil had been spread, Lieutenant General Ken Gillespie stood to attention, saluted and said "Well done Corpy". It was done in such a profound and respectful way, it brought tears to all our eyes, a beautiful and kind gesture from such an important man.

It was an amazing day, so full of emotion and extremes, we were very sad to leave Herbert but felt glad too that he was lying with all his mates, it made our departure just that little bit easier.

Fulfilment: In 2012, my husband and I, and our entire family, visited Fromelles again. It was very emotional for us to actually stand on the site of the mass grave at Pheasant Wood. It was such a very special and profound moment. It was also wonderful having Herbert's 2 grand-daughters (myself and my sister Mary) his 5 great grandchildren and their spouses, and 13 great great grandchildren all together there to pay their respects to our very brave ancestor. What a beautiful and very emotional visit it was for us all. I truly feel that after reading the daily entries he wrote in his diary, which he began the day he sailed from Sydney until his last entry on the 30th June when he arrived in France, that he was a remarkable and very brave young man who sadly made the ultimate sacrifice and was taken from us too early.

CLINGAN, ALEXANDER STANLEY

with thanks to Gwenda Stockton]

Alexander Stanley Clingan was born at Waterloo, Sydney on 28 November 1893. He lived in Union Street Newtown, within walking distance of his school, the Newtown Superior Public School. He lived with his parents, Ruth and William Clingan. He was their 6th child, having four brothers and two sisters. They were a strong, Methodist family. Unfortunately, Alexander's father, William, died at Camperdown NSW on 18 June 1907.

Alexander was a promising young sportsman in cricket and football. He played 2nd grade rugby league for the Newtown 'Bluebags' Rugby League team. Alexander had the highest individual score in junior grade cricket in 1915 at Moore Park in Sydney. After leaving school, he worked as a boilermaker's labourer for the NSW Railways.

Alexander was 21 years old when he enlisted for military service on 8 August 1915, Regimental number 3168. He embarked from Sydney, on 8 October 1915, a heart-wrenching time for his mother Ruth, saying goodbye to her youngest child.

He went on to train in Egypt before sailing with the 53rd Battalion to join the British Expeditionary Force at Marseilles, France, disembarking on 28 June 1916. His last letter to his dear mother was dated 12 June 1916 from France. His battalion were close to the firing line by then as he described what he saw and heard.

Seven days later on 19 July 2016, Alexander was killed in the Battle of Fromelles.

He was reported as missing in action on 19 July 1916, less than a year after joining the AIF.

On 13 February 1917, Alexander's identification disc was received from Germany. The War Office reported on 24 February, 1917: 'No particulars afforded except that this soldier is deceased. To be reported killed in action, 19 July 1916' on the authority of Assistant Adjutant General, ANZAC Section, War Office.

[Picture 57: AWM No. P05301.032]

The report from the Australian Red Cross, Wounded and Missing Bureau, 15 October 1920, showed: 'Clingan A.S. Regimental number: 3168, 53rd Battalion, A.I.F. Fell 19.7.1916 near Fromelles. Final German Death List, S'.

On 1 August 1921, Base Records wrote to the 'mother of Private Clingan', and asked her if she had any letters of communications that contain any reference to the circumstances surrounding his death, particularly the exact locality at which it occurred or where he was last seen alive.

Mrs Clingan replied, 'I have never received definite information as to whether my son was missing after Fromelles or Fleurbaix. The military informed me that he was missing on July 19, 1916. His mates have told me that they saw him go over on that date but did not see him again. It was through the efforts of the Red Cross that I found out that his name appeared in a German newspaper, but whether the death was due to wounds or otherwise, I do not know.'

In 2010, world-wide news had spread of the Army's request for the DNA of family members who believed their loved ones had died at Fromelles, France. Gwenda Stockton's husband

Donald Stockton was the great nephew of Alexander and his sister Valerie was a great niece on Alexander's mother's side. On Alexander's father's side, Graham Clingan, and Scott Clingan, contacted the Australian Army and were each given a DNA kit.

On 16 March 2010, both Valerie and Graham received phone calls from a Major in the Army at Duntroon, Canberra to tell them the wonderful news. Their DNA was a perfect match with the DNA taken from the remains of Alexander Stanley Clingan. Alexander was one of the first Australian soldiers to be positively identified.

Gwenda Stockton gives the following account of events following the receipt of that wonderful news:

The Australian Army invited families of those lost in France and those discovered and identified in the mass grave, to attend the Commemoration of the new Military Cemetery at Fromelles on 19 July 2010. My husband Donald had passed away in 2006, so our son Peter represented Donald at the Fromelles Commemoration Service.

Prior to attending the Commemorative Ceremony at Fromelles, Peter and I were in London. We visited the Imperial War Museum, as we were told there was a Fromelles Exhibition and an exhibit about Alexander.

We did this and our joy bubbled over when a huge photo of Alexander filled the glass window at the entrance to the Fromelles Exhibition. Peter stood to the side of Alexander's picture while I took his photo beside his Great, Great Uncle Alexander.

Nine members of our family then travelled to France for this very special occasion. My husband's cousin, Robyn and her husband, Bob Hanna, my son, Peter and other family members, were privileged to attend the Commemorative Ceremony for the opening of the new Military Cemetery at Fromelles in France on 19 July 2010. Alexander's face was projected onto the large screen at the commencement of the Ceremony and a family relative from Canberra, Margaret (Clingan) Wright, told Alexander's story and spoke of his mother Ruth's search for her beloved son. Our family felt such joy and pride as we watched this wonderful presentation.

It was a humbling experience and a privilege to be there. Homes were decorated with flags and French families gathered out the front of their homes to wave to us and welcome us. A special stand had been erected within the cemetery grounds for the families to view the ceremony.

During the ceremony, we could hear the clippety-clop of horses' hooves along the road outside the cemetery. On the screen in front of us, we could see a horse-driven First World War wagon arriving with the coffin of the last soldier who was to be buried during the ceremony. Walking behind the coffin, were Prince Charles and our Australian Governor-General, Quentin Bryce and members of the French, British and Australian Army.

Six Australian Army soldiers carried in the coffin of the Unknown Soldier, followed by the dignitaries."

At three o'clock, family members were seated near the graves of their loved ones in the cemetery. We were all addressed by Army personnel and the Governor-General, Quentin Bryce.

She spoke with such feeling and called us all 'friends' as she identified with us all and our deep feelings of emotions, both sadness and joy, as we all gathered together around the graves of our loved ones. She spoke these words at the Ceremony:

> *"Twenty five sets of brothers, two sets of father and son. All of them, loved ones.*
>
> *Allied by Kingship. Born of blood and Empire. United at the front line by a common humanity. Fine, willing, courageous men. Two hundred and fifty of them.*
>
> *Lost and now Found."*

Then the moment came when families were able to go to their soldier's graveside. What a moment that was. Each family of an identified soldier was given the opportunity to have their own service with an Army chaplain at the graveside.

This was a very special moment for us. The woman Army chaplain told us that she had been the chaplain at Alexander's burial. She spoke as if she knew Alexander personally. She said during his burial service, she read the 23rd Psalm and said she would like to read it with us together as a family. She read other scriptures and prayed and said the blessing. It was a very moving and emotional time.

The Army had written to family members of identified soldiers, giving them the opportunity to put together wording on their loved ones headstone. Alexander's head stone was engraved with the words:

ALEXANDER STANLEY

CLINGAN

Son of William & Ruth

Greatly Loved Never Forgotten

As Don's cousin, Robyn Hanna, later said "the ceremony provided some closure on a sad episode for our family. It was just an amazing day and most brilliantly done. We all cried all day, but we also felt like he finally had a home - that he had somewhere at last.

---ooo000ooo---

So, 100 Years On, when you ask this family member of a descendant of a soldier who fought at Fromelles what does Fromelles mean to them 100 years after that disastrous Battle, you receive this response:

The Pain from Him Never Returning: I remember meeting my husband's grandmother, Ethel Mary (Clingan) McDonald and her husband, Hubert Arthur, at their home in Earlwood, NSW. I noticed the name on the family home near the front door, 'Pozieres.' I wondered what the name meant. At that time, I didn't know any details about WW1 and the many Battles that took place in France. It wasn't until many years later, that I understood. Ethel was the sister of Alexander Clingan. I never heard Ethel talk about Alexander, but through her daughter, my husband's mother, I heard of the deep sadness and heartache of Ethel and her siblings and her mother and father, Ruth and William Clingan. They always lived with the hope that Alexander would one day, return home.

Alexander had said goodbye to his family to fight in the War that was raging throughout France. He never returned.

Ruth, never knew her son's fate. The family had been told by the War Office that Alexander was missing. Every day, Ruth would go to the waterfront to meet every troop ship returning to Sydney, hoping to see her beloved son again.

I now understand why Ethel and Herbert McDonald named their home, 'Pozieres.' They believed their son had died in the Battle at Pozieres in France.

Elation from the Identification: Alexander's DNA was one of the first of the soldiers positively identified with family DNA. I will never forget the joy our family felt at this incredibly wonderful news. For so long, we had lived with stories from Don's mother about Alexander. Now our beloved relative, once lost was now found.

Shared Passions: Fromelles has become a passion for myself and Robyn Hanna. Robyn is related to Alexander on the Clingan line. We've both had the privilege of meeting Lambis Englezos and hearing him speak about the discovery of the mass grave. We've heard others speak about Fromelles and both love to talk about Alexander's sacrifice with anyone willing to listen. We have such love and pride for him and honour him for the immense sacrifice he made that fateful day.[214]

I was in Sydney on Tuesday 19 July 2016 and attended the Commemorative Service of the Centenary of The Battle of Fromelles, which took place at the Anzac War Memorial in Hyde Park, Sydney, a very beautiful and meaningful service. During the service, I laid a wreath on behalf of my family, on the steps of the War Memorial. This service was held 100 years to the day since The Battle of Fromelles.

After the service, I travelled by train to Newtown and visited the Honour Roll of the students from the Newtown Superior Public School on the WW1 Memorial outside of Newtown Public School. I stood and looked up and read Alexander's name 'Clingan A.S.' on the France section of the Memorial. Reflecting on the enormous sacrifice of these brave students, I walked around the Newtown Public School. I felt I was walking in Alexander's footsteps.

He Has Not Been Lost: The importance of knowing Alexander's story and now knowing that he has been laid to rest in his own grave, with a beautifully engraved headstone in a the new Military Cemetery at Fromelles, surrounded by the many graves of his mates from the 53rd Battalion, gives our family, not only closure, but peace and much happiness in the knowledge Alexander has been found and has a place of his own that his family can visit. As Robyn Hanna said, "we felt like he finally had a home – that he had somewhere at last."

Through the death of Alexander Stanley Clingan, my family and Robyn Hanna's family, have a greater knowledge of the Battle of Fromelles. We are so proud of all our brave soldiers, their amazing courage, persevering to fight the enemy despite the impossible odds that confronted them. They are indeed heroes.

---

[214] Sadly, Robyn Hanna herself passed away since the material for this book was collated.

We will remember them. Lest we forget.

BOLT DESCENDANT MEETS WITH CLINGAN DESCENDANT

On 24 February, 2018, a memorial service was held for Herbert Thomas Bolt at the Australian War Memorial, Canberra. From that service, Gwenda Stockton and Josie Shelley met for the first time, years after they had become friends. They had realised both their relatives lived only streets from each other, that they went to the same Newtown school, that they both played football for the same Newtown 'Bluebags' football club, that both were killed at Fromelles, and that both are now buried in the Fromelles (Pheasant Wood) Military Cemetery at Fromelles. Their histories had become interwoven, each with the other's, 100 years after their soldiers, their family members, had been believed lost. The names of 'H. Bolt' and 'A. Clingan' are etched in the Memorial Column outside Newtown Public School.

BREGUET, JUSTIN HERCULES[215]

Private Justin Hercules Breguet enlisted at age of 18 on July 16, 1915 in his hometown of Geelong only to die almost exactly one year later on July 19 at the Battle of Fromelles.

The former bread carter turned soldier was one of 250 Australian and British World War One soldiers recovered from a mass burial site at Pheasant Wood in France in 2009 and reburied in the Fromelles Military Cemetery in 2010. He was only identified through DNA matching in May 2016.

[Picture 58: AWM. DA13759. 2017 Pte Allan James Eldridge, of Geelong, Victoria (probably left); 2059 Pte Ernest Charles Hopkinson, of Melbourne, Victoria (originally of Hertfordshire, England) (probably middle); 1983 Pte Justin Hercules Breguet, of Geelong (right)]

Private Breguet was one of six identified ahead of the centenary of the Battle of Fromelles, in 2016. Australia would go on to commemorate the centenary of the bloody Battle of Fromelles on Tuesday but for one relative, Ms. Sharyn Neil, *"It will be his first funeral, it's taken a while. I was so surprised when I was contacted because so many had been identified and Justin's name had never come up so I'd pretty much given up hope that he would be."*

Ms. Neil is one of a handful of descendants that have made the trek to the outskirts of the French village where in 1916 the Australian 5th Division suffered 5533 casualties in less than 24 hours.

215 Sourced from Charles Miranda, Australia Commemorates Centenary of Battle of Fromelles, online, July 18, 2016 2:55pm

Her relative's remains were only confirmed through DNA in May, 2016, allowing him to finally have a proper burial with full military honours.

Private Justin Hercules Breguet was Ms. Neil's grandfather's cousin who enlisted at age 18 on July 16, 1915 in his hometown of Geelong only to die almost exactly one year later on July 19 at the Battle of Fromelles.

Ms. Neil was first contacted in 2008 and asked to research her family tree and identify relatives from whom DNA could be matched. Thousands of descendants of those killed had been contacted after the discovery of the unidentified "lost" mass grave on the farmland.

Private Breguet's siblings had all died young or never had children so the best DNA was distant but it was Ms. Neil's father's cousin who provided a sample in September, 2015.

"It's been a long journey ... the whole family is really happy," she said of the identification.

"What's particularly special for me was my father was an avid family tree researcher and he passed away in 2006 and I know how much this would have meant to him. He knew Justin's story, he had visited Justin's older brother which was his great uncle and talked war stories. The older brother had survived World War One."

"I know dad would absolutely want me here."

---ooo000ooo---

Families have also conducted their own services of remembrance in Australia, even when their relative's remains have not been located. This was the case as recorded in the following report.

CRESSY, HENRY ALFRED

A DIGGER LAID TO REST[216] [with thanks, from Catherine Murray.]

My father, Alby Murray, served in the Air Force with Bomber Command and his proudest achievement came at the end of World War II when his Lancaster crew dropped food into the starving nation of Holland. My reflections on ANZAC Day have grown stronger over the years as I learned more of my family's military contribution.

My father's uncles, Henry Alfred Cressy (Harry) and his brother, Edward, grew up in Boolaroo. There is a treasured family photo taken outside the Cressy house in Boolaroo some time after August 1915 when both boys enlisted. It depicts Brougham (Bro) and Ann Cressy (the boys' parents), and other family members with the two boys standing proudly in uniform.

Australia was a different place then. England was 'the mother country' and when war broke out in 1914 there was no question that Australia would be there. The Cressy boys enlisted, Edward at 24 and Harry at 22.

The Cressy brothers trained at Liverpool and left Australia from Woolloomooloo aboard "Aeneas" on 21 December 1915. Harry Cressy kept a diary with descriptions of his training and

[216] http://mnnews.today/aurora/april-2017/17781-a-digger-laid-to-rest/

the effect of long hot marches, preparing for battles to come. His innocence and youthful enthusiasm come through; he described things that were new and exciting as 'bosker'. As he travelled north to the battlefields he noticed women in mourning, old men guarding the railway and the locomotives they passed on their journey north; these impressed him as a locomotive driver. He wrote, "It was a country worth fighting for."

[Picture 59: Edward Cressy (left) and Henry Alfred (Harry) Cressy (right) with their parents and siblings, at the family home, Creek Road, Boolaroo. From: Catherine Murray]

The boys arrived in Northern France where plans were at the ready for a major offensive. The plan was to attack the 'Sugar-loaf', a high ground near the village of Fromelles held by the Germans for at least 18 months. This would mean that the Germans would not be able to send extra troops to the Somme. Some of our force were seasoned Gallipoli veterans; others, like Harry and Edward, were to have a baptism of fire.

Gallipoli veteran from Mackay, Ronald McInnes, wrote at the battle of Fromelles, "We thought we knew something of the horrors of war, but we were mere recruits, and have had our full education in one day."

The Germans had had many months to prepare. They had comfortable reinforced dugouts, secure communication lines and reinforced concrete bunkers. If you go to Fromelles today, you will see the bunkers and realise that the German line was in a perfect position to see the plains below.

The battle began on 19 July. The 14th Brigade, to which Harry belonged, crossed about 400 yards of open ground to get to the German trenches in broad daylight. The continuous accurate fire of German machine guns, which had survived the allied bombardment earlier in the day, killed many.

Harry made it to the German trenches and some time after that a sniper bullet killed him. We know that Edward was wounded earlier in the battle. This probably saved his life.

The fate of those like Harry who fell behind German lines was not known for another 95 years. There is a poignant record of this time. Harry's parents wrote to the Australian Imperial Force from October 1916 up to 1922. They requested information about where Harry was and asked for Harry's effects, which they knew must be somewhere as they had received a letter from Private Wilson who had been present at Harry's death. His mother wrote that she knew you could have an inscription placed on the headstone, not knowing that Harry's resting place was unknown. They received a memorial plaque for Harry, and his effects finally arrived; a money belt, wallet, bible, cards, photos, letters and diaries which his descendants treasure as the record of an ordinary Australian involved in something greater than we could imagine. His parents were never to know where Harry lay.

Some 95 years later, Melbourne teacher, Lambis Englezos, focused his attention on the lost diggers of Fromelles. Through his tenacity, their remains were found in a mass grave in Pheasant Wood, Fromelles, in 2008. The Australian government launched a project to identify the diggers through DNA matching and Harry was identified in January 2010.

In 2009 my sister and I visited the quiet, peaceful village where Harry lay. In January 2010, soldiers from Britain and Australia reburied with honours the remains of soldiers killed during the battle. In July of that year a new cemetery was dedicated in the village of Fromelles to all our lost diggers. I was there and it was a hot, blue sky day, I imagine very like that day in 1916.

Last year Harry was honoured at Sandgate Cemetery on the 100th anniversary of the battle at the graveside of his parents, Brougham and Ann, with many Cressy descendants present. His inscription in the cemetery in that little village in northern France reads:

HENRY ALFRED CRESSY

Born 5 December 1893

Died 19 July 1916.

In Your Family's hearts

and minds forever.

CUCKSON, WILLIAM JOSEPH[217]

William Joseph Cuckson came from a line of William Cucksons, one of eight orphaned children of a British-born sailor who'd died in a wharf accident. Young William joined the army with his brother Victor in July 1915, leaving big sister Kathleen to look after the younger ones.

[217] Andrew Rule, **Lost and Found in Fromelles**, Sydney Morning Herald 20 July, 2010.

After training in Australia and Egypt, William landed at Marseilles in June, 1916, and reached the front in time to be killed on July 19, his 27th birthday.His family was told he was missing in action. Eight months later, his identification tag arrived from Germany. After that, there was silence for 90 years.

[Picture 60: William Joseph Cuckson standing with his brothers Victor (seated) & Ernie (inset). Photo from Geoffrey Cuckson]

Cuckson's body must have been one of those that Bill Barry had seen tossed into the pit, but no one knew that. In the sad reckoning after the war, his name was etched on a memorial with all the others, but he had no known grave. Victor Cuckson survived the war, married and named a son William in memory of his brother. Generations came and went, and branches of the family spread.

Geoff Cuckson, grandson of William's youngest brother, managed a business in Bairnsdale, East Gippsland. A second cousin he'd never met, (Kathleen Cuckson's granddaughter) Anne Bean, lived in Brighton, widowed matriarch of seven children and their offspring.

Geoff knew nothing of his grandfather's big brother except that he'd been killed and never found. But when a local man, Alan Marchant, wrote a piece in the Bairnsdale Anglican newsletter about names on the church honour board, he told Marchant what little he knew about his long-lost relative. Marchant's maternal grandfather, Sidney Green, had been killed at Fromelles, a fact that had overshadowed his mother's life, he said. Something of an amateur historian, Alan encouraged Geoff Cuckson to research his great uncle's life.

Geoff was hooked. When the mass grave revealed the remains of 250 diggers he registered the family's name with the Army.

When the call came for DNA samples to match against the skeletons, Geoff Cuckson provided a family tree, highlighting the female line because women are the best source of DNA. That's when Anne Bean and her brood in Brighton heard from their distant Gippsland cousin.

They found they had much in common — a growing sense that the missing soldier their elders had told them about would be found at last.

In March 2010, Anne Bean's telephone rang. An army major told her: "We have clear and unequivocal identification of your uncle William Joseph Cuckson."

"The hairs stood up on the back of my neck," Anne said later. "I called Geoff and his sister. We were both crying."

Anne Bean and one of her girls and Geoff Cuckson and his wife went to Fromelles in 2010, now friends as well as relatives, their scattered family pulled together by the man whose bones they've laid to rest at last. In finding him, they've found each other.

GREEN, ROBERT COURTNEY[218]

Robert Courtney Green, spent much of his childhood living in Budby, near Thoresby, England before he moved to Australia in 1908. Desperate and facing destitution because of his drought-ravaged farm outside Fremantle, Robert Courtney Green signed up with the Australian Army in 1916. He was sent straight to the front-line at Fromelles.

He was just 30 years old when he was hit in the chest with shrapnel and bled to death in the trenches on the night of July 19, 1916. Two Comrades, Corporal Allan Bennett and Private Percy Allengame were sent to bring him in, but they both perished.

He left behind his heartbroken fiancée, Nancie Pearson, a 25-year-old Scottish migrant. For years, Nancie penned letters to her missing love. "She writes to him, my Bob, my love, my handsome, I love you so much," his great nephew Mr. John Lockwood from Woolaton said. "When you read her letters, there is just a lot of love in them."

The Germans did return Robert's blood-splattered identity disc, which was to be used to identify him nearly 100 years later.

Robert's body was discovered in a mass grave in 2009, and he was finally laid to rest in 2010 along with 249 other British and Australian soldiers whose bodies were discovered in the same grave.

Due to extraordinary advances in DNA techniques, Robert was able to be identified. His great-nephews John Lockwood, 60, from Wollaton and David Lockwood, 69, from Farndon, both travelled to Fromelles to honour him.

For John Lockwood, 19 July, 2010 would be the day he shed tears for all his family who never spoke about their missing soldier, Robert Courtney Green. "I'm taking two handkerchiefs," he said of mourning his great-uncle.

John said: "It was incredible, very moving and respectful. It really made you appreciate what they did for all of us. At long last my great-uncle has been laid to rest."

By chance, John and David Lockwood found themselves sitting next to the family of Allan Bennett during the ceremony, one of those two soldiers sent to bring Robert in, but who were both killed while doing so. John said: "It was a really nice moment and such an amazing coincidence as well."

[218] See Picture 39 in Chapter 14.

Robert's niece Betty Savage, 95, from Wollaton, also attended the ceremony and was invited to a special reception where she met Prince Charles.

John said: "She's the only one left of that generation and, although she's not in the best of health, she insisted on coming to pay her respects to 'Uncle Bob', as she knew him."

More than 18 of Robert's family attended the event from Nottingham and further afield, and they were pleased he could finally rest in peace.

John said: "It's terrible that he died the way he did but it's fantastic that he's now had a dignified burial."

---ooo000ooo---

What might happen if you decide to investigate your family's possible involvement in The Great War, but your ancestry is more Baltic than English? This is the story of Colin Stalgis' inquisitiveness, and the outcome of his curiosity.

STALGIS, GREGORY FRANCIS [from, and with thanks to, Colin Stalgis]

The family name Stalgis originates from Lithuania according to family history and research my older brother did many years ago. With regards to where Gregory Francis Stalgis came into the story, well back in 2004, I was visiting The Imperial War Museum North during a lunch break when I was working in Manchester. The IWM have a data base located there and I typed in the name Stalgis. Lo and behold up came the details 2891 Cpl Gregory Francis Stalgis. I made a note of this information and began to ask around if anyone in the family had any information about him. At this point I drew a blank.

As he was listed as serving with the Australian Forces, I went ahead and posted a request for information about him on an Australian Web Site dealing with expats and family research and then I almost forgot all about it as I heard nothing for about 3 months.

What I learned about Gregory was that he was born in Newtown, Sydney, New South Wales. His parents were Mr Simon James Stalgis & Mrs Annie Stalgis, of 45 Edgeware Road, Enmore, New South Wales. His youngest sister was listed as Eileen Veronica Rose Stalgis.

He lived at Newtown and attended school at Christian Brothers, Sydney.

He had served for 4 years in the 34th Infantry (Enmore), Citizen Military Forces.

He worked as a Railway Porter, and on 26th July, 1915 at Liverpool, at the age of 20 he enlisted with the 20th Battalion, 6th Reinforcements.

He was allotted to and proceeded to join 56th Bn, Tel el Kebir, 16 February 1916. He was then transferred to 14th Machine Gun Company, 10 March 1916. He was promoted Corporal, 26 March 1916.

He embarked Alexandria to join the British Expeditionary Force, 19 June 1916; disembarked Marseilles, France, 25 June 1916.

He was reported missing, 20 July 1916.

His Disc was handed over by German authorities, and forwarded by Comite International de la Croix Rouge to His Majesty (sic) Minister at Berne.

Red Cross File No 8091731: '[Translation] This certifies that we hold a death card Ref: 111, which states that the Australian Prisoner of War 2898, Pte Gregory Francis Stalgis 14th Machine Gun Co., A.I.F. R.C., died on 20/7/16. Identity disc handed in by Inquiry Officer Army Corps 6, on 17.8.16. Stamp, Royal Prussian War Office, Central Inquiry Bureau.'

Previously reported Missing; he was officially now Killed in Action, 20 July 1916.

About three months after I posted to that Australian website, I received a phone call at home from some chap going on about some place called Fromelles in France and Gregory`s name being on some German burial list?

This man was Lambis Englezos and he was telling me about web pages and where to look and what to research. At first, I thought it was some sort of wind-up because I had posted on an open web site asking for information. But no, he was correct and I began to investigate further.

I became further involved and at Lambis' request I began to push for more involvement from sources [here] in the UK. I was emailing with the BBC, independent TV companies, The Times newspaper, basically anyone whom I thought might listen.

It was during this period I got my first breakthrough when I found out that a chap living in Sydney had a family connection with Gregory; his name is Jason Stalgis and his grandfather was named after Gregory. He said that the family never really knew the truth about what had happened to Gregory.

Late August 2006, I met with Martial Delebarre for the first time and he showed me Pheasant Wood, Fromelles. We viewed the ground where the burial pits were suspected of being located. We put some flowers down for the fallen who we believed were buried there.

Around 2008 after I had put together yet another email about the possible mass grave of WW1 Soldiers at Fromelles and sent it to ITV Channel 4 News, I got a phone call from my brother, Frank, telling me to look at the BBC text news page, it was the one I had sent to Channel 4! I travelled to France on Saturday 5th April and met with Lambis for the very first time on Sunday the 6th April. While we were in Fromelles there was supposed to be a ground non-invasive investigation to try to establish the presence of remains. However, due to the inclement weather this did not take place while I was there and I had to return home. Shortly after this time I received a call from Lambis to update me on the progress and the presence of remains had been established by Dr Tony Pollard and his team. While I was with Lambis for the first time I asked him: "Lambis, what do we do if there is nothing to be found and they are not here"? Lambis replied: "Me and you mate, we go and jump in the River Somme!"

In May 2009, I received a phone call from a reporter at The Times newspaper, Kaya Burgess, asking what exactly was happening and what progress had been made? I filled in the details as I had them to Kaya and he did a report in the Times newspaper, as a result of which interest in the story of the missing soldiers of Fromelles really took off. I was asked by our local TV

Company to give an interview and the story of the Fromelles missing was covered by local as well as national newspapers and appeared on the BBC TV News Channel. At this time both Lambis and Martial Delebarre were keeping me informed of progress.

The Australian Department of Defence wrote to me on a number of occasions asking if I would donate my DNA sample for testing. I wrote back giving them the details and contact information for Mr. Jason Stalgis. Around this time, I was fortunate enough to make contact with another distant relative on Gregory`s sister's side. At first she declined my request to become involved and did not wish to offer her DNA for analysis, however, after much persuading on my part she agreed. Later I heard back from Jason that he had submitted his DNA sample; just as well because I received a letter from DVA that it was male DNA they were looking for.

In January 2010 Jean and I travelled over to Fromelles to be present at the very first burial of the recovered soldiers, in the falling snow, the first of 250. Later in the year I was provided with a list of the first soldiers who had been identified but unfortunately Gregory`s name was not on it.

However, on Saturday 8th May 2010 all that changed. Jean and I were actually away for a few days and for some reason I cannot recall why, we had to return home during the day. During the time we were in the house I received a phone call, I noticed it was an international call but decided to answer it anyway, although I did assume it may be one of those nuisance calls you get. I was wrong fortunately; it was the Fromelles research group informing me that the remains of 2898 Corporal Gregory Francis Stalgis had been formally identified!

After waiting and hoping and praying for this day for more than 4 years I stood there, phone in hand with tears rolling down my face... One story sprang to mind. I heard from a family friend how Gregory's Mother, Annie, refused to believe her son was dead, after all she had heard some time after she got the news he was missing, that he had been seen in a Hospital in England. The fact that Annie would not believe Gregory was dead was borne out by the fact that each evening after the family meal Annie would go and stand at the garden gate and wait for her son to come walking up the road and back home to her. Annie passed away in 1941 never knowing what had happened to her son. As I stood there, phone in hand and tears rolling down my face all I could think was, "*Annie, your boy is finally home*".

He is now buried in Fromelles (Pheasant Wood) Military Cemetery (Plot I, Row D, Grave No 9), France.

---ooo000ooo---

Once Again, 100 Years On, when you ask this descendant of a soldier who fought at Fromelles what does Fromelles mean to them 100 years after that disastrous Battle, you receive this response:

If a complete stranger came up to me and asked me to express why it is so important, what would I say?

For myself personally I feel very much in touch with a person I never met, not just a person but a family member, who served his Country. As a person who also served his Country I feel very strongly that those who do such a thing should never be forgotten. I was one of the fortunate people who although I served in uniform someone was watching over me and I was never called upon to fire a weapon or put my life in danger in the defence of my Country.

There was a strong argument when the remains of the 250 missing soldiers of Fromelles were found, some people wanted them to remain where they were, undisturbed. I myself was of the opinion that they should rightly remain together, but be given a proper Christian burial with the honour, respect and recognition that they all deserved. Due to the logistics this meant the creation of a new Cemetery being constructed. This also in my opinion was right and fitting as it would be a beautiful place and would demonstrate how much these men who laid down their lives were truly cared for and would be remembered by the current generation.

In the case of my relative, Cpl Gregory Francis Stalgis, I know I am one member of the family who is in a position to be fortunate enough to 'be there for him' all these years later. Gregory's mother passed away never really knowing what happened to her son. His name is etched upon her grave in the Cemetery where she rests many miles away from her son. Now that the soldiers have been recovered and Gregory's remains have been formally identified, he has a fit and proper resting place where family such as myself can go and visit him and pay our respects. He is not forgotten.

All of the men and boys who were recovered must never be forgotten, no matter how many years have passed since they laid down their lives, they gave up their future so that we would have one. May they all rest in peace.

THE WILSON BROTHERS: SAMUEL, ERIC AND JIM

[from, and with thanks to Jim Munday]

George Wilson (from Ireland) and his wife Isabella Way (from Dapto) together with seven of their children moved from Majors Creek to Hamilton [now Hibbard, Port Macquarie] in 1886. George worked as an engine driver at the mill. The family lived in one of the old mill houses. A further five children were born in this house, including Ethel, my grandmother, and her brothers Eric, and Jim.

[Picture 61: 3 Wilson Brothers.

Photo: Jim Munday.]

At the age of 14 the boys left school to work with their dad in the village sawmill, the life blood of the village. Now with a couple of bob in their pockets they would head for town on the weekends to play football and enjoy the freedom of youth. Then on 6th August 1914 Australia declared war on Germany.

On the 17th July 1915 the three Wilson brothers, Sam aged 29, Eric aged 19, and Jim aged 18 along with a consent letter from his father in his pocket, presented themselves at Liverpool to join the AIF. Their country was at war and they were answering the call.

For reasons unknown, Sam and Jim sailed for Egypt together, but Eric sailed later on a separate ship.

On the 19th June 1916 the brothers embarked on the Royal George, disembarking in Marseilles on the 28th June 1916, their destination Fromelles. It was considered a quiet area on the western front and the normal process was to acclimatise new recruits in an area such as this - a so called nursery. Even the seasoned Anzacs had no experience in trench warfare in open country. The 53rd battalion arrived at Fromelles on the 10th July. On 13th July Sam wrote his last letter home to his mother:

France, July 13th 1916

Dear Old Mother,

Since I last wrote to you great things have happened. First, we left the land of the Pharaohs, and, later a beautiful trip arrived in France. I could not adequately describe the country in which we are; it is more lovely than I ever dreamed of. As you pass along the railway, there is nothing for miles but vineyards and orchards laden with fruit, and divided from each other by the most beautiful hedges. The houses look nice with their red tiled roofs, and little flower gardens in front. Further on we came to open fields, and for miles far as the eye could see were wheat and oat crops, and mingled with them were red poppies, blue and white corn flowers and white marguerites. At other places there were immense hills and lovely woods. One place in particular took my eye. It was a little chateau built close to the river. Behind it there was a beautiful wood, and along the river bank the most lovely garden I ever saw.

After a good train journey, we were quartered for a week in a little village not far from the firing line. We had a good time there undergoing final preparations and enjoying a good spell.

We are now right in the thick of it, and I can tell you frankly it is not one quarter as bad as I expected. Our boys are splendid lot, and a credit to Australia. We were under fire the other night, and they were as cool as cucumbers, each man at his allotted task like an old soldier.

All the boys are well, and as happy as song birds.

They were all on the adventure of a lifetime. Six days later, just 12 days after setting foot on French soil, Sam and Eric are dead. The Wilson brothers found themselves in the trenches at Fromelles. Nine days later they are in 14 hours of hell at Fromelles. Eric and Sam were reported missing, later confirmed killed in action.

Jim received a gunshot wound to the neck, reported missing before being found a couple of days later. After recovering Jim returned to the front, his records show him being hospitalized frequently during the rest of his service. He left England for home on 25th January1919 on board the Ceramic arriving in Australia on the14th March1919. His mother had died before his return.

In 2008 work started at the exhumation site at Pheasant Wood, Fromelles. Two days later the first remains were found. All the remains had been recovered by September2009.

As the work proceeded individual families registered with the Army and were advised of the probability of relatives being interned there. Eric and Sam Wilson were on the shortlist. The Army and GUARD then moved on to the delicate process of identification of the remains. Along with other soldiers' families, four members of the Munday family gave DNA samples. 96 of the 250 soldiers were identified, Eric and Sam being amongst them.

We then travelled to attend the opening of the cemetery at Fromelles on 19th July 2010. On that day I stood with my two daughters, my sister, son in law and eleven other family members, only three of whom I had met before.

We were at the dedication of the cemetery on a very hot summer's day in Fromelles, France. I had been honoured to have been selected to address the 5,500 people present and thanked all those that had made this day possible. It was a very moving ceremony in particular the burial of the 250th, and Unknown Soldier. Protocols were met and the politicians had their say, however the organizers kept them short, this was after all, about the soldiers and their families.

Later that day the family gathered for reflections at the graves of Eric & Sam buried side by side. What providence this was. Samples were taken from the remains of all soldiers recovered for identification purposes and the remains were then classified and reburied. The remains were then re-buried in sequence of presentation irrespective of rank. Names still unknown at that stage. It is some time after the burial that any sample was matched to a particular soldier. The researchers were amazed when they discovered that the two brothers Sam & Eric had been buried side by side. Truly Amazing. Sam and Eric arrived there together, they fought there together, they died there together, and forever more, they lie there together.

I contacted Martial Delebarre the curator of the Fromelles Museum and nine months later I went back to Fromelles to meet with him and obtain assurance of the artefacts display. There is little doubt in my mind that Martial would be the leading practical authority on The Battle of Fromelles, his knowledge is encyclopaedic. Whilst I was there Martial, a native of Fromelles - in fact his family has been there for 700 years - took me on a tour of the battlefield, a humbling experience indeed. During our walk across the fields I must have displayed a look of incredulity as I tried to take in Martial's narration because he stopped, looked at me and said "You know Jim the Australians were gladiators. They were gladiators."

---ooo000ooo---

Once Again, 100 Years On, when you ask this descendant of a soldier who fought at Fromelles what does Fromelles mean to them 100 years after that disastrous Battle, for Jim Munday the answer is a little different:

If you had asked me this question 10 years ago you would have received a different answer. My response would have focused on the direct impact and all the emotions revisited on my family and me. Whilst that position is still a given today, 4 visits to Fromelles have enlightened me to the suffering of others. What about all the other families of the fallen, and what about the families who call Fromelles home? After my first visit to Fromelles I have acquired a more holistic approach; I have an affinity for the people of Fromelles, a connection through my great uncles and the Battle of Fromelles.

Some Came Home, But They Are Now More A Part Of Fromelles.

OFFE, VICTOR RUDOLPH

[with thanks to Paul Fullston, Victoria Petho, & Nick Miller (SMH)]

Victor Rudolph Offe kept a diary as he went to the Western Front in 1916. The last line in his journal documents the beginning of the Battle of Fromelles. He was a private in the 32nd infantry battalion, embarking on HMAT Geelong from Adelaide on November 18, 1915.

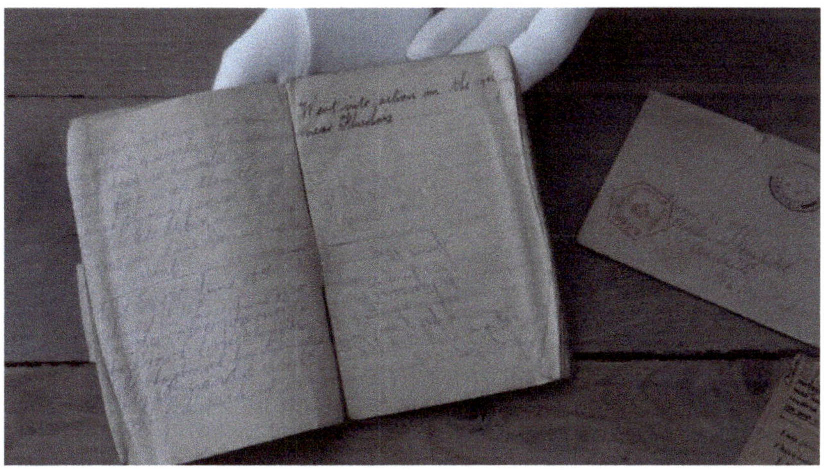

[Picture 62: Diary of Victor Offe. Photo: Nick Miller, Sydney Morning Herald.]

In a letter to his sister he wrote "My word our boys were splendid Mollie, as brave & fearless as you could hardly imagine." It is the calm after the storm. Behind the bold, defiant words was a man shocked by what he saw in the Anzacs' first, horrific major encounter on the Western Front.

For most of his life Victor kept his secrets from his descendants. He didn't want to talk about the war. He didn't want to talk about the time he was buried alive by an artillery blast, and rescued in the nick of time when someone saw the tip of his bayonet wiggling out of the Western Front mud - he only confided that story to his apprentice, who told it to his family after his death. He didn't want to talk about the extraordinary battles he saw, and somehow survived. He was at Armentieres in 1916, on the Somme for six months, then at Bullecourt in 1917 – where he caught trench fever that would invalid him out for a year or more.

And he didn't want to talk about his first fight at Fromelles – or Fleurbaix, as he knew it.

After four days of shelling from the Germans, at 6pm on Wednesday, July 19, 1916, Victor was sent to the front line, following his countrymen who had taken the first two German lines.

"Just as we got up a shell burst right over our heads, one was killed & two others were wounded I luckily got off without a scratch," he wrote.

Night fell as he made his way over No-man's land, "all torn & knocked about by shell fire with masses of barbed wire strewn all over the place".

He stumbled several times on the uneven ground. Almost as soon as he got across, he was told to return to the Australian trenches with a message for the OC.

Victor's notebook tells the tale in silence. This battered little pocketbook has been through the wars, literally. It is torn and stained, but the lines of ink and indelible pencil tell the tale of his enlistment, his travels to Egypt, Marseille and the north of France.

Then, at the top of one page: "Went into action on the 19th July near Fleurbaix," says a dry line, in stark black ink. The rest of the page is blank. The rest of the notebook is blank.

[Picture 63: Paul Fullston & Victoria Petho at Fleurbaix.]

"It's quite poignant," says Victor's granddaughter, Victoria Petho. "He found himself in the firing line when he made that entry. It feels like to me that there was no more time to write – or that was enough and he couldn't write any more. Victor kept a newspaper clipping of an address by Lieut.-Col. F. W. Toll, late commander 31st Battalion, published in Adelaide on 25 July 1919, mounted on a piece of cardboard, as a keepsake from his time at Fleurbaix. The family still have that original clipping.

"He didn't talk about the war, he didn't want to burden the family with that. It was only as an older man he started to tell us stories."

Victoria Petho, 59, and her younger brother Paul Fullston, 52, this week [19 July, 2016] donated the notebook and the letter, found among their grandfather's possessions after his death, to the museum at the Fromelles battlefield. They felt it was right for it to be there, where it would be treasured and displayed, rather than risking it fading and being forgotten in Australia.

They also donated a "Fleurbaix" sign, hammered by Victor onto a piece of metal in the trenches and then later placed, without explanation, above his front door back in Australia for the rest of his life.

Claire Fillassiez, director of the Museum of the Battle of Fromelles, held back the tears as she held the small, tightly-written letter and an old, stained and disintegrating notebook - a newly-discovered physical connection to one Anzac on the Western Front.

[Picture 64: Claire Fillassiez. Picture: Nick Miller, Sydney Morning Herald.]

She said "It is gifts like this that encapsulate what her museum wants to do: to remember the people, the soldiers, not just the cold cut-and-thrust of military tactics on this site.

"And families can fill in the happy endings. It's very emotional," she says. "I cried with Victoria yesterday. And now we have a name, and we have more history. I think families need to *partager* [share], to pass on stories, and this is a good site for it."

"We have more and more history," says Fillassiez, smiling as she reports the end of Victor's war story, and the start of his family's life.

---ooo000ooo---

The families did not forget.

Lest We Forget,

All those who fought at Fromelles!

---ooo000ooo---

20. Stop Perpetuating 100 Years of Myths and Lies

Abraham Lincoln is famously quoted as having said:

You can fool all the people some of the time,

And you can fool some of the people all of the time,

But you can't fool all of the people all of the time.

---ooo000ooo---

And so it is with Fromelles. For almost a century after the Battle of Fromelles had finished, it languished in the recounting of official histories of the Great War, more so in Britain than in Australia, but languish it did. Year after year, Gallipoli featured as the significant defining campaign of World War 1 for Australians. But slowly and surely, the key battles fought on the Western Front did emerge. Pozieres, Villers-Bretonneux, Bullecourt, Hamel, and Mont St. Quentin have all emerged out of the Gallipoli shadow.

The question is: Why didn't Fromelles?

Fromelles emerged into the modern era as a lost child, with limited information available to the general populace, until researchers and authors such as Robin Corfield, Paul Cobb and Peter Barton, and amateur historian Lambis Englezos gathered material together to open up what had been dormant for many decades.

German records of the Battle were also traced and translated, giving an even deeper appreciation of the events than was previously perceived.

But once the general populace became aware of the events of, and the significance of, the Battle of Fromelles, the myths and lies could not be concealed any further.

The First Casualty of War is Truth

In the fifth century BC, Greek dramatist Aeschylus said, "In war, truth is the first casualty." Quite so. Whenever national leaders decide to go on the warpath for the sake of their own ambition or self-aggrandisement, it's the citizenry that will pay the bloody price for their aspirations. Since war is rarely desired by the citizenry, it has to be sold to them. Some form of deception, exaggeration, or outright lies must be put forward to con the populace into getting on board with the idea.[219]

---

[219] [219]From an article by Jeff Thomas: http://www.internationalman.com/articles/the-first-casualty-is-truth

War, after all, represents a monumental failure of national leaders to serve the rightful national objectives of a citizenry – peace and prosperity. Of course, in the case of an *empire* going to war, this represents a monumental failure on steroids – the outcome may well be world war in such a case.

When an empire is nearing the end of its period of domination, war is almost always used by leaders as a last-ditch attempt to maintain order. During wartime, a populace tends to focus more on the war than the failure of its leaders. In addition, they're likely to tolerate the removal of freedoms by their leaders in order to be "patriotic".

This being the case, we might surmise that an empire in decline would be likely to display similar symptoms to a country at war. One of those symptoms might well be the loss of truth, not just as it relates to warfare, but as it relates to the society as a whole. A nation in decline might even *welcome* the disappearance of truth, as it would allow the people to continue to feel good about themselves at a time when a truthful outlook would be too unpleasant to be tolerable. Further, the closer to collapse the country may be (economically, politically, and socially), the more extreme the self-created loss of truth would likely be. [The] abandonment of basic truth in favour of whatever perception, is more pleasant.

When a ship is going down, the very *worst r*eaction is to pretend that everything's fine and that it will all turn out okay. Self-deception does not improve the situation; it exacerbates it.

---ooo000ooo---

Illustrations of these comments were manifestly evident during the Nazi domination of the Germans in World War II, particularly in the comments of the propagandist Joseph Goebbels, such as these: -

- *If you tell a lie long enough, it becomes the truth.*
- *The truth is the greatest enemy of the State.*
- *A lie told once remains a lie but a lie told a thousand times becomes the truth*

---ooo000ooo---

The first lie in relation to Fromelles came in the announcement that it was a raid, and not a Battle, without reference to the reality of the fighting. So, when reading reports from different sources in the preparation of this book, many of these quotes from Goebbels could have been reproduced and relied on to illustrate what had been found in those reports and other sources.

For example:

- *The government does not need war correspondents – it simply tells the people what it thinks will conduce towards winning the war. If truth is good for the war, it tells them the truth; if a lie is likely to win the war, it tells them lies. - Col. W. Tyrell, Allied Chief Military Intell, Gallipoli.*

And in the years that followed World War 1:

- *If the truth were told from the outset, there would be no reason and no will for war. - Arthur Ponsonby, British Politician, 1928.*

Whether or not a reader of the previous chapters, or the works of Corfield, Cobb, Barton and others would consider these comments to be applicable to Fromelles, it is my view that actions taken together illustrate a mindset of avoiding what was not tactically favourable, and being dismissive of the strategic gains made at Fromelles.[220] After 100 years, the myths can be exposed, and lies revealed. If not, they will only be perpetuated.

Myth: Fromelles 1916 Is A Unique Battle Site.

Many Australians still remain unaware that there were earlier attacks at Fromelles in December, 1914, and May, 1915. Or that 'VC Corner' originated as a result of the fighting in December, 1914. But it was here that much of the British Army's offensive action took place in 1915 – at Neuve Chapelle, Festubert and Aubers Ridge in the spring of 1915, and at Loos in the autumn.

And then this sector fell silent, as the fight moved south to the Somme and the Hindenburg Line, or north to Messines. It was to be undisturbed until the German offensive on the Lys in 1918 and the British Advance to Victory in the autumn of that year. Undisturbed, that is, except for the tragic battle of Fromelles in July 1916, almost ignored by British accounts of the war as the Somme battle raged, but of enormous significance to the Australians, representing as it did the first Australian battle on the Western Front.[221]

Fromelles 1916 became an extraordinarily faithful replica of Aubers 1915. What XI Corps elected to do in 1916 was what General Von Scanzoni, the German Commanding Officer of the 6th Bavarian Reserve Infantry Division and his staff had suspected, expected, and planned for. And so it became.[222]

The 1915 battle is seldom assigned more than a page or two in histories of Fromelles, and indeed remains shamefully un-commemorated at the place where it was fought. And while the action in 1915 is seldom assigned more than a few pages in histories of Fromelles, it was recognised with a Battle Honour – AUBERS 1915 - despite being a tactical failure. Fromelles fought in July 1916 received nothing.

Yet it played a monumental role in leaving the bodies of thousands of Diggers, Tommies, Indians and Germans so intertwined in death on the battlefield, that when fighting ceased in 1918 it was frequently impossible to distinguish one from the other.

The attacks of 1914, 1915 and 1916 are as inseparable as were those dead. It is also worth bearing in mind that many who survived both actions were to die long after the events as a direct result of the wounds they had sustained.

Myth: The Tactical .v. Strategic Purpose of Fromelles.

It is often repeated in reports dealing with Fromelles that its object was:

---

[220] Author Paul Cobb for example, believes there were none.

[221] See generally the contents of Chapter 2

[222] Barton, p.396.

- **to prevent the enemy (German troops) from moving southward to take part in the main battle** (on The Somme).[223] But what is usually not stated at this time, is that this object is actually the second listed objective in the order given to the soldiers. The first of the orders given was that:

1. "The XI Corps, assisted by portions of the Second Army, will carry out an offensive operation as early as possible **with a view to seizing the enemy's first line system of trenches** on the front between the Fauquissart-Trivelet road and La Cordonnerie Farm."
And further:

3. "For the period during which the operations last, the XI Corps **will continue to hold the front** as far north as point [map reference] N.10.c.7.8"
Signed: Major General G. de S. Barrow, General Staff, First Army, 15 July, 1916.

If we are to accord Fromelles its proper role in 1916, we cannot specify one stated objective in isolation. Thus Fromelles was set to be a confrontation involving soldiers, to assault the German trenches, and to seize the enemy's first line system of trenches.

As we have already read, a 'raid' is an active assertion with the objective of returning. Fromelles had the objective of active assertion and holding ground. It was an act of conquest. It can never in any realistic manner therefore, be called a 'raid'.

Myth: Superiority Regarding the Opposing Forces.

Only an optimist or a fool would discredit the strength of his opponent in a fight.

But according to two well-reported officers, one would be inclined to believe that there was not going to be any resistance to the soldiers moving into the First Line system of trenches. If the soldiers were to believe their commanding officers, then the artillery barrage which was to precede the assault on the trenches would have wiped out any opposition before the assault began.

Lt Gen Haking had the following note read out to all soldiers on 16 July:

When everything is ready, our guns, consisting of some 350 pieces of all descriptions, and our trench mortars, will commence an intense bombardment of the enemy's front system of trenches. After about half-an-hour's bombardment, the guns will suddenly lengthen range, our infantry will show their bayonets over the parapet, and the enemy, thinking that we are about to assault, will come out of his shelters and man his parapets. The guns will then shorten their range and drive the enemy back into his shelters again. This will be repeated several times. Finally, when we have cut all the wire, destroyed all the enemy's machine gun emplacements, knocked down most of his parapet, killed a large proportion of the enemy, and thoroughly frightened the remainder, our infantry will assault, capture and hold the enemy's support line along the whole front. The objective will be strictly limited to the enemy's support trenches and no more.

---

[223] See generally the contents of Chapter 4 and the Orders given.

The note was nothing less than a guarantee that resistance would be crushed before the assault.

But in the preparations for 'hopping the bags', and while the artillery was still sending shells across the German lines, it was becoming apparent to those in the front that the bombardment was not having anywhere near the promised damage to the enemy front-line defences, because of the inaccuracy of the Australian artillery.

The Germans themselves were responding with shrapnel, trench mortars, and heavy artillery. So much so that soldiers waiting in the Allied third and fourth waves in the reserve trenches were being killed or wounded.

Nevertheless, many of the Australians believed that judging by the noise, the dust and the smoke, their artillery was causing even more damage to the enemy, softening them up for the attack. No less than the much-loved Australian Brigadier Pompey Elliott was one of them. In the front trench with his men, he reassured them:

"Boys, you won't find a German in the trenches when you get there."[224]

When the shelling ceased, the Australians and the British were to find that there were indeed Germans still in their positions, heavily armed, and very much waiting for them.

Lie: Downgrading Fromelles to A Raid And Later to An Attack, And Not a Battle.

Immediately in the aftermath of the withdrawal of the troops to their own lines, the British High Command issued a statement which read:

> *'Yesterday evening, south of Armentieres, we carried out some important raids on a front of two miles in which Australian troops took part. About 140 German prisoners were captured.'*

The British BNC did not look further than the tactical result. It considered that *the rank of "battle" has been confined as a general rule to engagements of primary importance fought out between forces not smaller than the Corps*.[225] It did not properly look at the immediate strategic effect of the battle, or what was by their own definition a Battle.

So, it was an unsuccessful engagement, and not a success, it had to be an Attack. If so, contrary to all the indicia it used, and the explanation of its workings in coming to its classification of the various actions throughout the Great War, the BNC issued an assessment of Fromelles which is completely contrary to the justifications it gives for describing something as a Battle. We need only refer back to the material set out at length in Chapter 12.

In the ultimate analysis, and in an attempt to instill some consistency in defining the various fighting that occurred in the Great War, it is apparent from the background to the BNC Report that: -

1. if you had an engagement that resulted in a win, you had a 'Battle'; but

[224] Lindsay p.93
[225] BNC Report p.7

2. If you had an engagement that resulted in a loss, you had an 'Attack'.
And if you did not have a successful engagement, it is clear from the Battle Honours Committee's discussions, that you would not usually be a candidate for a Battle Honour.
Never mind the strategic effect of the win or the loss. Fromelles was thus cast as a loss, and was officially classified as such. It was never classified as a Battle. Technically, Fromelles was a Battle. As for the 'primary importance' that the BNC considered, it had arrived at the same result as the battle of Aubers in 1915 (a stalemate). But AUBERS 1915 is classified as a Battle with a Battle Honour, whereas 'Fromelles' is not.
Well before the BNC was to begin its deliberations, the soldiers of the 5th Division AIF erected their memorial in 1919, free of any political interference. Being fully aware of what was or what was not a Battle, they proudly listed Fromelles as the first battle in which they participated.

Myth: Fromelles Was Not A Success.

Peter Barton wrote[226] that: - After Fromelles, every commander – from Haig to Haking – made the claim that the sequence of summer actions in French Flanders had achieved the aim of holding German divisions in place. The statements were obligatory. Whether they were true, even in part, remains unclear, given that they were presented in the form of unsupported statement and unassailable claim.

One of those involved in the fighting, was Lt Col. Fred W. Toll, late commander of the 31st Battalion. In a letter published in the Brisbane Courier, Qld, on 18 July, 1919, and later, in the Journal, Adelaide on 25 July, he was vehement in his statement that it was successful. He wrote:

It may enlighten your readers to know that the special reason of the attack at Fleurbaix on July 19, 1916, was to draw off the overwhelming reserves of the enemy, who at this time threatened to smash our lines in the southern area, in which some of our own countrymen were fighting for dear life, that the Fifth Division was thrown into the vortex, with the result that it stopped the enemy's attacks, drew off his reserves, and caused considerable loss to his forces. With this knowledge in front of us at the beginning, also that little or no support would be available, and a very limited amount of reserve ammunition for our guns, can you appreciate why I place on record the Battle of Fleurbaix as being the finest exhibition of bravery of Australian troops in France or Belgium. All honour to our splendid dead. I salute your memory, my comrades.

Victor Offe (Chapter 19) kept a copy of the newspaper clipping of this speech until his death. As outlined in Chapter 7, it is this writer's view that the Battle at Fromelles was in fact strategically successful, if not a tactical success. Various comments are reproduced here: -

- Historian Michael Senior wrote that historians generally judged the attack to have failed to prevent German troops being moved south to the Somme.
- Wilfrid Miles, the British official historian writing that the IX Reserve and the Guard Reserve corps had been sent.

[226] Barton p. 394

- Peter Pedersen wrote that the Germans knew that Fromelles was a decoy and sent the reserves to the Somme.
- The Australian official historian, Charles Bean, wrote that the attack showed the Germans that they were free to withdraw troops.
- In 2007, Paul Cobb wrote that the Germans were not deterred from sending troops to reinforce the Somme.
- Michael Senior wrote that there was evidence that the transfer of troops to the south was delayed.

One aspect of the German forces remaining where they were, was a possibility which does not seem to have been explored by the various commentators referred to above. It may have been possible that the German forces considered they were holding the Somme area sufficiently so as not to require other reserves to be moved into the Somme and away from Fromelles.

Even so, a German intelligence officer of the 6th Bavarian Reserve Division wrote on 20 July, 1916:

There are no signs of any immediate repetition of the enemy attack. However, judging by the general situation, a new push is not impossible.

A Bavarian document discovered in 1923 related that:

An order was captured declaring that the object of the attack was to keep German troops engaged in the sector so as to keep pressure from the Somme...a repetition of these attacks is therefore to be expected.

And Bean wrote in 1930 that the Bavarians might have doubted that the British would sacrifice 7,000 men as a decoy.

But there is also evidence to show that the IX Reserve and Guard Reserve corps had been moved from the Souchez–Vimy area, less than 20 miles away (32 km) from Fromelles, well outside the XI Corps sector. Troops kept in the Loos–Armentières sector opposite the corps for four weeks after 19 July, remained there *as a precaution*.

German records show that eight divisions were between Loos and Armentières on 1 July and that two were sent to the Somme by 2 July, long before the Fromelles attack and the other six divisions stayed opposite XI Corps *for five to nine weeks after 19 July*. Had divisions moved earlier, the Battle of Pozières (23 July – 3 September), might have cost I Anzac Corps far more than 23,000 casualties.

Senior concluded that German troops **had** been retained opposite the XI Corps, because of the attack at Fromelles. On balance, the 2nd objective given to the troops had been achieved i.e.

to prevent the enemy from moving troops southwards, to take part in the main battle.

So while many commentators to this day describe Fromelles as a failure, there is also compelling evidence, available from contemporary accounts, that its objective had been partly successful.

Myth: You Can't Rely on The Tommies in A Fight.

It was a double misfortune that the UK 61st Division's first major action, at Fromelles on the 19th of July 1916, was a bloody failure, and that it involved the first Australian Division to go into battle on the Western Front. The attack and events surrounding the 5th Australian Division, on the left of the 61st Division, remain contentious, with the 61st Division subsequently the subject of criticism from both their own Corps commander and the Australians.

Australian criticism has been levelled at Australian commanders, particularly the Divisional commander, Major General Sir James McCay, Lt Gen Sir Richard Haking, the XI Corps commander, and at the commander of the 1st Army, General Charles Monro. For many Australian soldiers at the time they simply felt let down by a failure of "the Tommies" when they continued the attack and the 61st Division did not. See: Robin S Corfield, *Don't forget me, Cobber. The Battle of Fromelles*, for numerous instances. Sadly, Corfield does not give appropriate treatment of what amounted to poor communication with the 58th Battalion.

As we read in Chapter 6, the serious breakdown of communication led to Elliott continuing to make attempts at crossing No-man's land and onward to the German trenches. At 7.45pm, on receiving word that 15 Brigade was stalled, McCay authorized Elliott to use half of his third battalion, the 58th, to reinforce the line in No-man's land and hopefully carry it forward.

At 8.00pm, at XI Corps headquarters, General Haking ordered a fresh assault to take place in an hour's time. Having received divisional reports on the progress of the battle so far, his intention was to support the limited successes where they can be found. We know now that Haking was under the misapprehension that the Sugar-loaf salient had been dealt more damage than in fact it had. Plans were drawn up to launch an assault against the Sugar-loaf, which would be led by 61st Division to begin at 9.00pm.

But at 8.20 pm Haking was given a first-hand report on the beating the 61st Division had taken. He directed Mckenzie to abort the 9.00 pm attack and ordered that all troops were to be withdrawn after dark.

Whilst the order to delay did eventually make its way to McCay, Elliott who had been preparing his reserves to assault the Sugar-loaf had still not been informed of the change in plan and continued to prepare for the assault. The postponement of the attack failed to reach Elliott and the 15th Brigade ahead of the planned attack.

So at 9.00 pm, Elliott ordered the Australian 15th Brigade reinforcements to advance and assault the eastern face of the Sugar-loaf, with two companies of the 58th Battalion and some survivors of the 59th.

Unknown to Elliott and the 15th Brigade, the soldiers of the British 61st Division were not there to support the charge. They had already been told to withdraw.

Rising from cover, 200 metres from the Sugar-loaf, the 58th and 59th Battalions were immediately met by machine gun fire from the emplacements ahead of them. The result was a slaughter.

By 12.20am, Elliott learnt that the attack on the Sugar-loaf had failed disastrously and relayed the information to McCay. McCay's response was to order the attack abandoned and for the survivors to withdraw to their old positions and form a defensive line against an expected German counterattack.

It was not until 5.10am, that the order for a general retreat was finally given by General Charles Monro, Commander of the British First Army.

---ooo000ooo---

Not knowing the truth of what had happened to the 61st Division in their own attacks, the soldiers of the 15th Brigade, and in turn those from the 14th and the 8th Brigades, would form the belief that you couldn't rely on the Tommies in a fight. It was this criticism which condemned the Division to carry a reputation as a poorly performing fighting unit, even amongst its own men. But it was a criticism which was wholly unjustified.

As we read in Chapter 17, one vocal critic of 'the British' was Fr. Kennedy, Chaplain of the 14th Brigade, who gave a public address in Melbourne, which was published on 22 July, 1920 in The Advocate. In his address he accused the 61st Division of either cowardice or inefficiency:

> *Certain objectives were allotted to us, and also to the 61st. The 5th Division AIF, despite a tremendous barrage of artillery - hell let loose - never failed or quaked. They hopped over like heroes, and stopped not until they gained the second line of the enemy trenches. ... The 61st Division was allotted the Sugar-loaf as an objective, but only as a section got into the enemy trenches, and were compelled to retire. Whether they failed in their objective through cowardice or through inefficiency, does not matter very much to us, but it did matter to the shattered remnants of the 5th Division, which later came up to the cross roads to bury their dead.*

Fr. Kennedy then went on to accuse the 61st of failing to do their job:

> *Not only at Fromelles was it my misfortune, and the misfortune of the troops with whom I served, to be associated with English soldiers who failed to do their job, but right through our career, with very few exceptions, our experience was the same. ... This is not a mere statement from me. I have been associated with tens of thousands of Australians, and I say that during the years I served as chaplain I have not heard more than five Australian soldiers speak other than in terms of contempt and disgust of the English troops with whom they served. (Applause)*

As was stated in Chapter 17, Fr. Kennedy's comments on what happened to the UK 61st Division did not extend to advising his audience of their numbers being less than the Australians, or that the barbed wire in front of them had not been shelled and broken, or that the distance to their objective was much longer than some of the Australian units. Nor did Fr. Kennedy inform his audience that the 15th Brigade did not receive orders to postpone the attack until well after the Australians had left on an abortive charge against the Germans and the Surgarloaf.

Sadly, in the absence of critical examination of the role played by, and the manner in which the 61st Division were to carry out their assignment, it became all too easy to portray the British as being cowardly or inefficient, or having failed to do their job. The reality was that the 61st Division suffered severely and faced the same problems across a deeper front than the Australians.

The rumour had crossed the battlefields before the truth could catch up with it. And it was perpetrated throughout the years since.

Fr. Kennedy's portrayal became yet another spreading of mythology, which is not deserving of repetition other than to dispel it.

---ooo000ooo---

Let's now examine how the 61st Division came to be at Fromelles, and how they performed following their involvement there.

The 61st Division was made up following the decreasing supply of volunteers, which forced the government into the imposition of conscription in early 1916. This made the whole of the Territorial Force, including those who had elected not to go abroad, liable to full deployment.

The 61st Division came into being as the 2nd (South Midland) Division. Many of the Second - Line battalions making up the Division had been in existence since September and October 1914, the soldiers living at home and with little in the way of uniform or equipment, since priority was, understandably, given to equipping and training the First-Line battalions for overseas service. That overseas service would have foreshadowed service on The Somme and other sectors.

The Somme offensive was a joint initiative forced upon the French by the Germans, and upon the British by the consequent needs of the French for the allied army to undertake a "bataille d'usure" elsewhere on the Western Front. This, it was hoped, would reduce the Germans' ability to continue attacking at Verdun, and would also result in the destruction of the German reserves.

In his diary entry for the 29th March Haig, writing of a meeting with Kitchener, says:

"I have not got an Army in France really, but a collection of divisions untrained for the field. (The actual fighting Army will be evolved from them).

On 25th May, under its new General Officer Commanding, Major General Colin John Mckenzie, the 61st Division landed at Le Havre, the first of the Second–Line Divisions to arrive in France and the only one of the Second-Line Divisions that would be committed to battle on the Western Front in 1916.

The 61st Division moved rapidly to the front line once it reached France on the 25th May 1916. Six days later, on the night of the 31st, elements of the Division were in the front line, under instruction, to the left of Neuve Chapelle as part of the XI Corps.

To the south the great battle of attrition on the Somme started on 1st July 1916. The Fromelles operation, originally scheduled for the 17th of July, was designed as a demonstration with limited objectives, to take and hold the German front line in the area in front of the Aubers Ridge, between Fromelles and Aubers.

In 1915 Douglas Haig was GOC, First Army, and his conclusions on the outcome of the Aubers Ridge battle, noted in his diary on the 11th of May, included:

The defences in our front are so carefully (and strongly made), and mutual support with machine guns is so complete, that (in order to demolish them a) long methodical bombardment will be necessary (by heavy artillery [guns and howitzers] before infantry are sent forward to attack).
Whether the planners of First Army or of XI Corps bore these problems in mind in June 1916, as planning went forward for the Battle of Fromelles, is not clear.

Despite Haig's obvious concern in 1915, the supporting gun-fire on 19th July 1916 proved inadequate to destroy all the German front line defences and cut all the wire in front of them. In particular the machine guns were left largely intact and un-neutralised in two key locations, known as the Sugar-loaf and the Wick Salient. As in 1915 carefully sited machine gun fire would prove the decisive factor in British failure.

The defences on the Sugar-loaf, at the boundary between the British and Australian divisions were neither penetrated nor entered.

A second strong point, in front of the right-hand brigade of the 61st Division attack, the Wick Salient, also remained in action and was not entered by the attacking force. Able to enfilade the area attacked by the right hand two British brigades, it had an ideal tactical placement and could produce interlocking and overlapping fire in conjunction with the Sugar-loaf. Without the destruction of at least one, and preferably both, of these emplacements any advance by the five left hand battalions of the 61st Division attack were bound to take severe casualties as long as the German machine gunners could keep their guns in action.

In the event only on the extreme left and right of the whole attack front was any real success gained. In the centre, the front lines were nearly 400 yards apart, giving the Germans ample time to prepare once the preparatory barrage lifted. German riflemen and machine gunners, firing from the top of the breastworks, prevented any attacker who had survived the sustained machine gun fire sweeping No-man's land from entering the German front line.

On the left of the 61st Division attack both the 2/4th Royal Berks and the 2/1st Bucks regiments, which had been in the front line trenches since 9 a.m., had already incurred heavy casualties before the attack started.

Geoffrey Christie-Miller, an officer with the Buckinghamshire Regiment, recalled the battalion's experience:

Our front line was battered about and being packed with troops during this 7-hour counter bombardment (while waiting to attack), *the casualties were heavy and, in our case, estimated at over 100.*

… no effective destructive (sic) or neutralising of the Hun infantry, Artillery or M.G.s took place. The total effect of our artillery preparation on the Hun resistance was Nil.

When the time came … for the assaulting infantry to move out the sally ports were effectively covered by machine gun fire and progress was impossible.

The Berks (2/4th Royal Berkshire, on the immediate left) *lost a lot of men persevering, and few of the Battalion were seen to reach "No-man's land".*

The Australians went over the top of the breastworks. But The 61st Division were ordered to use sally ports, access tunnels broken through the breastworks onto No-man's land. Delays in breaking through and congestion as the engineers tried to get back as the infantry tried to get forward led to delays, while German machine gunning of some sally ports meant they could not be used, hence Haking's comment on the Division not going forward "like one man". The instruction to use sally ports however, was entirely Haking's.

At zero the Battalion advanced. The advance has been described to me as magnificent, not a man was seen to waver. The fire brought to bear was annihilating. Hardly a man if any, reached the German parapet.

This hardly supports Haking's suggestion that the Division lacked 'offensive spirit'.

On the 61st Division's centre sector the 2/4th Battalion, the Gloucestershire Regiment, attacked as the left-hand battalion of 183 Brigade. J D Wyatt, a regular officer from the Northamptonshire Regiment, wounded at Neuve Chapelle in March 1915 and posted to the 2/4th Gloucesters on his recovery in early July 1916 wrote:

Attacking Companies had started to go over the parapet when the bombardment lifted and were met immediately by a very heavy machine gun fire. The leading platoons were practically wiped out at once and no one could get any further. The other Regiments on our flanks got it worse if anything.

Most of the 61st Division deployed, as ordered, through the sally ports. The Australians, in contrast, went "over the top" by climbing over the breastworks, thus avoiding bunching.

183 Brigade would have been in the overlapping zone of fire from machine guns on both the Wick Salient and the Sugar-loaf.

Only on the extreme right of the attack, where the front lines were only 200 yards apart, and the German breastworks severely damaged, did a party of 2/7th Battalion, the Royal Warwickshire Regiment succeed in crossing the German front line and penetrating the German position.

The failure of the 2/6th Battalion, on their left, to penetrate the Wick Salient left them open to infiltration attacks by the defenders on both flanks. Once the Germans had retaken the front line the 2/7th men furthest forward were isolated and surrounded. All were subsequently killed.

These men are almost certainly those buried by the Germans three days later behind the village of Aubers, in what is now the Aubers Commonwealth War Graves cemetery. The first block of 100 graves on the left as the cemetery is entered are almost certainly those of the 2/7th Royal Warwicks killed in and beyond the German front line.

On the extreme left of the 5th Australian Division attack, where the front lines were also close together, the 8th Australian Brigade had succeeded in crossing the German front line only to find no second line trenches to consolidate on, only water filled ditches. With consolidation impossible the Australians on the extreme left, like the British on the extreme right, were vulnerable to flank attack and infiltration while the German domination of No-man's land made reinforcement impractical. Over the next twelve hours the Australians were driven back to their own lines.

The results of the 61st Division's first battle have always been overshadowed by a remark in the XI Corps Commander's after-action report to the GOC First Army, seeking to explain the failure of the attack, Lt Gen. Sir Richard Haking, wrote:

The 61st Division were not sufficiently imbued with the offensive spirit to go in like one man at the appointed time.

This was contained in Haking's XI Corps report to Advanced Headquarters First Army, NA WO 95/3055. It was this comment, which would have been seen by officers in the 5th Australian Division, which caused many Australians to jump to the conclusion that the 61st Division, by failing to take the Sugar-loaf, had also caused the failure of their part of the attack. Australian soldiers on the Western front were not noted for reticence and it may be assumed that the - less than justified - Australian opinion of the 61st Division would become widely known.

Haking's motive for attempting to blame the 61st Division for the failure at Fromelles may well have been informed by the events at Gommecourt surrounding the 46th Division on the 1st of July. The planning of that battle, as at Fromelles, was flawed and the Corps Commander, Lieutenant General Thomas D'Oyly Snow, saw fit to blame the 46th, reporting that "the 46th Division ... showed a lack of offensive spirit", then going on to criticise the Divisional Commander personally. As a result, the Army Commander, Allenby, sacked the 46th Division Commander, Major General Edward James Montagu-Stuart-Wortley, five days later. Haking did not name Mackenzie, but was very possibly seeking to avoid criticism of his planning and conduct of the battle, in a climate where sackings for bad performance led to considerable scope, on the part of senior commanders, to "scapegoat" subordinates.

Haking had, however, begun his report with what might be regarded as a defensive statement:

Briefly speaking, the Artillery preparation was adequate, there were sufficient guns and sufficient ammunition. The wire was properly cut and the assaulting Battalions had a clear run into the enemy's trenches.

He had good reason to be defensive. Not only had he repeatedly given assurances that his artillery preparations were adequate, he had personal experience of what was required of artillery on the same battlefield and certainly should have known that the battle he had promoted was based on flawed planning.

At no point in either of Haking's two after-action reports to Monro, (one on the performance of each of the divisions involved), and sent as covering letters to the Division's own report, do the words "machine guns" occur. This is a remarkable omission given that machine gun fire was probably responsible for the majority of British casualties during the attack, and for its ultimate failure.

A final comment on Haking's verdict against the 61st Division is that of casualty figures. It could be expected that men not intent on carrying through an engagement with the enemy would be intent on saving their own lives. It follows then that, if the men of the 61st Division were reluctant to do their duty they would take fewer casualties than those intent on following orders. This is simply not the case.

The casualty figures for the six attacking battalions of the 61st Division equate to 43% of their strength, three attacking battalions taking between 40-50% casualties and one 60%. The lowest casualties, 27-29% were taken by the three battalions unable to fully deploy forward of the sally ports into No-man's land. These were the three battalions between the Wick Salient and the Sugar-loaf, where machine gun fire prevented troops moving through the sally ports once the British artillery barrage lifted.

On the left the Australian casualties, if those taken as POWs are discounted, amount to about 5,000. The Australians eventually deployed 11 battalions, giving a casualty rate of 56%. No significant numbers of 61st Division men were taken prisoner. A small number may have been taken from the party of 2/7th Warwicks who penetrated the German positions on the left. This can be no more than 20 if the numbers lost and the numbers subsequently buried at Aubers by the Germans are taken into account.[227]

On the Somme the rates vary. Assuming normal battalion battle strength as 800, then 30 of the 203 infantry battalions present on the 1st of July had casualty rates of over 50%. Some were extraordinarily high, such as the 85% of the Newfoundland Regiment. Many were below 50%. Typical are the figures for the 8th (Service) Battalion, Kings Own Yorkshire Light Infantry at 45%, 15th (Service) Battalion, Durham Light Infantry at 44% and the 7th (Service) Battalion, Green Howards at 43%.

Neither the 61st Division, nor the 5th Australian Division, therefore, took 'remarkable' casualties at Fromelles. Instead, they took casualties commensurate with the level of tactical and operational development reached by the British Army in mid-1916.

Nor are the figures for the 61st Division indicative of any lack of resolution or discipline. If they had been then the Divisional Commander, Colin Mackenzie, would have been unlikely to avoid sacking. His survival in post shows that GHQ believed that any fault lay at other levels. Two battalion commanders and the Brigadier of 184 Brigade were sacked. Haking remained at Corps Commander level for the rest of the war and Monro's tenure as an army commander was short, being posted away from operational command after just seven months and posted to India.

There is therefore, compelling material to establish that it was a myth to say 'you could not rely on The Tommies in a fight.'

Official History .v. Official History?

Who Do You Believe? Bean Or Edmonds?

The corollary of *Truth being the first casualty of war*, is that *History is written by those who win*. This usually means that there will be some significant bias in the reporting. So too is the case when looking at the reporting of historians Captain Charles Edwin Woodrow Bean from Australia, and Sir James Edward Edmonds from Britain.

---

[227] See: Paul Cobb, *Fromelles 1916*, Tempus, 2007, p. 122, Account of 855 Sgt R Collett, 32nd Bn, AIF.

Edmonds, compiled material from information passed to him while he was at General Headquarters of the British Expeditionary Force. He wrote at the strategic macro level in broad terms.

Bean wrote his coverage of the war as it unfolded, having been universally recognised for having written and edited at the tactical micro level providing great details, and personal accounts.

Captain Charles Edwin Woodrow Bean won an Australian Journalists Association ballot and became official correspondent to the AIF.[228] He accompanied the first convoy to Egypt, landed at Gallipoli on 25 April 1915 and began to make his name as a tireless, thorough and brave correspondent. He was wounded in August but remained on Gallipoli for most of the campaign, leaving just a few days before the last troops.

He then reported on the Australians on the Western Front where his admiration of the AIF crystallised into a desire to memorialise their sacrifice and achievements. In addition to his journalism, Bean filled hundreds of diaries and notebooks, all with a view to writing a history of the AIF when the war ended. In early 1919 he led a historical mission to Gallipoli before returning to Australia and beginning work on the official history series that would consume the next two decades of his life.

Initially, Bean was a noted supporter of Generals White and Birdwood, but not so much of Monash. He was a fighter for conscription. He was educated in, and by his own writings, a lover of most things British.

He changed his views in some significant manners: he would become a supporter of an egalitarian Australian society; he was an opponent of the class system prevalent in Britain at the time; and he went on to make abundantly clear his lack of faith in the British Command.

Brigadier General Sir James Edward Edmonds CB, CMG (25 December 1861 – 2 August 1956)[229] was a British First World War officer of the Royal Engineers. Edmonds became the Director of the Historical Section of the Committee of Imperial Defence on 1 April 1919 and was responsible for the post-war compilation of the 28-volume History of the Great War. Edmonds wrote nearly half the volumes, including eleven of the 14 volumes dealing with the Western Front (Military Operations, France and Belgium). His task was not completed until the final volume was published in 1949.

Edmonds possessed a considerable intellect and was fluent in many European and Asian languages. In 1896, he entered the Staff College, Camberley, achieving the highest score of his class on the entrance exam. His classmates included Douglas Haig, who became Commander-in-Chief of the British Expeditionary Force in December 1915, during the First World War; Edmund Allenby, who led British forces in the Sinai and Palestine Campaign from 1917 to 1918, and William Robertson, who became Chief of the Imperial General Staff in 1916.

---

[228] See information accessible from Australian War Memorial website.
[229] Information from www.Wikipedia.org.

Edmonds passed the two-year staff course at the top of his class. He was understandably marked for work in military intelligence and was posted to the War Office Intelligence Department in 1899.

At the outbreak of the war in 1914, Edmonds was chief-of-staff of the 4th Division of the Expeditionary Force (renamed British Expeditionary Force on the arrival in France of the Indian Expeditionary Force) but the strain of the Great Retreat following the Battle of Mons (23 August), led to his replacement in September, within a month of the opening of hostilities. He spent the remainder of the war as a staff officer at GHQ of the BEF, during which time he gathered documents to be used in the Official History. Edmonds needed to demonstrate great diplomacy to obtain information. He told Bean, his Australian counterpart, that:

"I was on terms of friendship with all the British generals from Haig downwards. I never belonged to any party and since I was not competing for promotion, I enjoyed confidences I otherwise might not have had."

When comparing the manner in which each historian commented on Fromelles, two distinctly different approaches resulted.

Right from the beginning the British called the action at Fromelles, **an Attack.**[230] They never regarded the event as a battle and none of the regiments involved have it as a battle honour.[231] *The British Official History 1916* (compiled by Capt. Wilfred Miles: preface by Brig. General Sir James Edmonds) Chapter V has the title 'The Subsidiary Attack at Fromelles' which downgrades it even further, and extraordinarily there is no Fromelles or Fleurbaix in the index. And as it was the last official account written, this relegation can only be a direct comment by Edmonds on Bean's elevation of the event to a battle. The description occupies a mere 16 pages plus a full-page map and was first published in 1938, nine years after Bean, and eighteen years after Ellis: and the first Bavarian Regiment histories pre-dated the British effort by fifteen years, the last one by six. This should have surely meant that the British account would have been the last word, the final chance to deal properly with Fromelles.

Instead, we note considerable reliance on Bean and the Bavarian historians rather than digging deep into the British files, and that is not surprising when one examines the style and content of the British history. Unlike Bean it makes no judgments on people or events, and does not deal in individuals.

Bean for his part, supervised or wrote the Official History of Australia in the War of 1914-1918, spanning 12 Volumes. Chapters XII and XIII in Volume 3 deal exclusively with him having no hesitation in describing what was ***the Battle of Fromelles***.

---ooo000ooo---

[230] Corfield p.33

[231] Although McKenzie did in fact seek this, but it was not granted.

In reviewing the Official History of Australia in the War of 1914-1918 – Volume III Introduction, Historian P.A. Pedersen of the Australian War Memorial, Canberra wrote that:

The British War Historian, Sir James Edmonds, regarded Bean's summing up of the campaign as one of his finest chapters but held that his discussion of Haig's strategy dealt inadequately with its effect on the German Army.

Edmonds contended that as manoeuvre was impossible on the Western Front, there was no alternative to battles of attrition which wore down the opponent. On this ground Edmonds justified the policy pursued, asking Bean: "Would you say what the strategy of the Allies should have been? In a long war with sides very evenly matched… nothing but the slaughter of its men will bring one side to its knees".

Hence it was misleading to depict the battle as a struggle for "bits of ground" as Bean had done. Edmonds hoped to prove in his history that the Somme was the turning point of the war. Most of the best Germans were killed "but this could not be done without losing the flower of our manhood". He evidently assumed such a victory would be worth the sacrifice.

However, the whole question of Haig's intentions on the Somme was enmeshed with controversy. His apologists insisted that Haig had conceived the offensive as a campaign of attrition from the outset and vigorously denied that he had sought a breakthrough at all.

Their version was denounced by Liddell Hart as "one of the most elaborate perversions of historical truth that has been brought to light". Edmonds, too, emphasized the attritional aspects of the battle, although he did not actually deny that Haig aimed at a breakthrough.

Bean was in no doubt of the truth, nor did he attempt to mask it in any way. He annotated Edmonds' comments:

He [Haig] was a believer in attrition but not in attrition which loses two-and-one-half men to every one of the enemy. Moreover every operation order shows that he tried to break through. He put it to the government again and again that this break was close ahead and that he was wearing down the Germans faster than they him.

Again, The History of the Great War (1922–1948) produced by Edmonds has been criticised as propaganda, for being too lenient on the British generals.[232] It has been suggested that Edmonds' favourable portrayal of Haig was a counterpoint to the scathing criticism delivered by the former British Prime Minister, David Lloyd George in his memoirs. In 1991, Denis Winter, a critic of Haig, acknowledged Edmonds' comprehensive understanding of British operations during the war but said,

Only a profoundly knowledgeable man could have produced an Official History so misleading and yet with that ring of plausibility which has led to a general acceptance for so long.[233]

---ooo000ooo---

[232] See: Wikipedia.org generally, referencing 'Sir John Edmonds'.

[233] Michael Senior: Haig's Command: A Reassessment p.273

The two differing approaches to the recording of the history of Fromelles, and World War 1, can be reconciled to some degree. Bean and Edmonds were both historians, writing from very different national perspectives, and in different styles. They were like 'apples and oranges'. But while apples and oranges are both fruits, they are very different. Both their accounts are necessary, and ultimately, both played a vital part in assembling the history of events which made up the Great War, of which Fromelles was a part. We therefore need to read both, not one as opposed to the other.

Myth: We Should Preserve The Status Quo.

Fromelles has remained singularly absent from memorials which proclaim their homage to the fallen by reference to location. But there are some modern signs of change.

As we saw in Chapter 18, the process of honouring those who served in the Great War resulted in different treatment in different cities. In some instances, there were no mentions of specific battle engagements. In others there were. Some of the memorials combine places of battles with Battle Honours. It is doubtful whether the uninformed populace would be able to distinguish the difference if there are inscriptions on a remembrance monument.

The Principal Battle Areas inscribed in the France and Belgium Niche Wall 'Hall of Memory' ANZAC Memorial Sydney NSW are:

SOMME 1916	**BULLECOURT**	**MESSINES**	**YPRES**
AMIENS	**SOMME 1918**	**HINDENBURG LINE**	

In comparison, the Battle Names honoured on the 5th Division Memorial, Polygon Wood, Belgium, are:

FROMELLES (July – October 1916)	**BAPAUME, BEAUMETZ, DOIGNES, LOUVERVAL** (March – April 1917)
BULLECOURT (May 1917)	**3RD BATTLE OF YPRES, POLYGON WOOD, BROODSEINDE** (Sept - Oct 1917)
VILLERS-BRETONNEUX (April 1918)	**MORLANCOURT** (May – July 1918)
BATTLE OF AMIENS, 8th August, 1918	**VAUVILLERS** (August 1918)
PERONNE (September 1918)	**HINDENBURG LINE, JONCOURT** (October 1918)

There is little, if any, information available to show what steps have been take up until today, to have 'Fromelles' inscribed on those memorials where places of battles or Battle Honours are made prominent. The Anzac Memorial Sydney (AMS) was the subject of the presentation of an appeal to the Trustees to have 'FROMELLES' inscribed on the Western Front Niche wall, on 22 September, 2009. It was made by the Families and Friends of the First A.I.F.

The Trustees' case in refusing to have 'FROMELLES' so inscribed, and the case in rebuttal of those points are set out below:

1. ***The Building was handed down to them from the Original Trustees, has heritage value and is accordingly unalterable.***
Rebuttal:

(a) Australian 'Achievement and Sacrifice' are central to what an Australian Anzac Memorial should honour.

(b) The Anzac Memorial Building is charged by the Statute creating it to honour 'Achievement' of the AIF.

(c) 'Achievement' in military battles involves 'Sacrifice'.

(d) There was both extraordinary 'Achievement' and massive 'Sacrifice' at FROMELLES.

(e) The Battles Nomenclature Committee was concerned with an overall British High Command view. Its classifications were not concerned with or bound by Australian criteria for inscription.

(f) Dr Bean and the Original Trustees, who apparently took guidance from his advice as to appropriateness for Australian 'Achievement and Sacrifice', were in serious error in looking to the BNC classification to satisfy the obligation imposed by the Statute.

Comment: The inheritance/ heritage proposition is that the Building handed down to the Trustees from the Original Trustees has heritage value and is accordingly unalterable. Therefore, inscription of an additional battle such as FROMELLES would not be permissible, regardless of the merits of such inscription.

Were this reason to be sustained it would strongly favour the status-quo.

The building has been subject already to some alterations from the original but, more importantly, it has been contended that is in breach of its statutory obligation by not explicitly recognising Fromelles and is therefore "incomplete".

The FFFAIF's submission since is that until that breach is remedied the Trustees still have an obligation to complete the recognition of the principal battle areas and the Memorial does not have full and statute-conforming Heritage value.

2. ***The Original Trustees sought and were offered guidance as to inscriptions by Official Historian Dr CEW Bean who suggested they refer to the Battles Nomenclature Committee 1921 (BNC) whose report classifying major British Empire engagements in the Great War, was approved by the Army Council to King George V in 1922.***

Comment: While there does not appear to be a discrete rebuttal of this point, from material compiled by Historian Mr. Anthony Clancy (see next point), there is the following statement: -

The files held by the RSL concerning the ANZAC Memorial do not contain any information relating to the methodology or decision-making process relating to the choice of the inscriptions. In the Minutes of the Trustees' Meeting of 15 October, 1934 the following was listed:

40/34 Inscriptions

The Honorary Secretary submitted a report regarding the action taken by the Building Sub-Committee connected with the preparation of inscriptions to be incorporated into the Memorial. The matter was now in the hands of a special Sub-Committee consisting of Sir Mungo MacCullum, Dr. C.E.W. Bean and Mr. H.M. Green.

The former Honorary Architect, Bernard Connell, has confirmed verbally that he has a copy of these and/or additional minutes relating to the Building Committee and has promised to provide extracts. Verbally, he reported The Building Sub-Committee relied on the Official War Historian, C.W. Bean, and that Bean chose those names he considered most appropriate to New South Wales.

The four niches in Hall of Memory contain the names of "the most important battles in which the A.I.F. participated." It is therefore something of a conundrum that 'Fromelles' was not included given its treatment by Dr. Bean in his Chapters XII and XIII in Volume 3 of the Official War History.]

3. ***In refusing to remedy the non-recognition of FROMELLES by way of inscription within the Memorial the Trustees, relying in part on advice from Historian Mr Anthony Clancy MA (Hons), defended their decisions by arguing that:***
Consistent with the BNC classifying FROMELLES as part of The Somme Offensive (1 July – 18 November 1916) FROMELLES is honoured by the inscription SOMME 1916.

Rebuttal:

(a) This point i.e. that FROMELLES was actually honoured by the Inscription SOMME 1916 is really the crux of the matter. For if it were true the statutory obligation would have been discharged. But, it is not true.
(b) This advice may have misled the Trustees into being satisfied that all major battle areas involving Australians could be related to an in inscription. Sadly, the Trustees have been wrongly advised.
(c) The fact is that the 5th Australian Division was engaged in front line action at Flers in the final phase of the Somme Offensive as part of the Fourth Army commanded by Sir Henry Rawlinson from 20 October to 4 November 1916 and suffered casualties.
(d) Those engaged at Flers on The Somme, are honoured by the SOMME 1916 inscription.
(e) Those who fell or were put permanently out of action in July, at FROMELLES, some 80 kilometres to the north as part of the First Army, are not and could not possibly be honoured by that inscription. This was separate and distinct from the battle at FROMELLES. They were separate and distinct both geographically and in Army command structures.

4. ***Consistent with BNC classification and Wartime Cover-up, there never was a Battle of FROMELLES.***

This defence has since been abandoned by the Trustees.

5. ***A wreath placed annually in the France & Belgium Niche Wall of the Hall of Memory suffices.***

Rebuttal:

That the 5th Australian Division Association on its disbandment had to beg the Trustees to honour FROMELLES annually is abundant evidence that these Veterans knew that FROMELLES was NOT honoured by the SOMME 1916 inscription.

6. ***A successful plea for FROMELLES would lead to a plethora of such claims.***

Rebuttal:

This claim is unfounded:

(a) All claims for inscription are to be considered on their merit.

(b) Veterans of the Battle of POZIERES made no such plea. POZIERES is in the Somme arena and therefore honoured by the SOMME 1916 inscription.

(c) Representatives of only 1 of the 5 Australian Divisions of the First AIF have requested that the list of battle areas be revised. There are no known claims coming from representatives of 80% of the First AIF. Recently [in 2009] a search has been made, following a suggestion by a FFFAIF member that the representatives of the 5th Division AIF were happy with the current listing. No evidence was found to support this and on the contrary, the 5th Australian Division Association representatives, on its disbandment, were obviously aggrieved by the current listing when they requested that the Trustees honour FROMELLES annually.

(d) Of the 47 Battle Names incorporated in the Hyde Park Corner Australian War Memorial in London, opened in 2003, 10 relate the France – Belgium Theatre of War. 9 of the 10 clearly relate to the Battle Areas inscribed in the ANZAC Memorial niche – FROMELLES is the *only* Western Front Battle Name listed in London that is not recognised by inscription in the ANZAC Memorial niche.

Amiens (Amiens)	Passchendaele (Ypres)
Bullecourt (Bullecourt)	Polygon Wood (Ypres)
Fromelles	Pozieres (Somme 1916)
Messines (Messines)	Villers-Bretonneux (Somme 1918)
Mont St Quentin (Hindenburg Line)	Ypres (Ypres)

These Battle Names have been sanctioned by the Australian Government and its Department of Veteran's Affairs, so any new claims would need to demonstrate that 'Achievement and Sacrifice' would be comparable to these 10. So it is most unlikely that any new claims will be found of as much historical significance or as meritorious as to 'Achievement and Sacrifice' as FROMELLES. Clearly, the Australian Government and its Department of Veteran's Affairs has come to the view that the 'Achievement and Sacrifice' as FROMELLES ranks alongside or higher than the other 9 Battle Names listed.

(e) FROMELLES is proposed for inclusion on the new Great War Memorial proposed for ANZAC Parade, Canberra and is already listed on the 5th Division's battlefield monument at Polygon Wood.
(Note: the aforementioned Great War Memorial proposal has since been scrapped.)

Comment: The Office of Australian War Graves (DVA) consulted the 3 ADF Services. Chief of Army represented by Roger Lee, Head, Army History Unit nominated the most significant battle names for the Australian Army for WW1 and WW2. Fromelles was included.

The memorial, opened on Remembrance Day 2003, contains over 24,000 place names of enlistment and 47 Battle names are superimposed.

Thanks to individual persistence against official and other obstacles, millions of Australians through the discoveries at Pheasant Wood are now aware of the tragedy of Fromelles. As they gradually discover that 'FROMELLES' is missing from our principal Memorials they will be seeking to discover 'Why?'

The Trustees of the ANZAC Memorial Sydney in 2009 were: -

- The Premier of NSW – The Hon Nathan Rees MP
- The Opposition Leader – The Hon Barry O'Farrell MP
- The Lord Mayor of Sydney – The Hon Clover Moore MP
- The Public Trustee of NSW – Mr Peter Whitehead
- The President of the RSL (NSW Branch) – Mr Don Rowe OAM
- The President of the T.B. Sailors, Soldiers and Airmen's Association of NSW (Inc). – Mr Stan Poulsen.

The Case for inclusion of 'Fromelles' was presented by Mr. Russell Curley, President of the FFFAIF, and Mr. Neville Kidd, a self-published amateur historian who published a biographical history of Major Roy Harrison, 54th Battalion.

---ooo000ooo---

Between 2009 and 2013, further activity was undertaken by FFFAIF and Mr. Kidd. This did not result in any change to the outcome of 22 September, 2009. But of interest are the views expressed by the Trustees of other memorials for not inscribing 'Fromelles' on their memorials.

The National and State memorials that honour by inscription which are of interest are: The Australian National Memorial at Villers-Bretonneux (ANM), The Shrine of Remembrance, Melbourne (SRM) The Shrine of Remembrance, Brisbane (SRB) and the Anzac Memorial,

Sydney (AMS). The response from the Trustees of the Anzac Memorial Sydney were essentially as stated on 22 September, 2009. As for the other memorials Trustees: -

ANM: - The Minister for Veterans' Affairs responding in relation to the Australian National Monument at Villers-Bretonneux cited the following reasons: -

- Recognition within a wider honour: Fromelles was shown as a subsidiary Action to the Battles of the Somme. All twelve infantry battalions that fought at Fromelles were awarded the Battle Honour 'Somme 1916'.
  Comment: The 5th Division was moved to The Somme in October 1916. It is because of its service on The Somme at this time that its 12 Battalions qualified for the Battle Honour 'SOMME 1916', not because of its attack on Fromelles.
- Respect for decision makers at the time: Those who took part in the selection process were veterans of the Great War and the decisions they made should be respected. This is the principal reason the Government would not support any request to add the word Fromelles to the ANM.
- Aesthetics: There are secondary aesthetic reasons as well. The memorial was designed with the existing honours in mind and any addition would look out of place.
- Other recognition: Fromelles Memorial Park and 'Cobber' statue; In Victoria, Cobber's replica; Australian War Memorial in London; VC Corner Cemetery. Fromelles is the only battle where those Australians with no known grave have such a memorial at the battlefield; a new museum at Fromelles.

SRM: - The Trustees of the Melbourne Shrine of Remembrance: did not respond to correspondence

SRB: - Premier's Office and RSL Queensland cited:

- Battles of WWI on the Shrine have been grouped together in broad "campaigns". In this context, Fromelles is included under the inscription for the Hindenburg Line.
- We now believe Fromelles is included as part of Pozieres as it was the feint attack for Pozieres. Fromelles and Pozieres were part of the one battle.

  Comment: This response beggars belief. Two separate fields of war. Two separate battles. Part of one overall Campaign or Offensive, but in no way could Fromelles be included as part of Pozieres.

- There are many other notable battles that were excluded from the Shrine. Passchendaele was to be removed from the Shrine because there was not enough space to accommodate the name.
- It is believed that we should respect the past endeavours and the decisions that were made.

---ooo000ooo---

On 25 July, 2016: MP Mr. Paul Fletcher, in email to Neville Kidd stated:
"The Government's position continues to be that as the inscriptions of the Australian National Memorial reflect the decision-making processes of the responsible WW1 Commonwealth and military authorities, those decisions should continue to be respected."

---ooo000ooo---

In spite of the past defence of the issues raised by the FFFAIF, there is now some change of attitude at least on the part of the Trustees of AMS.
In the planning of the extensions of the AMS, incorporating a Hall of Service on the Liverpool Street side of the Memorial, a list of the *'100 most significant sites of NSW service spanning the State's military history'* will be created. FROMELLES is now one of them.

Myths, Lies, and the Evolution of Our History.

One of the many legacies of the First World War was the duty recognized by its several belligerent nations to provide both the participants and those who waited for word of success or failure, life or death, with an authoritative written context by which the events and final outcomes might be adequately understood. That context – which we call history – was also integral to grieving, remembering, healing and commemorating. But its creation required careful selection and construction. The primary vehicle was the *Official History*, a necessarily summarized version of events. The Official History was supplemented by a second influence: the slightly more personal and frequently more detailed also abridged unit histories.[234]

With a deal of military and political management, the catastrophe of Fromelles was played both up and down by the third and, from the public's point of view, most important agency, the press. Charles Bean first as a working journalist and then as the author of the Official History of Australia oin the Great War, perhaps did his best to portray the truth by injecting as much accuracy as the censors allowed, or indeed that he permitted himself, for he was already inextricably absorbed in the development of a concept that he and others had begun in Gallipoli: the Anzac legend. Giving birth to such a legend complicated his task because, once begun, myths under construction require not only continual supplementation, but protection. It was far from easy, for in the final analysis history and legend do not make comfortable bedfellows.

On 24 July, 1916, in a piece for the London *Times* and the *Daily Mail*, Bean labelled Fromelles as an 'attack' that had resulted in 'severe' casualties. The powers that be were not pleased, for the British official *communique* described it as, 'some important raids on a front of two miles in which Australian troops took part'. Raids, … not attacks. The Germans translated and circulated both the *communique* and the article, themselves noticing the contrast. To Digger and Tommy, this was a distinction of considerable significance: their 'battle' had been downgraded, and yet all the participants looked upon the clash as something far greater than the word 'raid' could ever articulate.

[234] Barton p.381.

The clash between those who are involved in creating the events which make the history, and those who analyse the circumstances under which the events took place, has been an ongoing process, and inevitably, a source of conflict.

Embedding of Myths

On one hand, we have those who give rise in the telling of events to what become widely understood truths, but which are in all reality, myths. Many parties have a vested interest in shaping the way we remember the Great War ahead of its centenary, but some are more equal than others.

The events surrounding Gallipoli for many decades exceeded the extent of remembrance for the rest of the areas at which Australians served in World War 1. Gallipoli was embedded into our national psyche. We came to idealise the bravery of our soldiers at Gallipoli, in order to exemplify idealized virtues that lie at the heart of our self-identity.

Thus images of war came to dominate conceptions of national identity. But many myths were continued in doing so, many falsehoods, a process which while not harmless is not seen as being harmful. Our beliefs become distorted and the truth is obscured for invalid reasons.

The centenary of the Great War resulted in a wave of events, reports, films, exhibits, books, television programs, research projects and many other contemporary forms of First World War commemoration. The existing myths were aided as never before by blogs, Wikipedia, Anzac supplements in the weekend papers, and bestselling popular histories not always based on archival research. There is no harm in commemorating loss. The real danger lies in having commemoration for valid reasons given almost carnival-like quasi celebrations.

Funding the Continuation of Myths

With the creation of myths, we need to ensure that those myths have some foundation.

Former Prime Minister Tony Abbott declared at Villers-Bretonneux that “no place on earth has been more densely sown with Australian sacrifice than these fields in France”. Australian attention focused again on the battlefields of Europe as a source of national inspiration.

Abbott pledged that Villers-Bretonneux would host an interactive museum, the centrepiece of the Australian Remembrance Trail stretching from Ypres in Belgium to Peronne in France, by 2018 as part of an estimated $AUS140 million in government funding allocated to the Anzac centenary.

This investment showed the importance the Australian government attaches to making the Great War accessible and comprehensible, linking Australians to people and places on the other side of the world. So, (in 2014) as we approach the centenary of the outbreak of the First World War, who or what is shaping the way we remember it?

The example of the Australian Remembrance Trail reminds us that even when governments seek to do less and ask citizens to do more “lifting”, a compelling national narrative is not something they can do without.

Many parties have a vested interest in shaping the way we remember the Great War, but some are more equal than others. Australia, and Australian bodies have been found to be the most 'aggressive' of the centenary commemorators. Aggressive in that they have spent in the vicinity of half a billion dollars on the commemoration.

That price tag was calculated by former Australian Army officer and UNSW alumnus James Brown in his book *'Anzac's Long Shadow'*. Brown, who has tours of duty in Afghanistan and Iraq, and is a Military Fellow at the Lowy Institute for International Policy, is a vocal critic of the Australian approach to the centenary. Brown estimated $325 million was being forked out by the Australian taxpayer for the string of "festivals". Add more than $300 million expected in private donations and what we will have, Brown predicted, was an Anzac centenary that risks fetishising war.

Unlike other countries in the Great War, Australia was never under serious threat of invasion, Military Historian Jeffrey Grey (UNSW Canberra) says. Even the number of Australians killed – around 60,000 – represents a tragic but small figure, compared with the 600,000 British deaths, more than one million Austro-Hungarian fatalities and almost 1.4 million French lives lost.

In commemoration of the centenary, Australia is not only spending more than twice what Britain is, we're even outspending the French. "And the French, you might think, have more reason than most to remember the Great War," Grey says.

By way of contrast: "At the War Memorial in Sydney's Hyde Park, inscribed words decree: 'Let silent contemplation be your offering'," Brown writes in his book. "Instead, Australians are embarking on a discordant, lengthy and exorbitant four-year festival for the dead."

Setting The Record Straight

We read previously of having found an article in the National Archives of the United Kingdom which had been classified and not released for 70 years, the letters of Maj. General Headlam to the Battle Honours Committee. Its non-disclosure led to a myth in relation to the selection process for Battle Honours.

One of our most cherished and enduring myths is the idea that Australia's military history, and by association our national identity, began at Gallipoli in 1915, despite Australia's military involvement in conflict extending at least to the 1899 Boer War.

Another myth is that the Australian Imperial Force was the only all-volunteer army in WWI (it wasn't), that its volunteer status made Australian soldiers inherently superior to their conscripted counterparts (there is no evidence their skills were inherent), and that Australian soldiers had higher ethics and morality (they demonstrably didn't).

For many Australians, prior to 1915, Australian identity was largely centred on the myth of the pioneering bushman. Thanks to Gallipoli, that image was replaced by the bronzed warrior – the ultimate fighter and larrikin prepared to die for his mates. But look more closely and a different picture emerges. It is a picture of ordinary men coping as well as they could in terrifying conditions.

The ordinariness of the Diggers is reflected in records that show the Australian Imperial Force had a higher rate of desertion than any other force on the Western front and that one in 10 Australian soldiers had some form of venereal disease. The honest accounts of the soldiers' lives are included in Professor Peter Stanley's book '*Bad Characters: Sex, Crime, Mutiny, Murder and the Australian Imperial Force*', which was jointly awarded the Prime Minister's Prize for Australian History in 2011.

"That book was made feasible by the release of the AIF's court martial files," Stanley says. Getting the tone right was critical. "I needed to be honest – this story had largely not been told – but also sympathetic to the situation of young men who in many cases suffered death and wounds in circumstances that none of us experienced. I remembered Madame de Stael's aphorism 'to know all is to forgive all', and that became the book's epigram."

The reviews of the book demonstrated the validity of this approach. "To my secret disappointment no one burned the book, and it sold reasonably well," Stanley says.

Despite the warts-and-all stories, one of the things that emerged from the historical accounts was that these ordinary people, when they entered battles and performed their duties in war zones, had a superb reputation.

"New Zealand soldiers were good. Canadians were good. Many Brit divisions were good but Australian divisions were very good and remarkably consistent," Stanley says. "Were they among the best fighting forces? Everybody at the time seems to agree they were."

Nevertheless, it is essential to continue to make the distinction between historical inquiry and mythology. The military is a reflection of Australian society and an agent of government policy, Jeffrey Grey stresses.

"For the vast majority of Australians their understanding of war and of our defence forces is refracted through their knowledge of the Great War. If that understanding is partial at best, by the time it is applied to today's defence force – one that bears no relation to the army of 1914 – it is going to be so skewed as to be deeply misleading."

Associate Professor Craig Stockings agrees: "It's one thing to believe in the idea of an invincible, seven-foot-tall Anzac soldier. It's another thing to understand these guys as normal people in harrowing circumstances who still achieved amazing things.

"They were human beings with human frailties. That's the reality; it's evidence from the source. Their legacy is far greater in truth than it is in myth," Stockings says.

The Backlash

On the other hand, we have the backlash from those who hold the myths comfortably close.

Attempts to set the historical record straight have attracted criticism and angered some.

Former James Cook University academic Dr Mervyn Bendle, writing in the conservative journal *Quadrant* distilled the criticism, accusing the UNSW historians and the Lowy Institute's James Brown, of declaring "a war on the Anzac legend", an accusation also seized on by *The Australian*.

According to Bendle, they have embarked on an "elitist project explicitly dedicated to destroying the popular view of these traditions". Ensconced in elite institutions (located mostly in Canberra, he adds), they exhibit a disdain for ordinary Australians and their beliefs.

"[Australians] should be allowed to honour the centenary without constant sniping from an anti-Anzac elite of obsessive academic leftists and disgruntled ex-officers," he concludes.

It's a cheap criticism but not entirely unexpected. After all, no one likes to have closely held beliefs challenged.

Who Owns the Myths and Lies?

As we move into the second century after the centenary of the outbreak of the First World War, who or what is shaping the way we remember it?

In a recently released book, *Nation, Memory And Great War Commemoration: Mobilizing The Past In Europe, Australia And New Zealand* [co-edited by Ben Wellings, and Shanti Sumartojo], contributors discussed the politics shaping First World War commemorations in the United Kingdom, New Zealand, France, Germany and Belgium in order to better understand Australia's commemorations in a comparative framework.[235]

The following points emerged:

1. These commemorations are global in nature. Similar pressures to create new narratives linking globalised states and national citizens generated demands for commemorative activity across the developed world in the past 30 years. Australia is no exception.
2. Despite common global themes (and maybe because of them), national and local circumstances loom large in the form and content of Great War commemorations.
3. Such activities are not organised in a political vacuum. They are performed with the crucial support – or at least the inevitable intrusion of – the state. Official organisation of commemorative ceremonies, government funding for memorials and education programs and the language our political leaders use, all shape our remembrance of the Great War.

This means that the themes and tones of commemorations must resonate at a personal as well as national level, and be politically useful in order to continue to enjoy official support.

A related point is that the use of "memorial diplomacy" is increasing as part of this trend and as a deliberate activity of government. Prime Minister Tony Abbott's attendance at D-Day remembrance ceremonies in June, 2014 along with leaders from the UK, France, Britain and Germany (among others), is a recent example. Such co-operative, multinational activity affects the tenor of domestic commemoration, but will also be an important stage for bilateral and multilateral relationships up to [2018].

---

[235] See the article published on 30 July, 2014: https://theconversation.com/who-owns-the-myths-and-legends-of-the-great-war-centenary-29348

As we know, remembering and forgetting are two sides of the same coin. But something else has become apparent in the run-up to the centenary: it has become extremely difficult – if not impossible – to remember the First World War without remembering the Second World War. This has been particularly the case in Germany but is no less true in Belgium. It is also difficult to imagine the Great War in Ireland, north and south, without memories of "The Troubles" intruding.

This blending of histories has its origins in specific national memories. In Germany, it occurs because the crimes of the Holocaust have permanently coloured German views of the entire 20th century. It exists in Belgium because accusations of Flemish collaboration with the Nazis are used to delegitimise Flemish nationalism. In Northern Ireland, there are direct organisational links between the nationalist and loyalist paramilitaries operative until the peace of 2007 and the civil wars fought in Ireland from 1916.

Finally, despite recent historical research stressing the transnational nature of the First World War, contemporary commemorative politics remain national despite, and also because of, its global scope. This is a very basic point but one that needs stating. The politics of nationalism has a profound impact on the shape of Great War commemoration.

In some places, nationalist politics may prevent a deep connection with state narratives, as happens in Flanders in relation to the rest of Belgium. In other multinational states, such as the United Kingdom, the politics of nationalism and multiculturalism produce state narratives that seek to include an almost impossibly wide national and international group of communities.

In France and Australia, where more unitary national narratives exist, commemoration provides an opportunity for governments to endorse values they hold dear.

So, who owns the national myth as it relates to the Great War?

The answer in Australia and Europe is that there are many players with vested interests – from individuals to organisations to governments. But some are more equal than others in shaping the way we remember the First World War, and even in a global era the self-weakened state still plays a dominant role in how and what we commemorate.

---ooo000ooo---

After 100 years, we can see behind the myths and the lies, and they can stop being perpetuated.

History will ultimately be recorded on the basis of all players' participation, not just those who want to perpetrate their own ideology.

Fromelles will therefore become the beneficiary of that process.

Again, after all is said and done,

1 person saying the wrong thing is saying the wrong thing.

1,000 people saying the wrong thing are still saying the wrong thing.

1,000,000 people saying the wrong thing are still saying the wrong thing.

---ooo000ooo---

21. Recognition

Babies CRY for it

Men DIE for it.

Those who served at Fromelles

have so far been DENIED IT.

--ooo000ooo---

John F. Kennedy, President of the USA said in rallying the people of the USA in the race to land a man on the moon:

We choose to go to the Moon! We choose to go to the Moon in this decade and do the other things, not because they are easy, but ***because they are hard****; because that goal will serve to organize and measure the best of our energies and skills, because that challenge is one that we are willing to accept, one we are unwilling to postpone, and one we intend to win, and the others, too.*

---ooo000ooo---

We can just as easily say the same about overturning 100 years of myths, lies and a failure to give proper recognition to the soldiers of Fromelles, 19 July, 1916.

For some of us, it is hard to accept that we should say something that has been done should be un-done. It seems insurmountable. For some of us it is something that should always have been the case. As John F. Kennedy's younger brother Robert 'Bobby' Kennedy is quoted as having said:

Some men see things as they are, and ask why.

I dream of things that never were, and ask why not.

---ooo000ooo---

Giving official recognition to the Battle of Fromelles has been treated with indifference for over 100 years. But if we do not do so, then we will only be left with a false history.

We must recognize Fromelles as a Battle. And when it is done, the soldiers of Fromelles will be given what they should always have had: **Recognition.**

Some Facts About The Battle of Fromelles.

Any analysis of the Battle of Fromelles needs to begin with some historical points being stated. Throughout the course of this book, these points have emerged:

1. There actually was a Battle of Fromelles, on 19/20 July 1916.
2. It was the first Battle by a Division of the AIF on the Western Front – the 5th Division.
3. The AIF went into Battle alongside the British 61st (2nd South Midland) British Division.

4. The orders given were to seize i.e. to capture and hold, the first line system of the enemy trenches.
5. There were no instructions on what would happen in the event the first line trenches were indeed seized.
6. The objective of the battle was to prevent the enemy from moving troops southwards to take part in the main battle (at the Somme).
7. The objectives expressly excluded an attack on Aubers.
8. The BNC footnote at page 17 of its Report contains a reference to 'On the Aubers Ridge', which is a geographical reference only.
9. The Battle on the Aubers Ridge was a battle which took place in 1915, for which a Battle Honour – 'Aubers' – was awarded.
10. Many German Trenches were captured and occupied by the AIF's 8th and 14th Brigades.
11. The 61st British Division withdrew from the Battlefield about two hours after the Infantry assault commenced, with disastrous consequences for the 15th Brigade AIF on its flank, affecting the duration and outcome of the Battle.
12. The orders to withdraw did not reach the 15th Brigade until after the 61st Division had begun their withdrawal.
13. The official tally of total Allied casualties – deaths, wounded, and prisoners of war - in a single night numbered 7,080. We now know that this is understated.
14. In official statements, AIF deaths and casualties numbered 5,533 in less than one day (a 24 hour period). In official statements, the number of deaths was 1917. The total number of deaths is higher when adding those who died from their wounds to those who were killed in action. The total who died due to or because of the Battle is 2678.
15. In comparison, the village of Pozieres was captured by the 1st Division on 23 July 1916. The 1st Division was relieved on 27 July having suffered 5,285 casualties in five days.
16. The 5th Australian Division received separate Laudatory Messages for their gallantry and achievement from the Commander-in-Chief, the Army Commander, the Corps Commander and the GOC 61st British Division. They were the first Australian Division to receive such recognition in France.
17. The Official High Command *communique* denied Fromelles its name and denied it the status of a Battle, deceptively reporting it as 'important raids', 'south of Armentieres'.
18. In turn the BNC styled the Battle as 'Attack at Fromelles', tabulating it as a 'subsidiary action' to the Battle of Bazentin Ridge (fought on the far-away Somme, 65 km from Fromelles), which had actually ended two days before the Battle of Fromelles commenced.
19. The Battalions of the 5th Australian Division took part in the Somme Offensive 1916. But they arrived there, and saw front line service there, and suffered casualties there in October

1916, before the official ending of the Offensive on 18 November 1916. They were thus entitled to claim the Battle Honour 'Somme 1916' for their service on the Somme.

20. 'Fromelles' is not inscribed by name on any of our Principal Memorials to the Great War which honour by inscription. Those which do, honour battles which received Battle Honours.
21. The Battle Honour and Inscription 'Somme 1916' have no application to the Battle of Fromelles 1916.
22. The (British) Berks Cemetery Extension Memorial at Ploegsteert, Belgium honours by separate inscriptions, FROMELLES 1915 and FROMELLES 1916. The latter honours the service of the 61st British Division at Fromelles.
23. 'FROMELLES' is listed as the first significant Battle on the (Australian) 5th Division Memorial, at Polygon Wood, Zonnebeke, Belgium.

The basis for making these points is set out in the previous chapters, and again in the comments which follow.

The Case For Recognizing Fromelles as A Battle.

The Forgotten Front

During the four years of the Great War, British division after division fought in that area of the Western Front which we now know as Fromelles and which forms part of what was until recently, referred to as "The Forgotten Front".

Prior to the discovery of the mass graves at Pheasant Wood, Fromelles was known only for being the place where the Australian Imperial Force had experienced its first, and disastrous, taste of action on the Western Front.

Whilst Fromelles hosts both the CWGC memorial to the AIF at VC Corner and the Australian Memorial Park, with the exception of a small private memorial to an officer of the Rifle Brigade, no memorial to the British casualties exists in this area.

British losses were numbered in their many thousands in and along this line, but the majority of the men killed in action in this area between 1914 and 1918 are commemorated on the Ploegsteert Memorial to the Missing in Belgium.

A charitable association, known as the British Memorial Association, Fromelles, has been set up to establish a memorial in Fromelles, dedicated to the thousands of British men who gave their lives in this small corner of France.

By educating people about the magnificent efforts of the British Army in this area of the 'Forgotten Front', the Association will endeavour to redress this balance.[236]

---

[236] See: http://en.wikipedia.org/wiki/Fromelles

Then Fromelles Was Rediscovered – 2007-2010

In July 2007, archaeologists from Glasgow University, relying on information provided by an amateur Australian historian, Lambis Englezos, tested the ground in order to determine if in a field on the outskirts of Fromelles, there existed the largest mass grave of allied soldiers discovered since the end of the Great War. There certainly was evidence of such.

Excavation of the mass graves was carried out May–September 2009, and the remains of 250 British and Australian soldiers were recovered.

The bodies were subsequently re-buried in the new Fromelles (Pheasant Wood) Military Cemetery, about 120 metres from the site of the original mass grave. 94 years later, on the 19th of July 2010, 94 years to the day after the original battle had taken place, the new cemetery and the service of those 250 soldiers was commemorated.

The Attack at Fromelles in July 1916 has been the subject of extensive research in the past two decades, most notably the matter of the unmarked mass grave. The story is now well known – see Wikipedia, for example: '*This unmarked mass grave at Pheasant Wood remained undisturbed for over 90 years. Its existence was discovered following research by retired Australian teacher Lambis Englezos. Following a visit to Fromelles in 2002, Englezos calculated a discrepancy in the numbers of unidentified Fromelles burials when compared to the lists of the missing. Following up on this, he investigated the possibility that the missing bodies had been buried in unmarked mass graves by the Germans, and never recovered since.*' The rest of the story, as they say, is history.

The recent literature on Fromelles 1916 has raised questions on how we should now judge this action. From the limited research I have conducted, that extended to questioning why the Attack at Fromelles does not have recognition as a Battle. Had it been classified as a Battle, then it is likely that it would also have been claimed for a Battle Honour.

You will recall that the intention of this book has been to present the case for Fromelles being granted recognition as a Battle. In raising the points which highlight that pursuit, there will invariably be points where the consideration of a Battle Honour, by way of comparison, will take place.

For example: We can draw a parallel between the Fromelles 'Feint' of 1916 and the Dieppe 'Raid' of 1942. Each action had specified objectives: pin German Forces down in order to prevent them being redirected [to the Somme (1916)], and prove the concept of seizing and holding a major port on the continent (1942). Each experienced significant casualties: 7,000 at Fromelles and 3,300 at Dieppe - the harsh reality is that while large scale casualties are never planned for or envisaged, such actions are a necessary evil of war. Though each proved to be tactical disasters, at the operational and strategic level they achieved their aims: no German forces were diverted to the Somme from the Fromelles front, and the lessons learned at Dieppe were put to good use in subsequent landings. However, while Dieppe is recognised with a Battle Honour, Fromelles is not.

The 'Battle of Fromelles'

A review of the Australian War Memorial website will identify that Fromelles:

…was the first major battle fought by Australian troops on the Western Front. Directed against a strong German position known as the Sugar-loaf salient, the attack was intended primarily as a feint to draw German troops away from the Somme offensive then being pursued further to the south. A seven-hour preparatory bombardment deprived the attack of any hope of surprise, and ultimately proved ineffective in subduing the well-entrenched defenders.

When the troops of the 5th Australian and 61st British Divisions attacked at 6 pm on 19 July 1916, they suffered heavily at the hands of German machine-gunners. Small parts of the German trenches were captured by the 8th and 14th Australian Brigades, but, devoid of flanking support and subjected to fierce counter-attacks, they were forced to withdraw.[237]

While it was the first major battle fought by Australian troops on the Western Front, and was 'intended primarily as a feint', and it was a 'complete failure', it is also patently obvious from the instructions to troops before the action that it was also designed to capture and hold enemy positions.

The British Battles Nomenclature Committee

Following the First World War, the British Battles Nomenclature Committee (BNC) was formed to decide on standard names for all engagements. Due to the sheer scale of the Great War, early on the BNC concluded 'that any attempt to classify the engagements in all theatres of war on hard-and-fast rules based on the numbers engaged, the casualties suffered, or any similar foundation must work unfairly in practice.' It was therefore decided, as a general rule, to classify engagements fought at corps level and above as 'battles', at divisional level as 'actions', and the remaining lesser engagements as 'affairs'.

Despite Fromelles having been fought at Corps Level, it was not given Battle recognition. And while it is clearly known that the soldiers and all who participated in the action at Fromelles saw it as a Battle, the only reason that I can assess that justified the BNC classifying it as an Attack was that it was seen as a *tactical* failure. There is now a significant body of opinion from commentators and researchers alike, that it was indeed, an *operational* or *strategic* success, which would give rise to its reassessment as a Battle in the first place.

Where there were prolonged operations the series of battles were arranged to form a Group of Battles. While some may say this exercise is an academic one, there is a need to bring some form of structure – particularly in a military environment – in order to create a hierarchy of engagements.

---

[237] https://www.awm.gov.au/military-event/E159/

http://www.diggerhistory.info/pages-battles/ww1/france/somme-1916.htm

By way of an example, the table below is an extract from the BNC Report, showing the hierarchy of 'engagements' with the overall 'Operation' listed first, followed by the various 'Battle' or Battles which were its major elements, followed next by the 'Tactical Incidents' and subsequently the various 'Actions'.

The 'Attack at Fromelles' is clearly identified as being linked to the Operations on the Somme 1916, and connected to / sequenced with the Battle of Bazentin Ridge. The Battles of Delville Wood and Pozieres followed after Fromelles.

The Anomaly: – Attack At Fromelles Linked to The Battle Of Bazentin Ridge

The BNC lists the 'Attack at Fromelles' as an Action as part of the Battle of Bazentin Ridge. At first glance this doesn't make sense.

The Battle of Bazentin Ridge was launched by the British Fourth Army (involving five divisions – none Australian) under the command of Lt Gen Rawlinson at dawn on 14 July,

1916. It is acknowledged as having been fought during the four day period 14-17 July.

The Battle of Fromelles was launched by the British First Army (involving two divisions – the 5th Australian Division and the 61st British Division) under the command of Lt Gen Monro at dawn on the 19 July 1916, It is acknowledged as having been fought as a 24 hour action 19/20 July 1916.

'Bazentin' is the name given to a resulting Battle Honour. How can the Attack at Fromelles be listed as an Action as part of the Battle of Bazentin Ridge? They occurred days apart, in different regions (65km apart), and were conducted by different British Armies.

Bazentin, located on the Bazentin Ridge, is a village in the Somme Department, Picardie Region of France, situated in the Somme River Valley. Fromelles, located on the Aubers Ridge, is a village in the Nord department, Nord-Pas-de-Calais Region of France, situated in the Lys River Valley. The distance by road from Fromelles to Bazentin is 88 km.

Nonetheless the fact remains that the Attack at Fromelles is grouped with the Battle of Bazentin Ridge. An explanation for this may be the aspect of the Attack at Fromelles being intended to be a diversionary attack in support of the BEF's main effort of the Operations on the Somme. So that although Aubers Ridge was in the Lille / Ypres sector, 65 km away from the Somme, because there was no 'Operations' in the Lille / Ypres sector in 1916 (as there had been in 1915 when The Battle of Aubers Ridge was part of the Summer Operations 1915, The Battles of Ypres 1915), and because commentary of the Attack at Fromelles is interwoven in the official records (UK and AUS) with the Operations on the Somme, it is logical to group it's Battles – it could logically have appeared in either the Battle of Bazentin Ridge, Deville Wood or Pozières Ridge as the dates for either of these fit with the Attack at Fromelles.

One reason I can postulate for Fromelles being considered *subsidiary* to the Battle of Bazentin Ridge, is that the fighting at Fromelles was due to be commenced two days prior to its actual commencement – i.e. on 17 July – but it was delayed on account of rain and inclement weather. Subsequently it was fought on the day that it was, and became a stand-alone engagement.

Another possible reason is that the BNC forgot to change it from being *subsidiary* to being *subsequent* as a result of the delay. As was the case with the next entry in the BNC table: *and subsequent Attacks on High Wood.*

Had it been commenced on the intended date, it would in all certainty have been a diversion for the purpose of holding troops in place. It still was a diversion, but for all the battles on the Somme, not just the Battle of Bazentin Ridge. It could just as easily have appeared in either Deville Wood or Pozières Ridge. Simply put, the German forces would have been fighting two actions and for tactical reasons there would not have been the possibility of moving troops from one site to another without significant risks.

Fromelles was thus a strategic success.

Military Cringe and Other Reasons for Denial of Fromelles as A Battle.

We have separately seen in Chapter 12, that when compared to other unsuccessful Battles, and other unsuccessful Attacks, that Fromelles appears to have been singly unrecognized for awards of both Battle recognition, and consequently unable to be the beneficiary of a Battle Honour.

The Battle of Fromelles involved the 5th Australian Division and the 61st British Division. It is often believed that the British did not press for Fromelles to be recognised as a Battle, let alone to receive any form of Battle Honour. But Major Christie-Miller did in fact press for a Battle Honour for the 2nd Bucks Battalion of the 61 Division at Fromelles. He was ultimately unsuccessful. The Australians it appears, if they sought a Battle recognition or a Battle Honour, were also unsuccessful. Five points which should be emphasised here are as follows:

1. Some Say: Covering-up a Disaster

The action is widely acknowledged as having been an unmitigated disaster. A communiqué regarding the action released to the press by British GHQ stated that: "*Yesterday evening, south of Armentières, we carried out some important raids on a front of two miles in which Australian troops took part. About 140 German prisoners were captured*". This brief description was arguably carefully worded to down play the perceived tactical failure of the action and the high number of British and Australian casualties – including 470 Australian prisoners. I submit that the British High Command, and in turn the BNC were disinclined to acknowledge this failure, and subsequently its true status was obscured.

2. Success versus Defeat

From an Australian perspective the action has been described as, 'the worst 24 hours in Australia's entire military history', and represents the largest number of Australian casualties ever experienced in a single day. Though viewed as a military failure, the defeat at Gallipoli was heralded as a great feat of arms by the ANZACs. Fromelles being the first action by an Australian Division on the Western front was viewed as another defeat. With the Australian success (albeit limited in terms of the 1916 Somme Offensive) at Pozières happening so quickly after Fromelles there was arguably a natural tendency to maximise the reporting of the success and minimise references to the defeat. I submit that in combination with the lack of British initiative to acknowledge Fromelles as a Battle, the Australian establishment may have been not prepared at the time to pursue the matter.

3. Maintaining Relations

The disastrous outcome of the action at Fromelles is reputed to have led to tension between the Australians and the British. Australian losses, doubts about the judgement of higher commanders and further doubts about the reliability of British troops spread in Australian units - the 61st Division being given the nickname '61st and worst' by the Australians. But the 61st Division did not just walk away from the battle. The message to the 5th Division to withdraw was held up in 5 Division HQ. I submit that at the executive level (both military and political), in the interest of maintaining relations, it was neither in Australia's interest, nor British interests to highlight the animosity that resulted from the action at Fromelles.

4. Lack of a Champion

The British XI Corps Commander in charge at Fromelles was Lt Gen Richard Haking. Maj.Gen Colin Mackenzie commanded the 61st Division while the 5th Australian Division was commanded by Maj Gen James McCay. Though Haking and McCay both survived the war, neither did so with a good reputation.

As the action at Fromelles was viewed as a disaster, I submit that none of the Generals may have been inclined to champion it being acknowledged as a Battle, preferring perhaps instead that it slip into obscurity. In his book, *Don't Forget Me, Cobber: The Battle of Fromelles*, the author Mr Robin Corfield goes so far to say that, "after the war McCay's powerful friends ensured that Fromelles was never examined in any depth".

But as stated prior to point 1 above, Major Christie-Miller did in fact press for a Battle Honour for the 2nd Bucks Battalion of the 61st Division at Fromelles. He was ultimately unsuccessful.

5. Military Mindset: 'Only a Feint'

As stated earlier, the official statement regarding Fromelles was that it was intended primarily as a feint to draw German troops away from the Somme offensive. The aim of holding German positions is often overlooked. I submit that because it was officially classified only as a 'feint', combined with the fact that the intention of capturing and holding German positions was a failure/disaster, the military mindset within the BNC may have been to simply overlook the action when it came time to consider Battle recognition.

Fromelles was a failure to capture and hold ground. The indications are that the military commanders preferred to emphasize that and it thus remained so. In spite of this, and the costly losses suffered there, it has been arguably shown to have been an operational success.

---ooo000ooo---

We read in Chapter 14 of the press release on 13 May, 2018 titled:

GALLANTRY AWARD FOR SERVICE DURING THE BATTLES OF CORAL AND BALMORAL.[238]

In that press release it was pointed out that the Unit Citation for Gallantry was created in 1991, along with the Meritorious Unit Citation. The Unit Citation for Gallantry recognises extraordinary gallantry in action.

Although a Unit Citation and a Battle Honour are different, they are arguably similar in nature. Whereas a Unit Citation is a device worn by all the members of the Unit, a Battle Honour is an award of a right by a government or sovereign to a military unit to emblazon the name of a battle or operation on its flags ("colours"), uniforms or other accessories where ornamentation is possible.

However, both recognise conspicuous service by the Unit as a whole.

The Australian Government's recent award of Unit Citations for Gallantry for the fighting at Fire Supports Bases Coral and Balmoral during the Vietnam War shows that retrospective review of past military conflicts and battles is possible.

Commentary From Historians, Academics, Researchers and Other Commentators Progressively Brought Fromelles Back into Public Knowledge.

Reviewing The Battle of Fromelles

In the past two decades a number of people have reviewed the 'Attack at Fromelles', producing well researched books and papers. The harsh assessments that were made in the past have in some cases been reaffirmed, but in others have been questioned. In his 2013 Thesis titled, *British Battle Planning in 1916 and the Battle Of Fromelles: A Case Study Of An Evolving Skill*, Historian Dr R.V. Lee (currently the Head of the Army History Unit) makes the following statement in his conclusion:

When the operational level objectives set for the Fromelles attack are compared with the outcome achieved, the attack can only be regarded as having met its objectives. The capture of ground was not, for this battle, the strategic and operational intent.

---

[238] See: - https://www.minister.defence.gov.au/minister/dchester/media-releases/gallantry-award-service-during-battles-coral-and-balmoral

The clearly articulated objective was to pin in place the enemy troops in the Lille area. On the evidence available, no further German troops were moved from Lille to the Somme after 19 July 16. This was the aim of the Fromelles plan and it succeeded. [Emphasis added.]

From the military position, it is of little issue whether, as some critics have claimed, the Germans had no intention of moving any more troops south. For the point of view of the commanders in July 1916, a threat had been identified and a successful solution applied.

If the 'feint' conducted by the 5th Australian Division and the 61st British Division at Fromelles on 19 July 1916 achieved its aim, then is that not worth Battle recognition, and if so, the accolade of a Battle Honour? As Dr Lee goes on to state: -

The problem is simple: too much emphasis is given to the tactical level of fighting and not enough to the strategic and operational levels. As the Red Army was to demonstrate so forcefully twenty-nine years later, battles could be lost while the war was won.

I submit that the Battle of Fromelles is viewed as a battle that was lost, but that nonetheless it was a Battle which contributed to winning the war.

---ooo000ooo---

Professor Joan Beaumont (internationally recognized historian of Australia in the two world wars, Australian defence and foreign policy, the history of prisoners of war and the memory and heritage of war) wrote on 18 July, 2016 that interest in Fromelles resurfaced in the 1990s. She wrote that[239]: -

Fromelles is now one of the most famous battles in which Australians fought during World War I, routinely remembered as the greatest disaster in Australian military history; 5533 casualties in 24 hours, all for an operation that manifestly had no strategic value.

But Fromelles was not always central to Australian memory of war. In the inter-war years, the battleground became the site of one of the thousands of cemeteries created by the Imperial War Graves Commission. Here, at VC Corner, the remains of 410 unidentified Australian dead were interred and the names of the 1299 Australian missing from Fromelles inscribed on the memorial walls. In Australia, meanwhile, survivors of Fromelles would gather on the anniversaries of 19 July, while families who had lost men in what they called Fleurbaix would also insert 'In Memoriam' notices in newspapers.

However, Fromelles was not then a national memory, in the sense of being a battle that was honoured in prominent national rituals of commemoration. Perhaps this was because it was soon eclipsed by other costlier battles on the Western Front, such as Pozieres on the Somme and the Third Battle of Ypres in Flanders in 1917. [See above: Covered in 'Success versus Defeat' and 'Covering-up a Disaster'.] *Perhaps it was because survivors of World War I preferred to see themselves as heroes, not victims, as is the vogue in war commemoration of the early twenty-first century.* [See: 'Military Mindset' referred to above.] *The 5th Division, for instance, when asked in 1919 where it wanted to install the memorial celebrating its wartime achievements chose not Fromelles but Polygon Wood, the site of one of the more*

[239] See article: Joan Beaumont: Why Australians Remember Fromelles. The Interpreter, Lowy Institute 18 July, 2016. Reproduced with permission.

successful actions during the Third Battle of Ypres. [See: 'Military Mindset' and 'Success versus Defeat' referred to above.]

All of this changed in the 1990s. As the extraordinary resurgence of interest in war memory occurred not only in Australia but around the globe, Fromelles was rediscovered. This was the result of both government intervention and individual initiatives. In 1998, the Australian government opened a memorial park at Fromelles on the eightieth anniversary of the end of World War I. At its heart was a statue of a man staggering under the weight of a wounded soldier draped across his shoulders. Called Cobbers, it immortalizes a Victorian farmer, Sergeant Simon Fraser of 57th Battalion, who in the days after the Fromelles battle joined small groups scouring the battlefield for the wounded. It was a story of compassion and mateship waiting to join other icons of Australian war memory: Simpson and his Donkey, and 'Weary' Dunlop. Indeed, the sculptor of Cobbers, Peter Corlett, had already created the statues of these two iconic figures that stand now outside the Australian War Memorial.

---ooo000ooo---

As Prof. Beaumont indicated, Fromelles was 'rediscovered' in the 1990s. By 19 July, 2002, it had certainly become more obvious in mainstream reporting when Ross McMullin (an award-winning historian and biographer) wrote an article – The Forgotten Fallen – which appeared in the Sydney Morning Herald. It was of necessity a relatively short article, being for mass consumption.

His article recounted the movements of Australian troops up until the Battle of Fromelles, and dealt with the simplistic faith of Lt General Richard Haking in believing an attacking infantry would be able to deal with the entrenched German positions.

According to Haking, even if a defending force was stronger than the unit attacking it, the attackers would win. Such absurdity during a war dominated by heavy artillery, machine-guns and barbed wire had been starkly demonstrated by mid-1916, by which time the English general had charge of the area near Fromelles recently occupied by the 5th Division. This sector had been relatively quiet, but Haking coveted the higher ground behind the German lines known as the Aubers Ridge.

Preparations for Haking's folly were rushed, inadequate and obvious to the enemy. The haste also played havoc with the effectiveness of the inexperienced artillery, whose role would be crucial.

The attackers never had a chance. Machine-guns at the Sugar-loaf, unsubdued by the artillery slaughtered much of Brigadier Pompey Elliott's brigade in a few minutes.

Hideous scenes were commonplace at the Western Front, but there was no more ghastly sight for Australians in the whole war than the scene in their front line after Fromelles. Red-splashed trenches were packed with dead and dying men. Another heart-rending spectacle confronted anyone looking out towards no-man's-land, where many Australians were raising limbs in pain or turning from side to side in obvious distress. The pitiful plight of helpless comrades inspired deeds of exceptional gallantry. Hundreds of wounded were saved. Even more could have been rescued if an impromptu truce arranged with the enemy by two members of the 8th Australian Brigade had been ratified. McCay disallowed it in accordance with Haig's order forbidding such agreements.

In the end, (official figures are that) 5,533 Australians were killed or wounded. German casualties were much fewer. McMullin recounted the prevailing view that:

'A cover-up was swiftly authorised.' [As noted previously, this is not the writer's view.] The official British communique referred to the battle as some important raids which had resulted in the capture of about 140 German prisoners. Australian survivors who read this were unimpressed. One contended that fierce fighting for eight hours under terrific gunfire might be considered a battle. Wartime censorship was an influential instrument in the clampdown. Unfamiliarity with Fromelles, compared with other Western Front battles and Gallipoli, was widespread, which increasingly annoyed participants.

The public's general knowledge of Fromelles remained low. Australians had remained unfamiliar with the calamity of Fromelles until comparatively recently. This was despite Brigadier Elliott delivering his 'pungent lecture in 1930 which generated headlines in London as well as Australia'. But it was known to relatives of those who served and especially those who never returned.

Even then [in 2002] there was a 92-year-old woman in a Victorian nursing home who had carried the name of Fromelles as a lifelong emotional scar. She could still remember her family being devastated by the news that her father and several uncles had been killed in the battle.

For thousands of Australian families like hers, the aftermath of the war that was supposed to end all wars was grim. Historian Bill Gammage summed it up: "Dreams abandoned, lives without purpose, women without husbands, families without family life, one long national funeral for a generation and more after 1918."

[Picture 65: 2016 Centenary Commemoration Service, Fromelles.Photo: Philippe Clerbout, 2016]

The family of the soldier who was introduced on page 1 of this book, kept a bible in which it was written: 'RIP William Polding Ryan. Killed at Fleurbaid (sic) France 16 July, 1916 aged 20 years 3 months'. His family and many others never forgot.

And yet, we were told to accept that it had been *some important raids*. The soldiers and their families knew that it could only ever have been properly described as a Battle, and always should have been.

---ooo000ooo---

And then the attitude toward remembrance took on a new facet altogether. Fromelles became front and centre in the news reports of 2008 as a result of the assiduous investigation and persistence of Lambis Englezos and a small group of like-minded supporters. As Prof. Beaumont continued:

Convinced that they could locate the burial site of the Fromelles missing who were not interred at VC Corner, they initially fought official scepticism. But in 2008 a geophysical survey commissioned by the Australian government identified a mass grave at Pheasant Wood near Fromelles. On excavation it was found to contain the remains of 203 Australians, 3 British and 44 soldiers of indeterminate nationality buried by the Germans in 1916.

There the remains might have lain undisturbed, with a new memorial erected to mark the location and the presumed dates of death. However, to leave the dead of Fromelles in a mass grave was unacceptable in early 21st century Australia.

The remains found at Pheasant Wood were reburied with great official ceremony in 2010 in a newly created cemetery, the first to be built by the Commonwealth (formerly Imperial) War Graves Commission in fifty years. Even more unusually, a search was launched to identify the missing individually. Where possible, DNA samples taken from Pheasant Wood remains were matched with those of the living descendants of missing soldiers in Australia and the United Kingdom.

No one questioned why many millions of dollars was spent on this exercise. No one, at least publicly, debated the precedent of using DNA to identify the missing. [It had been done recently in Belgium, so it was not going to be opposed in Fromelles.] Nor did any one query whether 'closure' was needed for relatives who had never known the missing of Fromelles, nor truly experienced grief.

The missing of Fromelles, it seems, spoke to phenomena that have fuelled the growth of war memory globally in recent decades: the explosion of genealogy and the desire to locate family stories in bigger, more universal narratives of the past. The new rituals of Fromelles also testified to the continuing salience of the implied contract between citizen and state that underpins a voluntary system of enlistment for military service. In a liberal democracy, where the rights of the individual are a core value, those who choose or are required to die in the defence of the state are considered entitled to be honoured individually. This was the ideal that informed the work of the Imperial War Graves Commission in the years after World War I, one of the first mass conflicts in which rank-and-file soldiers were granted their own grave and headstone carrying their name, age and date of death.

---ooo000ooo---

By 2010, on 24th April, the eve of Anzac Day, journalist Bridie Smith wrote in The Age and Fairfax press that a

PUSH GROWS TO HAVE 'FROMELLES' ON MEMORIALS.[240]

By that time, the mass graves at Fromelles had been revealed and the dedication of a new Commonwealth War Graves Cemetery at Fromelles was drawing closer. The public's knowledge of the battle had increased exponentially in the time since Ross McMullin's article in 2002. Smith wrote that: -

AUTHORITIES are facing a determined push to have the Battle of Fromelles inscribed on the nation's war memorials after the identification of 75 Diggers almost a century after they were buried in mass graves in northern France.

As the descendants of the soldiers prepare to mark their first Anzac Day knowing where their relatives lie - right down to the plot number at the cemetery outside Fromelles, soon to be opened - those pushing for the battle to be inscribed on war memorials are preparing for ''the third battle''.

''The first was the battle itself in July 1916, the second was the battle to get the investigation of [the mass graves at] Pheasant Wood under way and the third is to get Fromelles inscribed on World War I memorials,'' said Chris Bryett, president of Recovering Overseas Australia's Missing, or ROAM. His calls for the battle to be inscribed on memorials are backed by the Friends and Families of the First AIF, Friends of the 15th Brigade and amateur historian Neville Kid. They warn that the growing presence Fromelles has in the public consciousness will make it impossible for authorities to ignore.

''The number of descendants of Diggers who participated in Fromelles isn't getting smaller, it's getting larger,'' Mr Bryett said.

The name features on the Australian memorial in London's Hyde Park and on the 5th Division memorial in the Polygon Wood near Ypres in Belgium but not on any of the main World War I monuments in Australia.

Families and Friends of the First AIF president Russell Curley said the main memorials in Adelaide, Darwin, Perth and Hobart did not honour by inscription, while in Sydney, Melbourne, Brisbane and Canberra the battle is subsumed in honours such as 'Somme 1916'.

Melbourne's Shrine has a plaque for the 60th Battalion and the Cobbers sculpture in the grounds surrounding the Shrine.

''Our position is that anything other than inscription is ancillary,'' Mr Curley said. ''They honour by inscription and the best way to honour Fromelles is by inscription.''

But the chief executive of the Shrine Trust, Denis Baguley, doubts the name will be added to the walls of Melbourne's memorial.

''The original design of the Shrine is the original design,'' he said. ''I can't see that there's going to be any changes to the Shrine building.''

Curiously, Mr Bryett said the biggest obstacle in the push to have the battle inscribed on memorials had been the RSL, among whose founders was Frederick Toll, who led the 31st Battalion at Fromelles.

[240] ©: Bridie Smith, The Age: Push Grows to Have Fromelles On Memorials (reproduced with permission).

RSL national secretary Derek Robson said while Fromelles may be mentioned on some memorials, it was not mentioned on all as it was considered ''part of the Somme and the Western Front campaigns generally''. Interestingly, he speaks of Fromelles being part of the Western Front "campaigns" generally, i.e. not exclusively part of "the Somme Campaign".

He conceded the July 1916 battle was significant for its ''level of carnage'' - 5533 Australians were killed, wounded or reported missing in one night - and because Fromelles was the first engagement by the AIF on the Western Front. However, he said there was no need to rewrite history.

''[Fromelles] is a poignant part of the whole campaign itself ... would we really try to suggest that one action was more important than another? I'd let it lie.''

Mr Bryett can't make sense of the argument. He suggests that rather than being consigned to history, the story of Fromelles is still unfolding. Some of those killed in the battle were identified last month and more could be named next month after a second sitting of the Joint Identification Board in London.

The identification of remains continues despite initial estimates that it would only run until 2014. The 250 soldiers' remains would be reburied in 2010.

''This is a groundswell that's coming from citizens and the irony is that it's the citizens that are reminding the RSL to honour the Diggers,'' Mr Bryett said.

Sydney-based Neville Kidd said history had overlooked Fromelles.

''When these memorials were built, Fromelles was still under wraps,'' he said.

''Fromelles was such a tragedy that it was hidden in the initial reports from the high command ... [authorities] hid its casualties amid the slaughter of the Somme.''

But as Fromelles emerges in the public consciousness, Mr Kidd and those championing his cause hope that the push will become impossible for authorities to ignore, and the memorials changed.

''As people become aware that it's not on the memorials,'' he said, ''there will be a clamour which will become irresistible.'"

---ooo000ooo---

Later on 19th July, 2010, on the 94th Anniversary of the Battle of Fromelles, Ashley Ekins (then Head of the Military History Unit of the Australian War Memorial) gave an address in the Commemorative Area of the Australian War Memorial.

His speech came as part of the closing ceremony for the War Memorial, and recounted the background for the Battle, the numbers involved, and the failure of the Battle.

The extent of the slaughter was graphically stated when he pointed out that: -

the 15th (Victorian) Brigade was destroyed within fifteen minutes, entire companies of infantry being virtually annihilated. Their commander, Brigadier General Harold "Pompey" Elliott, was speechless with grief the following day, "the tears streaming down his face, as he shook hands with the returning survivors".

One of Elliott's battalions, the 60th, had gone into the attack with 887 officers and men. When the survivors gathered at brigade headquarters the following afternoon, only one officer and 106 men answered the roll call.

While there was success in capturing some 1000 metres of the German second line, the soldiers were unable to locate their final objective. Those who had not been able to make a retreat were either wounded, dead, or captured. Those who lay in No-man's land were lucky if they were able to be recovered. Many more died where they fell, unable to be retrieved.

Many were never recovered and [have] remained missing for decades.

Ekins recounted the recovery of fallen remains after the declaration of the Armistice on 11 November, 1918.

For over 90 years, 1,335 Australian soldiers remained 'missing' from the Fromelles battle, having no known grave. [Of which 1299 names were recorded at VC Corner Cemetery alone.] *Then in 2007, investigators began to uncover the remains of some 200 Australian and 50 British soldiers who had been buried in a mass grave at Pheasant Wood by German troops in 1916.*

Between 30 January, and 19 February, 2010, the remains of 249 soldiers were reburied with full military honours in Fromelles (Pheasant Wood) Military Cemetery, newly constructed by the Commonwealth War Graves Commission.

The last to be buried was interred during the ceremony at Fromelles on 19 July, 2010.

On its 94th anniversary, there was no hesitation in describing what had taken place as the Battle of Fromelles. The public have been told by their leaders that it was a Battle in every public broadcast since. While the public may give it the recognition it deserves as a Battle, sadly it has so far not been the subject of an ***Official*** Political or Military recognition.

---ooo000ooo---

Books dealing with the events at Fromelles, and the discovery of the burial pits were inevitable, and proliferated after 2008. One such book is that of Peter Barton – The Lost Legions of Fromelles.

A review of the book itself, by Paul Daley in The Guardian provided a positively glowing review, commending the extent of the research:

"this is meticulous, demanding dotted-i and crossed-t military history for the aficionado".

"Like most such Australian military moments, the barnacles and moss of myth, perpetuated by official history and politics, had for so long obscured the truth about what happened. Some scholarly work, not least by Peter Barton in his The Lost Legions of Fromelles, has further exposed the true folly and the horror of this Australian infantry charge."

When Barton came to Australia, he gave a talk on the book and its subject, the Battle of Fromelles, at Stanton Library, North Sydney. [241] Barton said:

[241] Jul 7, 2014 by Richard Jackson - Peter Barton Discusses "The Lost Legions Of Fromelles" At Stanton Library

"What you find in there (Fromelles) you can place anywhere along the Western Front, you can do this depth of research anywhere along the Western Front, it is inundated with memory and legacy, directly connected with you here in Sydney and all the other places I visited. It is the essence of Australia's memory."

Barton an archaeologist, historian and documentary filmmaker didn't so much discuss his book as present the equivalent of a university lecture. He talked passionately at length about the battle, key figures from both sides, and the battle's continued significance today. In fact he over ran the scheduled talk of one hour, leaving the audience unable to field questions.

Barton's research took him to various military archives across Europe drawing upon "the fixation that the Bavarians had for record keeping". He visited historical German accounts, including the Institut fr Zeitgeschichte (the Institute of Contemporary History), and the Volksbund Deutsche Kriegsgrberfrsorge (the German War Graves Commission).

Throughout he drew upon detailed information of Fromelles gathered by German intelligence, more detailed, accurate and self-critical than allied information.

The information showed Allied troop movements, the development of second trenches.

Some documents went so far as to record the precise spot allied troops had died on maps of the area, at the foot of one map an Australian soldier was named. Something that brought the whole thing home to the audience.

"Wherever you go in the Western Front, there are men to be found, you may be walking over these unknown graves."

Barton describes the action at Fromelles as a Battle, unless he quotes the official text of the BNC or from working documents otherwise. Fromelles' true significance is not and cannot be concealed under the descriptor of an 'Attack'.

---ooo000ooo---

Worthy of note is how Barton sympathetically recorded[242] the effect on families, when they are not aware of where they deceased soldier died. It serves to explain why so many families today strive to find just where 'their' soldier fell:

"After WW1 the Fromelles battleground remained a back-water, overshadowed by the majestic scenes of loss on both its flanks at Loos, Arras, Somme and Ypres. Almost a hundred years later, the reaction of the relatives of the soldiers occupying the Pheasant Wood pits suggests that regardless of the passage of time, a graveless death still represents a great deal more to grieving families than a man simply ceasing to exist."

---ooo000ooo---

Historian Professor Beaumont says[243] what occurred was simply a repeat of disastrous British assaults over the same ground in December 1914, and May 1915. But the events of May 1915

[242] At p.341

[243] As quoted in 'Fromelles, worst ever day for diggers', 9News National 12 July, 2016.

resulted in a Battle Recognition, while those of July 1916 did not. Neither was successful tactically.

Fromelles virtually destroyed the 5th Division - not until year's end could it stage even limited operations. Haunted by his experiences, Elliott took his own life in 1931.

The British high command won no digger friends, dismissing this debacle as "some important raids" in their official communique.

British General Richard Haking - Fromelles was his idea - blamed failure on untrained Australian troops and insufficient offensive spirit in British troops. He said this searing experience: 'had actually done both divisions a great deal of good.'

One factor overlooked in many accounts is that German defenders, soldiers from the 6th Bavarian Reserve Division, were actually very good.

British historian Peter Barton said they had created formidable defences, and troops were disciplined and experienced.

"The Germans were too well organised. It was a brilliant piece of long-term tactical planning," Barton said.

As an example of Germanic thoroughness, they mandated speedy burial of enemy dead for hygiene reasons. For this battle they dug eight pits, filling five, each with 50 dead.

Just how these were missed after the war isn't clear. Their presence was confirmed long afterwards in documents found in German archives in Munich.

Revising The First Rough Draft of History

I would guess that everyone reading this book has at some time read a newspaper, or listened to a news report on their radio, or picked up a report via television of some event happening around them or around the world. What they are experiencing is loosely called "news". News reports are prepared by journalists, plying what is known as their "craft" of journalism, reporting on events *du jour au jour*.

It is often said that journalism is the first rough draft of history.[244] Until becoming involved in writing this book, I had not had much cause to consider the wider impact of that expression, so just why is journalism the first rough draft?

We will often find later accounts of events, in which more information has become available, a witness has come forward to assist with enquiries, or evidence previously concealed has been revealed. The science of anthropology continues to fascinate and educate us. Journalism attempts to do so for matters close at hand in time.

[244] a phrase generally attributed to either Alan Barth in 1943 or to Philip L. Graham in 1953 – both are former writers or editors of the Washington Post.

So it is that the first rough draft of history becomes revised, updated, corrected, modified, or even negated altogether as wrong.

Yet among the community which investigates military reporting, and recording of events, and interpreting the events of the past in what is generally referenced as 'history' there are those who have on many occasions advised us to 'leave it alone', 'let sleeping dogs lie', 'you're opening Pandora's box', or similar. Or even to tell us that 'x-y-z' was done 'back then', and thus it always will be. 'It cannot be changed.' Or even 'don't fight the green'.

Those people are not wrong to do this. They are simply not prepared to deal with what may be revealed if they dig deeper than what they have already been comfortably exposed to. For them it may become un-comfortable, and their own thoughts may need to be revised.

Australians are renowned for their disdain of authority. A person is not respected for their position, but for the personal integrity and competence they bring to that role. We don't like hypocrites, show ponies, attention seekers and those who take themselves too seriously. We prefer to call people by their first name, and for them to be real, open, and relaxed. Is it any wonder then, that we do not accept anyone telling us to 'leave it alone', or that 'It cannot be changed.'

Even after the BNC and the BHC stated that there would be no more reviews of their decisions, cases continued to be presented to them for review. Those have been described in previous chapters. It should therefore come as no surprise that it is indeed possible to seek a review of a decision made in 1922 by the BNC with regard to the classification of Fromelles. Far from saying 'it can't be done', the truth is that 'it already has been done', for other units.

But Australian or not, it is a fundamental trait of being human to question what are perceived truths. At one time it was thought that the earth was flat. At one time it was though that all the planets and the sun revolved around the earth. Through scientific endeavour these and many other truths have been reviewed and updated. This process of questioning is not restricted to the sciences. Heinrich Schliemann's pioneering work in the field of archeology during the mid-1800s effectively lifted Homer's siege of Troy out of fictitious ancient mythology and into an actual historical event.

The search for the World War II Cruiser HMAS Sydney II was partly driven by setting the historical record straight. The ongoing and now successful [in 2017] search for the submarine AE1, which disappeared during the Australian Naval and Military Expeditionary Force capture of German New Guinea in September 1914, is similarly driven by a need to answer the question: 'what happened'?

In the course of this book, I have looked at what information has been left by and for past generations, and now for ourselves, concerning the events that took place at, and which subsequently involved that small part of France called Fromelles. It is indeed a small area, but it is filled with a large number of persons from many countries, friends and foe, who (on modern analysis at least) should not have ended their lives there.

And having left them there, our forebears have not given them their due recognition, for a range of reasons. But we do not need to perpetuate those past mistakes.

It has been stated that if no one back in the 1920/30s saw fit during the 'committee' phase of the discussion on what should be considered a Battle, or a Battle Honour for WWI, then why should we 90 to 100 years later question that process? In response I say not 'Why?', but rather: 'We must!'

Recognition of Fromelles as a 'Battle' is not simply a matter of accusing past Committees of getting it wrong, notwithstanding my comments to that effect in chapter 12. But it is a matter of suggesting that with the passage of time, and the ability to objectively consider the evidence - particularly the results of recent academic investigation, and the de-classification of official records, that the matter appears to be worthy of review and reconsideration.

There is a massive difference between interpreting the events of a century ago, with the hindsight of today, as opposed to considering those same events on the basis of standards applying at that time.

We have come to review the events of those first rough drafts of history, both civilian, and military, and to challenge what has been up until now 'left alone'.

We are not uncomfortable with what we now know, but we have been disturbed by what we have had to deal with. And it should not continue.

The populace accepts what is beyond dispute, and rejects that which is unfounded. In Australia we have evolved what is commonly known as 'the Pub Test': If it doesn't seem right, then it probably isn't. And if it is not right, we will reject it.

How would our current front line Defence Forces react to an equivalent *Official Communique* similar to that issued after Fromelles?

I suggest if the same process was foisted on us in this modern era, we would justifiably refuse to allow our soldiers to be left without due recognition at all.

For Fromelles not to be described as a Battle, does not pass the Pub Test.

To be classed as an Attack, and further, to be subsidiary to another action which has already taken place, not only does not seem right, but factually is not correct in any sense.

We have read earlier that there are in fact three Attacks that were recognized with Battle Honours. It is un-Australian and degrading to continue to condone the debasement of what was a Battle to the status of an Attack or a Raid.

Memorials are subject to alteration in light of the facts and a better understanding of them.

Our options now are to honour and inscribe them, or continue to dishonour their sacrifice and achievement by exclusion.

Conclusion: -

After due consideration it is clearly apparent that a strong case can be mounted to have the 'Attack at Fromelles' conducted 19/20 July, 1916 (as it was described by the BNC in 1922) retrospectively recognized as a ***Battle***, so that the work of all the units of the 5th Australian Division and 61st British Division that participated in it can be properly remembered.

For this to be progressed successfully it would be necessary and prudent to obtain cooperation and agreement with the British, as it was the British Battles Nomenclature Committee and subsequently their Battle Honour award system which Australia essentially followed.

The time for us to do so has arrived. We should therefore confront our Government, asking it to petition Her Majesty, Queen Elizabeth II, to commission a review of the determinations of the British Battles Nomenclature Committee and the Battle Honours Committee, with a view to exercise the Monarch's Royal Authority for the record to be amended.

The objective would be to record that their 'Raid' (initially by High Command) and 'Attack' (subsequently by the BNC) at Fromelles was actually a ***Battle***.

Separately, Fromelles should be considered for its own Battle Honour, thereby giving due respect and regard to those who served and fought there.

In the meantime, our Principal Memorials where honour is afforded by inscription, should be 'adjusted' to end the continuing dishonour of Fromelles by its exclusion on such memorials.

---ooo000ooo---

In the final analysis of the gravity of what took place at Fromelles, we can again look to the comments of Prof. Joan Beaumont: -

The Battle of Fromelles itself was reclaimed from its relative obscurity. It is no longer a 'raid' or even an 'attack' but 'the bloodiest twenty-four hours in Australia's military history before or since'.

---ooo000ooo---

There can be no doubt that Fromelles was a Battle in its own right. And nor can there be any doubt that those who fought there deserved a Battle Honour in recognition of their involvement there.

The time has come for Fromelles to be officially and correctly re-classified as a Battle.

The time has also come for 'Fromelles' to be inscribed on those memorials within Australia which are dedicated to those who died in the service of their country.

After all, and in adopting the Pub Test:

If it Walks like a Duck,
And it Swims like a Duck,
And it Quacks like a Duck,
Then it is probably a Duck!

---ooo000ooo---

FROMELLES WAS NOT AN ATTACK.

FROMELLES WAS A BATTLE.

ALWAYS WAS.

ALWAYS WILL BE.

---ooo000ooo---

LEST WE FORGET.

A Reflection: William Polding Ryan And Me

I never knew anything about William Polding Ryan while I was growing up. The first time I saw any reference to him was when I was 43 years of age. But he did exist. He was my mother's father's brother. He lived. He was part of a family. He felt the dawn. He felt the sunshine on his face. He heard birds in the trees and the sky above. He felt the rain from thunderstorms. And he was killed in France, in the Battle of Fromelles on 19th July, 1916.

I never heard my grandfather, his brother, mention him. I never heard my mother mention him. I only found his name for the first time after I started working on family history in 1999, after I received a copy of family records from an aunt, in which it was stated 'RIP William Polding Ryan. Killed at Fleurbaid (sic) France 19 July, 1916 aged 20 years 3 months'.

I went to the War Memorial in Canberra on 24 April, 2000 but I couldn't find a record of William, because without all his name written down, I couldn't find his records. I had guessed his name, and I got it wrong.

Seven years passed before I renewed any interest in finding out more about William, but in November 2007 I was able to find his service records through the Australian War Memorial website. I was surprised by what I found. I thought from the records that he was buried in a mass grave in VC Corner Cemetery, near Fromelles. And that was it. If I was ever going to travel to France again, I would go there and see the cemetery.

Then I heard about a teacher who had been working on tallying up the numbers of soldiers killed in France in WW1. The town of Fromelles was mentioned from time to time.

In 2008 I heard radio reports that spoke of an archaeological dig to take place, to see if soldiers were in fact buried at a site identified by the teacher and his friends.

I thought "what the heck? He might be buried there instead of VC Corner. There's nothing to lose if he is found there instead." So, I registered my interest with the Army. In May 2009, I received a letter advising me that the German war records did not show William as having been buried by them. He had not been included on the working list of soldiers whom the Army was trying to identify.

Later on, in November 2009 I was advised that my DNA had been identified as not suitable as a testing candidate. So, I arranged for my mother's brother, Cyril Ryan, at the time in his late 70s, a male with the same Y-DNA lineage, to provide a DNA sample and pass it on to the Army. I thought that might be the end of the line of inquiry for me.

Then on 16 March, 2010, I received a call at work, about 10 minutes before I was due to go out for lunch. A man named Christian, from the War Graves Commission told me that a Public Media Release was coming up soon. And that William Polding Ryan had been positively identified by use of the DNA sample given by my uncle. I began choking up, I had tears welling up in my eyes, and the scalp on my head was prickly from the news. I could not understand why these feelings were coming over me, but they did. And they still do. Something that seemed so simple, had helped to identify the remains of a soldier who had been killed in action nearly 94 years before, someone I had no idea even existed while I was growing up.

After my uncle was given the news, we spoke over the phone and he told me that I had to come up with the inscription on the headstone because "You started this." It took me a little bit of thinking, but I tried to come up with a connection between what I knew from records was the anguish of a concerned mother at home, and the generation who now know what happened to him so long ago. The inscription I came up with is:

LOST SO FAR FROM HOME, AND FOUND SO CLOSE TO OUR HEARTS.

William Polding Ryan is now buried in a marked grave, Plot I Row A, Grave 2 at the new Fromelles (Pheasant Wood) Military Cemetery. I stood with 6 others in front of his grave on 19th July, 2016, one hundred years after he was killed, and sprinkled some soil from near his home, at Dunbible.

There were 250 Australian soldiers' remains unearthed as a result of the actions of Lambis Englezos, a teacher with a passion for our history. More than half of them have since been identified. There were a lot of other people who travelled to Fromelles in 2016, all of whom in their own way would also say "Thank you" to Lambis Englezos for wanting to make sure our families' lost men were finally found and given the dignity of their own burial place.

---ooo000ooo---

Since those words for that short history of my connection with William Polding Ryan came to me nearly nine years ago, there have been many developments in the remembrance that families have shown for those who served at Fromelles. One such example has been immortalized in a life-size bronze statue of William holding the hand of a 5-year-old girl. It stands at Wollondilly Anglican College, Tahmoor NSW.

On Monday, 18 June, 2018, I was a guest of the College for the official dedication of their ANZAC Memorial Shelter, and garden adjoining, and the unveiling of the statue.

College students, and dignitaries including former Prime Minister John Howard, Chief of Defence Air Chief Marshal Mark Binskin, and Director of the Australian War Memorial Dr Brendan Nelson, joined with headmaster Dr Stuart Quarmby and veterans gathered at a ceremony to watch the unveiling of the College's Anzac memorial, and the official opening of their Anzac shelter.

The ceremony paid tribute and honoured soldiers who had died in World War I. Several speakers talked about the Battle of Fromelles, the horrors of war and the lives of particular soldiers and their families.

The statue is a representation of Acting Sergeant William Polding Ryan, who was 20 years old when he was killed in action at the Battle of Fromelles. The statue, made by sculptors Gillie and Marc Schattner, has a special meaning to the headmaster, and to myself.

He was Dr Quarmby's grandmother's uncle. He was also my maternal grandfather's brother. In a rare combination of relationships, Dr. Quarmby is also related to Alfred Tuck, who also perished at Fromelles, and is also buried in Pheasant Wood cemetery at Fromelles.

The statue was unveiled by me, and by 5-year-old Hannah. I had the honour of doing so as great nephew of William Ryan, and as Secretary of the Fromelles Association of Australia. Hannah was the model for the little girl in the statue, and is the granddaughter of Dr. Quarmby, and she was then a Year One student at the College. She is thus the great, great, great, great niece of Sergeant William Polding Ryan.

Dr Quarmby therefore spoke about his and my relatives Corporal Alfred George Tuck and Sgt. Ryan who died on their first day of battle on July 19, 1916 at Fromelles. Their bodies lay in mass unmarked graves until they were identified by DNA from surviving family members.

"The statue of the soldier represents all of the diggers who fought and died in the Battle at Fromelles in a 24-hour period," Dr. Quarmby said.

He went on to say in his dedication: "When looking at the statue, imagine every student saying: 'Thank you for paying such a price'."

"And imagine him saying 'Thanks for finding me'."

This event is in many ways, just one of the results of a phone call I received at work in March, 2010, to advise that my great uncle had been identified by use of another family member's DNA. I was, and still am very chuffed at what has come about because of that phone call.

Geoffrey Benn

12 March, 2019.

[Picture 66: William and Hannah.]

[Unveiled 18 June, 2018, Wollondilly Anglican College, Tahmoor NSW.]

APPENDIX I

Those Killed In Action, Died of Wounds, Taken Prisoner, Or Died From Wounds While Prisoner of War, Because Of Their Service at Fromelles

This Appendix records the names of Australian, British and Germn soldiers who

(a) Were killed in action in the Battle of Fromelles; or
(b) Died of wounds sustained in the Battle of Fromelles; or
(c) Were taken prisoner and survived the Battle of Fromelles; or
(d) Taken prisoner and died of wounds sustained in the Battle of Fromelles.

It also records six Australian prisoners of war who died from other causes (such as influenza), while they were prisoners, and who had been captured in the Battle of Fromelles.

The names of the casualties who were wounded, medically treated, and went on to serve are not recorded.

The numbers tell their own story. A tally of the numbers has been made in Chapter 9 of this book, in which I have set out the basis on which I assert that the Battle of Fromelles and its associated casualties should commence on 14 July, 1916, and not 19 July, 1916. This has led to a need to revise the hitherto officially-stated casualties of 5,533, which were restricted from noon on 19 July to 8 pm on 21 July. The correct number of Australian casualties should be increased by 102 for the period from 14 to 18 July, not previously included.

The correct number of ***casualties*** should be 5,635.

The names of the soldiers who now lie in the Fromelles (Pheasant Wood) Military Cemetery have been updated to reflect their final places of burial, instead of having been recorded on the Memorial Wall at VC Corner Cemetery, or the Villers Bretonneux War Memorial. Changes will take place into the future as more soldiers' names are announced that now lie at Pheasant Wood.

The number of soldiers 'killed in action' and those who 'died of wounds' after 21 July, have had their individual and Unit records reviewed to ensure their cause of death was related to their actual service under the Orders for the Battle of Fromelles. A number have been removed from the first edition of this book as a consequence. There were found to be no more killed in action beyond 26 July, but progressively more who died of wounds.

After July, 1916, the various units were in consolidation activities, or moving to other positions. Their circumstances had changed.

The numbers go on to include those who died of wounds even up until 1944. There are possibly more who died after lengthy and life-consuming illnesses after 1916, and if they become apparent, these numbers will be updated.

It is beyond this work to seek to document the number of German soldiers and the categories in which they have fallen, so the work from Robin Corfield dealing with the German numbers, has been reproduced without alteration.

What follows are the names of all the soldiers Australian, British, and German soldiers who died at or because of, or were taken prisoner at the Battle of Fromelles.

Lest We Forget.

KEY TO ABBREVIATIONS:

D	DIED – cause not stated – British Units on 19 & 20 July, 1916
POW Died	Died from other causes whle a Prisoner of War (Australian)
DPOW	DIED WHILE PRISONER OF WAR (British Regiments)
DOW	DIED OF WOUNDS
POW DOW	DIED OF WOUNDS WHILE PRISONER OF WAR
KIA	KILLED IN ACTION

ABBREVIATIONS FOR CEMETERIES AND MEMORIALS

FRENCH, BELGIAN OR GERMAN CEMETERIES

AA	Arras Road Cemetery, Rochincourt (No. Fr 1059)
AB	Abbeville Communal Cemetery Extension (No. Fr 52)
AN	Anzac Cemetery, Sailly-sur-la-Lys (No. Fr 255)
AR	Aubers Ridge British Cemetery (No. Fr 567)
BA	Bailleul Communal Cemetery & Extension (No. Fr 284-85)
BH	Bedford House Cemetery, Zillebeke (No. Bel. 165-67)
BO	Boulogne Eastern Cemetery (No. Fr 102)
BR	Brewery Orchard Cemetery, Bois Grenier (No. Fr 83)
BSW	Berlin South-Western Cemetery, Stahnsdorf, Germany
CA	Calais Southern Cemetery (No. Fr 8)
CB	Croix-du-Bac British Cemetery, Steenwerck (No. Fr 1092)
CC	Canadian Cemetery No. 2, Pas de Calais.
CD	Canadian Cemetery, Sailly-Sur-La-Lys
CI	Cite Bonjean Bilitary Cemetery, Armentieres (No. Fr 922)
CN	Couin British Cemetery, Pas de Calais (No. Fr 203)
CO	Corbie Communal Cemetery (No. Fr 22)

CR	Cabaret Rouge British Cemetery, Souchez (No. Fr 924)
CS	Canadian Cemetery, Sailly-sur-la-Lys (No. Fr 254)
CSC	Cologne Southern Cemetery, Cologne
CT	Contal Maison Chateau, Somme
DA	Daours Communal Cemetery, Daours, France
DC	Douai Communal Cemetery (No. Fr 1277)
DUL	Prisoner of War Cemetery, Dulmen, Germany
ES	Estaires Communal Cemetery (No. Fr 768)
ET	Etaples Military Cemetery (No. Fr 40)
FL	Flatiron Copse Cemetery, Mametz, Somme (No. Fr 453)
GZ	Gezaincourt Communal Cemetery Extension, Somme (No. Fr 168)
HA	Hazebrouck Cemetery (No. Fr 200)
HB	Hamburg Ohlsdorf Cemetery (No. Ger 2)
HS	Heilly Station Cemetery, Mericourt (No. Fr 833)
LA	Laventie Military Cemetery (No. Fr 1887)
LG	La Gorgue Communal Cemetery (No. Fr 354)
LI	Lille Southern Cemetery (No. Fr 1027)
LM	Loos Memorial (No. MR 19)
LO	Longuenesse (St Omer) Cemetery (No. Fr 134)
LS	Lijssenthock Military Cemetery (No. Bel 11)
LT	Le Trou Aid Post Cemetery, Fleurbaix (No. Fr 566)
MM	Meteren Military Cemetery (No. Fr 324)
MO	Morlancourt British Cemetery No. 1, Somme (No. Fr 633)
MV	Merville Community Cemetery Ono. Fr 345)
NC	Neufchatel Cemetery
NI	New Irish Farm Cemetery (No. Bel 96)
NW	Niederzwehren Cemetery (No. Ger 3)
OV	Ovillers Military Cemetery (No. Fr 393)

PH	Pont du Hem Military Cemetery, La Gorgue (No. Fr 705)
PM	Ploegsteert Memorial (No. MR 32)
PU	Puchevillers British Cemetery, Somme (No. Fr 74)
PW	Fromelles (Pheasant Wood) Military Cemetery, Fromelles
PZ	Pozieres British Cemetery (No. Fr 832)
RB	Rue-du-Bois Cemetery, Fleurbaix (No. Fr 348)
RC	Roiseol Communal Cemetery Extension (No. Fr 528)
RD	Rue David Cemetery, Fleurbaix (No. Fr 347)
RF	Ration Farm Military Cemetery (No. Fr 82)
RI	Royal Irish Rifles, Laventie (No. Fr 706)
RP	Rue Petillon Military Cemetery (No. Fr 525)
RTA	Returned to Australia
SS	St Sever Cemetery, Rouen (No. Fr 145-146)
TA	Trois Arbres Cemetery, Steenwerck (No. Fr 297)
TC	Tyne Cot British Military Cemetery (No. Bel 125)
TD	Thistle Dump Cemetery, High Wood, Longeval.
TM	Thiepval Memorial (No. MR21)
VA	Valenciennes Communal Cemetery (No. Fr 1142)
VB	Villers-Bretonneux Memorial (No. MR16)
VC	VC Corner Cemetery Memorial, Fromelles (No. MR 7)
VR	Varennes Military Cemetery (No. Fr 41)
WA	Warloy-Baillon Communal Cemetery Extension, Somme (No. Fr 43-44)
WI	Wimereux Communal Cemetery (No. Fr 64)
YF	Y Farm Military Cemetery, Bois Grenier (No. Fr 276)

UNITED KINGDOM CEMETERIES

UA	Fort Pitt Military Cemetery, Kent
UB	Kensal Green Cemetery, London

UC	Cambridge Borough Cemetery, Fenditton, Cambridgeshire
UD	Tottenham & Wood Green Cemetery, Middlesex
UE	Manchester Southern Cemetery
UF	All Saints Churchyard Extension, Orpington, Kent
UG	Brighton (Bear Road) Borough Cemetery, Sussex
UH	Nunhead Cemetery, London
UI	Keighley (Morton) Cemetery, York
UJ	Southend-on-sea (Sutton Road) Cemetery, Essex
UK	Brookwood Military Cemetery, Surrey
UL	Welford Road Cemetery, Leicester
UM	Melcombe Regis Cemetery, Dorset
UN	Stoke on Trent Cemetery, Staffordshire
UO	Epsom Cemetery, Surrey
UP	Arnos Vale Cemetery, Bristol
UQ	Lodge Hill Cemetery, Birmingham
UR	Nottingham General Cemetery, Nottinghamshire
US	Wandsworth Cemetery, London
UT	Ingham Churchyard, Suffolk
UU	Chelmsford Borough Cemetery
UV	Clacton Cemetery, Essex
UW	Sheffield (Burngreave) Cemetery, Yorkshire
UX	Portsmouth (Kingdom & Milton Cemeteries) Hants
UY	Bilton (St Mark's) Cemetery, Warwickshire
UZ	Warwick Cemetery, Warwickshire
WW	Wandsworth Cemetery (Australian Section), London
XA	Leeds (Lawn Wood) Cemetery, Yorkshire
XB	Little Wrattling (St Mary) Churchyard, Suffolk
XC	High Wycombe Cemetery, Buckinghamshire

XD	Adstock (St. Cecilia), Buckinghamshire
XE	Edinburgh Eastern Cemetery, Scotland
XF	Hastings Cemetery, Sussex
XG	Hetten Cemetery, Durham
XH	Greenwich Cemetery, London
XI	Norwich Cemetery, Norfolk
XJ	Northampton (Towcester Road) Cemetery
XK	Birmingham (Lodge Hill) Cemetery
XL	Cambridge City Cemetery
XM	Ipswich Cemetery, Suffolk
XN	Shorncliffe Military Cemetery, Kent
XO	Polmont Church, Stirlingshire, Scotland
GSS	Guildford (Stoke) Cemetery, Surrey
SAJ	St. Andrews & Jesmond Cemetery, Newcastle-On-Tyne, Northumberland.

BRITISH ROLL OF HONOUR

61st (2nd SOUTH MIDLAND) DIVISION

182nd BRIGADE:

2/6th ROYAL WARWICKSHIRE REGIMENT

o.	Rank	NAME	FATE	DATE	BURIED Or LISTED
6152	Pte	Adam, John	KIA	1916-07-19	LM
241891	Sgt	Allan, Andrew	KIA	1916-07-19	LM
3504	Pte	Allan, Leonard Walter	KIA	1916-07-19	LM
241890	Pte	Anderson, George	DOW	1916-09-10	LM
2177	Pte	Badger, Alfred	KIA	1916-07-19	LM
6296	Cpl	Baldwin, Percy Charles	KIA	1916-07-19	LM
4679	L/Cpl	Ballinger, George	DOW	1916-10-08	LM
240808	A/L/Sgt	Band, Ernest	KIA	1916-07-19	LM
()	2Lt	Barron, Louis	KIA	1916-07-19	LM
6054	Pte	Bignell, William Henry	KIA	1916-07-19	LM
240420	L/Sgt	Billers, William	KIA	1916-07-19	LM
241352	Pte	Bliss, Percy Charles	KIA	1916-07-19	LM
3872	Pte	Brittle, Bertram	KIA	1916-07-19	LM

4865	Pte	Brooks, John William	KIA	1916-07-19	LM
240903	Cpl	CHANCE, JOHN R.	KIA	1916-07-19	LM
6061	Pte	Charman, A W	DOW	1916-07-21	MV
6206	Pte	Clark, Henry James	KIA	1916-07-19	LM
2219	Pte	Clifford, W H	DOW	1916-09-26	LA
1636	Cpl	Collins, Edwin John	KIA	1916-07-19	LM
26361	Pte	Cook, T J	DOW	1916-09-05	RI
3037	L/Cpl	Cooke, W	DOW	1916-10-26	LG
2349	Pte	Cooper, Victor Frederick	KIA	1916-07-19	LM
()	Capt	Coulton, Aubrey Ewan	KIA	1916-07-19	PH
6066	Pte	Dale, Albert	KIA	1916-07-19	LM
3222	Pte	Darney, James Alfred	KIA	1916-07-19	LM
398	CSM	Davinson, Frank	KIA	1916-07-19	LM
6068	Pte	Dennett, William	KIA	1916-07-19	LM
20563	Pte	Dimbleby, Fred	DOW	1916-10-26	LA
6162	Pte	Donald, Tom	KIA	1916-07-19	LM
3263	Pte	Downing, John	KIA	1916-07-19	LM
2333	Pte	Evans, John	KIA	1916-07-19	LM
2828	Pte	Farr Harold John	KIA	1916-07-19	LM
6307	Pte	Field, Percy Frederick	KIA	1916-07-19	LM
242012	L/Cpl	Fielder, Edward Cecil	KIA	1916-07-19	LM
241945	Pte	Fitzgerald, Gerald William	KIA	1916-07-19	LM
()	2Lt	Flesher, A	DOW	1916-09-27	XA
3074	Pte	Frederick, G	KIA	1916-07-19	AR

6226	Pte	Gamblin, Ernest Frank	KIA	1916-07-19	LM
3277	Cpl	Garbett, F	KIA	1916-07-19	AR
2825	Pte	Glover, A W	DOW	1916-07-25	LO
202493	Pte	Gray, Robert George Freeman	DOW	1916-08-08	LM
2200	Pte	Green, Oliver	KIA	1916-07-19	LM
6079	Pte	Greenfield, Ernest Alfred	KIA	1916-07-19	LM
4841	Pte	Grieves, Austin	KIA	1916-07-19	LM
4395	Pte	Griffiths, Harry	KIA	1916-07-19	LM
3743	A/L/Sgt	Hadley, Horace Hewin	KIA	1916-07-19	LM
()	2Lt	Harper, Hal	KIA	1916-07-19	PH
2569	Sgt	Harris, P J	DOW	1916-07-20	MV
3268	Pte	Hemming, B J	DOW	1916-07-22	LO
6085	Pte	Hewitt, W	KIA	1916-07-19	AR
2867	Sgt	Hillman, G	KIA	1916-07-19	AR
3355	Cpl	Hollier, W A	KIA	1916-07-19	AR
4874	Pte	Hopkins, Benjamin	KIA	1916-07-19	LM
2093	L/Cpl	Hughes, William	KIA	1916-07-19	LM
()	2Lt	Hurdman, Cyril	KIA	1916-07-19	PH
1861	Dmr	Hyett, Charles Edward	KIA	1916-07-19	LM
6170	Pte	Ireland, David L	KIA	1916-07-19	LM
6246	L/Cpl	James, J C	DOW	1916-08-26	UX
6314	Pte	Jenner, Albert Ernest	KIA	1916-07-19	LM
4825	Pte	Jones, James William	KIA	1916-07-19	LM

2847	Pte	Jones, Robert Ellesmere	KIA	1916-07-19	LM
5515	L/Cpl	Joyce, A H	DOW	1916-07-22	BO
3545	Pte	Kenny, G	DOW	1916-07-21	MV
1660	Pte	King, C	DOW	1916-07-20	LG
241019	L/Cpl	Lamb, William Owen	KIA	1916-07-19	LM
4876	Pte	Law, William John	KIA	1916-07-19	LM
6175	Pte	Leishman, R	KIA	1916-07-19	AR
3725	Pte	Lloyd, Alfred	KIA	1916-07-19	LM
4935	Pte	Lucking, Maxwell	KIA	1916-07-19	LM
3354	Cpl	Luckman, C	KIA	1916-07-19	AR
6251	Pte	Luxton, Bertram Edward	KIA	1916-07-19	LM
3615	Pte	Major, W H	KIA	1916-07-19	AR
6182	Pte	McMain, Robert Kirk	DOW	1916-10-03	LA
36285	Pte	Minchin, George Victor	DOW	1916-09-03	AN
2937	Pte	Minett, George Henry	KIA	1916-07-19	LM
241904	Pte	Mitchell, Maxwell	KIA	1916-07-19	LM
2741	Pte	Morris, Norman Leslie	KIA	1916-07-19	LM
3716	Pte	Newby, W A	KIA	1916-07-19	AR
6256	Pte	Norman, Thomas George	KIA	1916-07-19	LM
5519	Pte	Peat A R	DOW	1916-07-22	MV
1794	Pte	Perrins H	DOW	1916-07-20	MV
3486	Pte	Peters, J H	KIA	1916-07-19	AR
6324	Pte	Pitman, Edgar Victor	KIA	1916-07-19	LM
6113	Pte	Pollard, Arthur	KIA	1916-07-19	LM

3821	Pte	Powell, William George	KIA	1916-07-19	LM
6266	Pte	Powley, Alfred George	KIA	1916-07-19	LM
5523	Pte	Ralph, Arthur James	KIA	1916-07-19	LM
6117	Pte	Randall, E W	KIA	1916-07-19	AR
3558	Pte	Reeves, A E	DOW	1916-07-20	LG
4882	Pte	Roberts, C J	DOW	1916-10-03	LA
3904	Pte	Rowson, Harry	KIA	1916-07-19	LM
6123	Pte	Ruttens, Edward	KIA	1916-07-19	LM
3783	Pte	Sanders Charles Edward	KIA	1916-07-19	LM
2885	Sgt	Saunders, Walter Joseph	KIA	1916-07-19	LM
6125	Pte	Searle, John James	KIA	1916-07-19	LM
()	Capt	Simms, W	KIA	1916-07-19	AR
240688	Pte	Smallwood, Ernest	KIA	1916-07-19	LM
3497	Pte	Smith, B	KIA	1916-07-19	AR
240909	L/Cpl	Starling, Frederick	KIA	1916-07-19	LM
2853	Sgt	Stephens, Sydney	KIA	1916-07-19	LM
241873	A/Cpl	Stevens, Harry	KIA	1916-07-19	LM
6276	Pte	Surgenor, Leslie	KIA	1916-07-19	LM
240749	L/Cpl	Taylor, William Thomas	KIA	1916-07-19	LM
6277	Pte	Teague, George Edwin	DOW	1916-09-10	LM
3578	Pte	Teasdale, B L	KIA	1916-07-19	RI
4885	Pte	Tew, W E	DOW	1916-09-13	LO
6278	Pte	Thompson, Harold Frederick	KIA	1916-07-19	LM
6135	Pte	Tilley, E S,	KIA	1916-07-19	LM

6136	Pte	Tipper, W	KIA	1916-07-19	AR
3868	Pte	Tonks, W	KIA	1916-07-19	AR
6285	Pte	Waterman, Clinton Stewart	DOW	1916-07-19	LM
()	Capt	Wathes, Thomas Sidney	KIA	1916-07-19	PH
20498	Pte	Watkin, Frederick John	DOW	1916-10-24	MV
1997	Dmr	Watkins, Tom	KIA	1916-07-19	LM
3799	Pte	Webster, Frederick	KIA	1916-07-19	LM
241763	Pte	Wellen, Albert	DOW	1916-08-18	LM
20413	Pte	Wells, Frank	DOW	1916-10-27	MV
6339	Pte	Wells, Harold George	KIA	1916-07-19	LM
240991	Pte	Weston, Leslie	KIA	1916-07-19	LM
241996	Pte	Whitmore, Henry	KIA	1916-07-19	LM
6336	Pte	Winter, F	DOW	1916-09-12	LA
5528	Pte	Witten, Thomas	KIA	1916-07-19	LM
2835	Pte	Wood, George Cyril	KIA	1916-07-19	LM
241887	Pte	Woodhatch, William	KIA	1916-07-19	LM
6288	Pte	Woodward, George William C	KIA	1916-07-19	LM

2/7th ROYAL WARWICKSHIRE REGIMENT

	Pte	Adkins, Frederick	POW	1916-07-19	
	Pte	Alford, Ernest Charles	POW	1916-07-19	
3660	Cpl	Allen, William John	KIA	1916-07-16	LA

		Atkins, Thomas	POW	1916-07-19	
266487	Pte	Austin, Harry	KIA	1916-07-19	LM
3224	Sgt	Bai1ey, Christopher	KIA	1916-07-19	LM
3755	Pte	Bannister, W	DOW	1916-07-20	MV
266144	Pte	Barlow, Henry	KIA	1916-07-19	LM
3367	Pte	Barnett, Stephen	KIA	1916-07-19	LM
3654	Pte	Barnett, William Henry	KIA	1916-07-19	LM
	Pte	Basley, Wilfred	POW	1916-07-19	
3435	Pte	Bates, Harold	KIA	1916-07-19	LM
	Cpl	Baxter, Horace	POW	1916-07-19	
3456	Pte	Beckett, Harry	KIA	1916-07-19	LM
266208	Pte	Beesley, Albert	KIA	1916-07-19	LM
	Pte	Bell, David W	POW	1916-07-19	
4628	L/Cpl	Berry, H	KIA	1916-07-19	CR
()	Capt	Bethell, Thomas Henry	KIA	1916-07-19	PH
5488	Pte	Birchall, John	KIA	1916-07-16	LA
267143	L/Cpl	Birchall, Thomas	KIA	1916-07-19	LM
2922	Sgt	Blakemore, Albert James	KIA	1916-07-19	LM
	Pte	Blogg, Edward H	POW	1916-07-19	
	Pte	Breengan, Allan W	POW	1916-07-19	
8285	L/Cpl	Brown, Frank Llewellyn John	DOW	1916-09-30	DC
3469	Pte	Calloway, A J	DOW	1916-07-21	LA

	Pte	Calloway, George	POW	1916-07-19	
5591	Pte	Campbell, F	DOW	1916-08-08	MV
265148	L/Cpl	Chatwin, Alfred	KIA	1916-07-19	LM
267150	Pte	Christie, Frank	KIA	1916-07-19	LM
4648	Pte	Clarke, John Alfred	KIA	1916-07-19	LM
4645	L/Cpl	Clarke, W E	POW	1916-07-19	
????	Pte	Colledge, Frank	POW	1916-07-19	????
267217	Pte	Collins, Mitchell	KIA	1916-07-19	LM
5598	Pte	Coull, George	D	1916-07-19	LM
268583	Pte	Crompton, Harold	KIA	1916-07-19	LM
267218	L/Cpl	Cunningham, Robert	KIA	1916-07-19	LM
5499	Pte	Dare, L	DOW	1916-08-08	MV
3858	Pte	Davenport, Charles	DOW	1916-08-28	LM
3492	L/Cpl	Davenport, Henry Charles	D	1916-07-19	LM
5949	Pte	Davies, James Ernest	KIA	1916-07-19	LM
267414	L/Cpl	Davies, Robert Clifford	KIA	1916-07-19	LM
3231	Pte	Davis, Thomas Henry	KIA	1916-07-19	LM
266259	Pte	Desborough, Thomas Henry	KIA	1916-07-19	LM
	Cpl	Dilworth, Ernest	POW	1916-07-19	
3000	Pte	Dobbins, J	KIA	1916-07-19	AR
()	Capt	Donaldson, Geofffrey Boles	DOW POW	1916-07-19	PH

265990	Pte	Drinkwater, Frederick James	KIA	1916-07-19	LM
267224	Pte	Dryburgh, Alexander	KIA	1916-07-19	LM
	Pte	Dudley, Joseph	POW	1916-07-19	
	L/Cpl	Easton, Thomas W R	POW	1916-07-19	
	Pte	Enoch, Daniel	POW	1916-07-19	
	Pte	Evans, Gordon	POW	1916-07-19	
	Pte	Evans, Samuel	POW	1916-07-19	
265409	Pte	Farren, Joseph	KIA	1916-07-19	LM
20851	Pte	Foley, William	D	1916-07-19	PH
3764	L/Cpl	Furnival, Alfred	DOW	1916-09-20	LA
267233	Pte	Galloway, George I.	KIA	1916-07-19	PH
	Pte	Gibb, Leslie	POW	1916-07-19	
2914	Sgt	Gilbert, J T	DOW	1916-08-01	UY
3909	Pte	Godsell, H	KIA	1916-07-19	AR
3930	Pte	Goode, Absolom	KIA	1916-07-19	LM
267165	Pte	Grady, George Andrew	KIA	1916-07-19	LM
267239	Pte	Gray, Alex	KIA	1916-07-19	LM
20860	Pte	Griffith, John McPherson	KIA	1916-07-19	LM
5958	Pte	Griffiths, John	KIA	1916-07-19	LM
	Pte	Gusley, John William	POW	1916-07-19	
	Pte	Hanlin, Thomas	POW	1916-07-19	
5728	Pte	Harries, J	KIA	1916-07-17	LA

5970	L/Cpl	Harris, Daniel	KIA	1916-07-19	LM
2216	Pte	Hayes, Frank Hinde	KIA	1916-07-19	LM
265933	Pte	Hazelwood, James	KIA	1916-07-19	LM
	Pte	Hobbs, Leonard	POW	1916-07-19	
5525	Pte	Holly, Robert Ballington	KIA	1916-07-19	LM
	Pte	Horn, William	POW	1916-07-19	
	Pte	Hughes, David G	POW	1916-07-19	
268591	Pte	Hulligan, M	KIA	1916-07-19	AR
267174	Pte	James, Benjamin	KIA	1916-07-19	LM
	Pte	Jenkins, John	POW	1916-07-19	
	Pte	Jenkins, William J	POW	1916-07-19	
266263	Pte	Jephcott, John	KIA	1916-07-19	LM
266351	L/Sgt	Jocelyn, Ernest	KIA	1916-07-19	LM
266341	Sgt	Johnson, Albert Joseph	KIA	1916-07-19	AU
5983	Pte	Jones, John Thomas	KIA	1916-07-19	LM
4624	Pte	Kendall, A	KIA	1916-07-20	MV
3206	Pte	Lamb, Alfred	KIA	1916-07-19	LM
5727	L/Cpl	Lewis, Thomas Stephen	KIA	1916-07-16	LA
267179	Pte	Lewis, W J	KIA	1916-07-19	AR
	Sgt	Lole, William Adams	POW	1916-07-20	
	Sgt	Lunn, Leonard	POW	1916-07-20	
5658	L/Cpl	Macdonald, Arthur	D	1916-07-19	LM

3896	Pte	Malin, W	DOW	1916-07-28	UZ
4209	Pte	Mapley, A W	DOW	1916-08-21	MV
	Pte	Mark, Edward C	POW	1916-07-19	
267261	L/Cpl	Marshall, David	KIA	1916-07-19	LM
	Pte	Marshall, Ernest	POW	1916-07-19	
5666	Pte	Maxwell, G	DOW	1916-07-25	XO
	L/Cpl	Mayor, Harry W	POW	1916-07-19	
	Pte	McDonald, Henry	POW	1916-07-20	
5664	Pte	McNab, W	DOW	1916-10-02	LA
267272	L/Cpl	Melville, John	KIA	1916-07-19	LM
	Sgt	Moncrieff, Alexander	POW	1916-07-19	
2132	Cpl	Moore, Harold	KIA	1916-07-19	LM
266167	Pte	Moore, Thomas	KIA	1916-07-19	LM
5984	Pte	Morgan, R	KIA	1916-07-19	AR
	Cpl	Muir, George	POW	1916-07-19	
266506	Pte	Neale, Alexander Grant	KIA	1916-07-19	LM
2851	CQMS	Nicholl, Duncan	DOW	1916-12-16	WA
2943	Sgt	Noakes, Robert Foster	KIA	1916-07-19	LM
5683	Pte	Paton, Ernest	KIA	1916-07-19	LM
2991	Pte	Pearman, A W	DOW	1916-07-29	UI
266488	Pte	Portman, Albert	KIA	1916-07-19	LM
5687	Pte	Pratt, George	KIA	1916-07-19	LM

6000	Pte	Price, H G	DOW	1916-09-04	LO
5688	L/Cpl	Pryde, John	D	1916-07-19	LM
	Pte	Rees, Ivor	POW	1916-07-19	
5962	Pte	Rees, William Simon	KIA	1916-07-19	LM
	Pte	Reynolds, Arthur	POW	1916-07-19	
267290	L/Cpl	Richardson, William	KIA	1916-07-19	LM
5693	Cpl	Robertson, W F	KIA	1916-07-20	LA
20871	L/Cpl	Ryan, Christopher	KIA	1916-07-19	LM
268586	L/Cpl	Sanderson, David	KIA	1916-07-19	LM
266353	Pte	Sharlott, Thomas Charles	KIA	1916-07-19	LM
	Sgt	Shaw, Thomas	POW	1916-07-19	
4618	L/Cpl	Sheffield, Sidney	D	1916-07-19	LM
268573	Pte	Shepherd, Charles	KIA	1916-07-19	LM
265192	Pte	Sheppard, Herbert Benjamin	KIA	1916-07-19	LM
267298	Cpl	Simpson, David	KIA	1916-07-19	LM
3859	L/Cpl	Smith, Alec James Victor	KIA	1916-07-19	LM
5704	Pte	Smith, John	KIA	1916-07-19	LM
265295	Cpl	Stanway, Herbert William	D	1916-07-19	LM
5711	Pte	Storrar, D	DOW	1916-09-20	LA
4206	Pte	Stuart, A C	DOW	1916-08-22	???
265915	Pte	Swift, John Charles	KIA	1916-07-19	LM
4233	Pte	Tallis, Thomas	KIA	1916-07-19	LM

3343	Pte	Tams, H F	DOW	1916-07-20	MV
4369	Dmr	Taylor, George Albert	KIA	1916-07-19	LM
	Pte	Tew, Frederick	POW	1916-07-19	
5712	Pte	Thom, David Low	D	1916-07-19	LM
4386	Pte	Topp, Aubrey	KIA	1916-07-19	LM
4638	Pte	Townley, Frederick	KIA	1916-07-19	LM
	Pte	Townson, John	POW	1916-07-20	
5715	Pte	Troupe, William	KIA	1916-07-16	MV
267305	Pte	Turnbull, Henry	KIA	1916-07-19	LM
266675	Pte	Twamley, Leonard	KIA	1916-07-19	LM
	Pte	Unith, Thomas A	POW	1916-07-20	
266674	Pte	Wale, W	KIA	1916-07-19	CR
	Pte	Walker, Arthur	POW Died	1918-11-21	CSC
	Pte	Wallace, James	POW	1916-07-20	
	Pte	Wallsgrove, Harry	POW	1916-07-19	
3354	L/Sgt	Ward, H	KIA	1916-07-19	LA
266345	Pte	Ward, Joseph	KIA	1916-07-14	LM
5948	L/Cpl	Watkins, Henry Percival	KIA	1916-07-19	LM
5718	L/Cpl	Welsh, Robert	D	1916-07-19	LM
2987	L/Sgt	Whateley, J R	KIA	1916-07-19	CR
266485	Pte	Whitcroft, Tom	KIA	1916-07-19	LM
2497	Sgt	White, William Harvey	D	1916-07-19	LM

2/8th ROYAL WARWICKSHIRE REGIMENT

5689	Pte	Anderson, J	DOW	1916-09-28	LA
4999	Pte	Bacon, John	DOW	1916-08-20	LM
5461	Pte	Bermingham, James	DOW	1916-10-20	MV
3261	Pte	Berry, Harold	DOW	1916-10-28	LA
5779	Pte	Bowers, J	DOW	1916-08-25	PH
5742	Pte	Bradbury, Alfred	DOW	1916-10-16	LA
3713	Pte	Brazier, W	DOW	1916-10-23	LA
306648	Pte	Bright, James	DOW	1916-08-27	LM
306669	Pte	Collins, Ernest Alfred	DOW	1916-08-27	LM
1570	Pte	Cotterill, Howard	KIA	1916-07-18	LM
3646	Sgt	Cowie, J D	DOW	1916-09-05	LA
3647	Dmr	Croshaw, Frank Herbert	DOW	1916-10-14	LA
2742	Sgt	Downard, C H	DOW	1916-08-16	MV
4160	Pte	Farmer, Arthur	KIA	1916-07-18	LM
306093	Cpl	Foster, Harry	DOW	1916-08-27	LM
()	2Lt	Guest, R V	DOW	1916-08-28	LO
306819	Pte	Hodges, Sydney Phillip	DOW	1916-08-27	LM
6417	Pte	Hough, W	DOW	1916-09-06	LA
306638	Pte	Ingram, William	DOW	1916-08-27	LM
1788	L/Cpl	Kidner, Walter James	DOW	1916-07-21	LA
306795	Pte	Langston, Samuel Elmer	DOW	1916-08-27	LM

306406	Pte	Lawrence, Percy Hayward	DOW	1916-08-27	LM
4428	L/Sgt	Leedham, George Edward	DOW	1916-08-14	PH
5407	Pte	Lewis, A V	KIA	1916-07-20	AR
306418	Pte	Miller, Christopher Sydney	DOW	1916-09-07	LM
306708	Pte	Miller, William John	DOW	1916-08-27	LM
2184	Pte	Milner, Horace Brooks	DOW	1916-08-18	PH
3040	Pte	Milward, Ernest	DOW	1916-07-21	LO
306685	Pte	Morgan, David William	DOW	1916-08-27	LM
5413	L/Cpl	Reynolds, William Arthur	DOW	1916-10-29	MV
3752	Pte	Rose, H	DOW	1916-08-25	PH
5684	Pte	Salmond, H	DOW	1916-10-23	LA
3229	CSM	Scrivener, J H	DOW	1916-09-19	BO
306803	Pte	Smith, Percy William	DOW	1916-08-27	LM
()	2Lt	Stone, Docksey	DOW	1916-07-21	LA
5388	Pte	Strickland, H	DOW	1916-07-30	LO
4434	Cpl	Topliss, Arthur	DOW	1916-09-05	LA
5601	Pte	Turner, J	DOW	1916-08-12	PH
5665	Pte	Watson, Frank	DOW	1916-07-31	BO
5391	L/Cpl	Watson, James Percival	DOW	1916-09-09	LA
5523	Pte	Wells, A S	DOW	1916-08-19	PH
306923	Pte	White, George	DOW	1916-09-07	LM

182nd MACHINE GUN COMPANY

14236	L/Cpl	Connors, Martin	KIA	1916-07-20	L M
	Sgt	Elliott, Houry Jas	POW	1916-07-19	
30877	Pte	Felsted, Charles	D	1916-07-20	LM
30508	Pte	Fisher, A	D	1916-07-19	CR
28681	Pte	Holdsworth, John	D	1916-07-20	LM
		Mahoney, K. Jas	POW	1916-07-19	
28697	Pte	Marshall, Albert Edward	D	1916-07-20	LM
28398	Pte	Rothwell, Joseph S.	D	1916-07-20	LM
28575	Pte	Scudamore, E F F	KIA	1916-07-20	CR
28683	Pte	Thornton, A	KIA	1916-07-20	AR
30525	Pte	Tipple, H	KIA	1916-07-19	CR
28579	Pte	Waller, Albert	D	1916-07-19	LM

183rd BRIGADE

2/4th GLOUCESTERSHIRE REGIMENT

201924	Pte	Bailey, Frank R	DOW	1916-09-04	LM
200801	Pte	Baker, Henry	D	1916-07-19	LM
201299	Pte	Barker, William	KIA	1916-07-19	LM
6328	Pte	Barnard, B	DOW	1916-07-28	BO

200979	Pte	Bees, F W	D	1916-07-19	LM
3291	Pte	Bees, Frederick W	KIA	1916-07-19	LM
6386	Pte	Bennett, Raymond P	KIA	1916-07-19	LM
201251	Pte	Bidgood, Frederick William	KIA	1916-07-17	LO
6320	Pte	Bland, Edward John	KIA	1916-07-19	LM
5201	Pte	Bleaken, Pte Albert F	KIA	1916-07-19	LM
2869	Pte	Bown D B	KIA	1916-07-19	LA
4026	Pte	Burrows, Walter E	KIA	1916-07-19	LM
()	Capt	Byers, R H K	KIA	1916-07-19	TM
6362	L/Cpl	Cave, C A	DOW	1916-07-22	LO
6249	Pte	Cawston H J	DOW	1916-07-30	LA
6235	Sgt	Chaplin, Robert W	KIA	1916-07-19	LM
5183	Pte	Chitty, A	KIA	1916-07-19	RB
3247	Pte	Colbourne, W	DOW	1916-08-11	MV
201735	Pte	Cooper, Frederick	KIA	1916-07-19	LM
1580	Pte	Cox, H F	KIA	1916-07-19	LA
201462	Pte	Creed, Arthur	DOW	1916-09-06	LM
6408	Pte	Critchley, C	DOW	1916-08-02	MV
3709	Pte	Cryer, Herbrt Henry	DOW	1916-10-04	MV
4435	Pte	Daniels, W	DOW	1916-07-19	TM
4533	Pte	Darke, Harold	KIA	1916-07-19	LM
5117	Pte	Davey, William E	KIA	1916-07-19	LM

5139	Pte	Deverell, Walter J	KIA	1916-07-19	LM
5053	Pte	Driver, V	KIA	1916-07-19	LA
3339	Pte	Duggan, W	KIA	1916-07-19	RB
201814	Pte	Durrant, Walter W	DOW	1916-09-06	LM
5109	Pte	Edwards, Arthur T	KIA	1916-07-19	LM
6358	Pte	Enefer, John	DOW	1916-08-09	LA
3620	Pte	Evans, Percival Gilbert Henry	KIA	1916-07-19	LM
5419	Pte	Gibbons, John B	DOW	1916-09-06	LM
2827	Pte	Hale, P D	KIA	1916-07-19	LA
202767	Pte	Hampson, Edward C	D	1916-07-19	LM
20715	L/Cpl	Harper, T	DOW	1916-10-16	PH
()	Lt	Hart, R M	DOW	1916-07-25	CT
4364	Pte	Hartrey, W	KIA	1916-07-18	LA
3808	Sgt	Holley, F	DOW	1916-08-09	LA
5249	Pte	Howard, F A	DOW	1916-08-07	LA
()	Lt	James, J C	KIA	1916-07-19	WI
4620	Pte	Jay, A E	KIA	1916-07-19	RB
202775	Pte	Mason, Herbert	D	1916-07-19	LM
4718	Pte	Mead, S	DOW	1916-08-25	PU
3597	L/Cpl	Miller, H W	KIA	1916-07-19	LA
2829	Cpl	Milliner, Wilfred Ernest	DOW	1916-10-24	PH
3331	Pte	Moore, C O	DOW	1916-07-29	WA

6233	Cpl	Moss, J N	KIA	1916-07-19	LA
20726	Pte	Newport, T	DOW	1916-08-26	PH
201838	Pte	Norman, Arthur	DOW	1916-09-04	LM
201256	Pte	Pagram, Arthur	DOW	1916-09-06	LM
3752	Pte	Paver, William	KIA	1916-07-19	LM
201152	Pte	Phillips, Ronald	D	1916-07-19	LM
20729	Pte	Pocock, H	KIA	1916-07-19	RB
()	Capt	Pratt, Ralph Lewis	DOW	1916-10-18	PH
1918	Pte	Richings, Albert Victor	KIA	1916-07-19	LM
()	Lt	Scrase, R G	KIA	1916-07-19	TM
()	2Lt	Stagg, A C	KIA	1916-07-19	MO
4623	Pte	Sullivan, H G	DOW	1916-09-02	CN
1857	Pte	Taylor, Walter	KIA	1916-07-19	LM
200277	L/Cpl	Thomas, William H	KIA	1916-07-19	LM
201745	Pte	Tiley, Reginald	DOW	1916-09-04	LM
6384	Pte	Tuke, W A B	KIA	1916-07-19	RB
5353	Pte	Tuplin, C	DOW	1916-09-06	HS
6294	Pte	Wakeling, J	DOW	1916-08-09	LA
4873	Pte	Watkins, E H	KIA	1916-07-18	LA
3832	Cpl	Wheeler, Sidney	KIA	1916-07-19	LM

2/5th GLOUCESTERSHIRE REGIMENT

4651	Pte	Agg, R J	DOW	1916-09-25	PH
5800	Pte	Angus, Leslie Dean	DOW	1916-07-27	???
3777	Cpl	Askew, A H	DOW	1916-08-31	LA
4837	Pte	Bishop, Richard B	DOW	1916-09-04	LM
2801	L/Cpl	Brain, H P G	DOW	1916-09-25	PH
5938	Pte	Clarkson, A	DOW	1916-09-25	PH
5869	Pte	Coe, W	DOW	1916-10-16	PH
3195	Pte	Dartnall, W J	DOW	1916-09-25	PH
1943	Cpl	Davis, Walter George	DOW	1916-08-19	LA
3342	L/Cpl	Godsell, Alec Charles	DOW	1916-09-25	PH
5910	Pte	Hornsey, H W	DOW	1916-08-17	MV
5888	L/Cpl	Howland, S P	DOW	1916-10-25	MV
5097	Pte	Jenman, Alfred	DOW	1916-09-24	LO
2096	Pte	Lafford, J	DOW	1916-09-25	PH
4853	Sgt	Murrell, O	KIA	1916-07-15	MO
4824	Pte	Page, Albert E	DOW	1916-09-04	LM
3552	Sgt	Palmer, John Edmund	DOW	1916-08-23	LA
2460	Pte	Rowe, S A	DOW	1916-07-21	TM
2128	L/Cpl	Saunders, W H	DOW	1916-08-19	LA
5881	Pte	Skinner, J W	DOW	1916-07-31	PH
3370	Cpl	Stephens, E L	KIA	1916-07-15	MV

5780	Pte	Townsing, Stanley Gordon	DOW	1916-07-24	???
4500	Pte	Ursell, Alfred	DOW	1916-09-04	LM
4915	Pte	Van Santen, S S	DOW	1916-09-01	LA
3172	Pte	Wasley, A G	DOW	1916-07-27	???
5142	Pte	Williams, William	DOW	1916-08-24	LA
5785	Pte	Wilson, A	KIA	1916-07-18	LA
4556	Pte	Windsor, Herbert Josiah	DOW	1916-09-23	MV

2/6th GLOUCESTERSHIRE REGIMENT

267172	Sgt	Bacon, Charles E	D	1916-07-19	LM
3718	Pte	Bailey, Frank Walter	KIA	1916-07-19	LM
4886	Pte	Baker, A W	KIA	1916-07-19	LA
267229	Pte	Barnes, George	D	1916-07-19	LM
4806	Pte	Beth , W	KIA	1916-07-19	LA
()	Capt	Bird, Ernest Walter	DOW	1916-07-27	GZ
267166	Sgt	Bird, Herbert	D	1916-07-19	LM
266731	Pte	Bond, Arthur G	D	1916-07-19	LM
5916	Pte	Broomfield, John P	KIA	1916-07-19	LM
4837	Pte	Brown, E J	KIA	1916-07-19	LA
4810	L/Cpl	Butler, C G W	DOW	1916-07-20	MV
5940	Cpl	Cater, Frederick	KIA	1916-07-19	LA
4832	Pte	Chapman, Arthur Gooding	KIA	1916-07-17	LA

266742	Pte	Charlwood, Frederick John	DOW	1916-07-19	LM
265346	Pte	Chippett, Arthut F	D	1916-07-19	LM
3185	L/Sgt	Chivers, Alfred Bertie	KIA	1916-07-17	LA
265959	Pte	Clements, Bertie A	DOW	1916-09-04	LM
5993	Pte	Cobb, H	DOW	1916-07-30	LO
266205	Pte	Collins John	DOW	1916-09-04	LM
3152	Pte	Cook, Alfred John	KIA	1916-07-19	LM
4984	L/Cpl	Cooke, Wilfred	DOW	1916-09-10	CR
5937	Sgt	Cooper, Sidney Joseph	DOW	1916-09-30	MV
5977	Pte	Cubitt, Stanley Arthur	DOW	1916-07-22	BO
267253	Pte	Cundy, Edward	KIA	1916-07-19	LM
4502	2Lt	Derrick, William	DOW	1916-07-22	LM
267207	Pte	Dixon, Charles	D	1916-07-19	LM
5018	Pte	Dore, H T	DOW	1916-07-20	MV
4856	Pte	Eason, Edward S	KIA	1916-07-19	LM
4615	Pte	Edwards, William S T	KIA	1916-07-19	LM
267247	Pte	Elmer, Charles Edgar	DOW	1916-08-04	LM
()	Capt	Eyre, H W	DOW	1916-07-29	LO
2176	Pte	Fegan, James	KIA	1916-07-17	LA
266280	Pte	Feltham, Frederick C	KIA	1916-07-19	LM
()	Lt	Fison, Frank Henry	KIA	1916-07-19	AR
3748	Pte	Foxon, L T	KIA	1916-07-19	LA

6022	Pte	Frost, B	KIA	1916-07-19	AR
()	2Lt	Fry, Leonard	KIA	1916-07-19	RP
3494	Pte	Gawler, Charles S	KIA	1916-07-19	LM
2108	Pte	Godfrey, John Robert Charles	DOW	1916-07-23	BO
4951	Pte	Griffiths, Cyril G	KIA	1916-07-19	LM
3876	Pte	Gwyther, C R	DOW	1916-07-20	LG
4966	Cpl	Hall, A	DOW	1916-09-01	PH
4946	Pte	Hancock, W S	KIA	1916-07-19	PH
2743	Pte	Hawkings, Arthur G	KIA	1916-07-19	LA
3085	Cpl	Hemmings, Royston Albert	DOW	1916-09-28	PH
267263	Pte	Hickmott, Henry	D	1916-07-19	LM
266295	Pte	Highnam, James H	KIA	1916-07-19	LM
265291	L/Cpl	Hobbs, William R	D	1916-07-19	LM
4947	Pte	Hodges, Edwin E	KIA	1916-07-19	LM
266662	Pte	Holden, Bertram John	D	1916-07-19	LM
6152	Pte	Hooper, A R	KIA	1916-07-19	CR
3592	Pte	Howell, Edward Coulston	KIA	1916-07-19	RB
6064	Pte	Hubbard, P	DOW	1916-08-22	PH
2515	Pte	Humphries, Albert Edward John	DOW	1916-09-06	LM
4957	Pte	James, C H	KIA	1916-07-19	PH
2439	Cpl	Johnson, Ernest	KIA	1916-07-19	LA

()	2Lt	Johnston, Leslie	DOW	1916-07-25	MV
4945	Pte	Jones, W	DOW	1916-08-02	LA
3675	L/Cpl	King, Walter A	KIA	1916-07-19	LM
266604	Pte	Lacey, Ernest	KIA	1916-07-19	LM
5958	Pte	Lambert, William Arthur	DOW	1916-07-23	MV
()	Lt	Langford, William John	KIA	1916-07-19	LM
4669	Pte	Lanham, Lawrence Edward	DOW	1916-09-04	LM
()	Lt	Lewis, Clifford Stanley	DOW	1916-09-19	MV
3858	Pte	Lewis, H	KIA	1916-07-19	LA
267236	Pte	Lupson, George Frank	D	1916-07-19	LM
2308	Pte	Manning, William George	KIA	1916-07-14	LA
266764	Pte	Marshall, William John Samuel	D	1916-07-19	LM
4887	Pte	Matthews, H C	KIA	1916-07-19	LA
266236	Pte	McCormick, William	D	1916-07-19	LM
267285	L/Cpl	Miles, Edward A	KIA	1916-07-19	LM
4802	Pte	Nelmes, A E	DOW	1916-08-07	LG
5981	Pte	Nicholson, W I	KIA	1916-07-19	LA
266132	Pte	Parsons, Arthur Arthur	D	1916-07-19	LM
4674	L/Sgt	Pearce, Charles W E	KIA	1916-07-19	RB
5914	Pte	Pennington, Thomas	KIA	1916-07-19	RB
266620	Pte	Perring, William J	DOW	1916-09-06	LM
3448	Pte	Pike, A	KIA	1916-07-19	RB

3382	Pte	Potter, C E	DOW	1916-08-28	PH
4797	Pte	Powell, T	KIA	1916-07-15	LA
4220	Pte	Putnam, H	DOW	1916-09-28	PH
3986	Pte	Raisey, Albert V	D	1916-07-19	LM
4932	Pte	Ralph, Claude Edward	DOW	1916-07-20	MV
1962	L/Cpl	Rich, Frank	KIA	1916-07-19	LA
4744	Pte	Robbins, George Henry	KIA	1916-07-19	LA
3711	Pte	Rolfe, Stanley	KIA	1916-07-19	CR
3959	Pte	Routley, Joseph	KIA	1916-07-19	LM
()	Capt	Rudman, Harold Ewart	KIA	1916-07-19	LM
2685	L/Cpl	Sampson, H C	KIA	1916-07-19	AR
4668	Pte	Sharp, C	DOW	1916-08-26	PH
265283	Pte	Sheppard, William J	KIA	1916-07-19	LM
267208	L/Sgt	Shreeve, Alan Robert	D	1916-07-19	LM
266289	Pte	Sims, Charles	DOW	1916-09-06	LM
4860	Pte	Smith, Walter J	D	1916-07-19	LM
267287	Pte	Spooner, Peter T	D	1916-07-19	LM
5941	Sgt	Stallworthy, W J	DOW	1916-07-20	LA
266734	Pte	Stockham, William	D	1916-07-19	LM
266695	Pte	Stoker, Albert Tom	D	1916-07-19	LM
266591	Pte	Sweetman, Alfred Henry	D	1916-07-19	LM
5014	Pte	Thomas, F	DOW	1916-08-26	PH

2659	Pte	Thorne, Frank	KIA	1916-07-19	LM
3043	Pte	Wall, Henry William	DOW	1916-08-28	PH
4071	Pte	Whitlock, E E	KIA	1916-07-19	LA
4923	Pte	Williams, J	KIA	1916-07-17	LA
5956	Pte	Willimott, Harry Leonard	KIA	1916-07-19	LM
265708	Cpl	Willis, Frederick J	D	1916-07-19	LM
4940	Pte	Willis, William Russell	D	1916-07-19	LM
4113	Pte	Wilmott, Caleb	DOW	1916-09-04	LM
5983	Pte	Winterborn, James Edwin	D	1916-07-19	LM

183rd MACHINE GUN COMPANY

28509	L/Cpl	Wass, R T	KIA	1916-07-19	LA
28606	L/Sgt	Whincop, H M	KIA	1916-07-19	LA

184th BRIGADE

2/4th ROYAL BERKSHIRE REGIMENT

()	2Lt	Abbott, G S	KIA	1916-07-19	LA
3376	Pte	Barber, R	KIA	1916-07-19	LA
3569	Pte	Bates, C	KIA	1916-07-19	LA
()	Lt. Col	Beer, J H	KIA	1916-07-19	LA
6670	Pte	Beresford, F	KIA	1916-07-19	LA

6004	Pte	Blades, A	DOW	1916-07-20	MV
6610	Pte	Brazier, E V	KIA	1916-07-19	LA
R	Pte	Brown, A	KIA	1916-07-19	LA
202104	Pte	Buckingham, H	DOW	1916-08-14	TM
2923	Pte	Cannon Pte W	DOW	1916-07-20	TM
2Lt	()	Cecil, Charles	KIA	1916-07-16	LA
3760	L/Cpl	Church, F R	KIA	1916-07-19	LA
6611	Pte	Cole, H V	KIA	1916-07-19	LA
2963	L/Cpl	Cooper, A D	DOW	1916-08-15	WA
3398	Pte	Couzens, W E	KIA	1916-07-19	LA
5334	Pte	Davis, C	KIA	1916-07-19	LA
3328	Pte	Day, A	DOW	1916-07-20	MV
6558	Pte	Fleet, C T	KIA	1916-07-19	LA
3974	Cpl	Fountain, H E	KIA	1916-07-19	LA
5277	Pte	Gibbons, L	DOW	1916-07-20	MV
202023	Pte	Guy, J F	DOW	1916-08-14	TM
3391	Pte	Hallum, H	KIA	1916-07-19	LA
200319	Pte	Hester, James William	DOW	1916-07-30	CR
3380	Pte	Hollands, J A	KIA	1916-07-19	LA
5302	Pte	Hooper, W E	KIA	1916-07-19	LA
5267	Pte	Lainsbury, F	DOW	1916-07-30	LO
5306	Pte	Martin J W	KIA	1916-07-19	LA

6563	Pte	Mayes, F	DOW	1916-08-05	XB
6574	Pte	Mayes, J	DOW	1916-10-07	LG
6675	Pte	McGhan, J	KIA	1916-07-19	LA
2143	Pte	Mulford, F	KIA	1916-07-19	LA
3975	Pte	North, R	DOW	1916-08-10	LA
3483	Pte	Painter, T C	KIA	1916-07-19	LA
3545	Pte	Rivers, A T	KIA	1916-07-19	LA
3390	Pte	Rose, J	KIA	1916-07-19	LA
5496	Pte	Rouse, J V	DOW	1916-08-18	VR
2507	Pte	Savage, F E	KIA	1916-07-19	LA
3619	Pte	Savory W J	KIA	1916-07-19	LA
5502	Pte	Searle, W H	DOW	1916-08-28	LA
5807	Pte	Sellwood, William George	DOW	1916-09-02	PU
6289	Cpl	Smith, F	KIA	1916-07-19	LA
3800	Sgt	Snow, G	KIA	1916-07-19	LA
5036	L/Cpl	Stallard, H	DOW	1916-08-25	ET
5312	Pte	Swain, A T	KIA	1916-07-19	LA
2769	Pte	Tull, C A	DOW	1916-09-02	LA
6591	Pte	West, F V	KIA	1916-07-19	LA
()	2Lt	Williams, F C D	KIA	1916-07-19	LA
6673	Pte	Williams, R H	KIA	1916-07-19	LA
3547	Pte	Wooldridge, C	DOW	1916-08-18	LG

2/1st BUCKINGHAMSHIRE REGIMENT

266591	Pte	Aldridge, Arthur	KIA	1916-07-19	LM
265211	Pte	Andrews, Frank Herbert	KIA	1916-07-19	LM
()	Lt	Atkinson, Geoffrey William	KIA	1916-07-19	LM
810	Sgt	Austin, Charles	DOW	1916-07-22	LM
3460	Pte	Avis, Stanley Harold	KIA	1916-07-19	LM
6559	Sgt	Bailey, Reginald David	DOW	1916-07-27	LA
4390	Pte	Baldwin, Horace	KIA	1916-07-19	LM
267121	Pte	Barkaway, Frederick William	KIA	1916-07-19	LM
3316	Cpl	Battams, Edward Charlie	KIA	1916-07-19	LM
266401	Pte	Bennett, Horace	KIA	1916-07-19	LM
3469	Pte	Biggs, H W	DOW	1916-07-25	ET
3428	Pte	Birch, S A	KIA	1916-07-19	LA
6013	Pte	Blissett, E	DOW	1916-10-08	PH
5731	Pte	Boddy, Evan Marlett Nepean	DOW	1916-08-14	LM
2290	Pte	Bond, J R	DOW	1916-07-19	MV
266226	Pte	Bone, William Joseph	DOW	1916-07-19	LM
4131	Pte	Bowie, Pte John Lumsden	KIA	1916-07-19	LM
5396	Pte	Boyer, Robert	KIA	1916-07-18	LG
()	2Lt	Brewin, H R N	KIA	1916-07-19	LA
3149	CSM	Brown, Arthur	KIA	1916-07-19	LM

2518	Pte	Brown, E J	KIA	1916-07-19	RB
2982	CSM	Brown, R W	DOW	1916-07-20	MV
266515	L/Cpl	Bruce, William Henry Victor	KIA	1916-07-19	LM
3558	Pte	Bunce, John	KIA	1916-07-19	LM
265764	L/Cpl	Burnard, James	KIA	1916-07-19	LM
4303	Pte	Burrows, A T	KIA	1916-07-19	LG
4304	Pte	Burrows, Albert	KIA	1916-07-18	LG
201856	Pte	Butler, George Henry	DOW	1916-08-14	LM
5415	Pte	Butt, J	KIA	1916-07-19	LA
3324	Cpl	Callaway, Charles Mavies	DOW	1916-08-27	LA
201717	Pte	Carpenter, George	DOW	1916-10-07	LM
4214	Pte	Carter, A G	KIA	1916-07-19	LA
266618	Pte	Carter, R	KIA	1916-07-19	LM
4229	Pte	Carter, Roland	KIA	1916-07-19	LM
267077	Pte	Carver, George Edward	KIA	1916-07-19	LM
266097	L/Cpl	Casmore, William	KIA	1916-07-19	LM
266612	Pte	Castle, George Alfred	KIA	1916-07-19	LM
()	Lt	Chadwick, D G	KIA	1916-07-20	MV
()	Capt	Church, H	KIA	1916-07-19	LA
266638	Pte	Clark, Ernest Arthur Percy	KIA	1916-07-19	LM
5354	Pte	Coleman, Ernest	KIA	1916-07-19	LM
265265	L/Cpl	Cook, Albert	KIA	1916-07-19	LM

266006	Pte	Cook, George	KIA	1916-07-19	LM
5412	Pte	Cook, H J	DOW	1916-07-19	MV
2585	Pte	Cox, J	DOW	1916-07-25	MV
2255	Pte	Cox, N	KIA	1916-07-18	ES
266713	Pte	Cox, Thomas Atkins	KIA	1916-07-19	LM
2737	Sgt	Crane, Joseph	DOW	1916-10-15	MV
1579	Sgt	Crump, J C	KIA	1916-07-19	AR
1580	Pte	Crutchfield, Frederick	KIA	1916-07-19	LM
3571	Pte	Cummins, Frederick	DOW	1916-07-23	LM
4265	Pte	Damant, S R	DOW	1916-07-25	MV
4228	Pte	Daveney, Thomas	KIA	1916-07-19	LM
6961	L/Cpl	Davies, John Eleazar Garnant	DOW	1916-10-14	PH
1614	Pte	Davis, Albert Edward	KIA	1916-07-19	LM
1718	Sgt	Davis, George Ellis	KIA	1916-07-19	LM
266125	L/Cpl	Dell, Cecil Clark	KIA	1916-07-19	LM
4345	Pte	Dimmock, F J R	DOW	1916-07-25	WI
5324	Pte	Doyle, A L	DOW	1916-07-24	MV
5416	Pte	Durrant, E	DOW	1916-07-19	MV
3294	Pte	Dwight, Sidney	KIA	1916-07-19	LM
1665	Sgt	East, A H	KIA	1916-07-19	AR
3585	Pte	Eley, William George	DOW	1916-10-07	LM
4151	L/Cpl	Evans, A	DOW	1916-07-19	MV

202513	Pte	Felce, Herbert	DOW	1916-10-15	LM
1477	Cpl	Fincher, Albert Victor	KIA	1916-07-19	LM
266604	Pte	Fowler, Ernest Henry	KIA	1916-07-19	LM
1810	Pte	Frith, J	KIA	1916-07-18	MV
1686	Pte	Gibbs, G	KIA	1916-07-18	MV
265824	Pte	Gibson, Harry Reginald	DOW	1916-07-19	LM
3182	Cpl	Golder, Reginald James	KIA	1916-07-14	CR
1804	Pte	Gray, William	KIA	1916-07-19	LM
1518	Pte	Groves, E W	KIA	1916-07-19	LA
266634	Pte	Guntrip, Richard William	KIA	1916-07-19	LM
3281	Pte	Hance, George James	DOW	1916-08-26	LA
266656	Pte	Harding, Harry	KIA	1916-07-19	LM
3289	L/Cpl	Harrall, R	DOW	1916-09-08	XC
4386	Pte	Harris, A J	KIA	1916-07-19	LA
4377	Pte	Haverley, C C	DOW	1916-07-26	LO
2971	L/Cpl	Hayward, Edgar Stewart	DOW	1916-08-31	MV
265926	Pte	Henley, Samuel	KIA	1916-07-19	LM
4376	Pte	Hill, Arthur	KIA	1916-07-19	LM
266221	Pte	Hodgson, Frank Sydney	KIA	1916-07-19	LM
266715	Pte	Hogston, Herbert William	KIA	1916-07-19	LM
266682	Pte	Holt, W	KIA	1916-07-19	LM
3352	L/Cpl	Holdom, William John	KIA	1916-07-19	LA

266682	Pte	Holt, Willie	KIA	1916-07-19	LM
()	2Lt	Hudson, Roland Burton	KIA	1916-07-19	LM
5302	Pte	Hughes, T S	DOW	1916-07-22	LG
2722	Pte	Humphrey, Charles James	KIA	1916-07-19	LM
2920	Pte	Jarvis, A G	KIA	1916-07-19	LA
5644	Pte	Jefferies, Christopher S A W	DOW	1916-10-07	LM
5423	Pte	Jefferies, Ernest	DOW	1916-08-27	LM
4219	Pte	Jeffs, S W	DOW	1916-07-24	BO
5303	Pte	Johns, Owen Llewellyn	KIA	1916-07-19	LM
4311	L/Cpl	Johnson, William	KIA	1916-07-19	LM
2707	Pte	Johnston, L	DOW	1916-08-13	LA
266393	Pte	Jones, James William	KIA	1916-07-19	LM
4325	Pte	King, Thomas Edward	KIA	1916-07-19	LM
3019	Pte	Kipping, J	KIA	1916-07-19	LA
5313	Pte	Knights, William	KIA	1916-07-19	LM
1996	Pte	Lambourne, T H	DOW	1916-07-24	WI
265854	L/Sgt	Lance, Frederick Clarence	KIA	1916-07-19	LM
265930	Pte	Lewis, Edgar	KIA	1916-07-19	LM
6678	Pte	Leyland, F W	KIA	1916-07-19	LA
267090	Pte	Line, Francis William	KIA	1916-07-19	LM
2631	Pte	Litchfield, J S	DOW	1916-09-08	MV
265225	L/Cpl	Luff, Cecil George	KIA	1916-07-19	LM

4201	Pte	Malcolm, F A	DOW	1916-07-22	BO
5376	L/Cpl	Mantripp, George William	KIA	1916-07-19	LM
266236	Pte	Marshall, Robert	KIA	1916-07-19	LM
266690	Pte	Marslen, Reginald Victor	KIA	1916-07-19	LM
5391	Pte	McClements, Alexander	KIA	1916-07-19	LM
2697	Pte	Mead, Arthur William	DOW	1916-09-08	MV
265298	Pte	Morris, Albert Henry	KIA	1916-07-19	LM
3375	Pte	Morris, Arnold William	KIA	1916-07-19	LM
265787	Pte	Munday, Cyril Rupert	KIA	1916-07-19	LM
3788	Pte	Naylor, Henry George	KIA	1916-07-14	???
5358	L/Cpl	Newson, L F N	KIA	1916-07-19	LA
1191	Cpl	Oxlade, Alfred Edward	KIA	1916-07-19	LM
3940	Pte	Parker, Thomas Edward	KIA	1916-07-19	LM
2083	Pte	Payne, E	KIA	1916-07-19	LA
2705	L/Cpl	Pearce, W C	KIA	1916-07-19	AR
267093	Pte	Perry, John William	KIA	1916-07-19	LM
4336	Pte	Phillips, F L	DOW	1916-07-20	XD
()	Lt	Phipps, Charles Percy	KIA	1916-07-19	LM
3234	Pte	Piddington, William John	DOW	1916-07-29	LA
266561	Pte	Plowman, Charles George	KIA	1916-07-19	LM
4626	L/Cpl	Pocock, T	DOW	1916-08-18	LA
265385	Pte	Pratt, George Edward	KIA	1916-07-19	LM

2910	Pte	Prentice, H	KIA	1916-07-19	LA
266714	Pte	Price, Henry Thomas	KIA	1916-07-19	LM
266657	Pte	Price, Percy	KIA	1916-07-19	LM
20865	Pte	Radwell, L	DOW	1916-10-22	PH
4314	Pte	Savin, G E	KIA	1916-07-19	AR
4276	Pte	Sayell Arthur	KIA	1916-07-19	LM
267112	Pte	Scull, James Thomas	KIA	1916-07-19	LM
1689	Pte	Sexton, G J	DOW	1916-07-19	MV
3631	L/Cpl	Sharp, Edward	KIA	1916-07-19	LM
2549	Pte	Sharpe, h	DOW	1916-10-16	PH
2273	Pte	Shaw, William Thomas	DOW	1916-08-18	LA
266619	Pte	Simmonds, Joseph William	KIA	1916-07-19	LM
266692	Pte	Sinclair, Robert	KIA	1916-07-19	LM
4109	Pte	Smith, Harry	KIA	1916-07-19	LM
265388	L/Cpl	Stevens Arthur Frederick	KIA	1916-07-19	LM
4222	Pte	Stone, Gilbert Charles	KIA	1916-07-19	LM
4340	Pte	Symonds, George	KIA	1916-07-19	LM
1634	Cpl	Tapper, J E	DOW	1916-08-29	LO
266175	Pte	Tavenor, Frederick	KIA	1916-07-19	LM
3430	Pte	Taylor, Seth	KIA	1916-07-19	LM
1963	Pte	Taylor, Tom	KIA	1916-07-19	LM
3387	Pte	Thomas, Samuel Augustus	KIA	1916-07-19	LM

3527	L/Cpl	Thorpe, A	KIA	1916-07-19	LA
4308	Pte	Tomes, Harry	KIA	1916-07-19	LM
4257	Pte	Upson, William Thomas	KIA	1916-07-19	LM
1726	L/Cpl	Walker, A	KIA	1916-07-19	LA
2028	Pte	Wallace, W A	DOW	1916-07-20	MV
4331	Pte	Warr, Benjamin William Green	KIA	1916-07-19	LM
4732	Pte	Weaver, H J	DOW	1916-09-08	MV
3614	Pte	Wellman, David John	KIA	1916-07-16	LA
4189	Pte	White, J	KIA	1916-07-19	LA
3601	Pte	Whittick, Frederick William	DOW	1916-07-29	LA
4309	Pte	Williams, R J	DOW	1916-07-19	MV
4207	Pte	Willis, Frederick Thomas	KIA	1916-07-19	LM
5430	L/Sgt	Wood, Gordon Ewart Douglas	KIA	1916-07-19	LM
5342	L/Cpl	Wood, Stanley	DOW	1916-07-19	LM
266138	Pte	Wood, Stewart Sidney	KIA	1916-07-19	LM
167082	L/Cpl	Wright, Chester Harry	KIA	1916-07-19	LM

2/4th OXFORD & BUCKS LIGHT INFANTRY

5417	Pte	Bateman, Frederick William	KIA	1916-07-19	LM
1327	L/Cpl	Brain, T	DOW	1916-10-17	MV
2275	Pte	Broughton, Albert Walter	DOW	1916-09-28	PH

5148	Pte	Bryden, Charles	KIA	1916-07-19	LM
202028	Pte	Butler, Sidney	KIA	1916-07-19	LM
5373	Pte	Butler, W C	DOW	1916-07-21	LA
4107	L/Cpl	Clarke, I J	DOW	1916-09-12	PH
5342	Pte	Clements, J	DOW	1916-07-22	LO
4288	Pte	Durham, John William	DOW	1916-10-11	MV
6720	Pte	Gale, Albert Thomas	KIA	1916-07-14	MV
5979	Pte	Gardner, Thomas Henry	DOW	1916-08-22	PU
3460	Cpl	Grierson, P W	DOW	1916-07-28	LA
2191	Cpl	Harding, R	KIA	1916-07-19	LA
202292	Pte	Hicks, Ernest	DOW	1916-08-06	LM
202291	Pte	Higgins, John	DOW	1916-08-06	LM
2990	Pte	Jones, G	KIA	1916-07-19	LA
6736	Pte	Jones, W J	KIA	1916-07-19	RB
1538	Pte	Light, James Henry	DOW	1916-08-21	LA
3560	L/Cpl	Lunn, A	KIA	1916-07-19	LA
6762	Rfn	Reynolds, E E	KIA	1916-07-19	LA
5816	Pte	Simmons, T	DOW	1916-08-16	WA
4317	Pte	Simpson, G E L	KIA	1916-07-19	LA
3357	Bglr	Sims, T	KIA	1916-07-20	LG
3619	Pte	Smith, J W	DOW	1916-07-23	???
4145	Pte	Smith, Leonard	DOW	1916-07-25	PH

2715	Rfn	Spratt, H F	KIA	1916-07-19	LA
202327	Pte	Stone, Frank	DOW	1916-08-06	LM
202380	Pte	Strong, Albert	DOW	1916-08-06	LM
4167	Pte	Taylor, WA	KIA	1916-07-19	LA
3553	Pte	Thomas, Henry	DOW	1916-07-25	MV
()	Lt	Tiddy, Reginald John Elliott	DOW	1916-08-10	LA
3022	Pte	Tolley, G E L	KIA	1916-07-19	LA
1973	Cpl	Upfold, F H L	KIA	1916-07-14	MV

6th BATTALION, OXFORD & BUCKS LIGHT INFANTRY

20413	L/Cpl	Barden, F	KIA	1916-07-19	RP
9259	CSM	Bright, A H	KIA	1916-07-19	RP
11961	L/Cpl	Broad, T	DOW	1916-09-06	ET
13167	Pte	Broadbent, A	DOW	1916-07-22	RP
12607	Pte	Chappin, Harry	DOW	1916-08-03	CN
12223	Pte	Collett, William	DOW	1916-09-05	CO
21022	Pte	Cooper, C	KIA	1916-07-18	BA
13428	Sgt	Duncuff, Arthur Francis	DOW	1916-08-03	CN
()	2Lt	Fussell, J W H	KIA	1916-07-19	RP
12173	Pte	Griffiths, Arthur	DOW	1916-09-04	HS
16358	L/Cpl	Hardington, Henry James	DOW	1916-09-11	CO
17328	Cpl	Hunt, Walter Jonathan	DOW	1916-09-06	CO

2270	L/Cpl	Lenton, W H	KIA	1916-07-19	RP
10108	Sgt	Mansell, David George	DOW	1916-09-10	CO
12116	Sgt	Monk, Hope Jubilee Victor	DOW	1916-08-29	CO
21237	L/Cpl	Morris, H	DOW	1916-08-30	HS
12010	Pte	New, H	DOW	1916-09-04	CO
14991	Pte	Porter, F W (MM)	KIA	1916-07-19	RP
17308	Pte	Rackstraw, Joseph	DOW	1916-09-11	CO
11655	Pte	Sippitts, A J	DOW	1916-10-01	HS

1/5th BATTALION, DUKE OF CORNWALL'S LIGHT INFANTRY

2184	Pte	Andrews, Pte	KIA	1916-07-18	MV
()	2Lt	Brett, Ernest Hugh William	DOW	1916-08-02	LA
2416	Pte	Dalley, Albert	DOW	1916-08-08	LA
2981	Pte	Hoskin, W	DOW	1916-07-20	MV
3159	Pte	Hunn, N	KIA	1916-07-19	AR
2297	L/Sgt	May, R	KIA	1916-07-18	LG
2109	Pte	Murton, Edward James	DOW	1916-10-26	LM
2118	Pte	Pomery, J	KIA	1916-07-19	AR
()	2Lt	Ratcliff-Gaylard, E R	KIA	1916-07-20	LA
1438	Pte	Stephens, Arthur Edward	KIA	1916-07-17	LA
1305	Pte	Williams, FA	KIA	1916-07-19	AR

ROYAL FIELD ARTILLERY

61st DIV TRENCH MORTAR BATTERY

109681	Gnr.	Bridge, E	KIA	1916.07.19	AR

61st DIV AMMUNITION COLUMN

1041	Gnr	Bemrose, Fred	KIA	1916-07-20	MV

BRITISH SOLDIERS WHO SERVED, FROM OTHER UNITS

6th BATTALION, CAMERON HIGHLANDERS REGIMENT

S/22169	Pte	Cumming, John	D	1916-07-20	LM
S/22003	Pte	Loudon, Alexander	D	1916-07-20	LM
S/22150	Pte	MacLean, Allan Charles	D	1916-07-20	LM
S/21711	Pte	Marshall, Robert	D	1916-07-20	LM
S/16302	Pte	McGuire, Josep0h	D	1916-07-20	LM
S/21548	Pte	Millard, Joseph Gordon	D	1916-07-20	LM
S/22113	Pte	Robertson, William	D	1916-07-20	LM

7th BATTALION, CAMERON HIGHLANDERS REGIMENT

S/22023	Pte	Baxter, William	KIA	1916-07-20	LM

S/22021	Pte	Melvin, James Donaldson Frame	D	1916-07-20	LM
S/11243	Pte	Mitchell, James	D	1916-07-19	LM
S/22145	Pte	Stewart, Peter	D	1916-07-20	LM
S/22263	L/Cpl	Wallace, George	D	1916-07-20	LM

2nd BATTALION, CAMERONIAN REGIMENT

9121	Cpl	Howison, Harry	D	1916-07-20	LM

11th BATTALION, DEVONSHIRE REGIMENT

()	2Lt	Mathews, Anthony Edward	KIA	1916-07-19	LA
()	2Lt	Mitchell, George James	D	1916-07-19	LM
()	2Lt	Skewes, Artur Courtis	KIA	1916-07-19	LA

18th BATTALION, DURHAM LIGHT INFANTRY

28232	Pte	Ford, George	D	1916-07-20	PM
30967	Pte	Henderson, Norman	D	1916-07-20	PM

1/4th GLOUCESTERSHIRE REGIMENT

()	Lt	Down, Robert Hayward	DOW	1916-08-17	PU

1/6th GLOUCESTERSHIRE REGIMENT

()	2Lt	Smith, Arthur Roughton	DOW	1916-07-22	WA

HERTFORDSHIRE REGIMENT

5436	Pte	Shephard, Robert	D	1916-07-20	LM

NOTTS & DERBYSHIRE REGIMENT
17th SHERWOOD FORESTERS

31220	L/Cpl	Beeby, William W.	KIA	1916-07-20	LM

9th BATTALION, ROYAL DUBLIN FUSILIERS

23042	Pte	Kavanagh, William	D	1916-07-20	LM

11th FIELD COMPANY, ROYAL ENGINEERS

32870	Spr	Eldered, Thomas Mart Griffin	D	1916-07-20	PM

1st BATTALION, ROYAL MUNSTER FUSILIERS

10399	Pte	McCarthy, Patrick	D	1916-07-20	LM
5891	Pte	Miller, William	D	1916-07-21	LM

8th BATTALION, ROYAL MUNSTER FUSILIERS

3431	Pte	Corbett, Michael	D	1916-07-20	LM

3rd BATTALION, SOUTH LANCASHIRE REGIMENT

()	2Lt	Wood, Leslie William	D	1916-07-19	LM

10th BATTALION, SOUTH LANCASHIRE REGIMENT

()	2Lt	Jefferies, Maurice Arnold	KIA	1916-07-16	LM

2/7th WORCESTERSHIRE REGIMENT

4500	L/Cpl	Pratt, Jesse	D	1916-07-19	LM

3rd WILTSHIRE RGIMENT

()	2Lt	Parker, Frederick Richard	D	1916-07-19	LM

14th BATTALION, YORK & LANCASTER REGIMENT

14/501	Pte	Shepherd, Horace	D	1916-07-20	PM

---ooo000ooo---

AUSTRALIAN ROLL OF HONOUR

5th DIVISION DIVISIONAL SUPPORT UNITS

10th FIELD ARTILLERY BRIGADE

7162	Dvr	Walker	William Douglas Wilson	DOW	1916.07.18	ES

11th FIELD ARTILLERY BRIGADE

991	Cpl	Johnson	Thomas Norman	KIA	1916.07.20	RF

12th FIELD ARTILLERY BRIGADE

()	Capt	Thompson	Gordon Kingsley	KIA	1916.07.19	YF

13th FIELD ARTILLERY BRIGADE

2107	Gnr	Adams	James Daniel	KIA	1916.07.16	RP
565	Sgt	Bourne	Arthur Ernest Everard	KIA	1916.07.16	RP
3460	Gnr	Burling	Arnold	KIA	1916.07.18	RD
1238	Gnr	Carr	Cecil Gregory	KIA	1916.07.18	RD

3319	Pte	Everitt	James Samuel	KIA	1916.07.18	AN
3047	Gnr	Green	Blencowe Edward	KIA	1916.07.18	RD
3554	Gnr	Le Brocq	Daniel Tallan	KIA	1916.07.18	RD
3264	Gnr	Stevens	George Ernest	KIA	1916.07.18	RD
3702	Gnr	Wheeler	William Henry	KIA	1916.07.18	RD

1st FIELD COMPANY ENGINEERS

2173	Spr	Jones	William Henry	KIA	1916.07.17	ES

5th DIVISION PIONEERS

2825	Pte	Fitch (Ansell)	Alfred William	KIA	1916.07.19	PW
906	L/Cpl	Hollows	James	DOW	1916.07.14	ES
5587	Pte	McGregor	John Roberts	DOW	1916.07.20	RB
3181	Pte	Moore	John	DOW	1916.07.29	WI
2172	Pte	Pettigrew	Arcihbald Douglas	DOW	1916.07.22	WI
2796	Pte	Thompson	Irving	KIA	1916.07.21	RB

5th DIVISIONAL SIGNAL COMPANY

3648	Spr	Galgraith	Alfred George Finlay	DOW	1916.07.15	CD
2845	L/Cpl	Johnston	Thomas Atherton	DOW	1916.07.15	CD
1864	Spr	O'Donnell	Thomas Joseph	KIA	1916.07.20	RP

5th DIVISIONAL TRENCH MORTAR BATTERY

()	Capt	Hopkins	Clive Boyer	KIA	1916.07.20	RB
4824	Gnr	Kenning	John	KIA	1916.07.19	AN

2nd AUSTRALIAN TUNNELLING COMPANY

2462	Spr	Riley	Earl	KIA	1916.07.16	RF
2224	Spr	McDonald	Charles Parnell	KIA	1916.07.19	RB

8th BRIGADE

6th FIELD COY ENGINEERS

6105	Spr	Wemestad	Oscar Olaf	POW	1916.07.20	RTA

8th FIELD COY ENGINEERS

1305	Spr	Betts	Samuel	KIA	1916.07.19	RP
4408	Spr	Eadie	Alexander McKinley	KIA	1916.07.19	RD
4511	L/Cpl	Morrison	William John	KIA	1916.07.18	RP
4557	Spr	Reeves	James	KIA	1916.07.19	RD
509	L/Cpl	Smith	Harold Thompson	KIA	1916.07.19	PW
()	2Lt	Tenbosch	Christian Peter	KIA	1916.07.19	RP
6035	Spr	Thomas	Stanley	POW	1916.07.19	RTA
1153	Spr	Ward	Richard Dunstan	KIA	1916.07.19	RD
4483	Spr	Young	Arthur Henry	KIA	1916.07.19	RD

8th MACHINE GUN COY

94	L/Cpl	Bernays	Gerald Chisholm	POW	1916.07.20	RTA
104	Pte	Buckley	Frederick	DOW	1916.07.22	BA
16	Pte	Carnegie	William Leslie	KIA	1916.07.20	AN
913	Pte	Copley	Allan Wallace	POW DOW	1916.07.25	VA
802	Pte	Davies	John Edward Lawrence	KIA	1916.07.20	RP

100	L/Cpl	Driscoll	Thomas	POW	1916.07.20	RTA
99	Cpl	Holman (Devenish)	Walter Ernest	POW	1916.07.20	RTA
876	Pte	Huxley	Samuel Lawrence	KIA	1916.07.20	RP
909	Pte	Ings	Percy Edwin	KIA	1916.07.20	AN
229	Pte	James	Lawrence Mouland	DOW	1916.07.21	BA
98	Pte	Makeham	John Joseph	KIA	1916.07.20	RP
111	Pte	Marks	Charles Beresford	POW	1916.07.20	RTA
110	Cpl	McKinnon	John	KIA	1916.07.20	RP
31	Pte	Milton	Roy Clifford	KIA	1916.07.20	RP
267	Pte	Pearce	Walter Stanley	KIA	1916.07.20	RP
673	Pte	Regan	Ambrose	KIA	1916.07.20	VC
113	Pte	Ross	John Valentine	KIA	1916.07.20	VC
102	Pte	Stratford	Gordon John (Edward)	KIA	1916.07.20	VC
39	Pte	Turner	Arthur Leslie	KIA	1916.07.20	VC
117	Pte	Watson	Sinton	POW	1916.07.20	RTA
118	Pte	White	William John	KIA	1916.07.20	VC

29th BATTALION

1057	Pte	Amy	Edward John	POW	1916.07.20	RTA
1968	Pte	Antrobus	Aubrey William	POW	1916.07.20	RTA
1969	Pte	Antrobus	Harold	POW	1916.07.20	RTA
1272	Pte	Ballinger	Ernest	POW Died	1918.06.15	CSC
1062	Pte	Barnatt	George Robert	KIA	1916.07.20	PW
2514	Pte	Barry	William Charles	POW	1916.07.19	RTA
166	Pte	Batey	Lemuel Reginald Valentine Fitzpatrick	KIA	1916.07.20	VC
191	Pte	Bell	Henry	KIA	1916.07.19	PW
1074	Pte	Bolton	Tom Dalzell	POW	1916.07.19	RTA
1983	Pte	Breguet	Justin Hercules	KIA	1916.07.19	PW
1361	Pte	Brittain	William	KIA	1916.07.19-20	RP
1059	Pte	Brown	Joseph	POW	1916.07.20	RTA
1470	Pte	Brumm	Norman Leslie	KIA	1916.07.19	PW
182	Pte	Bryant	Alfred William	DOW	1916.08.07	BO
1093	Pte	Carter	Otway Vernon	POW	1916.07.20	RTA
1079	L/Cpl	Cheshire	Arthur Eric	KIA	1916.07.19-20	RP

1094	Pte	Childs	Edgar	POW	1916.07.20	RTA
500	Cpl	Chrisp	Glenwylln	KIA	1916.07.19-20	RP
2190	Pte	Clarke	Andrew	KIA	1916.07.25	RP
417	Pte	Cole	Reginald Maurice	KIA	1916.07.20	VC
1321	Sgt	Cole	Oliver Stanley	POW	1916.07.20	RTA
210	Pte	Cozens	James Alfred	KIA	1916.07.19	VC
735	Pte	Cunningham	James Euphrates	KIA	1916.07.26	RP
1102	Pte	Davidson	Percy Albert	POW DOW	1916.07.20	CR
2008	Pte	Deans	Albert Burchnell	KIA	1916.07.20	VC
222	Pte	Donovan	Patrick Charles	POW	1916.07.20	RTA
1097	Cpl	Downie	John Charles Gordon	KIA	1916.07.20	VC
2016	Pte	Edwards	William George	POW	1916.07.20	RTA
2015	Pte	Edwards	Joseph Buxton	DOW	1916.08.06	UA
1107	Sgt	Ellison	Norman George	POW	1916.07.20	RTA
2020	L/Cpl	Ennis	Leo Joseph	KIA	1916.07.20	VC
80	Pte	Farlow	Samuel	KIA	1916.07.19	PW
2584	Pte	Francis	Thomas William	KIA	1916.07.20	PW

2033	Pte	Gill	Robert	DOW	1916.07.20	BA
1130	Pte	Gordon	James	KIA	1916.07.19	VC
242	Pte	Goy	George	DOW	1916.08.09	UB
2043	Pte	Harding	Thomas Walter	POW	1916.07.20	RTA
253	Sgt	Harland	Ormond	KIA	1916.07.20	VC
1390	L/Cpl	Haslam	Herbert James	KIA	1916.07.19	PW
2050	Pte	Higgins	James Ingersoll	KIA	1916.07.20	VC
2203	Pte	Johnson	Arthur Joseph	KIA	1916.07.19	PW
278	Pte	Johnson	William Frank Mayne	DOW	1916.08.02	UC
288	Pte	Kearney	John	POW	1916.07.20	RTA
2677	Pte	Kennedy	Leslie Frederick	KIA	1916.07.19-20	VC
1163	Pte	Lamont	Frederick Harold	POW	1916.07.20	RTA
2546	Pte	Langmead	Roy Leslie	POW	1916.07.20	RTA
1168	Cpl	Livingston	David Frederick	KIA	1916.07.19	PW
1165	Pte	Lund	Horace	KIA	1916.07.19-20	RP
2100	Pte	Machen	Harry	POW	1916.07.20	RTA
1192	Pte	Martin	Sydney James	POW	1916.07.20	RTA
2081	Pte	Mayston	Sydney Robert	POW	1916.07.20	RTA
1173	Cpl	McIntyre	Robert	DOW	1916.08.08	WI

573	Pte	Merten	Stanley Reuben	KIA	1916.07.19	RP
2088	Pte	Middleton	Frank	POW	1916.07.20	RTA
1319	Sgt	Morshead	Ernest Davey	KIA	1916.07.19-20	RP
()	Capt	Mortimer	Kenneth Malcolm	KIA	1916.07.20	PW
307	CSM	Newman	Kenneth Fossey	POW	1916.07.20	RTA
1014	Pte	Noakes	Nelson William	KIA	1916.07.19-20	BO
1197	Pte	Noll	Ernest Hansen	POW	1916.07.20	RTA
314	Pte	O'Donnell	Stanley Richard	KIA	1916.07.19	PW
319	Pte	O'Donnell	William	KIA	1916.07.19	PW
316	Pte	Oliver	Ernest Robert	KIA	1916.07.19	PW
320	Pte	Parry	Frederick	KIA	1916.07.19	PW
1211	Pte	Parry	Reuben	POW	1916.07.20	RTA
323	Pte	Pilling	Edward	DOW	1916.09.18	ET
2119	Pte	Pinne	Norman Ernest	KIA	1916.07.20	VC
324	Pte	Pollard	Herbert George	KIA	1916.07.19	PW
335	Sgt	Reid	James Alexander	KIA	1916.07.20	PW
1216	Cpl	Ross	James Hugh	KIA	1916.07.19	PW

3890	Pte	Ross	Benjamin	POW	1916.07.20	RTA
1213	Pte	Russell	Thomas Horace	POW	1916.07.20	RTA
2137	Pte	Scrimgeour	Thomas	POW DOW	1916.07.23	VA
()	Capt	Sheridan	Thomas Francis	KIA	1916.07.20	PW
1240	Pte	Shiels	Robert Barrett	POW	1916.07.20	RTA
683	Pte	Smith	Norman	KIA	1916.07.14	BA
1245	Pte	Smith	Gordon Thomas	KIA	1916.07.19	PW
339	Cpl	Smith	Percival Kerson	KIA	1916.07.19-20	VC
1238	Cpl	Spooner	George William	KIA	1916.07.20	RP
682	Pte	Stone	Paul Vincent	KIA	1916.07.19-20	RP
949	L/Cpl	Strachan	Robert Ritchie	KIA	1916.07.19-20	RP
407	Pte	Taylor	John Michael	KIA	1916.07.20	VC
2131	Pte	Torbitt	Francis Albert	KIA	1916.07.19-20	RP
974	Pte	Trinder	William	KIA	1916.07.26	RP
1252	Cpl	Tuck	Alfred George	KIA	1916.07.19	PW
1255	Pte	Turner	Harry Allan	POW	1916.07.20	RTA
1257	Pte	Tyers	William	POW	1916.07.20	RTA

352	Pte	Wait	Edison	POW	1916.07.19	RTA
354	Cpl	Warnock	Arnold Edwin	POW	1916.07.20	RTA
358	Pte	Weir	Arthur Joseph	KIA	1916.07.19	PW
361	Pte	Westmoreland	Herbert John	KIA	1916.07.20	VC
365	Cpl	White	George Stephen	POW	1916.07.20	RTA
1314	Pte	Wilkin	Ernest Frank	KIA	1916.07.19	PW
1269	Pte	Wilkin	Harold John	POW	1916.07.20	RTA
652	Pte	Wills	George Charles	KIA	1916.07.19-20	RP
651	Pte	Wills	Robert Joseph	DOW	1916.08.02	UD
1262	Pte	Woodcock	Francis Stringer	KIA	1916.07.20	VC

30th BATTALION

1516	L/Cpl	Abbott	Ernest William	KIA	1916.07.20	VC
90	Pte	Anderson	Peter	KIA	1916.07.20	VC
340	Pte	Arkell	Eric Charles	KIA	1916.07.20	VC
341	Pte	Arkell	Max Kenneth Dick	DOW	1916.08.28	DA
1968	Pte	Arnott	Colin Henry	KIA	1916.07.20	PW
1971	Pte	Aulton	Vivian Henry	KIA	1916.07.20	VC
593	Pte	Austin	Edward George	KIA	1916.07.20	RP

2276	Pte	Balsdon	Russell Henry	DOW	1916.07.28	BO
2274	Pte	Balsdon	James	POW DOW	1916.09.01	VB
2287	Pte	Bennett	William Alfred	KIA	1916.07.20	VC
840	Pte	Benson	George Verner	KIA	1916.07.19	PW
1682	Pte	Bourke	Harold John	KIA	1916.07.20	PW
357	Pte	Brien	Harry	KIA	1916.07.17	RP
1986	Pte	Briscoe	Leopold Lionel	KIA	1916.07.20	VC
609	Pte	Burke	William	KIA	1916.07.20	PW
1995	Pte	Callaghan	John Patrick	KIA	1916.07.20	VC
631	Pte	Camps	Victor Alexander	KIA	1916.07.20	RP
397	Pte	Caswell	Douglas	KIA	1916.07.20	VC
635	Sgt	Church	Chester Cecil	KIA	1916.07.20	VC
382	L/Cpl	Churchill	Walter Edmund	KIA	1916.07.20	RP
137	Sgt	Coady	Edward	KIA	1916.07.20	VC
1057	Pte	Collier	Percy	KIA	1916.07.20	RP
633	Pte	Cooper	Samuel John	KIA	1916.07.20	VC
1251	Pte	Cooper	Ehba	DOW	1916.08.05	BO
1025	L/Sgt	Cracknell	Ralph Bertram Webb	KIA	1916.07.20	RP
2309	Pte	Cracknell	Donald	KIA	1916.07.20	AN
2006	Pte	Croft	George	KIA	1916.07.20	PW

2010	Pte	Croker	Harry	KIA	1916.07.20	PW
867	Pte	Dalton	Witton Kenworthy	KIA	1916.07.20	VC
649	Pte	Dobbie	James	KIA	1916.07.20	VC
869	Pte	Doyle	John Thomas	KIA	1916.07.20	VC
406	Pte	Dryburgh	Leslie	KIA	1916.07.20	VC
2021	Pte	Dykes	Cecil	KIA	1916.07.23	RD
2026	Pte	Easton	Thomas	KIA	1916.07.20	VC
882	Pte	Fenwick	Robert Gladstone	KIA	1916.07.20	PW
1550	Pte	Fiddling	Charles	KIA	1916.07.19	ES
889	Pte	Gardner	Henry	KIA	1916.07.20	PW
2038	Sgt	Garland	Charles Stewart	KIA	1916.07.19	RP
1553	Pte	Gillingham	William	POW	1916.07.20	RTA
1291	L/Cpl	Glenn	Frederick James	KIA	1916.07.20	PW
898	Pte	Hall	Robert Ephraim	POW	1916.07.20	RTA
190	Pte	Hart	Joseph	KIA	1916.07.20	RF
188	Pte	Hawcroft	Charles Henry	KIA	1916.07.20	PW
192	Pte	Hayman	William Francis	DOW	1916.08.28	UB
186	Pte	Hedges	Leonard Jack	KIA	1916.07.20	VC
2056	Pte	Hepple	Matthew	KIA	1916.07.20	PW

196	Pte	Higgins	William Bruce	KIA	1916.07.20	PW
189	Pte	Holmesby	Horace Arthur	POW	1916.07.20	RTA
1079	Pte	Hunter	Robert	DOW	1916.07.26	CA
686	Pte	Inglis	William	KIA	1916.07.20	VC
451	Pte	Jackson	Frederick Hillgrove	KIA	1916.07.20	VC
689	Pte	James	Arthur Mossop	KIA	1916.07.20	VC
1880	Pte	Jones	John	KIA	1916.07.20	VC
915	Pte	Kelly	John Reginald	DOW	1916.07.28	BO
1096	Pte	King	David	KIA	1916.07.20	VC
2512	Pte	Laing	David	DOW	1916.07.27	LO
221	Pte	Laurie	Alec Duncan Gordon	KIA	1916.07.20	VC
217	Pte	Law	Hilton Bradlaugh	POW	1916.07.20	RTA
1104	Pte	Leahy	Roy Jeremiah	KIA	1916.07.20	VC
1309	L/Cpl	Leask	Eric Graham	KIA	1916.07.20	RP
()	2Lt	Lees	John Stanley Havelock	KIA	1916.07.20	VC
81	Pte	Lording	Rowland Edward	DOW	1944.10.02	Sydney
467	Pte	Lucre	George Henry	KIA	1916.07.20	PW
483	Sgt	Malcolm	Edward Samuel	KIA	1916.07.20	VC

1584	Pte	Martin	Roy Leslie	KIA	1916.07.20	VC
722	Pte	Matthews	Arthur Frederick	KIA	1916.07.19	VC
3557	Pte	Maxwell	Arthur Reginald	KIA	1916.07.20	VC
1595	Pte	McCoullough	Hector Gordon	KIA	1916.07.20	VC
721	Pte	McGarvey	Sidney Ernest	POW	1916.07.20	RTA
711	Pte	McGregor	Thomas	KIA	1916.07.20	RP
938	Pte	McKillop	Donald John Duncan	KIA	1916.07.20	RF
4575	Pte	McWatt	John	KIA	1916.07.20	RP
709	Pte	Miller	James	DOW	1916.08.08	XE
()	Lt	Mitchell	Alan	KIA	1916.07.20	PW
930	Pte	Murray	Raymond George	KIA	1916.07.20	VC
1590	Cpl	Murray	Charles William	KIA	1916.07.20	PW
728	Pte	Nelson	Henry Oscar	KIA	1916.07.20	VC
2414	Pte	Osmond	Lionel Edgbert Clement	KIA	1916.07.20	VC
1326	Pte	Owen	William Whyte Stuart	KIA	1916.07.20	VC
()	2Lt	Parker	John	KIA	1916.07.20	PW
515	Pte	Payne	Samuel	POW	1916.07.20	RTA

801	Cpl	Pearce	Harold John	POW DOW	1916.07.21	CR
962	Pte	Peberdy	Frederick Thomas Smith	KIA	1916.07.20	VC
739	Pte	Perks	John	KIA	1916.07.20	VC
1816	Pte	Phelan	Patrick James	POW	1916.07.20	RTA
1601	L/Cpl	Powis	Charles Edward	POW DOW	1916.08.23	NW
2141	Pte	Pritchard	Herbert Ernest	KIA	1916.07.20	RP
2392	Pte	Rawnsley	Albert Arthur	KIA	1916.07.20	VC
2433	Pte	Reeves	John	KIA	1916.07.20	RP
1134	Pte	Rich	Douglas Chaplin	KIA	1916.07.19	VC
1023	Pte	Ridley	Ernest Robert	POW	1916.07.20	RTA
1312	Pte	Robinson	Horace Frederick	KIA	1916.07.20	VC
746	Pte	Robson	Charles Ernest	KIA	1916.07.20	RP
743	Pte	Ryan	Daniel Bernard	KIA	1916.07.20	PW
1606	Pte	Ryan	Patrick Edward	KIA	1916.07.20	VC
286	Pte	Scott	William Francis	DOW	1916.07.30	UJ
82	Pte	Sheehan	James	KIA	1916.07.20	RP
531	Pte	Shone	Edward	KIA	1916.07.20	VC
3255	Pte	Smith	Alfred John	KIA	1916.07.20	VC

4614	Pte	Spence	Malcolm	KIA	1916.07.20	PW
973	Pte	Stanborough	Edward Albert Leslie	KIA	1916.07.20	RP
755	Cpl	Steed	Frank Isaac	KIA	1916.07.20	PW
750	Pte	Stephens	Leonard Aspinall	KIA	1916.07.20	VC
1616	Pte	Stokes	Sidney	POW	1916.07.20	RTA
1143	Pte	Storey	David Burn	POW	1916.07.20	RTA
2163	Pte	Stuart	Rowland Hamilton	KIA	1916.07.20	RP
1146	Pte	Theisinger	Norman Lloyd	KIA	1916.07.20	VC
2466	Pte	Thomas	Henry Lamert	POW	1916.07.20	RTA
2170	Pte	Thompson	Landy Creel	POW	1916.07.20	RTA
2171	Pte	Tickner	Frederick	KIA	1916.07.20	VC
1623	Pte	Tisbury	Charles Frederick	KIA	1916.07.20	PW
767	Pte	Turner	John	KIA	1916.07.20	PW
776	Cpl	Vale	Tasman George	KIA	1916.07.20	VC
777	Pte	Vincent	Laurence Stanley	KIA	1916.07.20	PW
783	Pte	Wall	John Douglas	KIA	1916.07.20	VC
1624	Pte	Wallace	Thomas	POW	1916.07.20	RTA
1010	Pte	Ward	Patrick Michael	KIA	1916.07.20	RP

2184	Pte	Ward	Claude	KIA	1916.07.20	PW
1157	Pte	Warrington	James Dale	DOW	1916.07.22	HA
2475	Pte	Watson	Norman Leslie	POW	1916.07.20	RTA
582	Sgt	Watterson	Randal	KIA	1916.07.20	RF
1003	Pte	Watts	Thomas Langford	KIA	1916.07.20	RP
1311	Pte	Welling (Studdon)	Ernest Frank	KIA	1916.07.20	PW
1627	Pte	White	Russell	KIA	1916.07.20	RP
782	Pte	Whitton	John Donald Gordon	KIA	1916.07.20	VC
2188	Pte	Williams	Alfred Edward	KIA	1916.07.20	RP
804	Sgt	Willis	Edward Emerson	KIA	1916.07.20	VC
1004	Pte	Wishart	John Hay	KIA	1916.07.20	VC
781	Pte	Wood	Archie	KIA	1916.07.20	VC
2194	Pte	Woods	Charles Albert	KIA	1916.07.20	VC
70	L/Cpl	Wright	Eric Stanley	POW	1916.07.20	RTA
789	L/Cpl	Wymark	Reginald Douglas	KIA	1916.07.20	VC
2485	Pte	Wynn	John Cyril	KIA	1916.07.20	PW
321	Pte	Yardy	Richard Alfred	KIA	1916.07.20	RP
795	Pte	Yeo	Claude	KIA	1916.07.20	PW

31st BATTALION

1033	Pte	Alder	Robert	KIA	1916.07.19 -20	AN
570	Pte	Anderson	Leonard William	KIA	1916.07.19 -21	VB
1113	Pte	Aitken	Colin Robert	DOW	1916.07.21	BA
3751	Pte	Anthony	Lester	DOW	1916.07.27	BO
135	Pte	Bain	Walter Ernest	KIA	1916.07.20	VC
814	Pte	Baker	William Henry	KIA	1916.07.20	VC
134	Pte	Barnett	Henry Ernest	KIA	1916.07.20	RP
505	Pte	Barry	William	KIA	1916.07.19	RP
353	Sgt	Batteram	Robert	POW	1916.07.20	RTA
1523	L/Cpl	Baynes	Francis Lambert	KIA	1916.07.19	RP
719	Pte	Bean	Reginald Alexander	KIA	1916.07.19 -21	VB
1136	Pte	Benson (Harrison)	Charles Frederick	KIA	1916.07.19 -21	VB
()	Capt	Bernard	Vivian Dalzell	POW	1916.07.20	RTA
817	Pte	Beseler	George Henry	KIA	1916.07.19 -21	AN
605	Pte	Bills	Thomas Henry	KIA	1916.07.19	PW
474	Pte	Black	James Joseph	POW	1916.07.19	RTA
589	Pte	Bloomfield	Bertram Benjamin	KIA	1916.07.19 -21	VB

590	Pte	Bolton	John	POW	1916.07.20	RTA
1050	Pte	Brailsford	William Wood	KIA	1916.07.20	RP
582	L/Cpl	Bray	Frank Davies	KIA	1916.07.19 -21	VB
1522	Pte	Broom	Sidney	KIA	1916.07.19	PW
829	Pte	Brown	Leslie Charles	KIA	1916.07.19 -21	VB
585	Pte	Brumby	William Christopher	KIA	1916.07.19 -21	PW
29	Pte	Campbell	John	KIA	1916.07.20	RP
620	Pte	Cartwright	Arthur Leslie	KIA	1916.07.19 -21	RP
71	Pte	Castles	William Edward John	KIA	1916.07.19	RP
3784	Pte	Caughey	Rodger	KIA	1916.07.19 -21	VB
1004	Pte	Chadwick	Fred	KIA	1916.07.19 -21	RP
2293	Pte	Chalk	Henry Frederick	KIA	1916.07.20	RP
166	Pte	Chapman	John Edward	KIA	1916.07.20	RP
174	Pte	Charles	Leslie Stuart	KIA	1916.07.20	AN
1528	Pte	Christensen	James Walter	DOW	1916.08.12	XF
560	Pte	Clarke	Frank	POW	1916.07.20	Disch
()	Maj	Clements	Charles Edward	DOW	1916.07.22	BO

624	Pte	Coghlan	John	KIA	1916.07.20	VC
626	Pte	Collier	Alfred Arthur	KIA	1916.07.20	VC
414	Pte	Collings	John Vessey	POW	1916.07.20	RTA
()	Lt	Cox	Harold	KIA	1916.07.20	RP
623	Pte	Cullen	William Joseph	KIA	1916.07.19 -21	PW
171	Pte	Cumiskey	Michael Joseph	POW	1916.07.20	RTA
494	Pte	Curran	James Joseph	KIA	1916.07.19	VC
184	Pte	Daly	Leslie Richmond	KIA	1916.07.20	AN
320	Pte	Dank	Charles Conrad	KIA	1916.07.19 -20	VC
1517	Sgt	Davies	Bertram	KIA	1916.07.19	AN
187	Pte	Davison	Gordon	POW	1916.07.20	RTA
629	Pte	Denton	Reginald George	KIA	1916.07.19 -21	VB
642	Pte	Dodson	Horace Edward	KIA	1916.07.20	RP
1989	Pte	Dolan	Samuel Thomas	DOW	1916.07.22	LS
190	Pte	Downes	John Finlay	POW	1916.07.24	RTA
641	Pte	Dunn	Leslie Clark	KIA	1916.07.19 -21	PW
391	Pte	Esam	Harold	KIA	1916.07.19	PW

2130	Pte	Farrell	James Stephen	KIA	1916.07.19 -21	VB
198	Pte	Ford	Percy	KIA	1916.07.20	NI
795	Sgt	Francis	Arthur Ernest	KIA	1916.07.19 -21	AN
852	L/Cpl	Francis	William John	KIA	1916.07.20	VC
200	L/Sgt	Fraser	Arthur Kinross	KIA	1916.07.20	AN
367	L/Cpl	Furlonger	James Walker	POW	1916.07.19	RTA
333	Pte	Gaden	Neil	KIA	1916.07.19 -20	VC
482	L/Cpl	Gavin	James	KIA	1916.07.19	RP
2005	Pte	Goodwin	Louis John	KIA	1916.07.17	VB
555	Pte	Goulding	John Joseph	POW DOW	1916.07.19	PW
1537	Pte	Grace	Walter Allen	KIA	1916.07.20	PW
685	Pte	Grant	William Joseph	POW	1916.07.20	RTA
2192	Pte	Green	Frank Henry	DOW	1916.07.18	ES
683	Pte	Greening	Joshua George	KIA	1916.07.20	RP
1584	Pte	Griffiths	Edward Beynon	KIA	1916.07.20	AN
702	Pte	Hale	Norman Arthur	KIA	1916.07.19 -21	PW
1546	Pte	Hall	Stephen	POW	1916.07.20	RTA
2138	Pte	Hamilton	Robert	KIA	1916.07.20	VC
345	Pte	Harker	Robert Herd	KIA	1916.07.20	RP

434	Pte	Heaney (Hart)	George Henry	POW DOW	1916.07.25	CR
344	Pte	Henry	Robert	POW	1916.07.20	RTA
706	Pte	Hibbert	Albert George	KIA	1916.07.19 -21	VB
870	Pte	Hicks	William Stanley	KIA	1916.07.19	VC
1591	Pte	Hill	Robert	KIA	1916.07.20	VC
2141	Pte	Holland	Thomas	KIA	1916.07.20	VC
538	L/Cpl	Holley	Clifford	KIA	1916.07.20	AN
693	Sgt	Hopkins	Leslie Ralph	KIA	1916.07.19 -21	RP
2340	Pte	Houghton	Richard	DOW	1916.08.06	XG
1619	Pte	Howard	Robert James	KIA	1916.07.19	RD
()	2Lt	Hudson	Arthur	KIA	1916.07.19	VC
342	Pte	Humphrey	Frederick William	KIA	1916.07.20	VC
1054	Pte	Hunt	Thomas Vincent	KIA	1916.07.19	VC
2028	Pte	Hyams	Sydney Joseph Walter	KIA	1916.07.19	VC
2144	Pte	Jamieson	William Alexander	KIA	1916.07.19 -21	PW
3549	Pte	Jeffery	William George	KIA	1916.07.20	VC
883	Pte	Jenkins	Albert Charles	KIA	1916.07.20	RP

715	L/Cpl	Johnson	Ralph	KIA	1916.07.19 -21	PW
888	Pte	Johnson	Albert John	KIA	1916.07.20	VC
4231	Pte	Jolly	Wesley Clarence	KIA	1916.07.19 -21	RP
891	Pte	Jones	Thomas Walker	KIA	1916.07.20	VC
1055	Pte	Joseph	Joseph Davis De Pass	KIA	1916.07.20	VC
377	Cpl	Keen	Enoch James	KIA	1916.07.20	RP
1706	Pte	Kelly	Frederick Charles	KIA	1916.07.20	RP
365	Cpl	Kendall	Haslem Marsden Harding	KIA	1916.07.20	PW
516	Pte	Kerle	Franz Ferdinand	POW	1916.07.20	RTA
900	Pte	King	Frederick John	KIA	1916.07.19 -21	RP
988	Pte	King	Henry	KIA	1916.07.19 -21	VB
76	Pte	King	Stanley Byfield	KIA	1916.07.20	VC
34	Pte	Kippie	John Murray	KIA	1916.07.19	AN
902	Pte	Kirkaldy	Thomas David	KIA	1916.07.20	AN
2038	Pte	Knight	James	POW DOW	1916.07.27	DC
2148	Pte	Larkins	John Joseph	KIA	1916.07.20	AN

1000	Pte	Law	Edward Manning William	KIA	1916.07.19 -21	VB
247	Pte	Leigh	Arthur Donald	KIA	1916.07.20	VC
2481	Pte	Lewis	John	POW	1916.07.19	RTA
454	Cpl	Linton	Alan	DOW	1916.07.22	BA
250	Sgt	Mackenzie	Alan	POW DOW	1916.07.21	CR
732	Pte	Marcus	Reuben	KIA	1916.07.19 -21	VB
254	Pte	Marrinon	Joseph Patrick Vidal	POW	1916.07.20	RTA
2058	Pte	Martin	Joseph Archibald Loraine	KIA	1916.07.19 -20	VC
470	Pte	Mason	James Lowther	POW DOW	1916.07.19	VB
515	Pte	Mates	Harold	KIA	1916.07.20	RP
2160	Pte	McDonald	Charles Gordon	KIA	1916.07.19 -20	VC
1553	Pte	McDonald	Archie	KIA	1916.07.20	PW
1140	Pte	McKenzie	Owen Alexander	DOW	1916.07.20	BA
1716	Pte	McLeod	Hector McDonald	KIA	1916.07.20	VC
3846	Pte	McPherson	John Joseph	POW DOW	1916.08.23	NW

379	Pte	McReynolds	Francis	DOW	1916.07.26	XH
252	L/Cpl	Millard	George Probert	KIA	1916.07.20	VC
()	Capt	Mills	Charles	POW	1916.07.20	RTA
2055	Pte	Morgan (Meyers)	Cecil (aka Colin)	KIA	1916.07.19	PW
258	Pte	Morley (Howard)	John	KIA	1916.07.20	PW
269	Pte	Nevill	Joseph Howard	KIA	1916.07.20	PW
136	Pte	Oliver	Stanley	KIA	1916.07.19 -21	RP
421	Pte	Olsen	Andrew	KIA	1916.07.19	VC
464	Pte	Otterspoor	Francis Jacobus	KIA	1916.07.19	RP
1735	Pte	Page	Percy William	DOW	1916.08.05	WI
2508	Pte	Paton	Albert Alexander	KIA	1916.07.19 -20	VC
739	Pte	Peak	Sydney George Barker	KIA	1916.07.19 -21	RF
1114	Pte	Pearson	George Charles	KIA	1916.07.19 -21	VB
1089	Pte	Perrin	Harold Edgar	KIA	1916.07.19	VC
330	Pte	Power	Timothy	KIA	1916.07.20	AN
2162	Pte	Purcell	John William	KIA	1916.07.19	AN
288	Pte	Radford	John Charles	KIA	1916.07.20	VC

916	Pte	Rawlings	Frederick	KIA	1916.07.20	PW
1067	Pte	Redfern	Santley Irving	KIA	1916.07.20	VC
407	Pte	Reoch	Robert	POW	1916.07.19	RTA
3891	Pte	Robertson	Robert	KIA	1916.07.19 -21	VB
2411	Pte	Robertson	William Charles	KIA	1916.07.20	RF
()	Capt	Robertson	Gregor Gordon	KIA	1916.07.20	TC
919	Pte	Robinson	Stanley Lewis	KIA	1916.07.20	RP
1604	Pte	Rodda	Benjamin Patrick	DOW	1916.07.26	LO
1877	Pte	Rowlands	Wesley Richard	KIA	1916.07.19 -21	VB
921	Pte	Ryall	Henry	KIA	1916.07.20	RP
1520	A/Sgt	Ryan	William Polding	KIA	1916.07.19	PW
542	Pte	Sadler	Herbert	KIA	1916.07.19	VC
301	L/Cpl	Sexton	James	KIA	1916.07.20	NI
23	Pte	Shadforth	George Stanley	KIA	1916.07.19	RP
1561	Pte	Sheppard	Roy William	DOW	1916.10.23	WW
929	Pte	Simmons	William Thomas	KIA	1916.07.19 -21	RP
1569	Pte	Sinclair	Walter Burgh	KIA	1916.07.20	AN
2090	Pte	Slimon	Robert McRae	KIA	1916.07.20	RP

59	Pte	Smith	Charles James	KIA	1916.07.20	RP
3983	Pte	Smith	James Robert	KIA	1916.07.20	PW
()	Lt	Spreadborough	Ernest William	KIA	1916.07.19	VC
293	Pte	Statham	George Ernest	KIA	1916.07.20	VC
944	Pte	Stiles	Albert	KIA	1916.07.19 -21	RP
802	Cpl	Still	Harry Andrew	POW	1916.07.20	RTA
1144	Pte	Stoddart	William Bede	KIA	1916.07.19	AN
946	Pte	Strong	James Albert	DOW	1920.01.14	11 AGH Vic.
567	Pte	Sutherland	George	KIA	1916.07.19	VC
430	Pte	Tanner	Richard	KIA	1916.07.19	VC
2491	Pte	Taylor	William	KIA	1916.07.19 -21	AN
2012	Pte	Tedford	Thomas Henry	KIA	1916.07.19 -20	VC
956	Pte	Terrens	Albert James	POW DOW	1916.07.21	CR
959	Pte	Thomas	William	POW DOW	1916.07.22	DC
803	Cpl	Thompson	Samuel	KIA	1916.07.19 -21	RP
962	Pte	Thomson	Robert	KIA	1916.07.20	VC
3547	Pte	Thomson	George Leslie	POW	1916.07.20	RTA

763	Pte	Thornburrow	Edward James	POW DOW	1916.08.02	CR
325	Pte	Toohey	Michael	POW	1916.07.20	RTA
60	L/Cpl	Tozer	Ernest William	KIA	1916.07.19	RP
973	Cpl	Treble	Arthur Gilbert	KIA	1916.07.20	VC
97	Pte	Vincent (Minchington)	John Kenneth (James Victor)	KIA	1916.07.20	RP
310	Pte	Vowles	Thomas	POW	1916.07.20	RTA
311	Pte	Walsh	Leslie Gordon	KIA	1916.07.20	PW
770	Pte	Walsh	Thomas	KIA	1916.07.20	RP
2436	Pte	Warby	Charles Francis	KIA	1916.07.19 -21	VB
318	Pte	Weakley	Percy	KIA	1916.07.19	PW
3551	Pte	Welsh	Alfred Joseph John	KIA	1916.07.20	VC
28	Pte	White	William	POW	1916.07.19	RTA
983	Pte	Willis	Henry Victor	KIA	1916.07.19 -21	PW
799	Pte	Wilson	Alexander	KIA	1916.07.19 -21	AN
794	CSM	Woodman	Harold George	KIA	1916.07.19 -21	PW

32nd BATTALION

1966	Pte	Allengame	Percy Henry David	KIA	1916.07.20	VC
1590	Pte	Ambler	Jack Lawrence	KIA	1916.07.19	VC
1205	Pte	Anderson	Percy Michael	KIA	1916.07.20	VC
1209	Pte	Ash	Cyril Nelson	POW DOW	1916.07.22	CR
261	L/Cpl	Backler	Albert Henry	KIA	1916.07.20	VC
1967	Pte	Bamblett	Albert Harold	DOW	1917.04.17	Adelaide SA
346	Pte	Barber	William	KIA	1916.07.20	PW
2283	Pte	Bartholomew	Harold Jack	POW	1916.07.20	RTA
881	Pte	Bartlett	Jack	KIA	1916.07.20	VC
3014	Pte	Bartlett	Claude Thomas	POW	1916.07.20	RTA
352	Pte	Batt	Alfred George	KIA	1916.07.20	PW
10	Sgt	Baumann	Oscar Eric	KIA	1916.07.20	PW
1973	Pte	Bayes	James Frank	POW	1916.07.20	RTA
4	Sgt	Beare	Bertie	KIA	1916.07.20	VC
442	Pte	Bellinger	Archibald Sydney Richard James	DOW	1916.07.20	VC
1602	L/Cpl	Bennett	Allan	KIA	1916.07.20	PW
()	2Lt	Benson	James	KIA	1916.07.20	PW

42	Pte	Black	John Henry Alonza	KIA	1916.07.20	VC
3023	Pte	Blacksell	William Richard	KIA	1916.07.20	VC
43	Pte	Blann	William	DOW	1916.07.22	BA
44	Pte	Bonney	George Edward	KIA	1916.07.19	VC
45	Pte	Bontoft	Percy George	KIA	1916.07.20	BH
891	Pte	Boswell	John Henry	KIA	1916.07.20	VC
1983	Pte	Bottrall	Joseph	KIA	1916.07.19	VC
885	Pte	Bourke	John Patrick	KIA	1916.07.20	VC
1218	Pte	Boyce	William Edwin	KIA	1916.07.20	PW
3034	Pte	Brooks	Arthur Edmond	KIA	1916.07.20	VC
50	Pte	Brooks	Robert Matthew James	POW	1916.07.20	RTA
51	Pte	Brough	Norman John	KIA	1916.07.20	RF
2167	Pte	Brown	Lewis George	KIA	1916.07.19	RP
411	Sgt	Burbidge	Herbert John Charles	KIA	1916.07.20	VC
450	Pte	Burchell	Clive Harold	KIA	1916.07.20	VC
3042	Pte	Burford	Anthony Ray Edward	KIA	1916.07.20	VC
57	Pte	Burgess	Edwin	POW	1916.07.20	RTA

1226	Pte	Burney	Edward Nason	KIA	1916.07.20	PW
1609	Pte	Burrows	Henry	POW DOW	1916.07.24	LI
1611	Pte	Butler	James	POW	1916.07.19	RTA
1229	Pte	Byrne	James	KIA	1916.07.20	VC
1230	Pte	Cahill	Peter	POW	1916.07.20	RTA
1989	Pte	Cahill	James William	POW DOW	1916.08.03	DC
233	Pte	Cameron	Peter	KIA	1916.07.19	VC
917	Pte	Cameron	Ian Alistair	KIA	1916.07.19	VC
457	Pte	Cann	Samuel Isaac	KIA	1916.07.19	VC
63	Pte	Carr	Norman Teale Pearce	POW	1916.07.20	RTA
452	Pte	Carson	David George	KIA	1916.07.20	RP
1235	Pte	Cartwright	Thomas Clift	KIA	1916.07.20	VC
()	Lt	Chinner	Eric Harding	DOW	1916.07.19	PW
66	Pte	Choat	Archibald Percy	KIA	1916.07.20	RP
67	Pte	Choat	Raymond Hadden	KIA	1916.07.20	VC
68	Pte	Choat	Wesley Paul	POW	1916.07.20	RTA
1995	Pte	Clarke	Thomas James	KIA	1916.07.20	VC
1239	Pte	Claxton	Maurice James	KIA	1916.07.20	PW
1238	Pte	Claxton	Theodore	POW	1916.07.20	RTA

71	Pte	Clayer	Edmund Archibald	KIA	1916.07.15	RP
73	Pte	Cochrane	Walter John Leslie	KIA	1916.07.20	VC
2004	Pte	Coffey	Ambrose James	POW	1916.07.20	RTA
915	Pte	Colledge	John Robert (Frank)	POW	1916.07.20	RTA
1194	Sgt	Colless	Frederick Joseph	POW DOW	1916.07.23	LI
855	Pte	Collett	Harold Reginald	POW	1916.07.20	RTA
1241	Pte	Cope	Walter Vernon	KIA	1916.07.20	VC
2011	Pte	Corigliano	Maurice	KIA	1916.07.20	PW
1242	Pte	Cormack	Hugh William	DOW	1916.07.21	BA
2007	Pte	Corrigan	Alfred	POW	1916.07.20	RTA
3073	Pte	Cotching	George Albert	KIA	1916.07.20	RF
2009	Pte	Cowland	John James	KIA	1916.07.19	VC
77	L/Cpl	Crawford	John Alexander	KIA	1916.07.20	PW
3072	Pte	Creedon	John Vincent	KIA	1916.07.20	VC
1997	Pte	Crisp	William Glen	KIA	1916.07.19	VC
79	Pte	Crocker	John Edward	KIA	1916.07.20	PW
2588	Pte	Croser	Willie Gordon	KIA	1916.07.19	VC

80	Cpl	Crossman	Gordon Llewellyn	KIA	1916.07.19	VC
465	Pte	Curtin	Frederick Norman	KIA	1916.07.20	VC
1522	Pte	Daniels	Leslie Gordon	KIA	1916.07.19	VC
473	Pte	Davids	Thomas John Herbert	POW	1916.07.20	RTA
469	Pte	Davidson	Douglas Howard	KIA	1916.07.20	RP
1249	Pte	Davies	Bryan Pellew	POW	1916.07.20	RTA
2015	Pte	Davy	David William	KIA	1916.07.19	RP
925	L/Cpl	Dean	Jack	KIA	1916.07.19	VC
1252	L/Cpl	Dennis	Sydney William	KIA	1916.07.19	VC
475	Pte	D'Herville	Frederick Alexander	POW	1916.07.20	RTA
84	Pte	Dillon	Michael Norman Harold	KIA	1916.07.20	VC
477	Pte	Dolphin	Frank Reuben	POW	1916.07.20	RTA
87	L/Cpl	Dubois	Bernard	KIA	1916.07.20	VC
2022	Pte	Dyke	Walter	POW	1916.07.20	RTA
484	Pte	Dyke	Thomas Donald	POW DOW	1916.07.20	CR
487	Pte	Edgley	Bert	KIA	1916.07.19	RP

1618	Pte	Elmer	Thomas William	POW	1916.07.20	RTA
13	L/Sgt	Emery	Edgar Sydney	KIA	1916.07.19	RP
1257	Cpl	Featherstone	George Henry	POW	1916.07.20	RTA
929	Pte	Feldt	Jens Jensen	KIA	1916.07.20	VC
2027	Pte	Findlay	Peter	POW	1916.07.20	RTA
93	Pte	Finey	Cecil Lloyd	KIA	1916.07.20	RF
1259	Pte	Fisher	Christian Clyde	KIA	1916.07.20	VC
494	L/Cpl	Fisk	Livingstone Thomas	KIA	1916.07.19	VC
3112	Pte	Fox	Frank William	POW	1916.07.20	RTA
931	Pte	Fraser	Alfred Hugh	KIA	1916.07.19	VC
98	Pte	Fursland	Frederick	POW	1916.07.20	RTA
1265	Pte	Furze	Reginald Arthur	KIA	1916.07.20	VC
3115	Pte	Fuss	Lyal Hasalby	POW	1916.07.20	RTA
1083	Pte	Gibney	Guy Sommerville Dion	KIA	1916.07.20	VC
2035	Pte	Gibson	George	POW	1916.07.20	RTA
284	Pte	Gillespie	Frank	DOW	1916.09.11	XI
937	Pte	Gillett	Albert Ernest	KIA	1916.07.19	VC
2036	Pte	Gladwell	Heber	KIA	1916.07.20	BA

351	Pte	Good	James McMillan	POW	1916.07.20	RTA
1272	Pte	Gradwell	James	POW	1916.07.20	RTA
20	Pte	Gray	Edwin Charles	KIA	1916.07.20	PW
1274	Cpl	Green	Robert Courtney	KIA	1916.07.20	PW
1275	Cpl	Greenfield	Bertie	KIA	1916.07.20	VC
3115	Pte	Greenwood	Percy John	KIA	1916.07.20	VC
()	Lt	Griffen	Reginald Theodore	KIA	1916.07.20	PW
3117	Pte	Griffin	John James	KIA	1916.07.19	RP
1276	Pte	Griffiths	Gilbert Allen	POW DOW	1916.07.20	PW
941	Pte	Gully	Thomas	KIA	1916.07.20	VC
508	Pte	Haddy	Horace Bertram	POW	1916.07.20	RTA
()	Lt	Hagan	Thomas Percival	KIA	1916.07.19	VC
1525	Pte	Hall	Melville Alfred	KIA	1916.07.20	VC
513	Pte	Hall	Frederick	POW	1916.07.20	RTA
1281	Pte	Hard	Gordon Jeffery	POW	1916.07.19	RTA
2041	L/Cpl	Harland	Cecill George	KIA	1916.07.20	VC
865	Cpl	Hart	Leslie William	KIA	1916.07.20	PW

511	Pte	Hazell	Henry Haworth	DOW	1916.07.24	CA
110	Pte	Hennessy	William John	KIA	1916.07.20	VC
1284	Pte	Henry	John Alexander	POW	1916.07.19	RTA
()	Maj	Higgon	John Arthur	KIA	1916.07.19	RF
1288	Pte	Hill	Ernest John	POW	1916.07.20	RTA
2050	Pte	Hoffman	Charles Rhody Swan	KIA	1916.07.20	VC
1402	Pte	Hollings	William Craven	KIA	1916.07.19	VC
955	Pte	Holmes	Arnold	KIA	1916.07.19	PW
114	Pte	Holzberger	Herman William	KIA	1916.07.19	RF
1291	Pte	Honey	George	KIA	1916.07.19	PW
525	Pte	Hook	Roy Samuel William Henry	POW DOW	1916.07.22	CR
674	Pte	Hopper	Charles Leonard	KIA	1916.07.19	RP
522	L/Cpl	Hounslow	Alfred Beryl	KIA	1916.07.20	VC
()	Maj	Hughes	John James	POW	1916.07.20	RTA
()	Lt	Hulks	Frederick	KIA	1916.07.20	VC
531	Cpl	Humphris	Bryan	KIA	1916.07.20	VC
347	Pte	Hunt	Thomas	KIA	1916.07.20	VC
962	Pte	Hutchinson	James Wesley	KIA	1916.07.19	VC

3158	Pte	Ingram	Victor Robert	KIA	1916.07.19	VC
()	Lt	Ion	John	DOW	1916.07.25	BO
1528	Pte	Irving	Allan William James	KIA	1916.07.20	PW
972	Pte	Jamieson	Roy Wilfred	KIA	1916.07.19	VC
866	Cpl	Jarman	Ernest Hewitt	KIA	1916.07.20	VC
2354	Pte	Jergens	Edward William	KIA	1916.07.20	VC
120	Pte	Johnson	George Ethelbert	POW	1916.07.20	RTA
969	L/Cpl	Johnston	Robert	POW	1916.07.19	RTA
1299	L/Cpl	Johnstone	Francis David	KIA	1916.07.20	VC
3177	Pte	Jolly	Denzil Cyril	KIA	1916.07.20	VC
121	Pte	Jose	Harold Newlyn	KIA	1916.07.20	VC
1624	Pte	Joyce	Jack	KIA	1916.07.20	VC
971	Pte	Judge	Clarence Frank	KIA	1916.07.20	VC
974	Pte	Keating	William	KIA	1916.07.20	VC
()	Capt	Keay	Robert Allyth	POW	1916.07.20	RTA
1603	Pte	Knable	Adolf Thompson	KIA	1916.07.20	PW
3190	Pte	Knight	Alfred Donald Yorke	KIA	1916.07.17	RP
547	L/Cpl	Korber	Andrew	KIA	1916.07.20	VC

977	Pte	Lacey	Sydney	KIA	1916.07.19	VC
123	Pte	Lampe	Norman Elliott	POW	1916.07.20	RTA
2062	Pte	Lamprey	Frank Augustine	POW	1916.07.20	RTA
124	Pte	Lanyon	Cuthbert John	KIA	1916.07.20	VC
126	Pte	Lawlor	Daniel Michael	KIA	1916.07.20	PW
982	Pte	Lee	James Thomas	KIA	1916.07.19	VC
407	Pte	Lewis	Clement Francis	POW	1916.07.20	RTA
3179	Pte	Liddle	Richard	KIA	1916.07.19	RP
2064	Pte	Loader	Frank Oliver	KIA	1916.07.20	PW
307	Pte	Lodge	Wallace Salisbury	KIA	1916.07.19	RP
553	Pte	Low	David	DOW	1916.08.04	UD
856	Pte	Luly	Alfred Esmond	POW	1916.07.20	RTA
985	Pte	Lydiate	Frederick William	KIA	1916.07.19	VC
3234	Pte	Mackenzie	George Hector	KIA	1916.07.20	VC
1009	Pte	Mackness	Herbert	KIA	1916.07.20	VC
3209	Pte	Magor	Reuben Harold	KIA	1916.07.20	VC
1536	Pte	Mahoney	Samuel Nicholas	KIA	1916.07.20	VC

1195	Pte	Malsch	Henry Arthur	DOW	1916.08.26	BO
135	Pte	Martin	Arthur Percy	KIA	1916.07.20	VC
993	Pte	Mason	Edward Isaac	DOW	1916.07.21	BO
560	Pte	Matthews	Percival Roy	KIA	1916.07.20	LT
994	Pte	Matthews	Harry Karn	DOW	1916.08.24	BO
137	Pte	Maudsley	Robert Thomas	KIA	1916.07.19	PW
1019	Pte	McArthur	Dugald James	KIA	1916.07.19	RP
1011	Pte	McCulloch	Alexander	KIA	1916.07.19	PW
18	Cpl	McDonald	Robert	POW	1916.07.20	RTA
2081	L/Cpl	McGregor	Peter	KIA	1916.07.19	VC
571	Pte	McIntyre	Terence Michael	POW	1916.07.20	RTA
151	Pte	McKenzie	John Gordon	KIA	1916.07.20	PW
1797	Pte	McKenzie	Alexander McGregor	KIA	1916.07.20	PW
1330	Pte	McKinnon	Daniel	POW DOW	1916.07.28	DC
293	Pte	McLean	Hugh	KIA	1916.07.19	PW
119	Pte	McPhee	William James Oliver	KIA	1916.07.20	VC
572	Pte	Messenger	John Arnold	POW	1916.07.20	RTA
141	Pte	Millar	George Hood	DOW	1916.07.25	BO
576	Cpl	Miller	Harry	KIA	1916.07.19	VC

2073	Pte	Miller	Harold George James	KIA	1916.07.19	VC
144	Pte	Minter	Charles Herbert	KIA	1916.07.19	RP
998	Pte	Mitchell	Albert James	KIA	1916.07.19	VC
3282	Pte	Momplhait	Alfred Victor	KIA	1916.07.19	PW
1539	Pte	Moore	Tom Dalzell	POW	1916.07.20	RTA
581	Pte	Moore	William Robert	DOW	1916.07.28	BA
340	L/Cpl	Morphett	Arthur Hurtle	KIA	1916.07.20	VC
22	Pte	Mortimore	Percy	KIA	1916.07.20	VC
415	L/Sgt	Mules	Robert James	POW	1916.07.19	RTA
147	Pte	Murdock	Samuel	POW	1916.07.20	RTA
1744	Pte	Murphy	James Thomas	KIA	1916.07.19	VC
3227	Pte	Murphy	James Parsons	KIA	1916.07.19	VC
3859	Pte	Neill	Bernard Peter	POW	1916.07.20	RTA
1547	Pte	Neill	Daniel Edward	POW	1916.07.20	RTA
587	Pte	Nelligan	Adrian Thomas	POW	1916.07.20	RTA
2084	Pte	Nolan	William	POW	1916.07.20	RTA
158	Pte	O'Niell	Owen Rowe	DOW	1916.08.05	CA
1025	L/Cpl	Oates	Clifford Nicholas	KIA	1916.07.20	VC
2087	Pte	O'Leary	Humphrey	KIA	1916.07.20	VC

1082	Pte	Page	Alexander Russell Robert	KIA	1916.07.20	PW
593	Pte	Palmer	William John	KIA	1916.07.20	VC
2092	Pte	Parham	Edgar William	KIA	1916.07.20	PW
2091	Pte	Parish	Leslie Walter	POW	1916.07.20	RTA
()	2Lt	Paterson	Alexander	KIA	1916.07.20	LT
672	Pte	Pearce	Frank Stanley	KIA	1916.07.19	RP
859	Sgt	Pearse	John	KIA	1916.07.19	RP
2095	Pte	Perry	Andrew Murray	KIA	1916.07.20	VC
161	Pte	Pflaum	Raymond Holstein	KIA	1916.07.19-20	PW
595	Pte	Pitt	Harold Charles	KIA	1916.07.20	PW
597	Pte	Plunket	Charles Francis	KIA	1916.07.20	VC
2387	Pte	Powell	Reginald Lester	POW	1916.07.20	RTA
1556	Pte	Pretty	Walter Hainsworth	KIA	1916.07.20	PW
3869	Pte	Putland	Allan Stewart	DOW	1916.07.25	BA
2391	Pte	Quintrell	Richard Hugh	KIA	1916.07.20	VC
1558	Pte	Randall	Howard James	KIA	1916.07.20	PW
167	L/Cpl	Redman	Thomas Henry Wright	POW	1916.07.20	RTA

3256	Pte	Reid	Maurice Leslie	KIA	1916.07.20	PW
604	Pte	Rice	William Henry	POW	1916.07.20	RTA
1036	Pte	Ridler	Samuel James Thomas	KIA	1916.07.20	PW
602	Pte	Riedel	John Bert	DOW	1916.07.21	BA
1356	Cpl	Roberts	Alfred William	KIA	1916.07.20	VC
417	Pte	Robinson	William	POW	1916.07.20	RTA
1040	Pte	Ross	Mervyn	KIA	1916.07.20	VC
605	Pte	Rostron	George Henry	KIA	1916.07.19	VC
607	Pte	Rowe	Jesse	KIA	1916.07.20	VC
349	Pte	Russell	George Patrick	POW	1916.07.19	RTA
1043	Pte	Ryan	Eugene Joseph	KIA	1916.07.19	RP
1046	Pte	Scott	Robert Grieve Moncrieff	KIA	1916.07.20	PW
1639	Pte	Searle	Henry	KIA	1916.07.20	VC
177	L/Cpl	Secker	Francis Marrett	POW	1916.07.20	RTA
3324	Pte	Sedunary	Percival Charles	KIA	1916.07.19	RP
1047	Pte	Shaw	Harold	KIA	1916.07.20	VC
1363	Pte	Shepherd	Charles Francis	KIA	1916.07.19	VC
296	Pte	Shields	Victor Forest Stanley	POW	1916.07.19	RTA
180	Pte	Siggins	Henry	POW	1916.07.20	RTA

2107	Pte	Simms	Clarence George	POW	1916.07.20	RTA
1516	Pte	Simon	Victor George	KIA	1916.07.20	PW
16	Cpl	Sinigear	Alfred George	KIA	1916.07.19	VC
1054	Pte	Smith	William James	KIA	1916.07.19	VC
1640	Pte	Smith	Arthur George	KIA	1916.07.20	VC
621	Pte	Smith	Percy Edward	DOW	1916.07.21	BA
2111	Pte	Smith	Cyuril Gordon George	DOW	1916.12.15	GSS
2110	Pte	Snook	Godfrey	KIA	1916.07.20	VC
622	Pte	Soans	Boris	KIA	1916.07.20	VC
1367	Pte	Solomon	Morton	KIA	1916.07.19	VC
1571	Pte	Spain	John	POW	1916.07.20	RTA
1591	Pte	Sporsen	Robert Charles	POW	1916.07.20	RTA
3283	Pte	Standley	Frank William Hone	KIA	1916.07.20	VC
187	Pte	Stead	Joseph Raymond	KIA	1916.07.20	PW
624	Pte	Steele	James Norman	KIA	1916.07.20	VC
1375	Pte	Stone	Charles Walter	POW	1916.07.20	RTA
1404	Pte	Stone	Allison Davie	POW	1916.07.20	RTA
165	Cpl	Stranger	Arthur Thomas	KIA	1916.07.20	RF
1057	Pte	Stuart	John Henry	KIA	1916.07.20	VC

1575	Pte	Summerly	Michael	KIA	1916.07.20	VC
631	Pte	Swanton	Henry Theodore	DOW	1916.07.24	BO
2126	Pte	Thorpe	Sidney Hamilton	KIA	1916.07.20	VC
345	Pte	Tilling	Sydney	POW	1916.07.20	RTA
1580	Pte	Train	Stanley Norman	POW	1916.07.20	RTA
1581	Pte	Tucker	William Charles	KIA	1916.07.20	PW
1064	L/Cpl	Turton	Forrest Raymond	POW	1916.07.20	RTA
408	Sgt	Tymons	James Patrick	KIA	1916.07.20	VC
341	Pte	Vardon	Arthur Edward	POW	1916.07.20	RTA
1582	Pte	Vincent	Edwin Aloysius	KIA	1916.07.20	VC
2134	Pte	Voitkun	Andreas	POW	1916.07.20	RTA
2135	Pte	Wahlstedt	Charles	KIA	1916.07.20	VC
1186	Sgt	Walker	John	KIA	1916.07.20	VC
200	Pte	Wall	Leonard George	KIA	1916.07.20	VC
1382	L/Cpl	Waller	William Archer	KIA	1916.07.17	RP
2837	Pte	Wallis	Vincent	KIA	1916.07.20	VC
636	Pte	Wallis	Edward	POW	1916.07.20	RTA

201	Pte	Walsh	Richmond Ambrose	KIA	1916.07.20	VC
420	Cpl	Waltke	Harold William	POW	1916.07.20	RTA
202	Pte	Warner	Ernest	POW	1916.07.20	RTA
203	Pte	Warner	Edgar Albert	DOW	1916.07.22	ES
1384	Pte	Warren	Clarence Hampton	KIA	1916.07.20	VC
634	Pte	Watson	John Elliott	POW	1916.07.20	RTA
640	Pte	Watson	James Horace	POW	1916.07.20	RTA
1385	Pte	Watterson	William	KIA	1916.07.20	VC
204	Pte	Webber	Albert Roy Bennett	KIA	1916.07.19	RP
206	Pte	Westlake	George	KIA	1916.07.20	VC
421	Sgt	Wilson	Henry	KIA	1916.07.19	VC
211	Pte	Wilson	Howard Rupert	KIA	1916.07.20	VC
2144	Pte	Wilson	William Wahlers	KIA	1916.07.20	VC
2143	Pte	Wilson	Victor Francis	DOW	1916.07.21	BO
649	Pte	Winter	Albert Victor	KIA	1916.07.20	VC
213	Pte	Wolff	Edward Alfred	POW DOW	1916.09.05	DC
653	L/Cpl	Zachariah	David Henry Monty	POW DOW	1916.09.10	NW

14th BRIGADE

14th FIELD COMPANY ENGINEERS

5363	Spr	Ashdown	Cecil Paarker	KIA	1916.07.20	AN
5376	Spr	Ettingshausen	Herbert Vaughan	KIA	1916.07.20	VC
()	Lt	Ferguson	John Stephen	DOW	1916.07.27	UB
5385	L/Cpl	Mason	Arnold Blandeston	POW	1916.07.20	RTA
5651	Spr	Murray	Robert Norman	KIA	1916.07.20	RP
1606	Spr	Norton	Harry	KIA	1916.07.19	VC
2426	Spr	Sainty	Frederick Leslie	KIA	1916.07.20	VC

14th LIGHT TRENCH MORTAR BATTERY

1981	Pte	McDowell	John Sydney Peter	KIA	1916.07.20	RB

14th MACHINE GUN COMPANY

2779	Pte	Atwell (Lee)	Sydney Harold (Sydney Thomas)	KIA	1916.07.20	VC

()	2Lt	Briggs	Henry Francis	KIA	1916.07.20	VB
()	Lt	Burns	Robert David	KIA	1916.07.20	PW
3252	Pte	Carr	Harold	KIA	1916.07.20	AN
()	Lt	Cox	Albert Charles	KIA	1916.07.20	RB
2640	Pte	Elliott	John Colin	POW	1916.07.20	RTA
3513	Pte	Gardner	Norman Leslie	POW	1916.07.20	RTA
3323	Pte	Hart	Frederick William	POW	1916.07.20	RTA
()	Lt	Hoddle-Wrigley	Theodore	KIA	1916.07.20	VB
3347	Pte	James	Frederick	KIA	1916.07.20	PW
2934	Pte	Jordan	Raymond	KIA	1916.07.20	VC
2722	Pte	Lauder	James	DOW	1916.07.20	AN
2869	Pte	Lowe	Albert Edward	KIA	1916.07.20	AN
2856	Pte	Luke	Cyril Spencer	KIA	1916.07.20	VC
3110	Pte	McCawley	George	KIA	1916.07.20	AN
3182	Pte	McKay	Francis Walter	KIA	1916.07.20	VC
3118	Pte	McMaster	George Stewart	POW	1916.07.20	RTA
2392	Cpl	Moore	Charles James	KIA	1916.07.20	VB

2878	Pte	Newman	Charles Melville	POW	1916.07.20	RTA
2972	Sgt	Richards	Samuel	KIA	1916.07.20	VA
2899	Pte	Rogers	George Wesley	POW	1916.07.20	RTA
3163	Pte	Rogerson	George	KIA	1916.07.20	CR
2806	Pte	Ryan	Daniel	POW	1916.07.20	RTA
6831	Sgt	Saunders	Samuel Archie	DOW	1916.08.14	BO
2898	Cpl	Stalgis	Gregory Francis	KIA	1916.07.20	PW
3454	Pte	Timperley	Richard Ernest	POW	1916.07.20	RTA
2805	Pte	Walker	Edwin James	POW	1916.07.20	RTA
3469	Pte	Wotton	Cyril Rupert	KIA	1916.07.20	RB

53rd BATTALION

4898	Pte	Abbott	David Roylstone Leslie	KIA	1916.07.19	VC
()	2Lt	Allan	George Ernest	KIA	1916.07.19	VC
4727	Pte	Allanson	Thomas Henry	KIA	1916.07.19	VC
3134	Pte	Allen	George William	KIA	1916.07.19	VC

4728	Pte	Allen	Frederick James	POW	1916.07.20	Died
2101	Pte	Anderson	Frederick Charles	KIA	1916.07.19	VC
()	Capt	Arblaster	Charles	POW DOW	1916.07.24	DC
3226	Pte	Arrowsmith	Clarence	POW	1916.07.20	RTA
4729	Pte	Ashton	Ernest	KIA	1916.07.19	VC
3230	Pte	Aspinall	Frederick John Stevens	KIA	1916.07.19	VC
4731	Pte	Austin	William James	KIA	1916.07.19	VC
3004	Pte	Aysom	Frank Herbert	DOW	1916.07.20	LO
3234	Pte	Back	Richard Clarence	KIA	1916.07.19	VC
3236	Pte	Bacon	Walter John	KIA	1916.07.19	VC
3486	Pte	Badmington	Richard Charles	POW Died	1916.10.14	CSC
2468	Pte	Bailey	Gilbert	KIA	1916.07.19	VC
3235	Cpl	Bailey	Leo Frank Joseph	POW	1916.07.20	RTA
1927	Pte	Bainbridge	John Andrew	KIA	1916.07.19	VC
1623	Pte	Baker	Thomas Norman	KIA	1916.07.19	VC
2116	L/Cpl	Baldwin	Eric Victor	KIA	1916.07.19	VC
3010	Pte	Bancroft	Harry Walter	DOW	1916.07.20	AN

3492	Sgt	Barrack	Roy Gordon	KIA	1916.07.19	VC
2104	CQMS	Bartley	Herbert Henry	KIA	1916.07.19	VC
3239	Pte	Beck	William Sydney	KIA	1916.07.19	VC
4734	Pte	Bell	Charles Frederick	KIA	1916.07.19	VC
2331	Sgt	Black	Archibald John	KIA	1916.07.19	VC
3010	Pte	Black	Stanley William	POW	1916.07.20	RTA
4744	Pte	Blomfield	Fred Gordon	KIA	1916.07.19	VC
3509	Pte	Bolton	Alfred Robert	KIA	1916.07.19	VC
3251	Pte	Booth	William Stanley	KIA	1916.07.19	VC
()	Lt	Bowman	Albert William Montague	POW	1916.07.20	RTA
3243	Pte	Boyd	William	KIA	1916.07.19	VC
4735	L/Cpl	Breen	Reginald James	POW	1916.07.20	RTA
4744	Pte	Bromley	Albert Clive	KIA	1916.07.19	PW
4903	Pte	Bromley	Sidney Reginald	KIA	1916.07.19	RF
3007	Pte	Brown	Charles	KIA	1916.07.19	VC
3249	Pte	Bryant	Thomas Sydney	KIA	1916.07.19	VC
3494	Cpl	Bryson	John Beresford	KIA	1916.07.19	RB

3241	Sgt	Buckland	Frederick Arthur	KIA	1916.07.19	RF
3537	Pte	Bugg	Leslie James	KIA	1916.07.19	VC
3261	Pte	Bundy	William John	KIA	1916.07.19	VC
501	Pte	Burns	Austral Hunter	KIA	1916.07.19	VC
1664	Pte	Byrne	John Alphonsus	KIA	1916.07.19	VC
3273	Pte	Cairns	James David	KIA	1916.07.19	VC
3265	Cpl	Camp	John William	KIA	1916.07.19	VC
3515	Pte	Campbell	Alen	POW	1916.07.20	RTA
3266	Pte	Capper	Alfred Metcalfe	DOW	1916.07.20	BA
3480	Pte	Carey	Timothy Joseph	KIA	1916.07.19	VC
3481	Pte	Carey	John Stephen	KIA	1916.07.19	RB
1445	Pte	Carmichael	Roy Albert	KIA	1916.07.19	VC
3224	Pte	Clarke	John Bowman	KIA	1916.07.19	VC
3168	Pte	Clingan	Alexander Stanley	KIA	1916.07.19	PW
3267	Pte	Coady	Richard (Roland)	KIA	1916.07.19	VC
()	2Lt	Collier	Clarence Timbrell	KIA	1916.07.20	VC
4755	Pte	Comerford	Michael	KIA	1916.07.19	VC

2810	Pte	Connell	William John	KIA	1916.07.19	VC
4757	Pte	Cooke	Sydney Ernest	KIA	1916.07.19	VC
4747	Pte	Coote	Claude George	KIA	1916.07.19	VC
4751	Pte	Craig	George	KIA	1916.07.19	YF
3497	Pte	Crossley	Herbert Percy	POW	1916.07.20	RTA
3818	Pte	Crossman	William	KIA	1916.07.19	VC
3272	Pte	Croucher	William Roy Elwin	KIA	1916.07.19	VC
3265	Pte	Crozier	William Wilson	POW	1916.07.20	RTA
3282	Pte	Cruickshanks	Roy	KIA	1916.07.19	VC
4904	Pte	Davies	Richard David	KIA	1916.07.19	VC
4934	Pte	Davies	Frederick	KIA	1916.07.19	VC
3297	Pte	Davis	Frederick James	KIA	1916.07.19	VC
1943	Sgt	Davis	Frederick	POW	1916.07.20	RTA
1697	Dvr	Dixon	Frederick	KIA	1916.07.19	VC
3288	Pte	Dobson	Thomas Harrizon	POW	1916.07.20	RTA
2128	Pte	Dodd	Reginald	KIA	1916.07.19	VC
3290	Pte	Downie	Edward Leslie	KIA	1916.07.19	VC
1125	Pte	Drane	Sidney	KIA	1916.07.19	VC
2494	Pte	Duffecy	Thomas	KIA	1916.07.19	RB

4761	Pte	Duncan (Le Briton)	James Andrew	KIA	1916.07.19	VC
3551	Pte	Dunn	William	KIA	1916.07.17	RB
3527	Pte	Edwards	William Oliver	KIA	1916.07.19	VC
4772	Pte	Emerton	Stephen	POW	1916.07.20	RTA
4769	Pte	Evans	William Francis	KIA	1916.07.19	VC
4773	Pte	Evans	Frederick	KIA	1916.07.19	VC
2364	Pte	Feeney	Charles	POW	1916.07.19	RTA
3478	Pte	Flaus	Cecil Hamilton	KIA	1916.07.19	VC
1651	Pte	Fleming	Henry John	KIA	1916.07.19	VC
4779	Pte	Forland	Robert Carrington	KIA	1916.07.19	PW
3530	Pte	Foulser	Phillip Joseph	KIA	1916.07.19	VB
1356	L/Cpl	Freirat	Peter	POW	1916.07.20	RTA
3145	Pte	Fry	William	DOW	1916.07.26	WI
3298	Pte	Fryer	Charles Hollingshead	KIA	1916.07.19	RP
2721	Pte	Fuller	Frederick Alonzo	KIA	1916.07.19	VC
3530	L/Cpl	Fulton	Robert John	KIA	1916.07.19	VC
2961	L/Cpl	Gardiner	Cedric Dalmahoy	KIA	1916.07.19	VC

3048	Pte	Gardiner	Bertie Stephen Way	KIA	1916.07.19	VC
3502	L/Cpl	Gavel	Sidney John	KIA	1916.07.19	VC
3501	Pte	Gearing	Patrick	KIA	1916.07.19	CR
4787	Pte	Gilbert	Andrew	KIA	1916.07.19	VC
4788	Pte	Giles	John Albert	POW	1916.07.19	RTA
2636	Pte	Gill	Patrick Joseph	POW	1916.07.20	RTA
3522	Pte	Gilmore	James Elsworth	DOW	1916.07.24	BO
4051	Pte	Gollan	Sylvester James	KIA	1916.07.19	VC
2443	Pte	Goold	John	KIA	1916.07.19	VC
1657	Pte	Gordon	Aubrey Ernest	DOW	1916.07.23	ES
3239	Pte	Gould	Thomas Henry	KIA	1916.07.19	VC
2838	Pte	Green	Archibald John	KIA	1916.07.19	VC
2667	Pte	Greenwood (Smith)	George	KIA	1916.07.19	VC
4789	Pte	Grimmond	Robert Priory	KIA	1916.07.19	VC
3318	Pte	Hall	James	KIA	1916.07.19	VC
3341	Pte	Hall	Anthony	KIA	1916.07.19	VC
1669	Pte	Harden	William Henry	KIA	1916.07.19	VC

2825	L/Sgt	Harrison	Arthur Lawrence	POW	1916.07.20	RTA
4613	Pte	Hatchett	Percy	KIA	1916.07.19	VC
3538	Pte	Hawley	Michael	KIA	1916.07.19	VC
3336	Pte	Heaton	William Henry	KIA	1916.07.19	VC
3308	Pte	Herbert	Stanley George	KIA	1916.07.19	VC
3309	Pte	Herket	Thomas Henry	POW	1916.07.20	RTA
3485	Pte	Higgison	Sydney Percy	KIA	1916.07.19	VC
842	L/Cpl	Hill	Charles Stevens	KIA	1916.07.19	VC
2152	Cpl	Hillyar	Alexander Unwin	KIA	1916.07.19	VC
2351	Pte	Hindley	Arthur Redman	DOW	1916.07.25	BA
3056	Pte	Howatson	Colin	KIA	1916.07.19	RF
3327	Pte	Hubbard	Edwin Arthur	KIA	1916.07.19	VC
3327	Pte	Hungerford	George William	KIA	1916.07.19	PW
2860	Pte	Ingram	Sydney Edward	KIA	1916.07.19	VC
247	Sgt	Janes	Charles Henry	KIA	1916.07.19	RF
3331	L/Sgt	Jentsch	Ernest Augustus	KIA	1916.07.19	PW

5385	Pte	Johnston	Roy Besty	KIA	1916.07.23	AA
3096	Pte	Johnston	George Henry	KIA	1916.07.19	VC
3357	Cpl	Kearney	Percy Robert	KIA	1916.07.19	VC
2729	Pte	Keenan	Oscar	POW	1916.07.20	RTA
3368	Pte	Kemp	Leo Milton	KIA	1916.07.19	VC
3117	Pte	Kendall	William John	KIA	1916.07.19	VC
2862	Pte	Kennedy	Joseph Vincent	KIA	1916.07.19	VC
3337	L/Cpl	Kent (Kimpton)	Bertram	POW	1916.07.20	RTA
3541	Pte	Kentwell	George Stanley	KIA	1916.07.19	VC
2697	Pte	Kilburn	Wilfred	KIA	1916.07.19	VC
3339	Pte	Killalea	John Francis	POW	1916.07.20	RTA
3333	Pte	King	Christopher Charles	KIA	1916.07.19	VC
2973	Pte	Laing	William Lewis	KIA	1916.07.19	VC
3336	Pte	Landsler	Edwin Lawrence	KIA	1916.07.19	VC
4808	Pte	Larkin	John Patrick	KIA	1916.07.19	VC
5402	Pte	Latham	Albert Henry	KIA	1916.07.19	AA
3076	Pte	Lawler	Edwin	KIA	1916.07.19	VC
5404	Pte	Lawler	Thomas James	DOW	1916.07.23	ES
4820	Pte	Lawson	Samuel	KIA	1916.07.19	VC

1956	Pte	Lawson	Walter Raleigh	DOW	1916.07.29	WI
4809	Pte	Layton	Charles Henry	KIA	1916.07.19	VC
3341	Pte	Le Maitre	Herbert William	KIA	1916.07.19	VC
3366	Pte	Legate	Charles Martyn	KIA	1916.07.19	VC
2870	Pte	Levy	Lionel Harold	KIA	1916.07.19	VC
3367	Pte	Loudon	John Alexander	KIA	1916.07.19	VC
4816	Pte	Maher	Thomas James	KIA	1916.07.19	RF
3394	Pte	Main	Alfred Ernest	KIA	1916.07.19	VC
4954	Pte	Mainger	Nicholas	KIA	1916.07.19	VC
4815	Pte	Major	Roy	KIA	1916.07.19	VC
4815	Pte	Malsch	Henry Arthur	DOW	1916.08.26	BO
4822	Pte	Marshall	James	KIA	1916.07.19	RF
4939	Pte	Marshall	Thomas Joseph	KIA	1916.07.19	VC
2742	Pte	Martin	Walter Basil	DOW	1916.10.13	US
3090	Pte	Martin	Walton	KIA	1916.07.19	VC
3398	Pte	Martin	Francis Leo Joseph	POW	1916.07.20	RTA
3355	Pte	Martin	Cyril Henry	DOW	1916.08.08	WI
4811	Pte	Mayo	William Burns	POW	1916.07.20	RTA
3155	Pte	McCrossin	John Stratford	POW	1916.07.20	RTA

3536	Pte	McCue	Charles Joseph	POW	1916.07.19	RTA
3402	Pte	McGrath	Daniel	KIA	1916.07.19	VC
4201	Pte	McHugh	Roger Bede	KIA	1916.07.19	VC
4826	Pte	McKay	Archibald James	KIA	1916.07.19	VC
4828	Pte	McKell (Nixon)	Arthur Edward	POW	1916.07.20	RTA
3087	Pte	McKenzie	Donald Henry	KIA	1916.07.19	VC
3368	Pte	McLennan	Murdo	KIA	1916.07.19	VC
3121	Cpl	McLeod	William Edward	KIA	1916.07.19	VC
3156	Pte	McMahon	Claude Stanley	POW	1916.07.20	RTA
3361	Pte	Meloy	Sydney Alexander	KIA	1916.07.19	VC
3380	Pte	Milne	James Silver	KIA	1916.07.19	VC
2734	Pte	Mitchell	Colin Alexander	POW	1916.07.20	RTA
2616	Pte	Moate	Percy Gladstone	KIA	1916.07.19	VC
()	Lt	Moffitt	Harry Lowry	KIA	1916.07.19	VC
4814	Pte	Mollison	Arthur Cecil	POW	1916.07.20	RTA
3371	Pte	Molloy	John	POW	1916.07.20	RTA
3393	Pte	Moore	William	KIA	1916.07.19	PW
3553	Pte	Moran	John	KIA	1916.07.19	VC

3366	Pte	Morey	Gilbert Major	KIA	1916.07.19	PW
3137	Cpl	Morris	John Elwyn	POW	1916.07.20	RTA
2451	Pte	Morrison	Melton James Leslie	KIA	1916.07.19	VC
()	2Lt	Mudge	Charles Edward	KIA	1916.07.19	RF
3545	Pte	Murphy	Peter	KIA	1916.07.19	VC
3555	L/Cpl	Murray	Lindasy Bertram	DOW	1916.07.28	WI
2737	Pte	Myers	John Archibald	POW	1916.07.20	RTA
2401	Sgt	Neil	Reuben	KIA	1916.07.19	VC
()	2Lt	Nelson	Beresford Joseph	KIA	1916.07.19	VC
3412	Pte	Newman	Benedict Francis	KIA	1916.07.19	VC
3387	Pte	Nicholes	Frederick Alfred Scott	POW	1916.07.20	RTA
3587	Pte	Nicholl	John William	KIA	1916.07.19	VC
3388	Pte	Nichols	Frederick William	POW	1916.07.20	RTA
()	Lt	Noble	William Edward	KIA	1916.07.19	VC
()	Lt Col	Norris	Ignatius Bertram	KIA	1916.07.19	PW
3562	Pte	O'Donnell	William	KIA	1916.07.19	VC

3394	Pte	Ohlson	Axel Olof	KIA	1916.07.19	VC
3367	Pte	Outlaw	William Samuel	KIA	1916.07.19	VC
3391	L/Sgt	Outram	Leonard William	DOW	1916.08.04	UE
3395	Pte	Owen	Robert Herman	POW	1916.07.20	RTA
3392	Pte	Owston	Victor Maxwell	KIA	1916.07.19	RF
3095	Pte	Oxenham	James Malachi	DOW	1916.07.25	NC
4837	Pte	Pain	Ralph Oswald Cleveland	KIA	1916.07.19	VC
3540	Pte	Paling	Gerardus Johannes Verburgh	KIA	1916.07.19	VC
2892	Pte	Palmer	Phillip	KIA	1916.07.19	VC
3528	Pte	Parkinson	Edwin George	KIA	1916.07.19	RP
3148	Pte	Parton	Richard	KIA	1916.07.19	VC
()	Capt	Paulin	Harry	KIA	1916.07.19	VC
1650	Pte	Pearce	Alfred James	POW	1916.07.20	RTA
3098	Pte	Pearce	William Henry	POW	1916.07.20	RTA
3580	Pte	Peat	Walter Henry	KIA	1916.07.19	YF
2896	Pte	Percy	Richard James	KIA	1916.07.19	VC
3422	Pte	Phillips	John Patrick	KIA	1916.07.19	VC

3656	Pte	Phillips	William Henry	KIA	1916.07.19	VC
3407	Pte	Philp	Samuel	KIA	1916.07.19	YF
4840	Pte	Pike	Reginald	KIA	1916.07.19	VC
2815	L/Cpl	Plummer	Elver Gordon	KIA	1916.07.19	VC
3397	Pte	Porter	William Henry	KIA	1916.07.19	VC
3395	Pte	Potter	Walter Ernest	KIA	1916.07.19	VC
4838	Pte	Poulton	John Walter	KIA	1916.07.19	VC
2641	Pte	Pratt	Sidney Richard	KIA	1916.07.19	PZ
()	Lt	Pratt	Albert Ernest	KIA	1916.07.19	PW
4941	Pte	Quinn	William	POW	1916.07.20	RTA
3524	Pte	Quist	Rupert Gustav	KIA	1916.07.19	VC
3412	Pte	Rankine	Andrew Bruce	POW	1916.07.20	RTA
()	Capt	Ranson	Frederick Richard	POW	1916.07.20	RTA
3547	Pte	Raward	Leslie Harold	KIA	1916.07.17	AN
()	2Lt	Rickard	Thomas Nathaniel	KIA	1916.07.19	VC
2196	Pte	Rilen	Richard Norman	KIA	1916.07.19	VC
4843	Pte	Riley	Douglas	POW	1916.07.20	RTA
3421	Pte	Roach	Jack	DOW	1916.08.14	UP

2247	Pte	Roach (Hawke)	David (Edward)	KIA	1916.07.19	VC
3103	L/Cpl	Roberts	Claude Austral	KIA	1916.07.19	VC
2944	Pte	Robertson	William Hugh	KIA	1916.07.19	RF
3516	Pte	Robertson	Eric Lionel	POW	1916.07.20	RTA
329	Sgt	Robin	Geoffrey De Quetteville	KIA	1916.07.19	RB
4841	Pte	Robinson	William John Lawrence	KIA	1916.07.19	VC
3581	Pte	Rolls	George Henry	POW	1916.07.20	RTA
3102	Pte	Ronson	William	KIA	1916.07.19	PW
2900	Pte	Royan	William	KIA	1916.07.19	VC
3414	Pte	Ryan	Joseph	POW	1916.07.20	RTA
()	Maj	Sampson	Victor Horatio Buller	KIA	1916.07.19	PW
3442	Pte	Saunders	Henry	KIA	1916.07.19	RF
3568	Pte	Scott	John	KIA	1916.07.19	RF
3430	Pte	Seach	Charles Adam	POW	1916.07.20	RTA
3433	Pte	Shannon	Peter Paul	KIA	1916.07.19	PW
4872	Pte	Sharkey	Edward	POW	1916.07.20	RTA
4940	Pte	Sharp	James	KIA	1916.07.20	RB
4862	Pte	Shearin	Charles Ebenezer	POW	1916.07.20	RTA
3448	Pte	Shepherd	Thomas	KIA	1916.07.19	VC

3468	Pte	Sim	David	POW	1916.07.20	RTA
1822	Pte	Simms	Samuel Charles	POW	1916.07.20	RTA
2414	Pte	Simpson	Harold John	POW	1916.07.20	RTA
3447	Pte	Sims	Frederick Charles	KIA	1916.07.19	VC
2445	Pte	Sinclair	Alex Aikeman	POW	1916.07.20	RTA
4870	Pte	Slate	Edward Albert	KIA	1916.07.19	VC
523	Pte	Smith	Michael Noble	KIA	1916.07.19	RF
2910	Pte	Smith	Frederick William Alexander	KIA	1916.07.19	VC
3483	Pte	Smith	Roy Edward Norman	POW	1916.07.19	RTA
3106	Pte	Smith	William Walker	POW	1916.07.20	RTA
3439	Pte	Smith	Ernest	DOW	1916.07.23	BO
3112	Pte	Soper	Stephen	POW	1916.07.20	RTA
1165	Pte	Sparks	John	KIA	1916.07.19	VC
3449	Pte	Stacker	Samuel Stephenson	KIA	1916.07.19	VC
3427	L/Cpl	Stanton	Dougald Fittar	POW	1916.07.20	RTA
3440	Pte	Stapleton	Albert Edward	KIA	1916.07.19	VC
3414	Pte	Staunton	Dennis Joseph	KIA	1916.07.19	RF

4860	Pte	Steele	James	POW	1916.07.20	RTA
4891	Pte	Stephenson	Thomas William	KIA	1916.07.19	VC
4874	Pte	Steuart	Alfred Henry	KIA	1916.07.19	VC
5462	Pte	Stewart	Colin	KIA	1916.07.23	AA
4857	Pte	Stiles	Joseph Frederick	POW	1916.07.20	RTA
4922	Pte	Stokes	Alfred Charles	POW	1916.07.20	RTA
3421	Pte	Sullivan	James Joseph	KIA	1916.07.19	AN
3429	Pte	Sullivan	Michael	KIA	1916.07.19	RD
3494	Pte	Summers	Henry	KIA	1916.07.19	RF
3297	Pte	Swinfield	Jack Edward	KIA	1916.07.19	VC
2417	Pte	Taylor	Sydney Leslie	KIA	1916.07.19	VC
4878	Pte	Taylor	Albert Duncan	POW	1916.07.20	RTA
2034	L/Cpl	Thompson	Peter Alexander	KIA	1916.07.19	VC
4908	Pte	Thompson	Charles	KIA	1916.07.19	RD
4881	Pte	Thomson	Arthur Clement	KIA	1916.07.19	VC
4877	Pte	Thorpe	George Horace	KIA	1916.07.19	VC
3454	Pte	Tollefson	Oliver Bernt	KIA	1916.07.19	VC
4261	Pte	Tollis	Reuben Henry Creighton	KIA	1916.07.19	VC
3455	Pte	Turner	Harry	KIA	1916.07.19	VC

3462	Pte	Turner	Arthur	KIA	1916.07.19	VC
3531	Pte	Turner	Bernard Colin	POW DOW	1916.07.24	CR
4885	Pte	Verpillot	Aime Constant	KIA	1916.07.19	PW
4893	Pte	Walker	Edward Ernest	POW	1916.07.20	RTA
3472	Pte	Wall	Benjamin	KIA	1916.07.19	VC
3578	Pte	Walsh	Francis John	KIA	1916.07.19	VC
4932	Pte	Walton	Thomas	DOW	1916.07.20	BA
3515	Pte	Waters	Edgar	KIA	1916.07.19	VC
715	Pte	Watkins	John Edward	KIA	1916.07.19	VC
3576	Pte	Watson	John Henry	KIA	1916.07.19	VC
4924	Pte	Watson	Leonard Leslie	KIA	1916.07.19	VC
3449	Pte	Weaver	Richard Alex	KIA	1916.07.19	RF
3151	Pte	Webb	George	POW	1916.07.20	RTA
3578	Pte	Weiss	Frederick Alfred	KIA	1916.07.19	VC
4891	Pte	West	Walter Henry	DOW	1916.07.28	WI
3476	Pte	Whitbread	Eric Stanley	KIA	1916.07.19	VC
3467	Pte	Wilkins	Herbert Alfred	POW	1916.07.20	RTA
3605	Pte	Williams	Francis James	KIA	1916.07.19	VC
4897	Pte	Williamson	Walter Burt	KIA	1916.07.19	VC
3457	Pte	Willis	John Arthur	POW	1916.07.20	RTA

3474	Pte	Wilson	Arthur William	KIA	1916.07.19	VC
3534	Pte	Wilson	Samuel Charles	KIA	1916.07.19	PW
4887	Pte	Wilson	Eric Robert	KIA	1916.07.19	PW
4943	Pte	Wilson	Albert Edward	KIA	1916.07.19	VC
3535	Pte	Young	Albert Edward	POW	1916.07.20	RTA

54th BATTALION

()	2Lt	Ahern	Timothy Patrick	KIA	1916.07.20	AN
4627	Pte	Aldham	John Henry	POW	1916.07.20	RTA
2560	Sgt	Anderson	David Samuel	KIA	1916.07.19-20	PW
4430	Pte	Armstrong	Joseph	POW	1916.07.20	RTA
2569	L/Cpl	Artist	Philip	DOW	1916.07.20	BA
4128	Pte	Ausburn	Edward Harold	KIA	1916.07.19	RP
4432	Pte	Backhouse	Victor John	KIA	1916.07.19-20	VC
4156	L/Cpl	Balcomb	Horace James	POW	1916.07.20	RTA
4254	Pte	Balkin	Michael William	KIA	1916.07.19-20	PW

4730	Pte	Ballard	Harold Archurt Smith	KIA	1916.07.19-20	RB
4732	Pte	Barnett	Joseph	POW	1916.07.20	RTA
3006	Pte	Barr	Percy George Archibald	KIA	1916.07.19-20	PW
4435	Pte	Barrett	Thomas	POW	1916.07.20	RTA
4735	Pte	Bayliss	James	KIA	1916.07.19-20	AN
3478	Pte	Bell	William George	KIA	1916.07.19-20	AN
3007	L/Cpl	Bentley	William	KIA	1916.07.19-20	RB
1909	L/Cpl	Bilbow	Herbert William	POW	1916.07.20	RTA
4440	Pte	Billington	Marcus Edward	KIA	1916.07.19-20	VC
4441	Pte	Bingham	Ernest	POW	1916.07.20	RTA
3473	Pte	Bishop	William Harrold	DOW	1916.07.28	UO
3720	Pte	Blacka	Arnold Edwin	POW	1916.07.20	RTA
3625	Pte	Blackford	Charles Thomas	POW	1916.07.20	RTA
2782	Pte	Bonnett	Thomas Matthew	KIA	1916.07.19-20	VC
3482	Sgt	Bonnett	James Raymond	POW	1916.07.20	RTA
()	2Lt	Boone	Claude Arthur	KIA	1916.07.19	AN

2786	Cpl	Bowen	Arthur Donald	POW	1916.07.20	RTA
3010	Pte	Boyd	Arthur James	POW	1916.07.20	RTA
1888	Pte	Bradney (Wildman)	Reginald Raymond	KIA	1916.07.19-20	PW
4140	Pte	Brand	John James	KIA	1916.07.19-20	VC
4261	Pte	Bremner	George	POW	1916.07.20	RTA
3487	Pte	Brindal	Clifton Sydney	KIA	1916.07.19-20	PW
4134	Pte	Brown	John	KIA	1916.07.19-20	CR
4145	Pte	Brown	Walter Norman	POW	1916.07.20	RTA
2572	Pte	Brown	Robert William	POW	1916.07.20	RTA
2789	Sgt	Brown	James Powell	POW	1916.07.20	RTA
4143	Pte	Browne	Frederick Thomas	KIA	1916.07.19-20	RF
3463	Pte	Browne	Roy Berridge	POW	1916.07.20	RTA
4151	Pte	Burgess	Ernest	KIA	1916.07.19-20	VC
4450	Pte	Calder	Leslie William	POW	1916.07.20	RTA
495	Sgt	Campbell	Jack Marchmont	KIA	1916.07.19-20	PW
3631	Pte	Campbell	Robert Alexander	KIA	1916.07.20	RP

2793	Pte	Cardew	Thuillier Lake	KIA	1916.07.19-20	PW
154	L/Cpl	Carlin	Harry	KIA	1916.07.20	VB
3511	Pte	Carr	Wilfred Daniel	POW	1916.07.20	RTA
4748	Pte	Carter	Herbert Norman	POW	1916.07.20	RTA
3272	Pte	Carter-Ihnen	Lawrence Phillip	POW	1916.07.20	RTA
3630	Pte	Cawley	Herbert Thomas	KIA	1916.07.19-20	VC
4166	Pte	Clapperton	George	KIA	1916.07.19-20	VC
4155	Pte	Clark	Roy Allison	KIA	1916.07.20	VC
3510	Pte	Clarke	Benjamin	POW	1916.07.20	RTA
2797	Pte	Cobb	Ernest Joseph	POW	1916.07.20	RTA
4362	Pte	Collier	Leslie	KIA	1916.07.19-20	AN
4459	Pte	Collis	David Lewis	POW	1916.07.20	RTA
3585	Pte	Connolly	William Thomas	KIA	1916.07.19-20	PW
4460	Pte	Conroy	Charles William Gordon	KIA	1916.07.19-20	VC
4357	Pte	Conway	John	DOW	1916.07.21	BO
4154	Pte	Corr	George	KIA	1916.07.20	RB
4151	Pte	Costello	John Joseph	POW	1916.07.20	RTA

606	Pte	Cottle	Archibald Percy	KIA	1916.07.19-20	RP
2387	L/Cpl	Coxall	Thomas George	KIA	1916.07.19-20	RP
4420	Pte	Craigie	William Andrew	KIA	1916.07.19-20	PW
4179	Pte	Cressy	Henry Alfred	KIA	1916.07.19-20	PW
4171	Pte	Creswell	Harry Daniel	POW	1916.07.20	RTA
3032	Pte	Cuckson	William Joseph	KIA	1916.07.19-20	PW
4280	Pte	Cutting	Harold Richard	KIA	1916.07.19-20	RB
()	2Lt	D'Alpuget	Jacques Montague	KIA	1916.07.17	RB
3636	Pte	Daley	William	KIA	1916.07.19-20	VC
4763	Pte	Dalzell	George	KIA	1916.07.19	TC
4190	Pte	Davis	Keith Whitmee	DOW	1916.07.22	BO
2803	Pte	Deveigne	Henry George	POW	1916.07.20	RTA
4182	Pte	Devos	Jerome	KIA	1916.07.19-20	VC
4768	Pte	Dewar	Roderick	POW	1916.07.20	RTA
4474	Pte	Diamond	Alfred Henry Lindon	POW	1916.07.20	RTA
4183	Pte	Dibben	Edwin Henry	KIA	1916.07.19-20	PW

3505	Pte	Donaghy	Samuel	POW	1916.07.20	RTA
4181	Sgt	Donnellan	Roger Bede	KIA	1916.07.19-20	RF
4476	Pte	Dorey	William George	POW	1916.07.20	RTA
4477	Pte	Dorfman	Wolf	POW	1916.07.20	RTA
3557	Pte	Doust	Willie Hilton	KIA	1916.07.19-20	PW
3558	Pte	Doust	Roy	DOW	1916.10.30	XJ
4160	Pte	Doyle	Vincent Patrick	KIA	1916.07.19-20	AN
4479	Pte	Dryden	Harry Daniel	POW	1916.07.20	RTA
4189	Pte	Duff	Alexander Moffatt	POW	1916.07.20	RTA
3037	Pte	Duff	Charles McCrea	POW	1916.07.20	RTA
4481	Pte	Dunbar	Randolph	KIA	1916.07.19-20	RP
4482	Pte	Duncan	Robert Ernest	POW	1916.07.20	RTA
4186	Pte	Dunlop	Edwin Raymond	POW	1916.07.20	RTA
4483	Pte	Dunstan	Benedick John Arthur	KIA	1916.07.20	PW
2390	Pte	Dwyer (Duryer)	Valentine John	POW	1916.07.20	RTA
3560	Pte	Dyson	Frederick Arthur	KIA	1916.07.19-20	PW

3286	Pte	Eather	Albert Ernest	POW	1916.07.20	RTA
4166	Pte	Eaton	William Joseph	POW	1916.07.20	RTA
4165	Pte	Eaves	Joseph Henry	POW DOW	1916.08.14	VA
4487	Pte	Emery	James	DOW	1916.08.10	UK
4489	Pte	Evans	Hugh Colman	KIA	1916.07.19-20	VC
4192	Pte	Eyles	Roland James	POW	1916.07.20	RTA
4198	Pte	Fairweather	Richard William	POW	1916.07.20	RTA
4207	Pte	Ferguson	Thomas Charles	DOW	1916.07.23	ES
4776	Pte	Ferrett	Gabriel Alphonse	POW	1916.07.20	RTA
3044	Pte	Fletcher	Harold Percival	POW	1916.07.20	RTA
3494	Pte	Flude	Henry	KIA	1916.07.19-20	VC
4784	Pte	Foley	Joseph Edward	POW	1916.07.20	RTA
4205	Pte	Ford	Samuel William	POW	1916.07.20	RTA
3046	Pte	Forrest	John Robertson	KIA	1916.07.19-20	PW
4492	Pte	Fry	Francis Patrick	POW	1916.07.20	RTA
3048	L/Cpl	Fryer	Thomas Edward	KIA	1916.07.19-20	VC

4782	Pte	Furlong	James	POW	1916.07.20	RTA
4291	Pte	Gallard	Roy Marston	POW	1916.07.20	RTA
4787	Pte	Gates	Charles	KIA	1916.07.19-20	VC
2604	Pte	Gaunt	Edwin George	KIA	1916.07.19-20	VC
2817	Sgt	Gibbons	Bert Alfred	POW	1916.07.20	RTA
4947	Pte	Gibson	William Bernard	KIA	1916.07.19-20	AN
3299	Pte	Gigg	George	POW	1916.07.20	RTA
3543	Pte	Graham	Herbert James	KIA	1916.07.19-20	PW
4296	Pte	Gray	Sidney John	KIA	1916.07.19-20	PW
2927	Pte	Gray	George Robert	KIA	1916.07.20	VC
2406	Pte	Green	Charles	KIA	1916.07.19-20	AN
4221	Pte	Hales	Arthur Ernest	KIA	1916.07.19-20	VC
3474	Pte	Hall	William George	POW	1916.07.19	RTA
()	2Lt	Hall	Henry John	KIA	1916.07.19-20	RF
4226	Pte	Hanna	Lindsay Donald	POW	1916.07.20	RTA
4507	Pte	Hansen	Leslie Hamilton	POW	1916.07.20	RTA

4509	Pte	Harriott	Laurence	KIA	1916.07.19-20	PW
()	Maj	Harrison	Roy	KIA	1916.07.20	RP
3061	L/Cpl	Hartmann	Stanley Theo Victor William	KIA	1916.07.19-20	CB
4512	Pte	Haslam	Percy Arthur	POW	1916.07.20	RTA
3559	Sgt	Hatter	George Henry	DOW	1916.08.03	XK
4232	Pte	Hay	David	KIA	1916.07.19-20	VC
3511	Pte	Healey	Reginald Alfred Charles	KIA	1916.07.19-20	VC
4805	Pte	Heard	Roy Frederick	KIA	1916.07.19-20	RF
4514	Pte	Heath	William Frederick	KIA	1916.07.19-20	VC
3647	Pte	Henry	George Joseph	POW	1916.07.20	RTA
515	Sgt	Hewes	George Walter	KIA	1916.07.19-20	VB
3067	L/Cpl	Hewitt	Frederick	KIA	1916.07.19-20	VC
4176	Pte	Higgins	Anthony James	KIA	1916.07.19-20	VC
2425	Sgt	Hill	Ernest John	KIA	1916.07.19-20	RP
2821	Pte	Hill	Joseph Henry	KIA	1916.07.19-20	VC
4310	Pte	Hoare	Walter Martin	POW	1916.07.20	RTA

4801	Pte	Holliday	Clifford Dawson	KIA	1916.07.20	PW
3074	Pte	Holloway	Leonard James Warren	POW	1916.07.20	RTA
4305	Pte	Holmes	James Leslie	KIA	1916.07.19-20	VC
4188	Pte	Hope	Edwin James	KIA	1916.07.20	PW
4518	Cpl	Horne	Bertram Ernest	POW	1916.07.20	RTA
1137	Pte	Hughes	Walter Godfrey	KIA	1916.07.19-20	RF
4223	Pte	Hughes	Arthur John	POW	1916.07.20	RTA
3323	Pte	Hull	Christopher	KIA	1916.07.16	BA
2818	Cpl	Hurdis	John	KIA	1916.07.19-20	VC
4242	L/Cpl	Innes	John	KIA	1916.07.19-20	LT
4807	Pte	Irvin	David George	KIA	1916.07.19-20	PW
4524	Pte	Jackson	Alfred John	POW	1916.07.20	RTA
3328	Pte	Jarvis	William	DOW	1916.07.21	BA
4247	Cpl	Jenkins	John Alexander Whitmore	POW	1916.07.20	RTA
3548	Pte	Johnson	Joseph	KIA	1916.07.19-20	RF
3085	Cpl	Johnson	Noel Mobrey	POW	1916.07.20	RTA

3549	Pte	Johnston	Fred	KIA	1916.07.20	VC
4315	Pte	Johnston	Cyril Donald	KIA	1916.07.20	PW
3081	Pte	Jones	David	POW	1916.07.20	RTA
4528	Pte	Keats	Prosper Goodwin	KIA	1916.07.19-20	VC
4811	Pte	Keed	Alfred Edward	DOW	1916.07.21	BA
4812	Pte	Keirnan	Leonard	POW	1916.07.20	RTA
4318	Pte	Kelly	Rowland Humphrey	POW	1916.07.20	RTA
4813	Pte	Kentish	Alfred	KIA	1916.07.19-20	VC
4250	Pte	Key	Harold George	POW	1916.07.20	RTA
4815	Pte	Key	Sydney Norman	POW	1916.07.20	RTA
4816	Pte	King	Herbert James	KIA	1916.07.19-20	VC
2835	Pte	Kiss	Reginald Sylvester	POW	1916.07.20	RTA
4253	Pte	Krogman	Arthur	KIA	1916.07.19-20	VC
4532	Pte	Land	Frederick John	POW	1916.07.20	RTA
4370	Pte	Lawson	William Wallace	POW DOW	1916.07.23	LI
4533	Pte	Lee	James Joseph	KIA	1916.07.19-20	VC

4534	Pte	Lee	Percy Victor	KIA	1916.07.19-20	VC
4822	Pte	Lee	James	KIA	1916.07.19-20	VC
4398	Pte	Lenehan	John Robert	KIA	1916.07.19-20	VC
1777	L/Cpl	Little	Vernon Harcourt	KIA	1916.07.19-20	VC
3563	Pte	Lohmann	Herbert George	POW	1916.07.20	RTA
()	Capt	Lovejoy	Harold Redman	POW	1916.07.20	RTA
4537	Pte	Lukeen	George Mathew	POW	1916.07.20	RTA
3599	Pte	Mackin (Mackey)	Terence Joseph	DOW	1916.07.21	BA
4324	Pte	Maclaren-Webb	Phillip Mark	KIA	1916.07.19-20	VC
3102	Pte	Malcolm	William	KIA	1916.07.19-20	AN
4538	Pte	Manley	Lance	POW	1916.07.20	RTA
4830	Pte	Martin	George	KIA	1916.07.16	LA
2848	Pte	Martin	Albert James	POW	1916.07.20	RTA
()	Chpln	Maxted	Spender Edward	KIA	1916.07.20	RP
2850	Pte	Mayo	Bruce Selwyn	KIA	1916.07.19-20	AN

4270	Pte	McAulay	Roderick	KIA	1916.07.19-20	VC
4391	Pte	McAulay	Hector	KIA	1916.07.19-20	VC
2557	Sgt	McCurley	Robert Thomas	POW	1916.07.20	RTA
3161	Pte	McDonald	Lionel Sydney Roderick	KIA	1916.07.19-20	VC
4278	Cpl	McDonald	John	POW	1916.07.20	RTA
4557	Pte	McFarlane	Claude Joseph	POW	1916.07.20	RTA
4560	Cpl	McKee	Alexander	POW	1916.07.20	RTA
4565	Pte	McPherson	Duncan	KIA	1916.07.19-20	VC
4840	Pte	Meyer	Norman	POW	1916.07.19	RTA
3576	Pte	Milne	James Douglas	POW	1916.07.20	RTA
432	Pte	Moir	Frank Edward	POW	1916.07.20	RTA
2636	Pte	Mondy	Charles	KIA	1916.07.19-20	VC
3577	Pte	Moody	Bruce	POW	1916.07.20	RTA
4274	Pte	Moore	Valentine Harding	KIA	1916.07.19-20	VC
4847	Pte	Munn	George	POW Died	1918.10.15	BSW
4850	Pte	Myers	Peter Lawrence	KIA	1916.07.19-20	VC
4341	Pte	Nash	Clarence Ernest	KIA	1916.07.19-20	VC

4946	Pte	Needham	Arnold	KIA	1916.07.19-20	VC
4330	Pte	North	Frederick	POW	1916.07.20	RTA
3586	Pte	O'Donnell	Alexander Thomas	KIA	1916.07.20	VC
4333	Pte	O'Halloran	Charles Henry	POW	1916.07.20	RTA
2934	Pte	Osborne	Terence	POW	1916.07.20	RTA
2906	L/Cpl	Pagan	George	KIA	1916.07.19-20	PW
2859	Pte	Paterson	John	KIA	1916.07.19-20	BH
2012	Pte	Paton	Charles	POW	1916.07.20	RTA
2462	Pte	Pheasant	Walter	KIA	1916.07.19-20	PW
4862	Pte	Plater	Ernest	KIA	1916.07.19-20	VC
4578	Pte	Potter	Walter Arthur	POW	1916.07.20	RTA
4579	Pte	Powell	Thomas Henry	POW	1916.07.20	RTA
3597	Pte	Preston	Victor Henry Gwynne	KIA	1916.07.19-20	VC
3639	Pte	Pryor	Harry Asher	POW	1916.07.20	RTA
3087	Pte	Pybus	Arthur Coventry Leroux	KIA	1916.07.19-20	VC
4297	Pte	Quihampton	Wilfred	KIA	1916.07.19-20	VC

3127	Pte	Read	Frederick	KIA	1916.07.19-20	LT
3398	Pte	Reed	George Henry	POW	1916.07.20	RTA
4304	Pte	Reid	Alexander Henderson	POW	1916.07.20	RTA
2912	L/Sgt	Richardson	Harold	KIA	1916.07.19-20	VC
4581	Pte	Richardson	Benjamin	KIA	1916.07.19-20	VC
4301	Pte	Richardson	Russell Burton	POW	1916.07.20	RTA
4583	Pte	Riordan	William Edgar	KIA	1916.07.19-20	VC
4584	Pte	Rodda	James Ashley	KIA	1916.07.19-20	VC
4299	Pte	Russell	Arthur Joseph	KIA	1916.07.19-20	PW
2678	Pte	Ryan	Daniel	DOW	1916.07.25	BO
4233	Pte	Sadler	Alfred Arthur	KIA	1916.07.20	VC
4315	L/Cpl	Samuels	Alfred Ernest	KIA	1916.07.19-20	RB
4587	Cpl	Sargent	Foster Hartley	POW	1916.07.20	RTA
4872	Pte	Scally	Francis Patrick	KIA	1916.07.19-20	AN
4873	Pte	Scott	Joseph	KIA	1916.07.19	VC
4876	Pte	Shaw	Herbert Victor	DOW	1916.08.07	XL
4878	Pte	Shepherd	Albert John	POW	1916.07.20	RTA

4316	Pte	Singleton	Albert	KIA	1916.07.19-20	VC
4880	Pte	Skea	James	POW	1916.07.20	RTA
4885	Pte	Smith	Charles Frederick	KIA	1916.07.17	RB
2876	Pte	Smith	Fred	KIA	1916.07.19-20	VC
4526	Pte	Smith	Albert John Horace	KIA	1916.07.19-20	RF
4884	Pte	Smith	Charles Thomas Pickering	KIA	1916.07.19-20	RP
4886	Pte	Smith	Edward George	KIA	1916.07.19-20	VC
2688	Pte	Smith	William Henry	POW	1916.07.20	Disch UK
998	Pte	Smith	Edward	DOW	1916.07.21	BO
4239	Pte	Smyth	Arthur Montague	KIA	1916.07.19-20	AN
791	Pte	Speakman	John	POW	1916.07.20	RTA
3138	Cpl	Speer	Robert Allan	POW	1916.07.20	RTA
4308	Pte	Spike	Williaj	KIA	1916.07.19-20	AN
3382	Pte	St John	Thomas	POW	1916.07.20	RTA
4601	Pte	St Ledger	Waltger John William	KIA	1916.07.19-20	AN

4241	L/Cpl	Starr	Victor James	KIA	1916.07.19-20	AN
3219	Pte	Stevens	Raymond Edward	POW	1916.07.20	RTA
4600	Pte	Stewart	Donald Singleton	POW	1916.07.20	RTA
4634	Pte	Stone	John Gregory Ambrose	KIA	1916.07.19-20	VC
()	2Lt	Strangman	John Gerald	KIA	1916.07.19-20	VC
4892	Pte	Summerfield	Herbert	POW	1916.07.20	RTA
4377	Pte	Sweeney	Edward	KIA	1916.07.19-20	VC
4894	Pte	Swift	George Edward	KIA	1916.07.17	RB
4606	Pte	Tate	Charles William Sturdy	POW DOW	1916.07.24	DC
()	Capt	Taylor	Harold	KIA	1916.07.19-20	RB
4607	Pte	Thomas	Edward	POW	1916.07.20	RTA
4608	Pte	Thomas	William John	POW DOW	1916.07.21	CR
3145	Pte	Thomson	John Percival	KIA	1916.07.19-20	VC
4323	Pte	Tibey	James	KIA	1916.07.19-20	VC

4378	Pte	Tilley	Walter Edward	POW	1916.07.20	RTA
4526	Pte	Tilney	William	POW	1916.07.20	RTA
690	Cpl	Toole	John	KIA	1916.07.19-20	VC
4813	Pte	Towne	Harold Victor	KIA	1916.07.19-20	AN
2942	Pte	Turner	Alfred James	KIA	1916.07.19-20	AN
3606	Pte	Turner	William Henry	POW	1916.07.20	RTA
4379	Pte	Vincent	Sydney	POW	1916.07.20	RTA
2506	Pte	Waldron	Harry	POW	1916.07.20	RTA
4617	Pte	Wallis (Wailes)	Joseph Patrick	KIA	1916.07.20	PW
4346	L/Cpl	Walshe	Bernard Martin	KIA	1916.07.19-20	VC
239	Sgt	Wass	William	KIA	1916.07.19-20	VC
1846	Cpl	Watkins	Herbert Ernest	KIA	1916.07.19	RF
1644	Sgt	Watts	John Hamilton James	DOW	1916.07.22	BA
2890	Pte	West	George Annesley	POW	1916.07.20	RTA
3152	Pte	White	Ernest Charles	KIA	1916.07.19-20	VC
4384	L/Cpl	Wilkinson	Henry Mark	KIA	1916.07.19-20	VC
3539	Pte	Wilkinson	Norman	POW	1916.07.20	RTA

4249	Pte	Williamson	Albert	KIA	1916.07.20	PW
4913	Pte	Williamson	Alexander William	POW	1916.07.20	RTA
1196	Pte	Wilson	Alexander	KIA	1916.07.19-20	VC
3170	L/Cpl	Wilson	John Henry	POW	1916.07.20	RTA
3464	Pte	Wright	Frederick Llewellyn	POW	1916.07.20	RTA
4355	Pte	Yabsley	Throsby William	KIA	1916.07.19-20	VC

55th BATTALION

2983	Pte	Anderson	Esmond Lewis	KIA	1916.07.20	RP
3753	Pte	Anderson	George	KIA	1916.07.20	VC
2555	Pte	Arnold	John	KIA	1916.07.20	VC
3210	Pte	Ashcroft	Herbert Alfonso	POW	1916.07.20	RTA
3230	Cpl	Austin	Merton James Desmond	POW	1916.07.20	RTA
3475	Pte	Bagge	Robert Nelson	DOW	1916.07.20	AN
3007	Pte	Baker	Vinton Battam	KIA	1916.07.20	PW
3247	Pte	Baker	Vivian Ernald	KIA	1916.07.20	VC
2582	Cpl	Barber	Alexander Herbert	KIA	1916.07.20	VC
3007	Cpl	Barber	Richard	KIA	1916.07.20	VC

3216	Pte	Barnard	John Osmond	KIA	1916.07.20	VC
2562	Pte	Barnes	James Joseph	POW	1916.07.20	RTA
3031	Pte	Barrett	Rossiter Alfred	KIA	1916.07.20	PW
2382	Pte	Barry	John	POW	1916.07.20	RTA
3760	Pte	Bell	Sydney	KIA	1916.07.20	RF
1304	L/Cpl	Bell	Eric Walter	POW	1916.07.20	RTA
3761	Pte	Bishop	Raymond Charles	KIA	1916.07.20	PW
3657	Pte	Blackburn	Horace Samuel	KIA	1916.07.20	VC
4747	Sgt	Blunt	George Alfred	DOW	1916.07.19	BA
3489	Pte	Bolder	Horace Dudley	POW	1916.07.20	RTA
3490	Pte	Bolder	Alfred Vincent	POW	1916.07.20	RTA
3009	Cpl	Bolt	Herbert Thomas	KIA	1916.07.20	PW
1715	Sgt	Bowman	David McQueen	POW	1916.07.20	RTA
2574	L/Cpl	Boyle	Edward Francis	KIA	1916.07.20	AN
3492	Pte	Bransdon	Frederick McGregor	POW	1916.07.20	RTA
4741	Pte	Brennan	William Arnold	POW	1916.07.19	RTA
3013	Pte	Broadhurst	Leonard	KIA	1916.07.20	PW
3767	Pte	Bryan	Richard Alfred	POW	1916.07.20	RTA
3221	Pte	Cane	Arthur Samuel	POW Died	1916.11.23	CSC

3027	Pte	Casey	Michael John	POW	1916.07.20	RTA
3032	Pte	Chambers	Frederick Joseph	POW	1916.07.20	RTA
4775	Pte	Clifton (Leneard)	Edward William	KIA	1916.07.20	RF
2601	Pte	Coates	John William	POW	1916.07.20	RTA
3262	Pte	Crawford	Clarence Roy	KIA	1916.07.20	AN
3045	Pte	Crebert	Walter Henry	POW Died	1918.11.05	NW
2617	Pte	Crockett	William James	DOW	1916.08.04	BO
()	Lt	Cummins	George	POW	1916.07.20	RTA
2341	Pte	Davis	Robert	POW	1916.07.20	RTA
4792	Pte	Deans	James Frederick	POW	1916.07.20	RTA
3047	Pte	Dewar	Robert Arthur	KIA	1916.07.20	PW
2600	Pte	Duffy	Frank	DOW	1916.08.07	UN
3086	Cpl	Eads	Sydney	POW	1916.07.20	RTA
4802	Pte	Egan	Edwin	KIA	1916.07.20	AN
3057	Pte	Ellis	Ronald Charles	POW	1916.07.20	RTA
3228	Pte	Elrick	James Robertson	KIA	1916.07.20	VC
3060	Pte	Fahey	Patrick William	KIA	1916.07.20	PW

1291	Cpl	Fairweather	Andrew Carnegie Booth	POW DOW	1916.10.16	HB
3061	Pte	Ferguson	Peter	KIA	1916.07.20	VC
3957	Pte	Ferris	Edward	DOW	1916.07.21	BA
3310	Cpl	Fletcher	Frederick	KIA	1916.07.20	PW
()	Capt	Folkard	George D'Arcy	POW	1916.07.20	RTA
3055	Pte	Foreman	William	POW	1916.07.20	RTA
1741	Pte	Freeman	George Harold	POW	1916.07.20	RTA
3051	Pte	Freeman	Norman Lister	POW DOW	1916.07.22	CR
3520	Pte	Fryklund (Franklin)	Charles	KIA	1916.07.20	VC
2650	Cpl	Furniss	Joseph Ross	POW	1916.07.20	RTA
3067	Pte	Galloway	James McCulloch	POW	1916.07.20	RTA
225	Pte	Garrett	William	KIA	1916.07.20	RF
4811	Pte	Geason	Percy	KIA	1916.07.20	PW
()	Capt	Gibbins	Norman	KIA	1916.07.20	AN
3517	Pte	Glynn	Harry	DOW	1916.08.10	UQ
3810	Pte	Gorman	Raymond Sidney	POW	1916.07.20	RTA
2728	Pte	Gribbon	George	POW	1916.07.20	RTA
3237	Pte	Grimes	Joseph Henry	POW	1916.07.20	RTA
4828	Pte	Hammond	James Hugh	POW	1916.07.20	RTA

3240	Cpl	Hancock	Eric Floyd	KIA	1916.07.20	VC
3507	Pte	Hare	David Edward	KIA	1916.07.20	VC
3819	Pte	Harris	James John	KIA	1916.07.20	VC
3311	Pte	Harrison	Frederick Ernest	POW	1916.07.20	RTA
3243	Cpl	Hester	William	DOW	1916.07.22	BA
3090	Pte	Horsfield	Frederick	DOW	1916.07.20	CB
3088	Pte	Housego	Charles	POW	1916.07.19	RTA
3326	Pte	Hybenett	Alfred Frederic	POW	1916.07.20	RTA
3539	Pte	Ibbott	Albert Edward	KIA	1916.07.20	AN
3076	Pte	James	William John	POW	1916.07.20	RTA
4831	Pte	Jennings	Walter James	KIA	1916.07.20	VC
3253	Pte	Johnston	William Henry	POW	1916.07.20	RTA
4830	L/Cpl	Jones	Waldo Emerson	KIA	1916.07.20	AN
3075	Pte	Jones	John Edward Copeland	POW	1916.07.20	RTA
3097	Pte	Jones	Albert	POW	1916.07.20	RTA
3256	Pte	Kendell	Arthur George	POW	1916.07.20	RTA
3258	Pte	Lawrence	Francis John	POW DOW	1916.07.21	CR
2954	Pte	Leane	Albert Charles	POW	1916.07.20	RTA
4840	Cpl	Leister	Leslie	KIA	1916.07.20	PW
4229	Pte	Lennard	Samuel Robert	KIA	1916.07.20	RB

4837	Pte	Luland	Frederick Henry	POW	1916.07.20	RTA
2705	Pte	Lund	William	KIA	1916.07.20	VC
3536	Pte	Mackay	John	POW	1916.07.20	RTA
3885	Pte	Maher	Joseph	POW	1916.07.20	RTA
6732	Pte	Mandeville	John	POW	1916.07.20	RTA
3861	Pte	Marshall	Sydney	POW	1916.07.20	RTA
3375	Pte	Martin	William James	POW	1916.07.20	RTA
3385	Pte	Martin	Cecil George	POW	1916.07.20	RTA
1990	L/Cpl	Mason	Arthur Frederick	POW	1916.07.20	Disch UK
687	Capt	Matthews	John Hamilton	POW	1916.07.20	RTA
3135	Pte	Maude	Frederick William	POW	1916.07.20	RTA
2873	Pte	Mayer	Henry	KIA	1916.07.20	PW
1702	Pte	McCabe	William Jenkins	POW	1916.07.20	RTA
3124	Pte	McGoldrick	Charles Dominic	POW	1916.07.20	RTA
3873	Pte	McGuarr	Richard James	KIA	1916.07.20	PW
3874	Pte	McInnes	William John	POW	1916.07.20	RTA
3577	Pte	McKenzie	Robert	POW	1916.07.20	RTA
3273	Pte	McLeod	Hector	KIA	1916.07.20	AN
3567	Pte	McPherson	Leslie Allan	POW	1916.07.20	RTA

3355	Pte	McPherson	John	POW	1916.07.20	RTA
()	Lt	Mendelsohn	Berrol Lazar	KIA	1916.07.20	PW
3264	Pte	Meyer	Ernest Sydney Rudolph	KIA	1916.07.20	VC
3880	Pte	Millard	Charles Westbrook	KIA	1916.07.20	VC
4854	Pte	Montgomery	Robert	POW	1916.07.20	RTA
2739	L/Cpl	Moore	Eric Osborne	KIA	1916.07.20	VC
2860	Pte	Morrall	Leslie Alexander George	POW	1916.07.20	RTA
3371	Pte	Morris	Norman Sidney	POW	1916.07.20	RTA
()	2Lt	Munro	Stuart	KIA	1916.07.19	VB
3875	Pte	Murphy	Edward Peter	KIA	1916.07.20	RP
3883	Pte	Murray	Dick Clifford	POW	1916.07.20	RTA
2712	Sgt	Naylor	Joseph	DOW	1916.07.23	UJ
4860	Cpl	Nelson	Alexander Edward	KIA	1916.07.20	VC
2509	Pte	Nolan	Justin	POW	1916.07.20	RTA
2423	Sgt	Noldart	Albert Stanley	KIA	1916.07.20	VC
3890	Pte	O'Brien	Charles Percy	POW	1916.07.20	RTA
4865	Pte	O'Brien	Thomas Francis	DOW	1916.07.23	CA
4866	Pte	O'Neill	John Joseph	POW Died	1918.11.30	BSW

3143	Pte	Owens	Michael Frank	POW	1916.07.20	RTA
3279	Pte	Packer	Frederick Ryland	POW	1916.07.20	RTA
4869	Pte	Parsonage	Anson Ronald	POW	1916.07.20	RTA
3353	Pte	Pattison	Cyril William	POW	1916.07.20	RTA
2748	Pte	Payne	Edgar Ashton	KIA	1916.07.20	RP
3200	L/Cpl	Pearce	Oliver Austin Jr	POW	1916.07.20	RTA
2668	Pte	Penfold	Neil	KIA	1916.07.20	VC
3901	Pte	Purdon	Reginald Arthur	POW	1916.07.20	RTA
3902	Pte	Purdon	Sydney Victor	POW	1916.07.20	RTA
3154	Pte	Quirk	Thomas Alfred	POW	1916.07.20	RTA
2896	Pte	Ralph	Victor	POW	1916.07.20	RTA
4882	Pte	Ratcliffe	John	POW	1916.07.20	RTA
4885	Pte	Reay	John Gilbert	POW DOW	1916.10.01	CR
3419	Pte	Richards	Alfred	POW	1916.07.20	RTA
1168	Pte	Robertson	Alexander Christopher	KIA	1916.07.20	RF
1087	CSM	Rose	William Henry Christian	KIA	1916.07.20	PW
4961	Pte	Roth	Harold Wallace	KIA	1916.07.20	VC
3610	Pte	Rowledge	Albert	POW	1916.07.20	RTA

3913	Pte	Rue	Albert Edward	KIA	1916.07.20	VC
3178	Pte	Seinor	William Ernest	POW	1916.07.20	RTA
4899	Pte	Sellen	Donald Alexander	POW	1916.07.20	RTA
3920	Pte	Seymour	Aubrey Isaac	POW	1916.07.20	RTA
3294	Pte	Sherwin	Harold Austin	POW	1916.07.20	RTA
3305	Sgt	Shirley	George	POW	1916.07.20	RTA
3235	Cpl	Skelly	James	POW	1916.07.20	RTA
3327	Pte	Smith	George William	POW	1916.07.19	RTA
2784	Pte	Smith	Alexander Robert	KIA	1916.07.20	AN
2675	Pte	Smith	Percy Carlton	POW	1916.07.20	RTA
3169	Pte	Smith	William Roy	POW	1916.07.20	RTA
3170	Pte	Smith	Frederick	POW	1916.07.20	RTA
3175	Pte	Sparkes	Valentine Camps	POW	1916.07.20	RTA
3924	Pte	St Smith	Herbert Newey	KIA	1916.07.20	PW
1291	Pte	Starr	William Gordon	KIA	1916.07.20	VC
3171	Pte	Steavens	John Edward	KIA	1916.07.20	VC
2772	Cpl	Stephen	William John	KIA	1916.07.20	PW
3436	Pte	Stringfellow	George Henry James	POW	1916.07.20	RTA

2510	Pte	Taylor	Jason	KIA	1916.07.19	VC
3304	Pte	Taylor	Arthur	POW	1916.07.20	RTA
3188	Pte	Tharme	Herbert Sampson	POW	1916.07.20	RTA
2825	Cpl	Thompson	Alfred	KIA	1916.07.20	PW
1838	Pte	Vogt	Peter Christian	KIA	1916.07.20	VC
3148	Pte	Waldon	Harold Victor	POW	1916.07.20	RTA
3310	Pte	Walter	Arthur Barton	POW	1916.07.20	RTA
3311	Pte	Weddup	George	POW	1916.07.20	RTA
4921	Pte	Westerway	Henry	POW	1916.07.20	RTA
3308	Pte	Whitnall	Alexander Charles	KIA	1916.07.20	VC
3151	Pte	Whitney	Marshall	POW	1916.07.20	RTA
2802	L/Cpl	Wilson	David	KIA	1916.07.20	AN
3197	Pte	Wilson	Frank Alexander	POW	1916.07.20	RTA
2039	Pte	Wood	George	DOW	1916.07.21	BA

56th BATTALION

3041	Pte	Barker	Ernest	DOW	1916.07.20	BA
1615	Pte	Baxter	Eric James	DOW	1916.07.22	ES
3038	Pte	Beith	William	KIA	1916.07.20	VC

3012	Pte	Bussell	Rupert Duffhan	DOW	1916.07.25	UT
3052	Pte	Carey	Herbert Ernest Edward	KIA	1916.07.19	VC
1633	Pte	Churchill	Stanley James Webb	KIA	1916.07.19 -20	VB
2579	Pte	Comport	Albert	DOW	1916.07.21	RB
3254	Pte	Coulter	Walter Henry	DOW	1916.07.20	RP
4770	Pte	Dodd	Downie	KIA	1916.07.20	VC
3086	Pte	Finch	Frederick Arthur	KIA	1916.07.20	VC
5109	Pte	Gardner	Walter	KIA	1916.07.16	RB
4789	Pte	Geldard	William Henry	KIA	1916.07.20	VC
4802	Pte	Hall	Horace	DOW	1916.07.28	CA
3555	Pte	Holmes	Hector	POW	1916.07.20	RTA
3119	Pte	Horsington	Archibald Henry	KIA	1916.07.20	VC
3127	L/Cpl	Hurley	George	DOW	1916.10.15	XI
1687	Pte	James	Francis Henry	KIA	1916.07.20	AN
2705	Pte	Johnston	Charles William	KIA	1916.07.20	VC
1690	L/Cpl	Kelly	Joseph	KIA	1916.07.20	AN
3074	Pte	Lonsdale	Norman	DOW	1916.07.20	AN
3430	Pte	Martin	Francis	KIA	1916.07.20	AN
4848	Pte	McCann	George	KIA	1916.07.20	AN

3609	Pte	Miskelly	James Patrick	KIA	1916.07.20	RP
3596	Pte	Oakley	Cecil Victor	DOW	1916.07.20	ES
10	Pte	Penfold	Percy James	DOW	1916.07.22	CA
5193	Pte	Poyitt	Leslie	DOW	1916.07.25	BO
1532	Pte	Purnell	Albert	KIA	1916.07.20	RP
3104	CQMS	Quick	Francis Henry	KIA	1916.07.20	RB
4889	Pte	Ross	Donald Ewen	KIA	1916.07.20	AN
3242	Pte	Rourke	Harold George	KIA	1916.07.20	AN
2804	Pte	Rudd	Percy Mansfield	KIA	1916.07.20	RP
2912	Pte	Sturgeon	William	KIA	1916.07.20	AN
4913	Pte	Sullivan	Arthur Thomas	KIA	1916.07.20	AN
4915	Pte	Sweeney	John Leonard	KIA	1916.07.20	AN
3457	Pte	Tuke	Richard	KIA	1916.07.20	AN
3902	Pte	Walsh	Thomas	KIA	1916.07.20	AN
3288	Pte	Watt	James	DOW	1916.07.20	AN
4955	Pte	Webber	Frederick John	KIA	1916.07.20	AN
3289	Pte	Wilson	Arthur Robert	DOW	1916.08.09	WI

15th BRIGADE

15th FIELD COY ENGINEERS

2	Spr	Elliott	Robert Bruce	KIA	1916.07.20	VC

5578	Spr	Gates	George Kevin	KIA	1916.07.21	VB
5648	Spr	Lawler	Francis Michael	KIA	1916.07.19	RF
6095	Spr	Lyell	John	KIA	1916.07.20	RB
6096	Spr	Newton	William	KIA	1916.07.20	VC
6219	Spr	Price	Edmond Shaw	KIA	1916.07.19	RB
6036	Spr	Tickner	Walter	DOW	1916.07.22	BO

15th MACHINE GUN COY

2567	Pte	Barber	George Francis Thomas	DOW	1916.08.05	BO
2795	Pte	Borneman	John Thomas	KIA	1916.07.19	RB
2345	Pte	Brodie	William Reid	KIA	1916.07.19	LA
2567	Pte	Cox	Graham Rodgers	KIA	1916.07.20	RB
3136	Pte	Horwood	Claude Harold	KIA	1916.07.20	RB
2637	Cpl	Jennings	George William	KIA	1916.07.19	RB
3073	Pte	Lewis	Henry Herbert	KIA	1916.07.20	RB
2733	Pte	Price	Arthur Spalding	KIA	1916.07.19	RB
3248	Pte	Shearer	Alexander	KIA	1916.07.19	RB
3475	Pte	Sutherland	Lisle Ben	KIA	1916.07.19	RB

57th BATTALION

CASUALTIES FROM 14-18 JULY

3041	Pte	Baxter	Herbert Russell	KIA	1916.07.15	RB
4464	Pte	Crozier	Frank	DOW	1916.07.17	ES
2612	Pte	Daniel	John Moore	KIA	1916.07.15	RB
3502	Pte	Edwards	William George	KIA	1916.07.15	RB
3563	Pte	Henderson	Ludovic	DOW	1916.07.17	BA
978	Cpl	Hogan	Arthur Henry	DOW	1916.07.18	BO
3142	Pte	Ireland	Valentine	KIA	1916.07.15	RB
3041	Pte	Kenny	Jep Francis	KIA	1916.07.17	AN
4147	Pte	King	Lawrence Vivian	KIA	1916.07.15	RB
3168	Pte	Leed	David Ray	KIA	1916.07.15	RB
3822	Pte	McDonald	Duncan	KIA	1916.07.15	RB
481	Pte	Pryse	Alexander Leslie	KIA	1916.07.15	RB
2782	Pte	Rosendale	John	KIA	1916.07.15	RB
2784	Pte	Ryan	Patrick Joseph	KIA	1916.07.15	RB
2872	Pte	Shepherd	Francis Charles	KIA	1916.07.15	RB
3547	Pte	Temple	Henry	KIA	1916.07.16	VB

57th BATTALION:

CASUALTIES, MAIN BATTLE 19-20 JULY

2555	Pte	Abrahamson	Anton	KIA	1916.07.20	RB
()	2Lt	Anderson	Edward Handfield	KIA	1916.07.20	VC
4428	Pte	Anglin	Michael	KIA	1916.07.20	RB
4429	Pte	Arbon	Roy Edward	DOW	1916.07.19	UR
4445	Pte	Buchan	William Ferguson	KIA	1916.07.21	RB
3036	Sgt	Bull	Walter Westlake	KIA	1916.07.20	RB
185	Sgt	Davies	Herbert Martyn	DOW	1916.07.22	ES
4963	Pte	Eldershaw	Athol Gale	KIA	1916.07.20	RB
2359	Pte	Fernance	Alexander	KIA	1916.07.20	RB
4484	Pte	Ferrari	John William	KIA	1916.07.20	RB
2928	L/Cpl	Forsyth	Ernest James	KIA	1916.07.23	RB
4500	Pte	Grant	Stanley William	KIA	1916.07.20	RB
3840	Pte	Ingleton	Charles Henry	DOW	1916.07.24	XM
171	Pte	Jenkin	George Ernest	KIA	1916.07.19	VC
3573	Sgt	Kerr	Andrew	KIA	1916.07.20	RB
4181	Pte	McConnell	William James Norwood	DOW	1916.07.22	CA

4855	Pte	Morrison	Henry Herbert	DOW	1916.07.27	BO
2744	Sgt	Moulton	Lewis Leeds	DOW	1916.07.26	CA
4846	Pte	Mullins	Alfred James	KIA	1916.07.21	RB
2882	Pte	Robertson	Archibald Farquhar	KIA	1916.07.20	RB
4228	Pte	Savolainan	Arthur John	KIA	1916.07.20	RB
3608	Pte	Simons	John Joseph	KIA	1916.07.20	RB
2867	Sgt	Strahan	Arthur	DOW	1916.07.20	BA
3653	L/Cpl	Williams	Ernest Charles	DOW	1976.08.11	XL

58th BATTALION
CASUALTIES FROM 14-18 JULY

2862	Pte	Aitken	Josephus Alexander	KIA	1916.07.15	VB
4427	Cpl	Anderson	Albrt Richard Grant	KIA	1916.07.15	VB
4429	Pte	Atkinson	Robert Peter	KIA	1916.07.14	RP
3682	Pte	Baker	John Henry	KIA	1916.07.15	RP
3023	Pte	Bogue	Thomas	KIA	1916.07.15	RP
4436	Pte	Bray	Stanley	KIA	1916.07.14	RP
2326	Pte	Bray	William Roy	KIA	1916.07.15	RP
4822	Pte	Broomhall	John	KIA	1916.07.15	RP
3026	Sgt	Bruce	George Henry	KIA	1916.07.15	RP

3027	Cpl	Bryant	Norman Hertley	KIA	1916.07.15	RP
262	Pte	Butcher	Gordon Leslie Fleuret	KIA	1916.07.26	RB
3489	Pte	Campbell	William	DOW	1916.07.17	ES
2597	Pte	Cartledge	Leslie Albert	DOW	1916.07.17	ES
3485	Pte	Cations	William John	KIA	1916.07.15	VB
2595	Sgt	Challis	George David	KIA	1916.07.15	RP
3483	Pte	Christie	James Skene	DOW	1916.07.16	ES
3709	Pte	Clayton	John Hume	KIA	1916.07.15	RP
2790	Pte	Clinch	Edward	KIA	1916.07.15	RP
3498	Pte	Douglas	Frederick Charles	KIA	1916.07.14	RP
3539	Pte	Dunn	Albert Henry	POW	1916.07.15	RTA
3506	Pte	Everett	Geoffrey James	KIA	1916.07.15	RP
3515	Pte	Farrar	Reginald Charles	KIA	1916.07.15	RP
3532	Pte	Gannon	Joseph Francis	KIA	1916.07.15	RP
4497	Pte	Gordon	Percival Leslie Dallas	KIA	1916.07.15	RP
1770	Pte	Grundy	Albert	DOW	1916.07.18	LO
2650	Pte	Gye	Sydney Charles	KIA	1916.07.15	RP
3550	Pte	Hambling	Robert Horace	KIA	1916.07.15	RP
3035	Pte	Hewitt	Rupert	KIA	1916.07.15	RP

2847	Pte	Johnston	Joseph Henry	KIA	1916.07.15	VB
3561	Pte	Johnston	Percy Albert	KIA	1916.07.15	RP
3570	Pte	Kelsall	Joseph	KIA	1916.07.15	VB
3057	Pte	Kidston	Ewen Burns	KIA	1916.07.15	RP
3166	Sgt	Lewis	Harold Clifford	KIA	1916.07.15	RB
3167	Sgt	Lochhead	William	KIA	1916.07.15	VB
3588	Pte	Lodge	George Lemuel	KIA	1916.07.15	RP
3562	Pte	Looney	Lincoln	KIA	1916.07.15	RP
()	Capt	Mair	Edward Hugh	KIA	1916.07.15	RP
4157	Pte	Maloney	Michael Frederick	KIA	1916.07.15	VB
3603	Pte	Mayo	George Albert	KIA	1916.07.15	RP
3187	Pte	Miller	Alex	KIA	1916.07.14	RP
2720	Pte	Morgan	Edward James	KIA	1916.07.15	VB
4549	Pte	Muir	Andrew Eric	DOW	1916.07.18	RB
4860	Pte	Mulconry	Samuel Joseph	KIA	1916.07.15	RP
3196	Pte	Newton	Thomas Joseph	KIA	1916.07.15	RP
3515	Pte	O'Connor	John Reginald	KIA	1916.07.15	RP
2531	Pte	O'Shannassy	Alan Neil	KIA	1916.07.15	RP
3618	Pte	Parker	Ivan Leslie	KIA	1916.07.16	VB
3621	Pte	Pilven	Hamilton	KIA	1916.07.15	RP

3900	Pte	Quinn	Victor William	KIA	1916.07.15	VB
2868	Pte	Semple	James	POW	1916.07.15	RTA
3253	L/Cpl	Simmons	Edward Yarram Winchelsea	KIA	1916.07.15	RP
3449	Pte	Skewes	Joseph	POW	1916.07.15	RTAOW
3420	Pte	Steers	William Robert	DOW	1916.09.05	ES
3954	Pte	Thompson	John Alexander	KIA	1916.07.17	VB
3276	Pte	Twitt	Frederick Thomas	KIA	1916.07.15	RP
2697	Pte	White	Henry George	DOW	1916.07.16	ES
3668	Pte	Wootton	Alfred John	KIA	1916.07.15	RP

58th BATTALION
CASUALTIES, MAIN BATTLE 19/20 JULY

401	Pte	Allen	Desmond Claude	KIA	1916.07.19	VC
3458	Pte	Allen	Ralph Sinclair	KIA	1916.07.19	VC
2777	Pte	Apperley	Frank Colin	KIA	1916.07.19	VC
3010	Pte	Aspinall	William Harold	KIA	1916.07.19	VC
2560	Pte	Baines	Charles Vincent	KIA	1916.07.19	VC

3022	Pte	Baker	John Leslie Bryan	DOW	1916.07.22	BA
()	Lt	Barnfather	Andrew Ernest Ralph	KIA	1916.07.19	VC
2785	CSM	Bassett	Edward Alfred	KIA	1916.07.19	VC
4751	Pte	Batten	Albert Thomas	DOW	1916.08.03	BO
2572	Sgt	Bird	Cecil George	KIA	1916.07.19	RP
2575	Pte	Bond	William Arthur	KIA	1916.07.19	AR
4296	Pte	Bond	William Edward Ernest	KIA	1916.07.19	VC
4442	Pte	Bowden	Ernest	DOW	1916.07.22	LO
3239	Pte	Briant	Calvin Ellis	POW	1916.07.20	RTA
2565	Pte	Bright	Edgar Henry	KIA	1916.07.19	VC
3043	Pte	Browne	Phillip	KIA	1916.07.19	VC
602	Pte	Bruce	George Owen	KIA	1916.07.19	VC
3311	Cpl	Bunn	Edwin	KIA	1916.07.19	RB
2571	Pte	Butler	John Thomas	KIA	1916.07.19	VC
4452	Pte	Carr	William Chambers	DOW	1916.07.31	BA
2597	Pte	Chandler	Thomas	KIA	1916.07.19	VC
2598	Pte	Chapple	Robert Charles	KIA	1916.07.19	RF
3055	Pte	Clarke	Thomas Leslie	KIA	1916.07.19	VC
3063	Pte	Collins	Charles	KIA	1916.07.19	RB

3496	Pte	Connor	Adrian Ambrose	DOW	1916.07.30	CA
3064	Pte	Coogan	Michael Thomas	KIA	1916.07.19	VC
4465	Pte	Cooper	Thomas Matthew	KIA	1916.07.19	VC
4643	Pte	Daly	James Fitzagibbon	KIA	1916.07.19	RB
4474	Pte	Daly	William Warren	DOW	1916.07.22	BA
4471	Pte	Davis	Frank Herbert	KIA	1916.07.19	RB
2616	Pte	Deakin	George Robert	KIA	1916.07.19	VC
4470	Pte	Dial (Dakin)	George Joseph Farrell	KIA	1916.07.19	RB
2148	Sgt	Dobson	William John	DOW	1916.07.20	ES
2622	Pte	Duncan	William	KIA	1916.07.19	VC
3500	Pte	Dupe	Charles Stanley	KIA	1916.07.19	RB
4187	Pte	Eade	Frederick	DOW	1916.07.31	SS
1698	Pte	Ellis	Robert Porter	KIA	1916.07.19	VC
3125	L/Cpl	Ellis	Arthur Henry	KIA	1916.07.19	VC
4484	Pte	Fair	Frederick Rowland	KIA	1916.07.19	VC
1750	Pte	Ferguson	William	KIA	1916.07.19	RB
3038	Pte	Ferguson	Alexander Edgar	DOW	1916.08.02	UE

3098	Pte	Fleet	Artrhur Gunn	DOW	1916.07.23	BA
639	Sgt	Flynn	Eugene Joseph	DOW	1916.07.20	AN
4797	Pte	George	William Eden	DOW	1916.08.12	UF
3535	Pte	Gill	Maurice William	KIA	1916.07.19	NI
3531	Pte	Glew	Clarence Rupert	KIA	1916.07.19	VC
()	Lt	Gray	Clark Maxwell	KIA	1916.07.19	VC
3307	Sgt	Hannan	Edward	KIA	1916.07.19	VC
3016	L/Cpl	Hansen	Herbert	KIA	1916.07.19	VC
3542	Pte	Hargreaves	Ernest George	KIA	1916.07.19	VC
2503	Sgt	Harvey	Robert Marines	KIA	1916.07.19	VC
2599	Pte	Hearle	Edward Robert	DOW	1916.07.21	UG
2857	Cpl	Hegarty	Albert	DOW	1916.07.20	AN
3775	Pte	Hills	George Robert	DOW	1916.08.01	CA
4510	Pte	Hollioake	Percy	KIA	1916.07.19	VC
2948	Pte	Holten	Richard Frederick	KIA	1916.07.19	RP
4816	Pte	Hunter	John	KIA	1916.07.19	VC
3808	Pte	Hutcheson	William Thurman	KIA	1916.07.19	VC
()	Maj	Hutchinson	Arthur Justin Sanford	KIA	1916.07.19	VC

108	Pte	James	Frank	KIA	1916.07.19	RB
3378	Pte	Jasper	Bernard	KIA	1916.07.19	VC
726	Cpl	Johnston	William	KIA	1916.07.19	VC
3159	Pte	Keen	George Culverwell	DOW	1916.08.25	LO
3158	Sgt	Kenney	Arthur Ralph	KIA	1916.07.19	VC
4814	Pte	Knight	Samuel	KIA	1916.07.19	VC
3574	Pte	Lamb	John James	KIA	1916.07.19	VC
3581	Pte	Landy	William Joseph	KIA	1916.07.19	VC
()	Lt	Latta	Charles Harold	DOW	1916.07.23	CA
3586	Pte	Leary	Alfred Bernard	KIA	1916.07.19	VC
3164	Sgt	Leech	William	KIA	1916.07.19	PH
4539	Sgt	Lepp	Albert Edward	KIA	1916.07.19	RF
2707	Pte	Lock	Albert	DOW	1916.07.21	BO
2875	Pte	Madgwick	Henry William	DOW	1916.07.20	BO
2726	Pte	Maxwell	Richard	DOW	1916.07.21	BO
3199	Pte	McDonald	Leonard James	KIA	1916.07.19	VC
2213	Sgt	McGregor	Francis Alexander	DOW	1916.07.21	BO
3890	Pte	McLean	Huey Walter	KIA	1916.07.19	VC
3118	Pte	Millich	Thomas Como	DOW	1916.07.23	LO
1865	Pte	Nash	Charles Buxton	POW DOW	1916.09.18	CR

3201	Sgt	Nolan	William	DOW	1916.07.23	BO
3610	Pte	O'Neill	James Francis	KIA	1916.07.19	VC
3126	Pte	Partridge	Sydney Carton	DOW	1916.09.01	LO
4569	Pte	Payne	Alfred Roy	KIA	1916.07.19	VC
1808	Pte	Pearson	William John	KIA	1916.07.19	VC
4567	Pte	Phillips	William Charles	KIA	1916.07.19	RB
2420	Pte	Plozza	James	KIA	1916.07.19	VC
4576	Pte	Ryan	Joseph Ernest	KIA	1916.07.19	YF
()	2Lt	Scott	George Norman	KIA	1916.07.19	VC
3256	Pte	Smith	Charles Henry	KIA	1916.07.19	VC
4189	Pte	Smith	Sydney	KIA	1916.07.19	VC
2454	Pte	Smithers	Robert Claude Neil	KIA	1916.07.19	RB
4595	Pte	Stevenson	Denholm Brown	KIA	1916.07.19	VC
3482	Pte	Thomas	James Russell	KIA	1916.07.19	AR
2812	Pte	Thomas	David Henry	DOW	1916.08.12	BO
3681	Pte	Trigg	Raymond Leslie	KIA	1916.07.19	VC
3288	L/Sgt	Watt	George	KIA	1916.07.19	VC
3281	Pte	Whitehill	John Alger	KIA	1916.07.19	VC
4615	Pte	Williams	Thomas Edward	KIA	1916.07.19	VC

4621	Pte	Williams	Edgar	KIA	1916.07.19	VC

59th BATTALION

3451	Pte	Affleck	Duncan Campbell	KIA	1916.07.19	VC
3003	Pte	Aitken	Thomas	KIA	1916.07.19	PH
4729	Cpl	Allen	Charles Prowse	DOW	1916.07.23	BA
4727	Pte	Alliss	George Sharp	DOW	1916.07.23	BA
3003	Pte	Anderson	James	KIA	1916.07.19	VC
1503	Sgt	Andrews	George	KIA	1916.07.19	VC
2778	Pte	Armstrong	Leonard	KIA	1916.07.19	VC
3453	Pte	Baker	Percy	KIA	1916.07.19	VC
2565	Pte	Balfour	Arthur Gordon	KIA	1916.07.19	VC
762	Pte	Bambury	Jacob	KIA	1916.07.19	VC
3012	Pte	Barber	William Ernest	DOW	1916.07.19	AN
3005	Pte	Barclay	Charles William	DOW	1916.08.03	BO
1308	Pte	Barker	James Grisby	KIA	1916.07.19	VC
2789	Pte	Bastin	William James Archibald	KIA	1916.07.19	VC
4146	Pte	Beard	Frederick Arthur	DOW	1916.07.30	UD

3239	Pte	Bell	Algernon Francis	DOW	1916.07.24	BO
2560	Pte	Bellingham	Phillip John	KIA	1916.07.19	VC
3240	Pte	Bennett	Clyde James	KIA	1916.07.19	VC
2569	Pte	Benton	Albert	KIA	1916.07.19	VC
415	Pte	Beston	Thomas Joseph	KIA	1916.07.19	VC
3672	L/Cpl	Bethune	Rupert Charles	KIA	1916.07.19	VC
4973	Pte	Bevan	George Alphonse	KIA	1916.07.19	VC
2570	Pte	Bicket	John Alexander	DOW	1916.08.28	BO
3022	Pte	Biggs	William Frederick	KIA	1916.07.19	VC
2792	Pte	Bill	Clive	KIA	1916.07.19	VC
3691	Pte	Bishop	Thomas Charles	KIA	1916.07.19	AR
4737	Pte	Blake	George Francis	KIA	1916.07.19	RF
3009	Pte	Blee	George Glouis	KIA	1916.07.19	VC
2559	Pte	Bond	Ralph Richard	KIA	1916.07.19	CR
3466	Pte	Bond	William Joseph	KIA	1916.07.19	VC
4736	Pte	Botterill	Frederick Charles	KIA	1916.07.19	VC
()	Lt	Bowden	John Charles	KIA	1916.07.19	VC
1664	Pte	Boyce	Albert John	KIA	1916.07.19	VC

2562	Private	Boyd	Hugh James	KIA	1916.07.19	VC
3689	Pte	Boyd	Allen James	KIA	1916.07.19	AR
4738	Pte	Brache	Melville William	KIA	1916.07.19	AR
4749	Pte	Bradshaw	Frederick James	KIA	1916.07.19	VC
1914	Pte	Breslin	Thomas	KIA	1916.07.19	RF
2327	Pte	Brinckmann	Hendrick Cornelius	KIA	1916.07.19	VC
2584	L/Cpl	Brown	Alexander Malcolm	KIA	1916.07.19	VC
4740	Pte	Brown	Frederick Henry	KIA	1916.07.19	VC
1843	Pte	Brownridge	Joseph	KIA	1916.07.19	VC
4745	Pte	Brunston	Henry	KIA	1916.07.19	YF
3038	Pte	Burden	Albert Herbert Arthur	KIA	1916.07.19	AR
4762	Pte	Burford	John Henry	KIA	1916.07.19	VC
3267	Pte	Butler	Arthur Ormond	KIA	1916.07.19	VC
3321	Pte	Byrne	Frederick Thomas	KIA	1916.07.19	VC
3281	Pte	Cahill	William Patrick	KIA	1916.07.19	VC
3037	Pte	Cain	Walter James	KIA	1916.07.19	VC
4748	Pte	Caines	Percy Colston	KIA	1916.07.19	VC

2618	Pte	Cameron	Norman	KIA	1916.07.19	VC
3042	Pte	Campbell	Arthur	KIA	1916.07.19	VC
2590	Pte	Cant	Charles Rutherford	KIA	1916.07.19	VC
2591	Pte	Cantwell	James William	KIA	1916.07.19	RF
()	Lt	Carr	Eric Thomas	KIA	1916.07.19	VC
3773	Pte	Charles	William Spencer	KIA	1916.07.19	RF
2827	Pte	Christian	Orlando Julius	KIA	1916.07.19	VC
3054	Pte	Christie	John Leslie	KIA	1916.07.19	VC
4751	Pte	Christie	William Jennings	KIA	1916.07.19	VC
3053	Pte	Clark	Stanley Talbot Claude	KIA	1916.07.19	VC
4978	Pte	Clarke	Owen James	KIA	1916.07.19	VC
3292	L/Cpl	Clayden	Arthur James	DOW	1916.07.25	UB
3055	L/Cpl	Clements	Vivian John	KIA	1916.07.19	RP
4076	Cpl	Clements	Harold George	KIA	1916.07.19	VC
4753	Pte	Cobbett	William	DOW	1916.08.01	CA
818	Pte	Coleman	William Joseph	KIA	1916.07.19	YF
4783	Pte	Connor	Eden Dennis	KIA	1916.07.19	VC
3482	Pte	Cooper	Henry Ackroyd	KIA	1916.07.19	VC
3715	Pte	Corin	Edward Arthur	KIA	1916.07.19	VC

2150	Pte	Cosgriff	Thomas Joseph	KIA	1916.07.19	PW
4462	Pte	Cotter	William Clarke	KIA	1916.07.19	VC
()	Lt	Cousins	Frank Leicester	KIA	1916.07.19	VC
2211	Pte	Cox	Edward Thomas	KIA	1916.07.19	VC
3277	Pte	Crow	William Frederick	KIA	1916.07.19	VC
2349	Pte	Cunningham	Alexander Joseph	KIA	1916.07.19	VC
3284	Pte	Curtis	Frederick Gillman	KIA	1916.07.19	VC
3718	Pte	Dale	Frederick William	KIA	1916.07.19	VC
3535	Pte	Daley	Claude Henry	KIA	1916.07.19	AR
3299	L/Cpl	Dalzell	Bertram Allen	KIA	1916.07.19	VC
2849	Pte	Daniel	Harry Gordon	KIA	1916.07.19	VC
864	Pte	Davies	Henry Gascoigne	KIA	1916.07.19	VC
3503	Pte	Davis	George Young	DOW	1916.07.25	WI
3027	Pte	Dewsnap	John William	KIA	1916.07.19	VC
2575	L/Sgt	Dixon	Frank Allen	KIA	1916.07.19	VC
3289	Sgt	Dobson	David	KIA	1916.07.19	RB
3091	Pte	Donahoo	Michael Francis	KIA	1916.07.19	AR

1694	Pte	Duffy	Joseph Cavan	KIA	1916.07.19	VC
4790	Pte	Duncan	George Robert	KIA	1916.07.19	VC
5375	Pte	Dunstan	Edwin William	KIA	1916.07.19	VC
3086	Pte	Dwyer	Frederick	DOW	1916.08.11	UB
2625	Pte	Edney	John Edward	KIA	1916.07.19	VC
4095	Pte	Ellis	Arthur William George	KIA	1916.07.19	VC
4776	Pte	Elsden	Henry Robert	KIA	1916.07.19	VC
3540	Pte	Emery	Harold Ernest	DOW	1916.07.28	UJ
3170	Pte	Farrelly	Martin	KIA	1916.07.19	VC
3727	Pte	Ferns	William Morton	KIA	1916.07.19	NI
4782	Pte	Fitton	William Reid	KIA	1916.07.19	AR
2474	Pte	Fitzgerald	John Lawrence	KIA	1916.07.19	VC
1706	Pte	Flack	Alfred Walter	KIA	1916.07.19	VC
1704	Pte	Fletcher	Japhet Money	KIA	1916.07.19	VC
3733	Pte	Fletcher	John Kearn	KIA	1916.07.19	VC
3732	Pte	Flowers	Wilfred Field Carlyle	KIA	1916.07.19	VC
3097	Pte	Forster	Harry	KIA	1916.07.19	VC
3282	Pte	Fraser	Roderick John	KIA	1916.07.19	VC
3519	Pte	Frew	Leslie George	DOW	1916.08.03	UH
3738	Pte	Fulton	Arthur Rupert	KIA	1916.07.19	VC

3314	Pte	Gaborit	Thomas Leopold	KIA	1916.07.19	VC
3979	Pte	Galloway	Thomas	KIA	1916.07.19	VC
3739	Pte	Gandy	John Frederick	DOW	1916.07.22	WI
4798	Pte	Garry	Walter	KIA	1916.07.19	VC
4821	Pte	Geal	Albert	KIA	1916.07.19	VC
()	Lt	Gibbs	Richard Horace Maconochie	KIA	1916.07.19	VC
2639	Pte	Giblett	Cecil Woods	KIA	1916.07.19	VC
1742	Pte	Gibson	Alfred Lewis	DOW	1916.07.21	LO
2641	Pte	Gilfoy	Herbert	DOW	1916.07.26	CA
1551	Cpl	Gill	William Alexander	KIA	1916.07.19	VC
3071	Pte	Gillies	William	DOW	1916.08.10	CA
3539	Pte	Ginman	George Leo	DOW	1916.08.10	BO
2213	Pte	Gittoes	Ernest Claude	KIA	1916.07.19	VC
3550	Sgt	Goldby	William	KIA	1916.07.19	VC
1553	L/Cpl	Goldie	Charles Edward	KIA	1916.07.19	VC
3063	L/Cpl	Goldsmith	Robert George Henry	KIA	1916.07.19	VC
2671	Pte	Goodwin	Alfred Basil Harrison	KIA	1916.07.19	VC
2798	Pte	Grace	George Henry	KIA	1916.07.19	VC

2799	Pte	Gray	John Millyard	KIA	1916.07.19	VC
3036	Cpl	Green	Sidney Frank William Harold	KIA	1916.07.19	VC
3110	Pte	Gregory	Charles Clapham	KIA	1916.07.19	AR
1811	Pte	Grenville	Vincent	KIA	1916.07.19	RB
3062	Sgt	Grumont	Charles Albert	KIA	1916.07.19	VC
3134	Pte	Halloran	Timothy	KIA	1916.07.19	VC
2173	Pte	Harbert	Gershun	KIA	1916.07.19	VC
2073	Pte	Harris	Robert	KIA	1916.07.19	VC
2811	Pte	Hart	William Henry Gordon	KIA	1916.07.19	VC
3766	Pte	Hart	John Ernest	KIA	1916.07.19	VC
3769	Sgt	Hayes	Henry Seymour	KIA	1916.07.19	VC
3348	Pte	Henderson	Frederick Arthur	KIA	1916.07.19	VC
3085	Pte	Henning	George Ernest	DOW	1916.08.06	UC
3364	Pte	Herriot	Walter Charles	DOW	1916.08.14	UV
3317	L/Cpl	Heward	Frank Horace	KIA	1916.07.19	VC
3066	Pte	Hind	James	KIA	1916.07.19	VC
4800	Pte	Hitzeroth	Glynn Rupert	KIA	1916.07.19	VC
3990	Pte	Hodson	Victor Henry Herbstone	POW	1916.07.19	RTA

4115	L/Cpl	Holmes	Edgar Charles	KIA	1916.07.19	VC
2844	L/Cpl	Hosie	Gustave George	KIA	1916.07.19	VC
1679	Pte	Hotham	Charles Joseph	KIA	1916.07.19	VC
()	Lt	Howard	Hubert Charles	KIA	1916.07.19	NI
1642	L/Cpl	Humphreys	Herbert	KIA	1916.07.19	VC
3357	Sgt	Hunt	William Henry	KIA	1916.07.19	VC
3046	Sgt	Hunt	Arthur	DOW	1916.08.13	UL
3784	Pte	Hunter	Reginald Arthur	KIA	1916.07.19	VC
4816	Pte	Hunter	David	KIA	1916.07.19	VC
2824	Pte	Hutcheson	David William	KIA	1916.07.19	VC
4522	Pte	Jackson	Albert Henry John	KIA	1916.07.19	AR
1773	Pte	Jacobs	Leslie John	KIA	1916.07.19	RP
2473	Pte	Jamieson	Young James	KIA	1916.07.19	AR
1810	Pte	Jennings	William West Anderson	KIA	1916.07.19	RB
1692	Pte	Jessop	Samuel	KIA	1916.07.19	VC
2957	Pte	Jewell	Herbert James	KIA	1916.07.19	VC
3367	Cpl	Johnson	Robert William	KIA	1916.07.19	VC
3524	Pte	Johnston	Albert	KIA	1916.07.19	RB
3372	Pte	Johnston	Andrew Archibald	DOW	1916.07.20	BA

3151	Pte	Jones	John	KIA	1916.07.19	VC
36	Pte	Joyce	Charles Edward	KIA	1916.07.19	RB
4823	Pte	Kennedy	James	KIA	1916.07.19	RF
3797	Pte	Kerr	John Cameron	KIA	1916.07.19	VC
3160	Cpl	Kiellerup	Otto James	KIA	1916.07.19	VC
2597	Cpl	King	Walter Tasman	KIA	1916.07.19	VC
1726	Pte	King	Charles	DOW	1916.07.21	BA
411	Pte	King	Arthur Wallace	DOW	1916.07.23	BO
3042	Pte	Kingston	William	DOW	1916.07.22	BO
4925	Pte	Ladd	William Henry	KIA	1916.07.19	VC
1712	Pte	Lake	Cecil Lancelot	KIA	1916.07.19	VC
3146	Pte	Lane	Thomas Arthur	KIA	1916.07.19	VC
4130	Pte	Lear	Isaac James	KIA	1916.07.19	VC
3385	Pte	Lee	Frank Albert	KIA	1916.07.19	VC
4227	Pte	Leigh	Cecil Thorold	KIA	1916.07.19	VC
3165	Pte	Letts	John	KIA	1916.07.19	VC
()	Capt	Liddelow	Aubrey	KIA	1916.07.19	VC
4832	Pte	Lidgett	Archie Fred	KIA	1916.07.19	VC
3812	Sgt	Lister	Robert Murray	KIA	1916.07.19	VC
3064	Pte	Love	James Richard	KIA	1916.07.19	RF

3816	L/Sgt	Lucas	Horace Stanley	KIA	1916.07.19	VC
3158	Pte	Lynch	Ellery Gordon	KIA	1916.07.19	VC
4828	Pte	Maddocks	John Henry	KIA	1916.07.19	RF
3358	Pte	Mann	Percy Roy	KIA	1916.07.19	VC
2723	Pte	Mathews	Albert Edward	DOW	1916.07.21	AN
1595	Pte	Matthew	Alec Duncan	KIA	1916.07.19	VC
3066	Pte	Maule	Alexander John	KIA	1916.07.19	VC
2212	Cpl	Mayo	Gilbert Joseph	KIA	1916.07.19	RF
4547	Cpl	McColl	Daniel Hugh	KIA	1916.07.19	AR
3184	Pte	McCulloch	Frederick Henry Alexander	KIA	1916.07.19	VC
1678	Pte	McDonald	Ronald	KIA	1916.07.19	VC
1743	Pte	McEwan	Thomas Michael	KIA	1916.07.19	VC
3416	Pte	McGhee	Donald	KIA	1916.07.19	VC
1791	Pte	McIlroy	Robert	DOW	1916.07.21	ES
3007	Pte	McInerney	William John	KIA	1916.07.19	VC
2420	Pte	McInnes	Donald	KIA	1916.07.19	VC
3408	Pte	McKeone	James Joseph	KIA	1916.07.19	VC
4636	Pte	McLean	Murdoch	KIA	1916.07.19	YF
794	Pte	McLear	Christopher Henry	KIA	1916.07.19	VC

3996	Pte	McMahon	Thomas Joseph	KIA	1916.07.19	VC
3120	Pte	McNamara	Christopher James	KIA	1916.07.19	VC
4153	Pte	McRae	Charles Robert	KIA	1916.07.19	AR
3840	L/Cpl	Meginess	Frank	KIA	1916.07.19	VC
3178	Pte	Merton	Charles	KIA	1916.07.19	VC
3575	Pte	Miles	Walter O'Malley	KIA	1916.07.19	VC
3356	Pte	Milham	Thomas Sidney	KIA	1916.07.19	VC
3582	Pte	Mills	Leslie Charles	KIA	1916.07.19	VC
2714	Pte	Mitchell	Sydney James	KIA	1916.07.19	VC
2746	Pte	Mitchell	Alfred Charles	KIA	1916.07.19	VC
4845	Pte	Mitchell	Albert Edward	KIA	1916.07.19	VC
2217	Pte	Montgomery	William Henry	KIA	1916.07.19	VC
3180	Cpl	More-Reid	George	KIA	1916.07.19	VC
2743	Pte	Morris	Leslie Charlton	KIA	1916.07.19	VC
()	Lt	Morrow	Andrew Duncan	DOW	1916.07.21	BA
2635	Pte	Mossenton	Harry Deering	KIA	1916.07.19	VC
3063	Pte	Mowbray	Arthur Charles	DOW	1916.07.31	UN
2729	Pte	Murphy	Christopher	KIA	1916.07.19	VC

2406	L/Cpl	Neil	Leonard John James	KIA	1916.07.19	VC
3126	Pte	Newey	Horace Roland	KIA	1916.07.19	VC
1586	Pte	Nicholls	Orton	DOW	1916.07.18	AN
2750	Pte	North	Ernest Frederick	KIA	1916.07.19	VC
1737	Pte	O'Day	William John	KIA	1916.07.19	VC
3214	Pte	O'Hara	James Augustine	KIA	1916.07.19	VC
2826	Pte	O'Keeffe	Walter Patrick	KIA	1916.07.19	VC
3096	Pte	O'Leary	Joseph	KIA	1916.07.19	VC
4570	Pte	Olston	John Augustus	KIA	1916.07.19	VC
3216	Pte	O'Meara	Patrick Julian	KIA	1916.07.19	VC
3871	Cpl	Orenshaw	William Henry	KIA	1916.07.19	VC
3867	Pte	O'Sullivan	William	DOW - Suicide	1916.07.19	AN
3873	Pte	Outen	Roy James	KIA	1916.07.19	VC
1404	L/Cpl	Palmer	William Edward	KIA	1916.07.19	VC
3860	Pte	Pamphlet	James George	KIA	1916.07.19	VC
3878	Pte	Parry	Robert Duncan	KIA	1916.07.19	VC
4864	Pte	Parsons	William	KIA	1916.07.19	VC
4564	Pte	Pearce	Roy Victor	KIA	1916.07.19	VC
2764	Pte	Peiper	Robert Frederick	KIA	1916.07.19	VC

2014	Pte	Perkins	Colin Ernest	KIA	1916.07.19	VC
2421	Pte	Perkins	Eric	KIA	1916.07.19	VC
2015	Pte	Perrett	Stanley Charles	KIA	1916.07.19	VC
3448	Pte	Perry	Robert Francis	KIA	1916.07.19	VC
3894	Pte	Phelan	John Francis	KIA	1916.07.19	VC
2877	L/Cpl	Phelps	William Thomas	KIA	1916.07.19	VC
1804	Pte	Phillips	Alfred Ernest	KIA	1916.07.19	VC
4868	Pte	Phillips	Arthur Egbert	KIA	1916.07.19	VC
4869	L/Cpl	Phillips	Clemons William	KIA	1916.07.19	VC
4911	Pte	Pitcher	Basil Bernard	KIA	1916.07.19	VC
2884	Pte	Polglase	Charles Alexander	KIA	1916.07.19	VC
1755	Pte	Powell	Clarence	KIA	1916.07.19	YF
2858	Pte	Powell	John	KIA	1916.07.19	VC
3895	Sgt	Power	Albert Chalmers	KIA	1916.07.19	RF
1757	Pte	Prescott	Hugh Henry Percival	KIA	1916.07.19	AR
1798	Pte	Rabinovitch	Bezelle	KIA	1916.07.19	VC
2759	Pte	Rankin	Stanley	DOW	1916.07.31	BO
4570	Pte	Ratcliffe	Frederick Edmund	KIA	1916.07.19	RB

3527	L/Sgt	Raw	Walter Whitfield	KIA	1916.07.19	VC
3231	Pte	Ray	Arthur William	KIA	1916.07.19	NI
3466	Sgt	Read	George Reginald	DOW	1916.07.25	BO
1811	Sgt	Reeves	Robert Edward	KIA	1916.07.19	VC
3897	Pte	Reid	Charles John	KIA	1916.07.19	AR
4515	Pte	Richardson	John Leslie	KIA	1916.07.19	VC
4882	Pte	Richardson	Frank	KIA	1916.07.19	VC
3224	Pte	Ricketts	Lawrence Arthur	KIA	1916.07.19	VC
4252	Pte	Roberts	Charles Edward	KIA	1916.07.19	VC
2844	Pte	Rogers	Harry Ernest	KIA	1916.07.19	VC
3132	Pte	Rosney	Peter Jeremiah	KIA	1916.07.19	VC
3568	Pte	Ross	Roy Leigh	KIA	1916.07.19	VC
3225	Pte	Ruff	Victor Henry	DOW	1916.07.28	UC
2672	Pte	Russell	Thomas Lippett	KIA	1916.07.19	VC
3129	Pte	Russell	Frederick George	KIA	1916.07.19	VC
2768	Pte	Russell	Alan James	DOW	1916.07.21	BA

3911	Pte	Samuel	Edward Ephraim Barnet	KIA	1916.07.19	VC
3912	Pte	Saunders	Horatio Norman Stewart	KIA	1916.07.19	RP
4605	Pte	Sawyer	Abraham Prosser	KIA	1916.07.19	VC
4183	Pte	Schimleck	Charles Stanley	KIA	1916.07.19	VC
2654	Pte	Scowcroft	Reginald James	KIA	1916.07.19	VC
4590	Pte	Serong	Emmanuel	KIA	1916.07.19	VC
4903	Pte	Seymour	James Allen	KIA	1916.07.19	VB
3922	Pte	Sibbin	George Stuart	KIA	1916.07.19	VC
4899	Pte	Simmons	Norman Thomas	KIA	1916.07.19	VC
3239	Pte	Simpson	Walter Stanley	KIA	1916.07.19	VC
1966	Pte	Slattery	Francis Patrick	KIA	1916.07.19	VC
3244	Pte	Sleigh	Stephen	KIA	1916.07.19	VC
2895	Pte	Sloan	Robert	KIA	1916.07.19	VC
3485	Pte	Slockwitch	Anthony	KIA	1916.07.19	VC
2268	Pte	Smith	Reginald	KIA	1916.07.19	RF
2786	Pte	Smith	James	KIA	1916.07.19	VC
3932	Pte	South	Frank Joseph John	KIA	1916.07.19	VC

3241	Pte	Statham	George William	DOW	1916.08.14	UC
3944	Pte	Stephens	Roy William	KIA	1916.07.19	RB
2000	Pte	Summers	Albert Victor	KIA	1916.07.19	VC
3476	CSM	Sydes	Charles Frederick	KIA	1916.07.19	RP
3915	Pte	Taylor	Norman Bowman	KIA	1916.07.19	VC
4609	Pte	Thexton	James Winwick	KIA	1916.07.19	VC
515	Pte	Thompson	Walter Herbert	KIA	1916.07.19	VC
2811	Pte	Thomson	Eroc Moslyn	KIA	1916.07.19	VC
3652	Pte	Thomson	George James	KIA	1916.07.19	VC
3269	Pte	Tolley	William Albert	KIA	1916.07.19	VC
959	Pte	Tracey	Murray	KIA	1916.07.19	VC
3956	Pte	Trost	Ludwic John	KIA	1916.07.19	VC
3491	Sgt	Tulloch	John David	KIA	1916.07.19	CC
1284	Cpl	Turner	Stanley George	KIA	1916.07.19	VC
2813	Pte	Usher	Frederick George	KIA	1916.07.19	VC
1928	2Lt	Vaile	Walter Hugh	DOW	1916.07.24	CA
2044	L/Sgt	Vass	James Johnston	DOW	1916.07.25	BA
3991	Pte	Vennell	William James	KIA	1916.07.19	VC
4951	Pte	Wade	Sydney Vivian	DOW	1916.07.26	CA

3497	Pte	Walker	Leslie George	KIA	1916.07.19	VC
2926	Pte	Wallis	Richard Ernest	KIA	1916.07.19	VC
2896	Pte	Walsh	Harry Raymond	KIA	1916.07.19	VC
2831	Pte	Watson	Roy	KIA	1916.07.19	YF
3500	Pte	Watt	Allan	KIA	1916.07.19	VC
4669	Cpl	Weller	Frederick Nicholas	DOW	1916.07.21	CB
3669	Pte	Wheeler	William James	KIA	1916.07.19	VC
3524	Pte	Whiston	Fred	DOW	1916.07.21	BA
3155	Sgt	White	Walter James	KIA	1916.07.19	RB
3651	Pte	Whitehouse	Charles Henry Benjamin	DOW	1916.09.01	BO
3436	Pte	Whylie	James Allan	KIA	1916.07.19	VC
3293	Pte	Wicking	Oswald Mallam	KIA	1916.07.19	NI
3984	Pte	Wiggins	Robert Thomas	KIA	1916.07.19	VC
4916	Pte	Wild	Ernest Victor	KIA	1916.07.19	VC
2830	L/Cpl	Wilkinson	William Bruce	KIA	1916.07.19	VC
1789	Pte	Willey	Josiah	KIA	1916.07.19	VC
3644	Dvr	Wilson	Bernard Francis	KIA	1916.07.19	RF
3279	Pte	Windram	Henry Charles	KIA	1916.07.19	VC
3966	Pte	Winfield	Ernest	KIA	1916.07.19	VC

4017	Cpl	Witham	Stanley	KIA	1916.07.19	RF
3969	Pte	Wood	Douglas	KIA	1916.07.19	VC
3968	Pte	Woodcock	Arnold	KIA	1916.07.19	VC
3297	Pte	Woodhouse	Raymond Lee	KIA	1916.07.19	VC
3660	Pte	Woodland	Charles William	DOW	1916.07.21	BA
2202	Cpl	Woods	Walter Wallace	KIA	1916.07.19	VC
2698	Pte	Yendle	George Roy Reginald	KIA	1916.07.19	VC

60th BATTALION

2452	Sgt	Anderson	Robert Douglas	KIA	1916.07.19	VC
3998	Pte	Andersen	Carl Christian	DOW	1916.07.31	XN
2554	Cpl	Andrews	William Henry	KIA	1916.07.19	VC
3681	Pte	Armstrong	John Hubert	KIA	1916.07.19	VC
1905	Pte	Arnold	William	KIA	1916.07.19	VC
2329	Pte	Arnold	Walter Smith	KIA	1916.07.19	VC
1901	Pte	Arnold	Edwin (Edward)	DOW	1916.07.22	BA
2328	Pte	Artis	Arthur	KIA	1916.07.19	RB
3009	Pte	Ashley	Alfred	DOW	1916.07.20	BA
1656	Pte	Ashworth	Percy Alexander	KIA	1916.07.19	VC

688	Pte	Bailey	Andrew John	KIA	1916.07.19	VC
2849	Pte	Bailey	Clarence York Teck	KIA	1916.07.19	VC
4065	Pte	Bannister	William Henry	KIA	1916.07.19	VC
3456	Pte	Bannon	Charles	KIA	1916.07.19	RB
3655	Pte	Barger	William Gabriel	KIA	1916.07.19	VC
4910	Pte	Barker	Phillip	KIA	1916.07.19	VC
2459	Pte	Barnes	Hurtle Edgar	KIA	1916.07.19	VC
3011	Pte	Barr	George Hamilton	KIA	1916.07.19	VC
3474	Pte	Barr	David	KIA	1916.07.19	VC
3479	Pte	Barr	Colin Campbell	DOW	1916.08.01	XH
4431	Pte	Barrett	Benjamin James	KIA	1916.07.19	VC
3768	Pte	Bartlett	Douglas Clarence	KIA	1916.07.19	VC
4734	Pte	Barton	John	KIA	1916.07.19	VC
1160	Pte	Beattie	Walter George	KIA	1916.07.19	VC
2562	Pte	Beirne	William Bernard	DOW	1916.07.25	WI
2578	Pte	Betancor	Royce John	KIA	1916.07.19	VC
3245	Pte	Bicket	Matthew Dunlop	KIA	1916.07.19	VC

2787	Pte	Bishop	John Henry	DOW	1916.07.25	WI
1921	Sgt	Bishop	Alan Stephen	DOW	1916.12.09	UW
2119	Pte	Bleazby	George	KIA	1916.07.19	VC
3248	Pte	Bloore	Leslie	KIA	1916.07.19	VC
3461	Pte	Bone	Richard Henry	KIA	1916.07.19	VC
4747	Pte	Bonetti	Ernest Edward	KIA	1916.07.19	VC
3250	Pte	Botting (Hopkins)	Dennis Benjamin	KIA	1916.07.19	VC
2932	Pte	Bowden	Frank Arthur	KIA	1916.07.19	VC
2799	Pte	Boyd	Cyril Tasman	KIA	1916.07.19	VC
3336	Pte	Boyd	Thomas Hector Cameron	KIA	1916.07.19	VC
761	Pte	Bradley	Edward Joseph	KIA	1916.07.19	VC
2564	Cpl	Bragg	Arthur Robert	KIA	1916.07.19	VC
3458	Pte	Brasher	William	KIA	1916.07.19	VC
4438	Pte	Brear	Charles Henry	KIA	1916.07.19	VC
3038	Pte	Brook	Arthur Stanley	KIA	1916.07.19	VC
3005	Pte	Brown	John Henry	KIA	1916.07.19	VC
3025	Pte	Brown	Frederick Leonard	KIA	1916.07.19	RD
3694	Pte	Brown	Percy	DOW	1916.08.07	BO
1668	Pte	Bruce	Walter Douglas	KIA	1916.07.19	VC

3256	Pte	Bull	Edward Leslie	KIA	1916.07.19	RB
3035	Pte	Bullen	Stephen	KIA	1916.07.19	VC
2333	Pte	Butler	William Wallace Cameron	KIA	1916.07.19	VC
2812	A/Sgt	Cahill	Richard Nicholas	DOW	1916.07.24	BO
2816	Pte	Caldwell	Edward Thomas	KIA	1916.07.19	VC
3056	Pte	Callen	Albert	DOW	1916.08.05	UH
3039	Cpl	Cameron	Claude Branscombe Vaughan	KIA	1916.07.19	VC
3040	Cpl	Cameron	Ewen Alexander	KIA	1916.07.19	VC
4769	Pte	Cameron	Leslie George	KIA	1916.07.19	VC
3062	Pte	Campbell	John	DOW	1916.08.09	CA
3060	Pte	Candy	George	KIA	1916.07.19	VC
1529	Pte	Carlin	George	KIA	1916.07.19	VC
1920	Pte	Carroll	Patrick	KIA	1916.07.19	VC
1883	Pte	Cassell (Meadows)	Allen (Robert)	KIA	1916.07.19	VC
1899	Pte	Catchpole	Ernest Edward	KIA	1916.07.20	RB
2596	Pte	Chamberlain	Frederick	KIA	1916.07.19	VC
3050	Pte	Charles	Henry	DOW	1916.07.22	BA

4454	Pte	Christensen	Dennis	KIA	1916.07.19	RB
2600	Pte	Claridge	Clarence Roy	KIA	1916.07.19	VC
3050	Pte	Clark	Archibald	KIA	1916.07.19	YF
2830	Pte	Clarke	Thornton Gainsborough	KIA	1916.07.19	VC
3266	Pte	Clarke	Alfred Raymond	DOW	1916.07.22	BA
850	Pte	Clayton	Charles John	KIA	1916.07.19	VC
2600	Pte	Clery	Albert	KIA	1916.07.19	VC
4965	Pte	Clout	Leslie Walter Francis	KIA	1916.07.19	VC
3583	Pte	Cohen	Cyril Sydney	KIA	1916.07.19	VC
2835	Pte	Collins	George Stanley	KIA	1916.07.19	VC
3082	Pte	Condon	John Charles	KIA	1916.07.19	VC
3063	Pte	Conley	John	KIA	1916.07.19	RF
988	CSM	Conway	John	KIA	1916.07.19	VC
3281	Pte	Cook	John William	DOW	1916.07.24	WI
3066	Pte	Cook (Trainer)	David Charles	KIA	1916.07.19	VC
3067	Pte	Cooper	Harry Edward James	KIA	1916.07.19	VC
3050	Pte	Cottren	Arthur Edwin Crisp	KIA	1916.07.19	RP
4305	Pte	Coubrough	Thomas John Victor	KIA	1916.07.19	YF
3019	Pte	Cowl	Robert Henry	KIA	1916.07.19	VC

2605	L/Cpl	Crammond	James Henry	KIA	1916.07.19	VC
3717	Pte	Currie	Joseph	KIA	1916.07.19	VC
3017	Pte	Curwen	Vernon Rudolph Adrian	KIA	1916.07.19	VC
2497	Pte	Dalton	Arthur Victor William	KIA	1916.07.19	VC
3715	Pte	Daly	John Michael	KIA	1916.07.19	VC
3087	Pte	Davies	George	KIA	1916.07.19	VC
3289	Pte	Davis	James Henry	KIA	1916.07.19	VC
3323	Pte	Dawson (Besley)	Thomas James	KIA	1916.07.19	VC
2350	Pte	De Lisle	Clarence Vernon	KIA	1916.07.19	VC
485	Sgt	De Melker	John Henry Wiltch	KIA	1916.07.19	VC
2470	Pte	Delaney	Patrick Anthony	KIA	1916.07.19	VC
3500	Pte	Dicker	Charles William	KIA	1916.07.19	RB
3717	Pte	Dixon	Frederick	KIA	1916.07.19	VC
2612	Pte	Docter	Thomas Cancel	KIA	1916.07.19	VC
3088	Pte	Donaldson	Frank Harold	KIA	1916.07.19	VC
3084	Pte	Doney	Alfred	DOW	1916.07.21	BA
3290	Pte	Donohue	William Pietrel	KIA	1916.07.19	VC

1674	Pte	Douglas	John Alexander	KIA	1916.07.19	RB
1150	CQMS	Drew	Edward John	DOW	1916.07.20	ES
465	Cpl	Dugdale	Benjamin Robert	KIA	1916.07.19	VC
4771	Pte	Easom	Woodley Cole	KIA	1916.07.19	VC
1846	Pte	East	Benjamin	KIA	1916.07.19	VC
3723	Pte	Eddy	Richard	KIA	1916.07.19	VC
2361	Pte	Edwards	Hugh Henry	KIA	1916.07.19	VC
3739	Pte	Edwards	John	KIA	1916.07.19	VC
()	Maj	Elliott	Thomas Patrick	KIA	1916.07.19	VC
3802	Pte	Elliott	Patrick William	KIA	1916.07.19	VC
1916	Pte	Engel	Arthur Francis	KIA	1916.07.19	RF
4803	Pte	Evans	Frederick Alfred	KIA	1916.07.19	VC
()	Capt	Evans	Ernest Andrew	KIA	1916.07.20	VC
1700	Pte	Fackrell	Leslie James	KIA	1916.07.19	VC
2022	Pte	Fargher	Philip Athol	KIA	1916.07.19	VC
376	Pte	Fletcher	William	KIA	1916.07.19	VC
4911	Pte	Fletcher	Edwin Douglas	KIA	1916.07.19	RD
3512	Pte	Flowers	Frank	KIA	1916.07.19	VC
2184	Pte	Fogarty	James Patrick	KIA	1916.07.19	VC

3676	Pte	Forrest	Matthew Thomas	KIA	1916.07.19	RB
3760	Pte	Forsyth	John Williamson	KIA	1916.07.19	VC
3102	Pte	Fowler	Walter Henry	KIA	1916.07.19	VC
3104	Pte	Franklin	Herbert Leyshon	KIA	1916.07.19	VC
3105	Pte	Franklin	Charles Wyndham	KIA	1916.07.19	VC
1703	Pte	Fraser	Alexander Gregory	KIA	1916.07.19	VC
4105	Pte	Frazer	Arthur Albert	KIA	1916.07.19	VC
1926	Pte	Fry	Alfred George	KIA	1916.07.19	VC
1133	L/Cpl	Gibson	Victor Robert	KIA	1916.07.19	RF
2133	Cpl	Gibson	Alexander Edmund	DOW	1916.07.25	UE
4494	Pte	Gilbee	Francis George	KIA	1916.07.19	VC
2929	Pte	Given	Andrew Moncrief	KIA	1916.07.19	VC
3549	Pte	Gleeson	David Joseph	KIA	1916.07.19	VC
2802	Pte	Gregory	Frederick	KIA	1916.07.19	VC
4793	Pte	Griffiths	Cecil Vincent	KIA	1916.07.19	VC
3114	Sgt	Grogan	Vincent Michael	KIA	1916.07.19	VC
3042	Pte	Groom	William Charles	DOW	1916.07.29	WI

()	Capt	Ground	Harry Oscar	DOW	1916.07.22	BA
1797	Pte	Gunn	Edwin Innes	KIA	1916.07.19	VC
2363	Pte	Hales	Harold George	DOW	1916.07.22	HA
4275	Pte	Hamilton	William	KIA	1916.07.19	VC
4505	Pte	Hamilton	Herbert Thomas	KIA	1916.07.19	VC
3124	Pte	Hammond	Wallace Hill	KIA	1916.07.19	VC
1162	Pte	Hannan	John	DOW	1916.07.30	BO
3006	Pte	Hansen (Andrews)	Jack	KIA	1916.07.19	VC
3118	Pte	Harrison	Mellor	KIA	1916.07.19	VC
3325	Pte	Harrison	Thomas	DOW	1916.07.28	UO
3313	Pte	Harry	Sydney Harold	KIA	1916.07.19	VC
3148	Pte	Hart	James John	KIA	1916.07.19	VC
1108	Pte	Havey	Vincent	KIA	1916.07.19	VC
3122	Pte	Hawkins	Richard	DOW	1916.07.23	BA
2815	Pte	Heather	Horace Gordon	KIA	1916.07.19	VC
3166	Pte	Heeps	Elton Sydney	KIA	1916.07.19	VC
3531	L/Cpl	Heggie	Allan Dow	KIA	1916.07.19	VC
2603	Pte	Henderson	Leslie Donald	KIA	1916.07.19	VC
3798	Pte	Henderson	Ernest Albert	KIA	1916.07.19	VC
2871	Pte	Hoberg	Thomas Gustave	KIA	1916.07.19	VC

2925	Pte	Holst	Frederick Giffen	KIA	1916.07.19	VC
4653	Pte	Hopper (Sherer)	Henry	KIA	1916.07.19	VC
1363	L/Cpl	Horsley	Roy	KIA	1916.07.19	VC
1948	Pte	Hosie	Russell William	KIA	1916.07.19	VC
1946	Pte	Howes	Ernest	KIA	1916.07.19	VC
4654	Pte	Humphries	Joseph Henry	KIA	1916.07.19	VC
2664	Pte	Hussey	Charles	KIA	1916.07.19	VC
35	Sgt	Incledon	Eryx Gabriel	KIA	1916.07.19	VC
4824	Pte	Irvine	John McDonald	KIA	1916.07.19	VC
4140	Pte	Jackson	Thomas	KIA	1916.07.19	VC
3340	Pte	Jacobs	Albert	DOW	1916.08.19	CA
3547	Pte	Jenkin	Herbert Francis	DOW	1916.07.25	BO
189	Sgt	Jewell	William Benjamin	KIA	1916.07.19	VC
194	Pte	Jones	Herbert Evan	KIA	1916.07.19	VC
798	Sgt	Jones	Geoffrey Ernest	KIA	1916.07.19	VC
1707	Pte	Jones	Cyril David	KIA	1916.07.19	VC
2392	Pte	Jones	George	KIA	1916.07.19	VC
3544	Pte	Jones	Rictor	KIA	1916.07.19	VC

3793	Pte	Jones	Mark Somerville	KIA	1916.07.19	VC
3155	Pte	Jourdain	Albert William	KIA	1916.07.19	VC
3794	Pte	Kann	Sydney	KIA	1916.07.19	VC
3141	Pte	Keegan	Walter William	KIA	1916.07.19	VC
2868	Pte	Kelly	John	KIA	1916.07.19	VC
3155	Pte	Kenny	William Alfred	KIA	1916.07.19	VC
2486	Pte	Kerr	Henry	KIA	1916.07.19	VC
3061	Pte	Kett	James Leo	KIA	1916.07.19	VC
1251	Pte	King	William John	KIA	1916.07.19	VC
1710	Pte	Kingham	James Henry	KIA	1916.07.20	RB
3162	Pte	Kingsford	Shrewsbury Frederick	KIA	1916.07.19	VC
3552	Cpl	Knight	Percy Thomas	KIA	1916.07.19	VC
1971	Pte	Lakeman	Stanley	KIA	1916.07.19	VC
3088	Pte	Lancaster	John William	KIA	1916.07.19	VC
2958	Pte	Latham	Peter James	KIA	1916.07.19	RF
3566	Pte	Layton	William Henry	KIA	1916.07.19	VB
207	Pte	Le Roy	Alfred	KIA	1916.07.19	VC
94	Cpl	Lee	Joseph Edward	KIA	1916.07.19	VC
3071	L/Cpl	Lee	William	KIA	1916.07.19	VC
3807	Pte	Leheny	George Daniel	KIA	1916.07.19	VC

997	Cpl	Lennox	Robert Thomas	KIA	1916.07.19	VC
158	Cpl	Letcher	John Gordon	KIA	1916.07.19	VC
4132	Pte	Little	Stanley	KIA	1916.07.19	VC
3866	Pte	Lock	Charles William	KIA	1916.07.19	VC
3400	Cpl	Lockett	Henry	KIA	1916.07.19	VC
2620	Pte	Luckhurst	Alfred	KIA	1916.07.19	VC
2626	Pte	Lukey	Charles William	KIA	1916.07.19	VC
219	Pte	Lynch	Michael	KIA	1916.07.19	VC
4834	Pte	Mackay	George	KIA	1916.07.19	VC
2631	Pte	MacKenzie	Robert Henry	KIA	1916.07.19	VC
2101	Pte	Maclaren	Frank Forbes	KIA	1916.07.19	VC
3137	Pte	Maidment	Horace	KIA	1916.07.19	VC
4929	Pte	Maning	Henry Anthony Beresford	KIA	1916.07.19	VC
2891	Pte	Mannall	Charles	DOW	1916.07.31	BO
1745	Pte	Marshall	Harry	KIA	1916.07.19	VC
3175	Sgt	Marshall	Albert Charles	KIA	1916.07.19	VC
2836	Pte	Martin	Claude Joseph	KIA	1916.07.19	VC
3183	Pte	Martin	Frederick William	KIA	1916.07.19	VC

3084	Pte	Mathieson	Clarence George	KIA	1916.07.19	VC
3174	Pte	Matthews	James Edward	KIA	1916.07.19	VC
2644	Pte	Mays	William Henry John	KIA	1916.07.19	VC
39	Pte	McArthur	Andrew	KIA	1916.07.19	RB
1980	Pte	McCaul	Donald Curteis	KIA	1916.07.19	VC
2884	Pte	McClaren	George Henry	KIA	1916.07.19	RF
3858	Pte	McClure	William Cyril	KIA	1916.07.19	VC
1631	Pte	McCooke	John / Jack	KIA	1916.07.19	VC
()	Maj	McCrae	Geoffrey Gordon	KIA	1916.07.19	RB
3415	Pte	McDermott	William Robert Patrick	KIA	1916.07.19	VC
3194	Sgt	McDowell	Athol Halliday	KIA	1916.07.19	PW
2643	Pte	McEwan	Alfred James	KIA	1916.07.19	RF
913	Sgt	McIntosh	James Donald	KIA	1916.07.19	VC
414	Sgt	McKenna	Robert John	KIA	1916.07.19	VC
1974	Pte	McKenzie	John	KIA	1916.07.19	VC
3201	Pte	McKenzie	James	KIA	1916.07.19	VC
3208	L/Cpl	McKenzie	Kenneth Fraser	KIA	1916.07.19	VC
2883	L/Cpl	McKenzie	Murdoch	DOW	1916.08.09	CA
1824	Pte	McKinery	John Patrick	KIA	1916.07.19	VC

()	Lt	McKinnon	Andrew Charles	KIA	1916.07.19	VC
495	CSM	McKinnon	John Alexander	KIA	1916.07.19	VC
3200	Pte	McLean	Alexander Leslie	KIA	1916.07.19	VC
3209	Pte	McLean	Victor Henry	KIA	1916.07.19	VC
2719	Pte	McManus	Samuel	KIA	1916.07.19	VC
3831	Pte	McManus	Patrick	KIA	1916.07.19	VC
4851	Pte	Mildren	William John	KIA	1916.07.19	VC
2735	Pte	Miller	William	KIA	1916.07.19	VC
2736	Pte	Miller	Robert Charles	KIA	1916.07.19	VC
4547	Pte	Milligan	William Henry	KIA	1916.07.19	VC
2901	Pte	Mogg	Leslie	KIA	1916.07.19	VC
3579	Pte	Molloy	Henry MacDonald	KIA	1916.07.19	VC
2648	Pte	Moore	James John	KIA	1916.07.19	VC
3562	Pte	Morgan	Robert Edwin	KIA	1916.07.19	VC
3190	Sgt	Morley	Edwin Joseph	KIA	1916.07.19	VC
3191	Cpl	Mortimer	Albert Edward	KIA	1916.07.19	VC
3059	Pte	Morton	Vincent Edward	KIA	1916.07.19	VC
3844	Pte	Mullin	Hugh	KIA	1916.07.19	VC
1725	Pte	Munro	Joseph Edward	KIA	1916.07.19	VC

171	Sgt	Murray	William	KIA	1916.07.19	VC
3194	Pte	Murrowood	George Thomas	DOW	1916.08.01	UN
2904	Pte	Murton	Harold Roy	KIA	1916.07.19	VC
3198	Pte	Nash	Felix	DOW	1916.07.24	BA
3436	L/Sgt	Nathan	Frederick Joseph	KIA	1916.07.19	VC
4970	Pte	Naylor	Joseph	KIA	1916.07.19	VC
3847	Pte	Neill	Thomas	KIA	1916.07.19	VC
3197	Pte	Nelson	Martin	KIA	1916.07.19	VC
2279	Pte	Newman	Albert Edward	KIA	1916.07.19	VC
146	Pte	Nitchie	James Louis	KIA	1916.07.19	VC
3595	Pte	Noble	Norman John	KIA	1916.07.19	VC
2098	Pte	Norris	Leonard	KIA	1916.07.19	VC
1736	Pte	Nott	Walter Ernest	KIA	1916.07.19	VC
3212	Pte	Nunn	Stanley Milton	KIA	1916.07.19	VC
4991	Pte	O'Connor	Gerald Augustine	KIA	1916.07.19	VC
4192	Pte	O'Grady	John	KIA	1916.07.19	VC
3423	Pte	O'Grady (McClare)	Patrick Aloysius	KIA	1916.07.19	VC
3869	Pte	Olney	Edwin Percival	KIA	1916.07.19	VC
4867	Pte	O'Neill	Frank	KIA	1916.07.19	VC
4875	Pte	Orchard	Martin Hosker	KIA	1916.07.19	VC

3100	Pte	Oriel	William Harrison	KIA	1916.07.19	VC
4268	Pte	Osmond	Walter William	KIA	1916.07.19	VC
2777	Pte	Paice	Victor Alfred Godfrey	KIA	1916.07.19	VC
3605	Pte	Pattison	Henry	KIA	1916.07.19	VC
2872	Pte	Peart	Roy Carlin	KIA	1916.07.19	VC
2233	Pte	Pegler	Joseph	KIA	1916.07.19	VC
2255	Pte	Pemberton	Thomas Alfred	KIA	1916.07.19	VC
3584	Pte	Penaluna	Edgar Briton	KIA	1916.07.19	VC
4865	Pte	Pengelly	Alfred	KIA	1916.07.19	VC
3440	Pte	Perry	Arthur James	KIA	1916.07.20	RB
1743	Pte	Phillips	Edwin John	KIA	1916.07.19	VC
3982	Pte	Pinniger	John Lewington	DOW	1916.07.23	CA
()	Capt	Plowman	Hugh McDonald	KIA	1916.07.19	VC
3615	Pte	Pollard	Frederick John	KIA	1916.07.19	VC
2749	Pte	Potts	Hedley Vasey	KIA	1916.07.19	RD
43	Sgt	Power	George	KIA	1916.07.19	VC
2893	Pte	Power	William Henry	KIA	1916.07.19	VC
3894	Pte	Power	Albert Patrick	KIA	1916.07.19	VC
3115	Pte	Pritchard	Horace Rudolph	KIA	1916.07.19	VC

3215	Pte	Prosser	Richard	DOW	1916.07.21	BA
2655	Pte	Purcell	John James	KIA	1916.07.19	YF
1748	Pte	Purvis	Robert	KIA	1916.07.19	VC
4889	Pte	Rafferty	Roy	DOW	1916.08.08	BA
4994	Pte	Rahilly	Patrick Vincent	KIA	1916.07.19	VC
2736	Pte	Rainsbury	George	KIA	1916.07.19	VC
4769	Pte	Rainsbury	William	KIA	1916.07.19	VC
1501	CQMS	Ralston	James	KIA	1916.07.19	VC
3131	Pte	Ramsay	George Seamer	KIA	1916.07.19	VC
1417	Pte	Rattray	Arthur Alexander	KIA	1916.07.19	VC
3458	Pte	Reeves	Alfred	KIA	1916.07.19	VC
3134	Pte	Reynolds	George William	KIA	1916.07.19	VC
3899	Pte	Reynolds	Charles Wakeman	KIA	1916.07.19	VC
()	Lt	Rhind	James Morison	KIA	1916.07.19	VC
3904	Pte	Richardson	Percy	KIA	1916.07.19	VC
3565	Pte	Roach	Edward	KIA	1916.07.19	RF
941	Pte	Roberts	Albert James	KIA	1916.07.19	VC
4579	Pte	Robertson	William	KIA	1916.07.19	VC
2889	Pte	Robinson	Albert Douglas	KIA	1916.07.19	VC
3139	Pte	Rogers	Stanley Arthur	KIA	1916.07.19	VC

3342	Pte	Rogers	Roy Charles	KIA	1916.07.19	VC
2786	Pte	Rojo	Ernest Francis	KIA	1916.07.19	VC
3906	Pte	Roney	George	KIA	1916.07.19	VC
3243	Sgt	Routley	Raymond Francis	KIA	1916.07.19	VC
3598	Pte	Rowe	William Alfred	KIA	1916.07.19	RD
3244	Pte	Ryan	William	KIA	1916.07.19	VC
3501	Pte	Ryan	Clarence Timothy	KIA	1916.07.19	VC
339	Pte	Salter	Roy Stuart	KIA	1916.07.19	VC
3134	Pte	Santwyk	Henry Mervyn	KIA	1916.07.19	VC
4589	Pte	Sawers	Stanley Sydney	KIA	1916.07.19	VC
2797	Pte	Scott	Archibald Allen	KIA	1916.07.19	VC
2300	Pte	Sergeant	Hector Thomas	KIA	1916.07.19	VC
3650	Pte	Sharp	Henry George Keller	DOW	1916.07.21	AN
3249	Pte	Shawe	William Charles	KIA	1916.07.19	RB
4210	Pte	Sheehan	Frank	KIA	1916.07.19	VC
1795	Pte	Sheen	James	DOW	1916.07.20	BA
2477	Pte	Sheldrick	Richard	KIA	1916.07.19	VC
2776	Pte	Shephard	George Arnold	KIA	1916.07.19	VC

2778	Pte	Shephard	Thomas Henry Phillip	KIA	1916.07.19	VC
2884	Pte	Shine	Henry Frank	KIA	1916.07.19	VC
4239	Pte	Shore	James	KIA	1916.07.19	RB
3885	Pte	Shurmer	Roy Stanley	KIA	1916.07.19	VC
2390	Pte	Silver	John	KIA	1916.07.19	VC
4909	Pte	Simpson	John James	DOW	1916.08.01	UP
4939	Pte	Sims	John Edward Theodore	KIA	1916.07.19	VC
452	Pte	Sitlington	Alexander Farndale	KIA	1916.07.19	AA
272	Pte	Slattery	Gerald Malyon	KIA	1916.07.19	RB
()	2Lt	Smith	John Huntingdon	KIA	1916.07.20	RB
2679	Pte	Smith	John Henry Adam	DOW	1916.07.21	BA
3263	Pte	Smith	John James	DOW	1916.07.22	BO
683	L/Sgt	Smyth	Henry Bowyer	KIA	1916.07.19	VC
3934	Pte	Smyth (Smythe)	Henry	KIA	1916.07.19	VC
3941	Pte	Spooner	James Edward	KIA	1916.07.19	VC
2663	Pte	Spooner	Edward Mason	DOW	1916.07.31	UJ
()	2Lt	Sterling	John Harold	KIA	1916.07.19	VC
2796	Pte	Steward	Frederick John	KIA	1916.07.19	VC

2902	Pte	Stobaus (Stobans)	George	KIA	1916.07.19	VC
823	Cpl	Stuart-Murray	Stuart Henry	DOW	1916.07.24	BA
2793	Pte	Sullivan	Arthur Joseph	KIA	1916.07.19	VC
3472	Pte	Sullivan	John Henry	KIA	1916.07.19	VC
635	L/Cpl	Summers	David	KIA	1916.07.19	VC
3640	Pte	Terry	Ernest Howard	KIA	1916.07.19	VC
3638	Pte	Thompson	Edward James	KIA	1916.07.19	VC
2033	Pte	Thomson	Charles William	KIA	1916.07.19	VC
3693	Pte	Tiedeman	Arthur Otto	KIA	1916.07.19	VC
4928	Pte	Tippett	Henry	KIA	1916.07.19	VC
4647	Pte	Tipton	James Larkin	KIA	1916.07.19	VC
4201	Pte	Toogood	Arthur William	KIA	1916.07.19	VC
3268	Pte	Towe	James	KIA	1916.07.19	VC
3271	Pte	Tredrea	Francis Stanley	KIA	1916.07.19	VC
2913	Pte	Trehearn	David Yorke	KIA	1916.07.19	RB
4598	Pte	Tully	Walter Allen	KIA	1916.07.19	VC
2691	Pte	Turnbull	William Robert	KIA	1916.07.19	VC
2929	Pte	Underwood	Albert Charles	KIA	1916.07.19	VC
1784	Pte	Wallmeyer	Francis	KIA	1916.07.19	VC

1446	Pte	Walters	Frederick Charles	KIA	1916.07.19	VC
4933	Pte	Ward	Rupert Charles	KIA	1916.07.19	VC
990	Pte	Wardell	Arthur Percy Montague	KIA	1916.07.19	VC
1767	Pte	Warren	Samuel Madden	KIA	1916.07.19	VC
3959	Pte	Warren	Robert Clyde	DOW	1916.07.24	WI
3286	Pte	Watkins	Frederick	KIA	1916.07.19	RD
3493	Pte	Watt	George Donald	KIA	1916.07.19	VC
3958	Pte	Waugh	Herbert Henry	KIA	1916.07.19	VC
2910	Pte	Webb	Thomas Richard	KIA	1916.07.19	PW
3642	Pte	Whelan	Francis Joseph	KIA	1916.07.19	VC
3647	Pte	Whelan	Victor Harold	KIA	1916.07.19	VC
2273	Pte	White	Walter	KIA	1916.07.19	VC
3323	Pte	White	Isaac	KIA	1916.07.19	VC
1449	Pte	Whitehurst	John	KIA	1916.07.19	VC
3514	Pte	Wickham	Alfred William Findlay	KIA	1916.07.19	VC
1774	Pte	Wilde	Kenneth Richard	KIA	1916.07.19	VC
295	CQMS	Wilkins	Thomas Gerismo	KIA	1916.07.19	RF

1800	Pte	Williams	Henry Egryn	DOW	1916.08.18	UU
3921	Sgt	Williamson	Harry Templeton	KIA	1916.07.19	VC
2225	Pte	Willis	John	KIA	1916.07.19	RF
727	CSM	Willox	James Simpson	KIA	1916.07.19	NI
1750	Pte	Wilson	William	KIA	1916.07.19	RF
2686	Pte	Windley	Harry	KIA	1916.07.19	VC
2833	Pte	Wood	Arthur Beaumont	KIA	1916.07.19	VC
()	Lt	Wright	Ernest Edward	KIA	1916.07.19	VC
2481	Pte	Wright	John Ralph	KIA	1916.07.19	VC
2482	Pte	Wright	Kensey	KIA	1916.07.19	VC
4635	Pte	Wright	Reuben	KIA	1916.07.19	VC
3299	Pte	Wyatt	Charles John	KIA	1916.07.19	VC
3663	Pte	Wynne	Walter Cornelius	KIA	1916.07.19	VC
4639	Pte	Young	William Victor	KIA	1916.07.19	VC

---ooo000ooo---

GERMAN ROLL OF HONOUR

16th BAVARIAN RESERVE INFANTRY REGIMENT

Name	Unit	Fate	Date	Age
Aicher, Pte M	9 Coy	KIA	19	22
Andessmer, L/Cpl C	8 Coy	KIA	20	25
Babl, Pte G	12 Coy	KIA	19	21
Bauer, Pte J	10 Coy	DOW	2/8	35
Benninger, Pte L	3 Coy	KIA	20	29
Bestler, Cpl O	1 Coy	KIA	20	31
Böck, L/Cpl O	1 Coy	KIA	20	28
Boedicker, Cpl R	10 Coy	KIA	19	37
Bratengeier, L/Cpl E	1 MG Coy	KIA	19	21
Braun, Pte A	11 Coy	KIA	20	33
Brenner, Pte J	10 Coy	KIA	19	22
Brummer, Pte M	6 Coy	DOW	22	19
Burger, L/Cpl J	11 Coy	KIA	19	21
Emerich, Cpl O	10 Coy	DOW	21	25
Englmann, Pte A	1 Coy	KIA	20	35
Enrainer, Pte J	9 Coy	KIA	19	24
Estner, Pte K	11 Coy	DOW	23	23
Fischhaber, Pte P	9 Coy	KIA	19	31

Frenzl, Pte G	12 Coy	KIA	20	29
Frommel, Pte M	8 Coy	KIA	20	29
Führer, Sgt Major N	11 Coy	KIA	19	28
Gartmeier, L/Cpl A	2 Coy	KIA	20	27
Gilles, Pte J	8 Coy	KIA	19	24
Gleissner, Pte F X	9 Coy	KIA	19	22
Grünwald, Pte J	9 Coy	KIA	19	23
Gumbert, Pte F	10 Coy	DOW	21/8	21
Gutbrod, Pte L	1 Coy	KIA	20	22
Härtle, Cpl A	1 MG Coy	KIA	19	25
Heiss, Pte J	3 Coy	KIA	20	22
Hitzenberger, Pte H	10 Coy	KIA	19	30
Hofmann, L/Cpl J	11 Coy	KIA	20	26
Holzner, Pte I	8 Coy	KIA	20	23
Hörath, Pte W	1 Coy	KIA	20	20
Huber, Pte J	10 Coy	KIA	19	23
Hund, Pte J	3 Coy	KIA	20	29
Ingold, L/Cpl M	1 Coy	DOW	27	22
Janssen, Pte E J A L	3 Coy	KIA	20	38
Jörg, Pte J	9 Coy	KIA	19	25
Kameter, Pte D	9 Coy	KIA	19	27
Kestel, Pte M	10 Coy	DOW	21	30

Kiebensberger, Pte K	8 Coy	DOW	23/11	22
Kling, Pte M	3 Coy	KIA	20	24
Koch, Cpl G	11 Coy	KIA	19	22
König, L/Cpl E	3 Coy	KIA	20	21
Kreutzer, Pte E	12 Coy	KIA	19	26
Kreuzpaintner, Pte A	1 MG Coy	KIA	20	23
Kronseder, Pte J	12 Coy	KIA	19	24
Kurzmaul, Pte J	1 MG Coy	KIA	19	21
Kuttroff, Lieut K	1 Coy	KIA	20	21
Lehr, Lieut M	11 Coy	KIA	19	24
Leib, Pte J	10 Coy	KIA	19	32
Leicher, L/Cpl J	3 Coy	DOW	30	30
Lerch, Pte X	12 Coy	KIA	19	21
Lösch, Pte L	1 Coy	KIA	20	22
Maas, Pte F	1 Coy	KIA	20	33
Machl, Pte J B	6 Coy	KIA	19	31
Maier, Pte M	12 Coy	DOW	20	30
Mayer, Pte L	12 Coy	KIA	20	20
Meier, Pte A	9 Coy	KIA	19	22
Möst, Pte H	6 Coy	KIA	19	23
Müller, L/Cpl F S	3 Coy	KIA	20	26
Müller, Pte M	9 Coy	KIA	19	21

Obermaier, Pte G	9 Coy	KIA	19	29
Petz, Lieut A	1 Coy	KIA	20	29
Preissel, Pte X	10 Coy	KIA	19	23
Pröller, Pte M	1 Coy	KIA	20	27
Reichenberger, Pte J	1 MG Coy	KIA	19	27
Reiser, Cpl A	6 Coy	KIA	20	24
Rieder, L/Cpl S	11 Coy	KIA	19	25
Risinger, Pte J	10 Coy	DOW	28	29
Roe, Pte J	9 Coy	KIA	20	23
Roiter, L/Cpl L	6 Coy	DOW	25	27
Ruckdäschel, Pte M	9 Coy	KIA	19	20
Ruckdäschel, Pte W H	11 Coy	KIA	20	40
Saumweber, Pte W	9 Coy	KIA	19	28
Schiessl, Pte J	6 Coy	DOW	29/8	28
Schmid, Pte J	12 Coy	KIA	19	22
Schmid, Pte J	9 Coy	DOW	20	24
Schmidhammer, Pte J	2 Coy	KIA	20	23
Schneider, Pte G	9 Coy	KIA	19	31
Schöll, Pte F	12 Coy	KIA	19	22
Schorer, Pte A	2 Coy	KIA	20	28
Schürf, Pte F	1 Coy	KIA	20	29
Seiling, Pte M	11 Coy	KIA	20	26

Sepp Pte L	10 Coy	KIA	19	21
Simon, Cpl E	8 Coy	KIA	20	32
Sollinger, Pte J	3 Coy	KIA	20	25
Specht, Pte J	9 Coy	DOW	20	22
Springer, Pte F	12 Coy	DOW	20	27
Steinle, Pte K	10 Coy	DOW	23	20
Stiller, Pte A	10 Coy	DOW	21	32
Strasser, Pte J (from Meitenkamm)	12 Coy	KIA	19	20
Strasser, Pte J (from Pleiskirchen)	12 Coy	KIA	19	29
Strötz, Cpl J	9 Coy	KIA	19	29
Stumpf, Pte G	7 Coy	DOW	20	21
Thaller, Pte G	9 Coy	KIA	19	26
Titscher, Pte F	9 Coy	KIA	19	29
Traxinger, Pte J	12 Coy	KIA	20	27
Treiber, Pte R	1 Coy	KIA	20	27
Vogl, L/Cpl S	2 Coy	KIA	20	25
Weber, Pte L	9 Coy	KIA	19	32
Weindl, Pte S	9 Coy	KIA	19	27
Wenl, Pte L	1 Coy	KIA	20	26
Wesche, Lieut G	9 Coy	KIA	19	26

Zahn, Cpl J	12 Coy	KIA	19	26
Zeitler, Pte K	1 Coy	KIA	20	22
Zeulmann, Lieut S	2 Coy	KIA	20	26

17th BAVARIAN RESERVE INFANTRY REGIMENT

3 Officers and 56 Ors unnamed = 59			
Buff, Cpl E	KIA	19	
Hehl, Sgt Major 2Coy	KIA	19	
Mayer, Lieut F	KIA	19	22

20th BAVARIAN RESERVE INFANTRY REGIMENT

1 Officer and 35 Ors unnamed = 36			
Eichhorn, Lieut 6 Coy	KIA	19	
Weglein, Pte W	KIA	19	23
Ziegel, Sgt Major R	DOW	27/8	21

21st BAVARIAN RESERVE INFANTRY REGIMENT

7 Officers and 288 Ors unnamed = 295			
Baumgärtel, Lieut	KIA	19	-
Bernhart, Sgt Major A	KIA	20	21
Fleishmann, I	KIA	19	31

Gerngross, J	DOW	20	33
Keim, Lieut	KIA	19	-
Keller-	KIA	19	-
Rosenhaupt, L/Cpl W	KIA	19	32
Schonberg, Sgt Major J	DOW	20	31
Wagner, Lieut	KIA	20	-
Wolf, Lieut	KIA	20	-

15 July Raid

Kriechbaum Lieut	KIA	15	
[and 5 Ors unnamed]			

104th SAXON REGIMENT

4 Ors unnamed			

HINDSIGHT

The German Roll of Honour as shown for 16th BRIR was taken from their history published in 1932, and shows 107 dead, The Divisional Returns dated I August 1916 gives 91 killed, 392 wounded, I missing, The difference is clearly accounted for by those who later died of wounds.

The Divisional Return for 17th BRIR was 59 killed, 116 wounded, 49 missing, Only the number of dead was available when this book (note: Robin Corfield's) was first published—as in the next two Regiments, 20th BRIR-36 killed, 118 wounded and 5 missing; 21st BRIR had 236 killed, 377 wounded and 111 missing,

The Divisional Return also gives the 17th BRIR Mining Coy as having 5 killed, 4 wounded, 5 missing, 20th BRIR Mining Coy having 3 killed, 21st BRIR Mining Coy, 6 killed, 8 wounded,

The artillery had 4 killed, 37 wounded. They lost I horse killed, and 10 wounded, The Pioneers, Tunnellers, Meinenwerfer crews, other miners and 2nd Bavarians add a total of 35 killed, 44 wounded and 49 missing,

Totals given:

Officers: 17 killed, 26 wounded, 2 missing,

NCOs & men: 452 killed, 952 wounded, 204 missing

[Total: 469 killed, 978 wounded, 206 missing,]

Horses: 1 killed, 10 wounded.

This was 606 itialed by Scanzoni and sent to Army Command, 6th Army.

Of the missing, it is believed that 146 of those were taken prisoner.

---ooo000ooo---

APPENDIX II

Soldiers' Reunions

The following pages are a record of the notices issued by or for the various Battalions, for reunions or the participation in commemorative services, by those which were involved in the battle at Fromelles. This material has been compiled in conjunction with that of Chapter 16. The records were identified by conducting a search using the National Library of Australia's 'Trove' Search Engine – https://trove.nla.gov.au.

In the years after 1918, the former soldiers who held observances and reunions on or near 19 July each year became known as 'the 19th of July men' the last reported meeting of the 19th of July men took place in 1966. Contrary to the suggestion of some historians, the soldiers' reverence of the memory of Fromelles, particularly by those who were involved did not go away. It carried on until the last of the survivors could no longer attend a service of remembrance.

In the boxes below will be set out the date of the publication of the event to be held, the source of the information, the heading of the article in which the information appeared, and the location at which the event would take place. In many cases, there is also a short extract following that information in the single box below those items, which gives the reader some insight into the manner in which the soldiers conducted their event.

Date of Event	Source of Notice	Heading of Article	Location of Reunion
25 May, 1917	The Register (Adelaide)	Reunion of Veterans	Prince of Wales Hotel Adelaide
Every other war had been put into the shade by the present great struggle of nations. The Allies had grown stronger every day in morale and fighting force. They all felt confident in the strength of the Empire.			
14 Feb., 1918	Daly Herald Adelaide	Reunion 10th & 60th Battalions	Cheer-Up Hut, Adelaide

<table>
<tr><td colspan="4">There was a large attendance, over which Colonel Weir, DSO, presided. Captain Blackburn, V.C., president of the Returned Soldiers' League, hoped that all the returned men would join it.</td></tr>
<tr><td>27 May, 1918.</td><td>The Advertiser, Adelaide</td><td>Veterans. Annual Reunion</td><td>Prince of Wales Hotel</td></tr>
<tr><td colspan="4">Today they were linked up with all the free democracies of the world, and an alliance like that must win victory in the end. German spread-eagle-ism had to go. The nation was called upon to bear and suffer much, but no more than what their forefathers had suffered for their freedom and security. The dial of the clock of destiny now pointed to the western front, and it was there that big battalions, as in the days of Napoleon, counted. They therefore had to put every ounce of their strength into the struggle, in men, money, and materials.</td></tr>
<tr><td>9 July, 1918</td><td>The Age, Melbourne</td><td>15th Infantry Brigade (57th, 58th, 59th & 60th Bns)</td><td>Orient Hotel, Melbourne.</td></tr>
<tr><td colspan="4">A reunion of all members of the brigade who have returned from active service, and for discussing general matters of interest.</td></tr>
<tr><td>24 July, 1918</td><td>Geelong Advertiser, Vic.</td><td>15th Brigade Assn.</td><td></td></tr>
<tr><td colspan="4">It is estimated that 3700 men of this unit, comprising the 57th, 58th, 59th and 60th battalions, have returned from the war. A reunion dinner is being arranged for Show Day.</td></tr>
<tr><td>24 Sep., 1918</td><td>Bendigo Advertiser, Vic.</td><td>15th Brigade Association</td><td>Athenaeum Hall, Collins Street, Melbourne.</td></tr>
<tr><td colspan="4">The objects of the association are to look after the interests of its members and to provide reunions.</td></tr>
<tr><td>23 Jan., 1919</td><td>Daily Herald, Adelaide</td><td>32nd Battalion Reunion</td><td>Leavitt Hall, Wakefield Street, Adelaide</td></tr>
</table>

<table>
<tr><td colspan="4">The first social reunion of members of the 32nd Battalion Club for 1919.</td></tr>
<tr><td>8 Apr., 1919</td><td>The Advertiser, Adelaide</td><td>32nd Battalion Reunion</td><td>Cheer Up Hut</td></tr>
<tr><td colspan="4">Took the form of a smoke social. The battalion was the first to go over the top in France. They went in 1,000 strong at 6pm, and the following morning at roll call only 52 answered. The Battalion numbers among its survivors Major Walk, V.C., formerly second in command. Colonel Davis is now brigadier-general commanding the Eighth Brigade. It was decided to arrange an annual meeting on July 19, in commemoration of the date in 1916, when the 32nd Battalion (the first South Australian battalion to go over the top in France) obtained its baptism of fire. It was resolved to obtain and present a 32nd Battalion flag to be hung in the Cheer-up Hut. The necessary sum was subscribed in the room.</td></tr>
<tr><td>7 Nov., 1919</td><td>The Argus, Melbourne</td><td>15th Brigade Reunion</td><td>Sargent's Café.</td></tr>
<tr><td colspan="4">In the absence of the president, Brigadier-General Elliott, who was unavoidably detained at Warragul, Lieut-Colonel H.M. Duigan presided. The usual toasts were honoured and an enjoyable musical programme rendered.</td></tr>
<tr><td>19 Nov., 1919</td><td>Daily Herald, Adelaide</td><td>32nd Battalion</td><td>Selborne Hotel</td></tr>
<tr><td colspan="4">The fourth anniversary of their departure from Australia (Nov. 16, 1915) was celebrated last night.</td></tr>
<tr><td>18 Dec., 1919</td><td>The Age, Melbourne</td><td>5th Division Meeting and Book.</td><td>Fifteen Brigade Association room, Swanston Street</td></tr>
<tr><td colspan="4">A divisional meeting will be held in conjunction with the souvenir war book "Some of the Fifth Division's Glory". General Stewart, Colonel Cass, Colonel Duigan, Colonel Layh, Colonel Marshall and Private Hornsey will be amongst those present. "Some of the Fifth Division's Glory" is backed by a strong military committee, with General H.E. Elliott, C.B., C.M.G., D.S.O., D.C.M., as president. Each brigade's actions and history will be given just and equal prominence, and be written by the men and officers who fought and fired them.</td></tr>
</table>

<table>
<tr><td>4 July, 1920</td><td>Sunday Times, Sydney</td><td>Fifth Division Reunion</td><td>Sargent Café, George Street, Sydney</td></tr>
<tr><td colspan="4">The members of the 5th Div. Artillery will hold a reunion in the form of a dinner and smoke concert.</td></tr>
<tr><td>8 Aug.,1920</td><td>Sunday Times, Sydney</td><td>53rd Battalion Reunion</td><td>Moonlight Excursion</td></tr>
<tr><td colspan="4">Those desirous of attending are requested to communicate at once with the secretary of the reunion committee, Mr. A. C. Elliott, Langhurst, Shirley-road, Roseville. Tickets will be: Gents. 4/6, ladies 2/6. The moonlight cruise was postponed until Wednesday 29 Sep.</td></tr>
<tr><td>23 July, 1921</td><td>Daily Herald Adelaide</td><td>32nd Battalion Reunion Dinner</td><td>Austral Gardens</td></tr>
<tr><td colspan="4">The function was attended by 155 members, and was presided over by Lieut. Colonel D.M.R. Coghill, V.D. The assembled company stood in silence for two minutes in honour of the honored dead of the battalion. Kindred clubs were represented. A unanimous motion was carried, that a letter of condolence be sent to Brigadier General Tivey, expressing deep sympathy with him in the loss recently sustained by him in the death of his wife.</td></tr>
<tr><td>29 May, 1922</td><td>The Brisbane Courier, Qld</td><td>31st Battalion Reunion</td><td>R.S.S.I.L.A, Elizabeth Street, Brisbane</td></tr>
<tr><td colspan="4">Held to make preliminary arrangements for a battalion reunion. The main object of the committee is to hold a big reunion in Brisbane during Exhibition week. A suggestion by Captain L.J. Kimber that a list of members should be prepared for the Defence Department, with a view to the presentation of 1914 stars general service medals, and victory medals, at the reunion during Exhibition week, was unanimously approved. The next meeting will be held in the league rooms on July 22, at 7.45 p.m.</td></tr>
<tr><td>26 July, 1924</td><td>Daily Mail, Brisbane</td><td>31st Battalion Reunion</td><td></td></tr>
</table>

<table>
<tr><td colspan="4">Arrangements are well in hand for the reunion dinner of the members of the 31st Battalion. The function has been fixed for the evening of Aug. 12. It is anticipated that there will be a good roll up.</td></tr>
<tr><td>15 Aug. 1924</td><td>Brisbane Courier, Qld</td><td>31st Battalion Reunion Dinner</td><td>Brunswick Café, Valley</td></tr>
<tr><td colspan="4">Amid the rejoicing, fallen comrades were not forgotten, and all stood in silence for a minute in honour of their memory. Then fleecy clouds of tobacco smoke and the amber ale, those two anodynes for most of the ills of life, dissipated gloomy thoughts; and every one settled down to enjoy the present, and cheerfully to reconstruct the past.

Colonel Toll said that he had had the honour of forming the battalion and therefore he had more than ordinary interest in it. He had seen it grow from an amateur machine to an efficient unit of the A.I.F. The splendid traditions of the earlier Queensland battalions had been before it. The work of a commanding officer could be made much easier by the co-operation of his officers. There were excellent officers in the 31st, and the result had been four splendid companies.</td></tr>
<tr><td>3 Apr. 1925</td><td>The St. George Call, Kogarah</td><td>4th & 56th Battalions Reunion</td><td>Highland Society Hall, 89 Phillip Street.</td></tr>
<tr><td colspan="4">The ex-Members of the 4th and 56th Battalions, (A.I.F.) will hold their 10th Annual Smoko on Wednesday, 15th April</td></tr>
<tr><td>22 June, 1925</td><td>Queensland Times, Ipswich</td><td>31st Battalion Reunion</td><td></td></tr>
<tr><td colspan="4">Arrangements are being made for a reunion of members of the 31st Battalion A.I.F. and other units of the 8th Brigade in Brisbane in Exhibition week. Col. Toll, the popular commander of the battalion, is keenly interested in the movement.</td></tr>
<tr><td>14 Aug. 1925</td><td>The Brisbane Courier, Qld</td><td>31st Battalion Reunion</td><td>Anzac Memorial Club, Brisbane</td></tr>
<tr><td colspan="4">To-day the scene was very different on the old fields of battle. All that remained were the lonely spots with the graves of many thousands who gave their lives that those at home</td></tr>
</table>

<table>
<tr><td colspan="4">might live in freedom. On the monuments on Gallipoli, France, Belgium, and Mesopotamia, the words were engraved: "Tell Australia you who pass this monument, we died for her, and we rest content." They were men their country could ill afford to lose, but with their lives they purchased a tradition beyond all human power to appraise. They set for all time the standard of conduct for the generations of Australia to follow. The lapse of years would not dim the splendour of their courage. But their deeds had been written a glorious page in the history of Australia. Of the whole battalion, only three officers and 110men were left to tell the tale. They were known as the men who could not be beaten,</td></tr>
<tr><td>28 Oct. 1925</td><td>Geelong Advertiser, Vic.</td><td>8th Australian Infantry Brigade Association</td><td>Manchester Unity Hall, Swanston Street, Melbourne</td></tr>
<tr><td colspan="4">The annual re-union and smoke night will be held on Cup Eve, Monday, Nov. 2nd. Members of units of the 8thBrigade visiting Melbourne for the Cup should take this opportunity of renewing old acquaintances formed during the war. Tickets available at the door.</td></tr>
<tr><td>24 Feb. 1926</td><td>News. Adeaide SA</td><td>Strathalbyn Sub-Branch of R.S.S.I.L</td><td>Forresters Hall</td></tr>
<tr><td colspan="4">Announcement: There will be a grand reunion in July on the occasion of the anniversary of the battle of Fleurbaix, 1916.</td></tr>
<tr><td>14 June, 1926</td><td>Newcastle Morning herald and Miners Advocate</td><td>30th Battalion Reunion</td><td>30 Battalion Drill Hall</td></tr>
<tr><td colspan="4">Over 200 ex-members of the 30th Battalion A.I.F. assembled at the Drill Hall on Saturday night on the occasion of the first re-union. Among those present were Colonels F. Street and W. Cheeseman, Major P. Charlton, and the gathering included a sprinkling of Sydney and country members. It was decided to meet on Nov. 9 each year, that being the day on which the Battalion sailed from Sydney in 1915.</td></tr>
<tr><td>15 July, 1926</td><td>News, Adelaide S.A.</td><td>32nd Battalion Reunion</td><td>Clubrooms of R.S.S.I.L.A.</td></tr>
</table>

<table>
<tr><td colspan="4">Members of the 32nd Battalion will hold their annual dinner on Monday night. It will be the tenth anniversary of the attack at Fleurbaix when the Fifth Australian Division received its baptism of fire. On that day, July 19, 1916, many South Australians lost their lives. The date has always been regarded as sacred by surviving soldiers. During the evening a number of congratulatory telegrams and messages were read by the secretary (Mr. P. A. Ohlstrom), among which was one from the former commander of the 8th Brigade (Major-General Tivey). The general sent the following reply from Melbourne to an invitation to be present at this dinner: —" I *received your letter, and regret that I am unable to accept the kind invitation. I shall not be able to leave Melbourne at that time. I would like you to convey a message to the boys, with my kindest regards and best wishes. I have not forgotten the gallant men who fought and died at Fromelles. In the war no soldiers were given a harder task than the men of the 32nd Battalion, who fought through that night of carnage on the left flank of the 8th Brigade. No men could have faced the ordeal with greater courage or braver hearts. No battalion could have ended its fighting career with a more glorious and successful operation than the 32nd Battalion at the battle of the Hindenburg Line.*"</td></tr>
<tr><td>20 Apr. 1927</td><td></td><td>4th & 56h Battalions</td><td>Maccabean Hall, Darlinghurst</td></tr>
<tr><td colspan="4">All ex-members of both Divisions are cordially invited to attend.</td></tr>
<tr><td>19 July, 1927</td><td>News, Adelaide</td><td>32nd Battalion Reunion</td><td>R.S.S.I.L. Clubrooms Adelaide</td></tr>
<tr><td colspan="4">Today is the eleventh anniversary of Fleurbaix when the battalion first went into action. Of 36 officers and 1,016 other ranks, six officers and 75 other ranks answered the roll call next day. As a result of a recent resolution of Federal congress regarding the segregation in mental hospitals of former service men a communication has been received to the effect that the work of building the required accommodation is now in hand.</td></tr>
<tr><td>13 Oct. 1927</td><td>The Age, Melbourne</td><td>60th Battalion Smoke Night on 5 Nov., 1927.</td><td>Manchester Unity Hall, Swanston Street.</td></tr>
<tr><td colspan="4">Colonel N. Marshall, D.S.O., will be visiting Melbourne during Cup week. Knowing that old members of the battalion who served under Colonel Marshall would like to meet him,</td></tr>
</table>

<table>
<tr><td colspan="4">the committee has decided to hold a smoke night at the Manchester Unity Hall, Swanston-street, on Saturday, 5th Nov..</td></tr>
<tr><td>9 Nov. 1927</td><td>The Newcastle Sun NSW</td><td>30th Battalion Reunion</td><td>Newcastle Drill Hall</td></tr>
<tr><td colspan="4">The 30th Battalion reunion will be the second since the Great War. The 30th Battalion. On July 19, 1916, it took part in the tragic battle of Fromelles, the first battle in which the A.I.F. was engaged In France. The honors of the Great War, are as follows: — Somme 1916-18; Bapaume, 1917; Bullecourt, 1917; Ypres 1917; Polygon Wood, 1917; Ancre, 1918; Hindenburg Line, 1918; St. Quentin Canal, 1918.</td></tr>
<tr><td>21 Jan. 1928</td><td>The Age, Melbourne</td><td>Reunion of Returned Soldiers for 24 Apr., 1928</td><td>To be held in Exhibition Buildings, Melbourne.</td></tr>
<tr><td colspan="4">The largest reunion of returned soldiers yet held in Australia will take place on Anzac Day eve. Thousands of men who will be in Melbourne to take part in the Anzac Day march will attend the function, and it will thus enable the men to renew friendships made during the four years of fighting. It is hoped that vice-royalty will be present, as well as members of the Naval, Army and Air Force leaders.</td></tr>
<tr><td>10 Apr., 1928</td><td>The Age, Melboune</td><td>58th Battalion Reunion</td><td>Melbourne</td></tr>
<tr><td colspan="4">It is proposed to hold a reunion of the former members of the 58th Battalion, A.I.F. in Melbourne during Show week of this year.</td></tr>
<tr><td>12 Apr. 1928</td><td>Sydney Morning Herald</td><td>4th & 56th Battalions Reunion</td><td>Maccabean Hall, Darlinghurst.</td></tr>
<tr><td colspan="4">About 200 members of the two regiments were present including the president, Brigadier-General J.G. Mackay, Brigadier-General J. Heane, Colonels D. Storey and A.J.G. Simpson, and the secretary, Mr. T. W. Hewitt. Colonel Estall and Colonel Murray, the commanding officers respectively of the 4th and 56th Battalions, C.M.F, were present by invitation, accompanied by junior officers and non-commissioned officers. It was resolved that the two battalions should place a wreath on the Cenotaph on Anzac Day.The 4th Battalion remained</td></tr>
</table>

<table>
<tr><td colspan="4">with the First Division, while the 56th helped to complete the newly formed Fifth Division. Both battalions distinguished themselves in France, the 4th at Pozieres, Le Bargue, Dernicourt, Bullecourt, and Hargicourt; whilst the principal engagements of the 56th were at Fromelles, the Somme, Bapaume, Bullecourt, Polygon Wood, and Amiens.</td></tr>
<tr><td>2 June, 1928</td><td>Sydney Morning Herald</td><td>55th Battalion Reunion</td><td>Cocoa Tree Cafe</td></tr>
<tr><td colspan="4">It has been decided to hold a reunion of all branches of the 55th battalion at the on Saturday, July 21. The committee will place a wreath on the Cenotaph on July 19, at 1.15 p.m.</td></tr>
<tr><td>11 July, 1928</td><td>Sydney Morning Herald</td><td>54th Battalion Tribute and Reunion</td><td>Cocoa Tree Café.</td></tr>
<tr><td colspan="4">July 19 will mark the anniversary of the Battle of Fromelles, and ex-members of the 54th Battalion will pay tribute to their comrades by placing a wreath on the Cenotaph at 5.30pm.</td></tr>
<tr><td>13 July, 1928</td><td>Sydney Morning Herald</td><td>55th Battalion Reunion</td><td>Cocoa Tree Café.</td></tr>
<tr><td colspan="4">Members of the 55th Battalion will hold their first reunion at 8 p.m. on July 21, at the Cocoa Tree Café.</td></tr>
<tr><td>17 July, 1928</td><td>Evening News, Sydney</td><td>14th Brigade Reunion</td><td>Sargent's Market Street, Sydney</td></tr>
<tr><td colspan="4">Diggers of the 14th Machine Gun Coy will hold their annual reunion on Thursday, the anniversary of the Battle of Fromelles, in which the 14th Brigade had its first engagement in France.</td></tr>
<tr><td>30 July, 1928</td><td>Sydney Morning Herald</td><td>54th Battalion Reunion</td><td>Cocoa Tree Cafe</td></tr>
<tr><td colspan="4">The first reunion of the 54th Battalion of the A.I.F. attracted nearly three hundred ex-service men. A wreath was laid on the cenotaph on July 19, the anniversary of Fromelles.</td></tr>
</table>

<table>
<tr><td>10 Aug. 1928</td><td>The Brisbane Courier</td><td>31st Battalion Reunion</td><td>Anzac Club, Brisbane 9 Aug., 1929</td></tr>
<tr><td colspan="4">The Padre recounted many details connected with the activities of his old battalion, referring particularly to that date inseparable from the Unit – July 19, 1916 - when the 31st received its baptism of fire. The toast was honoured in silence.</td></tr>
<tr><td>3 Sep. 1928</td><td>Sydney Morning Herald</td><td>30th Battalion Reunion on Sep. 1</td><td>Drill Hall, Miller's Point</td></tr>
<tr><td colspan="4">Members spent a happy evening renewing old friendships, and laughing over the fun they shaded amidst the tragedies of the "other side."</td></tr>
<tr><td>14 Sep. 1928</td><td>The Argus, Melbourne</td><td>57th Battalion Reunion and Smoke Night to be held on 25 Sep.</td><td>The Victorian Softgoods Club, Melbourne</td></tr>
<tr><td colspan="4">The second annual general meeting, will be held on 25 September.</td></tr>
<tr><td>15 Sep. 1928</td><td>The Argus, Melbourne</td><td>5th Divisional Australian Army Service Corps Assn</td><td>Sargent's Café, Elizabeth Street, Melbourne</td></tr>
<tr><td colspan="4">The Association will hold a reunion on Wednesday, Sep. 26.</td></tr>
<tr><td>21 Sep. 1928</td><td>The Age, Melbourne</td><td>59th Battalion Reunion Smoke Night</td><td>Clota's Cafe, Empire Arcade, Flinders St.</td></tr>
<tr><td colspan="4">General Elliott is again to be guest of honor.</td></tr>
<tr><td>14 Nov. 1928</td><td>Northern Champion, Taree</td><td>Diggers Reunion & Social Meet.</td><td>Russell & Smith's Rooms, Taree.</td></tr>
<tr><td colspan="4">The second reunion of Diggers held under the auspices of the Taree sub-branch of the R.S. & S.I.L. was held on the eve of Armistice Day. Lieut-Col. W. J. Cheeseman, the president of the branch occupied the chair. The evening was happily spent in toasts, lecturettes,</td></tr>
</table>

<table>
<tr><td colspan="4">choruses, vocal songs and general conversation during which many humorous wartime incidents were recalled. The idea of the reunion was to preserve the spirit of the A.I.F. which did not ask what a man's religion was or what he was in private life. A man was accepted for his soldierly qualities, and that spirit should be remembered at all Digger functions.</td></tr>
<tr><td>15 Mar. 1929</td><td>The Muswellbrook Chronicle</td><td>4th & 56th Battalion Reunion 3 Apr, 1929</td><td>Maccabean Hall, Darlinghurst.</td></tr>
<tr><td colspan="4">The function this year will be the usual Smoke Concert with a Lantern Slide display of views of the two battalions abroad.</td></tr>
<tr><td>22 Jun. 1929</td><td>Sydney Morning Herald</td><td>55th Battalion Reunion</td><td>Burlington Café 20 July, 1929 8 pm.</td></tr>
<tr><td colspan="4">The second reunion of the 55th Battalion, A.I.F. The members will meet at the Cenotaph at 7.30.</td></tr>
<tr><td>5 Jul. 1929</td><td>Sydney Morning Herald</td><td>30th Battalion Annual Reunion on 20 July</td><td>Drill Hall, Millers Point</td></tr>
<tr><td colspan="4">Prior to the reunion a wreath will be placed on the Cenotaph at 7.30 p.m.</td></tr>
<tr><td>9 Jul. 1929</td><td>Daily Examiner Grafton & 5 other newspapers</td><td>53rd Battalion Reunion on 20 July, 1929.</td><td>Marlborough Café (Miss Bishop's) King Street, Sydney</td></tr>
<tr><td colspan="4">A wreath will be placed on the Cenotaph at 6.30 p.m.</td></tr>
<tr><td>18 Nov. 1929</td><td>The Newcastle Sun (NSW)</td><td>30th Battalion Diggers fourth Reunion</td><td>Drill Hall Newcastle</td></tr>
<tr><td colspan="4">Held to commemorate the sailing of the battalion from Sydney on Nov. 9, 1915 on the steamer Beltana. Many of the diggers had not met since the days of the war. The 30th Battalion was originally formed in Newcastle, and, with the exception of one company, comprised men from Newcastle and Maitland districts.</td></tr>
</table>

<table>
<tr><td>14 May, 1930</td><td>Sydney Morning Herald</td><td>5th Division Artillery Reunion, on May 21</td><td>Blinded Soldiers' Gate. 30 Pitt Street</td></tr>
<tr><td colspan="4">Before the commencement of the function a wreath will be placed on the Cenotaph.</td></tr>
<tr><td>15 Apr. 1930</td><td>Dungog Chronicle and Gloucester Advertiser</td><td>4th & 50th Battalions' 14th Reunion Smoke Concert, 23 Apr..</td><td>Maccabean Hall, Cnr. Darlinghurst Road and Burton Street, Darlinghurst</td></tr>
<tr><td>20 Jun. 1930</td><td>Albury Banner and Wodonga Express</td><td>53rd Battalion Reunion Smoke Concert 19 July</td><td>Miss Bishop's Cafe, King Street, Sydney</td></tr>
<tr><td>5 Jul. 1930</td><td>The Canberra Times, ACT</td><td>Returned Soldiers' League</td><td>Canberra Sub-branch meeting</td></tr>
<tr><td colspan="4">Senator Elliott to Lecture on Battle Of Fleurbaix, on 17 July. This will be appreciated as it takes place on the eve of this battle which was the first big engagement of Australians in France on July 19, 1917. General Elliott was in charge of the15th Brigade, 5th Australians.</td></tr>
<tr><td>9 Jul. 1930</td><td>Goulburn Evening Penny Post</td><td>55th Battalion Third Annual Reunion</td><td>Sargent's Burlington Cafe, George Street, Sydney</td></tr>
<tr><td colspan="4">The function on 19 July will be preceded by a muster in Martin Place at 7.15p.m. to lay a wreath on the Cenotaph in honour of fallen comrades.</td></tr>
<tr><td>17 Jul 1930</td><td>Daily News Perth</td><td>32nd Battalion Reunion on 19 July, 1930</td><td>RSL Rooms, Perth</td></tr>
<tr><td colspan="4">Saturday next will mark the anniversary of the Battle of Fleurbaix on July 19, 1916, when the 32nd Battalion, portion of the 51st Division, had its baptism of fire and was practically wiped out. Shortly after noon on Saturday Mr. L. St. John Kennedy, president of the 32nd Battalion Association, will lay a wreath on the State War Memorial in King's Park.</td></tr>
<tr><td>16 Aug, 1930</td><td>The Brisbane Courier</td><td>31st Battalion Reunion</td><td></td></tr>
</table>

<table>
<tr><td colspan="4">In proposing the toast of "The 31st Battalion," Brigadier-General E. M. Ralph, D.S.O., said that the records of the 31st were such that every member of the battalion might be proud. It had done its job well and left behind a splendid reputation.</td></tr>
<tr><td>24 Sep 1930</td><td>The Age, Melbourne</td><td>58th Battalion Reunion</td><td>Drill hall, Albert & Gisborne Streets Eastern Hill</td></tr>
<tr><td colspan="4">The committee of the Assn invites any member of the battalion who has attended previous reunions but owing to the present financial depression has been unable to purchase a ticket for this year's function, to attend the reunion. It is only necessary for any such member to ask the Secretary, who will arrange matters - nobody else will know anything about it.</td></tr>
<tr><td>8 Apr. 1931</td><td>Sydney Morning Herald</td><td>5th Division Artillery, A.I.F., Second Annual, Reunion</td><td>Blinded Soldier's Tearoom, Pitt Street.</td></tr>
<tr><td colspan="4">More than 100 members of the division assembled at the Cenotaph earlier in the evening with three trumpeters, and a wreath was placed on the Cenotaph.</td></tr>
<tr><td>7 Jul. 1931</td><td>The Sun, Sydney</td><td>15th Brigade Reunion On 18 July.</td><td>Imperial Service Club, O'Connell Street</td></tr>
<tr><td colspan="4">The reunion dinner of the 57th, 58th, 59th and 60th Battalions, A.I.F., will be held at 7 p.m.on July 18.</td></tr>
<tr><td>7 Jul. 1931</td><td>The Sun, Sydney</td><td>53rd Battalion Reunion on 18 July.</td><td>Drill Hall, Addison-road Marrickville</td></tr>
<tr><td colspan="4">The annual reunion "smoko" of members of the above unit will be held on July 18. The function will serve to commemorate the 15th anniversary of the Battle of Fromelles, where the battalion received its baptism of fire. Colonel W. J. Cheeseman. M.C., will preside.</td></tr>
<tr><td>18 Oct. 1931</td><td>Sunday Times WA</td><td>32nd Battalion AIF</td><td>Annual General Meeting</td></tr>
</table>

<table>
<tr><td colspan="4">There was a splendid gathering of ex-members of the 32nd Battalion. Mr. A. Stone, the retiring president, stressed particularly the inauguration of the laying of a wreath on the State Memorial by the association in commemoration of Fleurbaix. This was to be an annual function, to take place on the Sunday afternoon following the reunion.</td></tr>
<tr><td>16 Sep. 1931</td><td>The Argus, Melbourne</td><td>59th Battalion AIF Assn</td><td>Central Café, Elizabeth Street on 22 Sep, at 3pm.</td></tr>
<tr><td colspan="4">In accordance with the Show-time custom the59th Battalion A.I.F. Association will hold a re-

union at the Central Cafe, Elizabeth street opposite McEwan's, on Tuesday, Sep. 22, at 3pm.</td></tr>
<tr><td>also</td><td></td><td>57th Battalion Assn. on 22 Sep, at 7.45pm</td><td>Temple Court Tea Rooms, 422 Collins Street.</td></tr>
<tr><td>also</td><td></td><td>32nd Battalion, City of Footscray Regiment</td><td>27 Sep, at 3pm West Footscray Oval.</td></tr>
<tr><td colspan="4">Trooping of the Colour in honour of their service battalion (the 32nd A.I.F.), as well as the 29th, 30th, and 31st Battalions, in the taking and holding of Polygon Wood, fought on Sep. 25, 26, and 27,1917.</td></tr>
<tr><td>10 June, 1932</td><td>The Sun, Sydney</td><td>54th Battalion Reunion on 16 July.</td><td>Miss Bishop's 117 Bathurst Street, Sydney</td></tr>
<tr><td>18 July, 1932</td><td>The West Australian</td><td>32nd Battalion Assn Reunion Dinner</td><td>Soldiers' Institute</td></tr>
<tr><td colspan="4">Every year the 32nd Battalion (A.I.F.) Association commemorates the battle of Fromelles, which was fought in France on July 19, 1916, when the battalion, at Fleurbaix, on the left flank of the attack, lost about 750 officers and men in less than 24 hours. 150 members of the unit were present. A feature of the attendance was the number of men who had come from the country for the function.</td></tr>
</table>

<table>
<tr><td>27 Oct, 1932</td><td>The Argus, Melbourne</td><td>14th Brigade Reunion on Cup Eve.</td><td>Manchester Unity Hall</td></tr>
<tr><td colspan="4">The annual smoke night and reunion of the 14th Field Ambulance and of the 14th Brigade (including members of the 53rd, 54th, 55th, and 56th Battalions) will be held at 8 o'clock on Cup Eve, Oct. 31.</td></tr>
<tr><td>4 Nov. 1932</td><td>The Newcastle Sun</td><td>30th Battalion Reunion</td><td>Memorial Institute, Newcastle.</td></tr>
<tr><td colspan="4">The seventh annual reunion of the 30th Battalion will be held on Nov. 19.</td></tr>
<tr><td>7 Apr. 1933</td><td>Sydney Morning Herald</td><td>4th & 56th Battalions</td><td>Maccabean Hall, Darlinghurst</td></tr>
<tr><td colspan="4">18th Annual Smoke Concert will be held on 19 April at 8pm.</td></tr>
<tr><td>2 May, 1933</td><td>Sydney Morning Herald</td><td>5th Divisional Artillery Reunion</td><td>Bishop House, 117 Bathurst Street</td></tr>
<tr><td colspan="4">To be held next Saturday night. Professor H Tasman Lovell will discuss "Patriotism."</td></tr>
<tr><td>27 June 1933</td><td>Sydney Morning Herald</td><td>53rd Battalion Reunion</td><td>Gum Blossom Cafe, 32 Pitt Street</td></tr>
<tr><td colspan="4">Ex-members will assemble at the Cenotaph at 7.45.</td></tr>
<tr><td>22 July 1933</td><td>The Mail, Adelaide</td><td>32nd Battalion Reunion Dinner</td><td>RSL Club</td></tr>
<tr><td colspan="4">More than 200 members were present at the annual reunion dinner.</td></tr>
<tr><td>27 July 1933</td><td>Western Mail, Perth</td><td>5th Division Memorial Service</td><td>War Memorial Kings Park</td></tr>
</table>

<table>
<tr><td colspan="4">July 19 was the 17th anniversary of the battle of Fromelles. The 5th Division (A.I.F.), of which the 32nd battalion was a unit, attacked the "Sugar-loaf salient," south of Armentieres, with great loss of life. Last Sunday, the president of the 32nd Battalion Association (Mr. R. Alexander) placed a wreath on the state war memorial in King's Park in memory of the members of the unit who fell and died overseas.</td></tr>
<tr><td>27 Sep. 1933</td><td>The Argus, Melbourne</td><td>60th Battalion Reunion</td><td>Victorian Softgoods Club</td></tr>
<tr><td></td><td></td><td>57th Battalion AIF Assn</td><td>Clota's Café, Melbourne</td></tr>
<tr><td colspan="4">The reunion of the 57th Battalion was attended by a larger number of members than at any previous reunion.</td></tr>
<tr><td></td><td></td><td>58th Battalion Reunion</td><td>Rumplemayers.</td></tr>
<tr><td>29 Sep. 1933</td><td>The Age Melbourne</td><td>60th Battalion Annual Reunion</td><td>Victorian Softgoods Club, Flinders Lane.</td></tr>
<tr><td colspan="4">Over 110 old soldiers attended, and spent an enjoyable evening with songs and stories of the war.</td></tr>
<tr><td>25 Oct. 1933</td><td>The Argus, Melbourne</td><td>14th Brigade and 14th Field Ambulance</td><td>Manchester Unity Hall, 339 Swanston Street. 6 Nov. 8pm.</td></tr>
<tr><td>23 Apr. 1934</td><td>The Argus, Melbourne</td><td>15th Brigade Reunion</td><td>Naval & Military Club</td></tr>
<tr><td>11 June 1934</td><td>The Sun (Sydney)</td><td>54th Battalion Reunion</td><td>Miss Bishop's Bathurst Street on 21 July.</td></tr>
<tr><td>30 June 1934</td><td>The Sun (Sydney)</td><td>30th Battalion Reunion</td><td>Miss Bishop's, 117 Bathurst Street, 21 July, 7.45 pm</td></tr>
<tr><td colspan="4">Ex-members are coming from Victoria and Newcastle.</td></tr>
</table>

<table>
<tr><td>1 July, 1934</td><td>The Argus, Melbourne</td><td>29th Battalion Reunion</td><td>Hall of Honour, Geelong, 3pm 14 July.</td></tr>
<tr><td colspan="4">A wreath will be placed in the memorial foyer at Johnstone Park at 10.15 a.m. on Sunday, and a church parade will follow at St. Matthew's Church of England. Motor-cars for Geelong will leave Anzac House at 3 p.m. on Saturday.</td></tr>
<tr><td>15 July 1934</td><td>Truth (Sydney)</td><td>To form an AIF 15th Brigade Reunion Assn.</td><td>7.30pm on 19 July at Junior Red Cross Café, Angus & Coote Bldg</td></tr>
<tr><td colspan="4">It is hoped that all members of Pompey's Brigade will attend.</td></tr>
<tr><td>18 July 1934</td><td>Sydney Mail, Sydney</td><td>55th Battalion Reunion</td><td>Sargent's Market Street on Saturday next.</td></tr>
<tr><td colspan="4">Members will assemble at the Cenotaph, Martin Place, at 7.15pm</td></tr>
<tr><td>21 July 1934</td><td>The Daily News, Perth</td><td>32nd Battalion Annual Reunion Dinner Tonight</td><td>Soldiers' Institute. Dinner at 7.00pm.</td></tr>
<tr><td colspan="4">The date of this reunion is set down in each year as the nearest Saturday to July 19 - On July 19, 1916, the first battle in which the 32nd took part was fought at Fleurbaix-Fromelles and history relates that this proved (while being a "disastrous show') the discipline and mettle of the new A.I.F. unit. A wreath will be laid at the State War Memorial in Kings Park at 8.30pm tomorrow.</td></tr>
<tr><td>4 July 1934</td><td>The Katoomba Daily</td><td>Leura Diggers Entertain</td><td>Leura Memorial Hall on 19 July.</td></tr>
<tr><td colspan="4">President E. K. Gillies was Master of Ceremonies, and in his opening remarks said that that day (July 19) was the anniversary of the battle of Fromelles — a conflict that would live long in the memory of members of the A.I.F. During that battle, the 5th Division of the A.I.F. lost 5533 officers and men in one night's sharp fighting. Six battalions, the 31st, 32nd, 53rd, 54th, 59th and 60th, which had attacked on the flanks, suffered 3879 casualties. The battle of</td></tr>
</table>

Fromelles was part of the great offensive in France, in 1916. "Of the 60th Battalion, which had gone into the fight with 887 officers and men, only one officer and 108 men survived," said the President.			
6 Aug. 1934	The Age, Melbourne	31st Battalion Reunion	The Cabin, Elizabeth Street.
A large number of men who served in the 31st Battalion in the war attended the smoke social held by the battalion association on Saturday night. When proposing the toast of the association, Mr. R. B. Wingate, a member of the committee, said that the organisation had formed a vigilance body, and its functions would embrace taking up the cases of men of the battalion who found themselves affected with recurring war disabilities and whose condition required to be placed before the Repatriation authorities with a view of obtaining assistance for them. In a number of instances that help was urgently needed, and when a Digger was unable to prepare his claim for consideration the association would show him how to present his claim. The association was also active in keeping alive the fellowship of the association. (Applause.)			
22 Sep, 1934	Sydney Morning Herald	15th Infantry Brigade AIF Reunion and Dinner	The Horse Shoe Cafe, Hosking-Place, on Sep. 29, at 6 p.m.
8 Oct. 1934	The Age, Melbourne	59th Battalion Assn. Annual Reunion	Unity Hall, Bourke-street, Melbourne, 23rd Oct. at 8 p.m.
2 Nov, 1934	The Age, Melbourne	58th Battalion Annual Reunion Dinner	Rumpelmayers, Howey Court, on Thursday next, at 7 p.m.
The date has been fixed to coincide as nearly as possible with the official opening of the Shrine of Remembrance by the Duke of Gloucester on Sunday next, 11th Nov.			
13 Apr. 1935	Sporting Globe, Melbourne, Vic.	58th Battalion Reunion	Hotel Australia on Apr. 24 (Anzac Eve), at 7 p.m.

<table>
<tr><td colspan="4">To give members who can only come to Melbourne for the Anzac celebrations an opportunity of being present at an official reunion, and to celebrate the 20th Anniversary of Anzac</td></tr>
<tr><td>8 June 1935</td><td>The Sun, Sydney</td><td>53rd Battalion Reunion and annual Smoke Night</td><td>Thorne's Cafe. 17 Castlereagh-street, Sydney.</td></tr>
<tr><td>12 June 1935</td><td>Sydney Morning Herald</td><td>54th Battalion Reunion</td><td>117 Bathurst Street on July 20, at 8 p.m.</td></tr>
<tr><td colspan="4">A wreath will be placed on the Cenotaph at 7.15 p.m.</td></tr>
<tr><td>12 June, 1935</td><td>Sydney Morning Herald</td><td>31st Battalion Reunion</td><td>Miss Bishop's café, Bathurst-street, on July 20.</td></tr>
<tr><td>10 July 1935</td><td>Daily Advertiser, Wagga Wagga</td><td>55th Battalion Annual Reunion</td><td>Sargents. 55 Market Street, Sydney, on Saturday, July 20, at 8 p.m.</td></tr>
<tr><td colspan="4">The ex-members of the 55th Battalion. A.I.F., many of whom come from this district, will hold their 8th annual reunion at, after assembling at the Cenotaph at 7.15.</td></tr>
<tr><td>11 July 1935</td><td>Sydney Morning Herald</td><td>54th Battalion Annual Reunion</td><td>117 Bathurst-street on July 20.</td></tr>
<tr><td colspan="4">A wreath will be laid on the Cenotaph at 7.15 p.m.</td></tr>
<tr><td>16 July 1935</td><td>The Argus, Melbourne</td><td>29th Battalion Annual Reunion and Smoke Night</td><td>8 p m on July 20 at Annie House Collins Street.</td></tr>
<tr><td colspan="4">More than 500 members are expected to attend.</td></tr>
<tr><td>17 July 1935</td><td>The Sun, Sydney</td><td>53rd Battalion Reunion Smoko</td><td>Thorne's Café next Saturday night.</td></tr>
</table>

Colonel W J. Cheeseman, and Detective-Superintendent Prior, of the C.I.B., will be present. Members will meet in Martin-place at 7.45 p.m. to lay a wreath on the Cenotaph. Superintendent Prior fought in the South African war and is president of the Returned Sailors and Soldiers' Police Association.			
11 Sep. 1935	The Age, Melbourne	59th Battalion AIF Assn Ninth Annual Reunion.	Tuesday, 24th inst., Unity Hall, 636 Bourke-street, 8 p.m.
29 Oct. 1935	The Argus, Melbourne	14th Field Ambulance and 14th Brigade (combined)	8 pm. on Nov. 4 at the Willow Cafe, Small Arms Buildings, 233 Elizabeth Street.
11 Nov, 1935	Newcastle Morning Herald and Miners' Advocate	30th Battalion Tenth Annual Reunion	Newcastle Drill Hall on 9 November.
150 members attended the reunion, to celebrate the 20th anniversary of the departure of the battalion for service in the Great War. The toast list included a. tribute to "departed comrades," with special remembrance of the 458 members of the battalion who fell in battle. The battalion sailed from Sydney in the steamer Beltana on Nov. 9, 1915, with Lieut. Colonel J. W. Clarke as Commanding Officer. The present Commanding Officer is Colonel F. Street, who assumed command in 1917.			
15 May, 1936	The Sun, Sydney	54th Battalion Ninth Annual Reunion	Miss Bishop's, 117 Bathurst Street, Sydney on 25 July
16 June 1936	The Sun, Sydney	56th Batt. First Annual Reunion	Legacy Rooms, Nock and Kirby's old building, George-Street North, on July 20, at 7.45 p.m.
The **first** annual reunion of the 56th Battalion will be held at the Legacy Club rooms.			

<table>
<tr><td>2 July 1936</td><td>Western Mail Perth</td><td>32nd Battalion & other 5th Divn units Annual Reunion</td><td>Anzac House Supper Room on 18 July.</td></tr>
<tr><td colspan="4">Ex-members are reminded of the annual re-union. This is the twenty-first anniversary of the formation of the battalion. Following the usual custom, a wreath in memory of fallen comrades will be placed on the State war memorial on Sunday, July 19, the anniversary of the battle of Fleurbaix (Fromelles). The battalion will parade at the Queen Victoria statue, in King's Park, at 3 p.m., and march to the memorial.</td></tr>
<tr><td>8 July 1936</td><td>News, Adelaide</td><td>Reunion of War Associates – 32nd Battalion Reunion</td><td>To be held 29 September.</td></tr>
<tr><td colspan="4">The 32nd Battalion of the A.I.F. is planning an important gathering of wartime associates as a centenary reunion in Adelaide. Visitors are expected to include Major Blair Wark, V.C., and Major-Gen. E. Tivey, who commanded the 8th Brigade (of which the 32nd Battalion was part) and later the 5th Division. Arrivals from the eastern States are expected in Adelaide on Monday, Sep. 28. The reunion has been brought about by the untiring work of Mr. Perdy S. Sanders, who has been honorary secretary for the 32nd since the cessation of hostilities in 1918.</td></tr>
<tr><td>17 July 1936</td><td>The Age, Melbourne</td><td>29th Battalion Reunion Smoko Night.</td><td>Masonic Hall, Little Ryrie Street, Geelong Saturday night 8pm.</td></tr>
<tr><td>17 July 1936</td><td>Daily News Perth</td><td>32nd Battalion Anniversary</td><td>Anzac House Supper Room tomorrow.</td></tr>
<tr><td colspan="4">At Anzac House supper room tomorrow, 18 July, the 32nd Battalion (A.I.F.) will hold a grand reunion, to celebrate its 21st anniversary. On the following day, 19 July at 3 p.m., members of the battalion will "fall in" at the Queen Victoria Statue in King's Park from whence they will march to the State War Memorial where a wreath will be placed in remembrance of fallen comrades. It has been the custom of this association to carry out this remembrance on the Sunday nearest the anniversary of the Battle of Fleurbaix-Fromelles July 19, 1916 in which "stunt" the great majority of the original battalion were wiped out.</td></tr>
</table>

<table>
<tr><td>19 July 1936</td><td>Truth, Sydney</td><td>54th Battalion Ninth Annual Reunion</td><td>Miss Bishop's, 117 Bathurst Street, Sydney, on July 25, 1936. at 8 p.m.</td></tr>
<tr><td>21 July, 1936</td><td>Sydney Morning Herald</td><td>56th Battalion First Annual Reunion</td><td>Legacy Club's Rooms 20 July.</td></tr>
<tr><td colspan="4">The first annual reunion of the 56th Battalion A.I.F., was a great success, there being more than 150 ex-A.I.F. men present. In years gone by the battalion had combined with the 4th Battalion in such functions, but it had been decided that they should hold separate reunions.</td></tr>
<tr><td>20 Nov 1936</td><td>The Age, Melbourne</td><td>30 Battalion Reunion</td><td>Softgoods Club, Flinders Lane, on Monday night, 23rd Nov.</td></tr>
<tr><td colspan="4">Reunion of Members of A Company 30th Battalion survivors of those ex-naval men who formed the original 30th Battalion A.I.F., at Liverpool camp in 1915. Major-General E. Tivey, the original brigadier of the 8th Brigade, will be the guest of honor.</td></tr>
<tr><td>17 Apr 1937</td><td>Sporting Globe, Melboune</td><td>58th Battalion Annual Reunion, on 2 April, 7pm.</td><td>Rondo's Café, Capitol House, Swanston Street</td></tr>
<tr><td>3 June 1937</td><td>The Sun, Sydney</td><td>29 Battalion Reunion</td><td>Railway & Tramway Institute</td></tr>
<tr><td>15 June 1937</td><td>Newcastle Morning Herald and Miner's Advocate</td><td>31 Battalion Reunion</td><td>Miss Bishop's Carrington Street, Sydney on 17 July, 7.30pm</td></tr>
<tr><td colspan="4">Members will meet at the Pitt-street corner, G.P.O., at 7 p.m.</td></tr>
<tr><td>2 June 1937</td><td>Western Mail, Perth</td><td>32nd Battalion Reunion, 17 July, from 7.30pm</td><td>Mrs. O'Connell's rooms, E.S. and A. Bank</td></tr>
</table>

			Chambers, corner Hay and Milligan Streets
On the day following the reunion (Sunday, July 18), the members will parade at the Queen Victoria statue in King's Park at 3 pm., and march from there to the State War Memorial, where the president will lay a wreath in memory of Fallen Comrades.			
15 July 1936	The Sun, Sydney	55th Battalion Veterans Assn 10th Annual Reunion	For 17 July, at 85 Market Street, Sydney
Members are asked to assemble in Macquarie Street, at the top of Martin-place, preparatory to marching to the Cenotaph with the Wreath of Remembrance.			
22 July 1937	The West Australian	32nd Battalion, 8th Brigade Annual Reunion Dinner	
About 120 survivors of the Battle of Fleurbaix of July 19, 1916, when the Eighth Brigade, A.I.F., had its first engagement, assembled at the annual reunion dinner last Saturday night. Fallen comrades were remembered when the lights were extinguished and an illuminated cross bearing the words "Lest we Forget" appeared on the wall. The observance culminated on Sunday in a brief ceremony at the State War Memorial. About 50 members paraded at the Queen Victoria statue in King's Park and marched to the State War Memorial, where the president laid a wreath in memory of fallen comrades. With the sounding of the last post, the men on parade doffed their hats and placed them over the heart; standing in silence for two minutes. At the call of reveille, they quietly dispersed.			
1 Oct. 1937	The Age, Melbourne	60th Battalion Annual Reunion	Richmond Memorial Hall, Church Street, last night.
Over 150 former members of the 60th Battalion, attended. Song, story and reminiscences of A.I.F. days were enjoyed.			
6 July 1938	The Sun, Sydney	31st Battalion Annual Reunion	Miss Bishop's, Carrington Street, city on 16 July

<table>
<tr><td colspan="4">Members will assemble at the Pitt Street corner of the G.P.O. at 7.15 p.m.</td></tr>
<tr><td>26 July, 1938</td><td>The West Australia Perth</td><td>32nd Battalion Reunion</td><td>The Stirling Social Rooms on the night of July 16.</td></tr>
<tr><td colspan="4">To mark the anniversary of July 19, the date in 1916 of the Battle of Fleurbaix (France), which was the first engagement in which the 32nd Battalion A.I.F., participated and which cost the unit 700 casualties in 12 hours, about 100 members of the 32nd Battalion Association held their annual reunion.

Plans were in train for a contingent of the 32nd Battalion, A.I.F. from Adelaide, and a party from the 32nd Battalion of the Commonwealth Military Forces stationed at Footscray (Victoria), to be entertained in Perth in 1940, when the 25th anniversary of the formation of the battalion would be celebrated. The proposal had emanated from the interstate reunion of the battalion held in Adelaide in 1936, when a large contingent of 32nd men from Western Australia had attended the 21st anniversary celebrations.

In memory of the late Lieut.-General Sir Talbot Hobbs, a ceremony was observed. The lights were put out, and an electric light sign depicting a white cross surrounded by a laurel wreath shone from one wall, towards which the gathering faced, standing in silence. The last post and reveille were sounded, the president reciting Laurence Binyon's famous lines between the bugle calls.

At 3 p.m. on July 17, several members of the association gathered in King's Park for a wreath-laying ceremony. They marched in file from the meeting-place, and made a fan formation in front of the State War Memorial, where the president placed a wreath. After the customary bugle calls and period of silence, the files were reformed and the party marched away.</td></tr>
<tr><td>7 Sep. 1938</td><td>Sydney Morning Herald</td><td>55th Battalion Annual Reunion</td><td>Sargents Cafe, Market Street, on Saturday, Oct. 1, at 8 p.m.</td></tr>
<tr><td colspan="4">Before the dinner, there will be a ceremony' at the Cenotaph, Martin Place. Those participating will, assemble at the eastern end of Martin Place at 7.15 p.m.</td></tr>
</table>

<table>
<tr><td>6 Apr. 1939</td><td>The Sun, Sydney</td><td>56th Battalion Annual Reunion</td><td>Legacy Club Rooms 199 George Street North, Saturday at 7.30pm</td></tr>
<tr><td>6 July, 1939</td><td>Sydney Morning Herald</td><td>31st Battalion Reunion</td><td>Saturday, July 22, at Miss Bishop's, Carrington Street</td></tr>
<tr><td colspan="4">Members are requested to assemble at the G.P.O. at 7.15 p.m.</td></tr>
<tr><td>28 July, 1939</td><td>The West Australian Perth</td><td>32nd Battalion Reunion commemorating the battalion's part in the Battle of Fleurbaix</td><td>Held on Saturday night at the Stirling Social Rooms</td></tr>
<tr><td colspan="4">An attendance of 95, nearly all being members of the unit, signified the success of the annual dinner of members. Formalities were restricted to give the maximum of opportunity for fraternisation. Fallen comrades were remembered in a special ceremony in which the lights were extinguished, and all present faced one corner of the room, where a cross of sacrifice was illuminated with the words, "Lest We Forget". Appropriate bugle calls were sounded. It was recalled that three years ago an interstate reunion of the battalion had been held in Adelaide, a contingent having gone from this State and another, including the band and colour party of the 32nd Battalion, C.M.F., from Victoria.

It was then announced that this procedure would be repeated in Apr. next in Perth. On Sunday afternoon 50 members of the battalion visited the State War Memorial. Marching from the Queen Victoria statue, headed by the president and vice-presidents, the column was met by the warden of the memorial, and a wreath was laid by the president. Bugler Hewens sounded the last past and reveille.</td></tr>
<tr><td>14 July 1939</td><td>The Sun, Sydney</td><td>54th Battalion 12th Annual Reunion will be held on Saturday 22 July at 8pm</td><td>At Miss Bishop's Crystal Ballroom, 500 George Street, Sydney</td></tr>
<tr><td>22 June 1940</td><td>The Daily News, Perth</td><td>32nd Battalion Annual Reunion</td><td>Saturday, July 20, at Mrs. O'Connell's Stirling Social</td></tr>
</table>

			Rooms at the corner of Milligan and Hay streets.
The ceremony of placing of a wreath on the State War Memorial, will take place on Sunday, July 21, at 3.15 p.m. Members will fall-in at the Queen Victoria Memorial in King's Park at 3 o'clock.			
23 July 1940	Daily Advertiser, Wagga Wagga	55th Battalion Reunion	On Saturday night last, the Union Club Hotel
Although hurriedly arranged, about 25 members attended the function. The occasion coincided with the annual reunion of the battalion in Sydney and opportunity was taken by the Wagga 'boys' to entertain at a smoke social their former commander. Major E. W.J. Stutchbury, M.C. (bar) and former padre, the Rev. H. Green. At the conclusion of the gathering it was unanimously decided to make the function an annual one.			
19 Apr. 1941	News, Adelaide	32nd Battalion	Majestic Hotel
19 Apr. 1941	News, Adelaide	5th Division Field Ambulance	Sturt Arcade Hotel, Grenfell Street
Country diggers who will flock to Adelaide next week for the Anzac Day march through the city on Friday will be welcomed by city comrades at battalion reunions. Most of the reunions will be held on Anzac Day eve. They will be in the form of socials at which there will be two minutes' silence for fallen comrades.			
3 July 1941	Sydney Morning Herald	55th Battalion 14th Annual Reunion on Saturday, July 19.	Sargents, Market Street
Members will assemble at the corner of Macquarie Street and Martin Place at 7 p.m. for a march to the Cenotaph.			
3 July 1941	Sydney Morning Herald	14th Field Company Engineers, A.I.F. Annual reunion	The Legacy Club, 188 George Street, Sydney, on Saturday. July 19.

Members are requested to attend at 6.45 p.m. at the G.P.O. for the ceremony at the Cenotaph.			
7 July 1941	Sydney Morning Herald	54th Battalion 14th Annual Reunion	The Legacy Clubrooms, 188 George Street, Sydney
9 July 1941	Sydney Morning Herald	31st Battalion, A.I.F., annual reunion dinner.	Miss Bishop's, Carrington Street, at 8 p.m. next Saturday week.
Members will assemble at the corner of Pitt Street and the G.P.O. at 7 p.m. to place a wreath on the Cenotaph.			
9 July 1941	Sydney Morning Herald	30th Battalion Annual Meeting	National Dining Rooms, 3rd Floor, 30 Ash Street
A wreath will be laid on the Cenotaph, Martin Place, at 7.50 p.m. on Saturday, July 19.			
13 July, 1941	Truth, Sydney	31st Battalion annual reunion dinner	Miss Bishop's, Carrington Street, City, at 8 p.m. on Saturday next.
Diggers will assemble at the corner of the GP.O. and Pitt Street at 1 p.m. to place a wreath on Cenotaph.			
22 July 1941	West Australian Perth	32nd Battalion annual reunion, and with 2/32 Battalion	
Members of the original 32nd Battalion and members of the 2/32 Battalion gathered at the Queen Victoria monument in King's Park on Sunday afternoon and marched to the State War Memorial. Lieut. L. D. Lobascher placed a wreath on the memorial in memory of fallen members. On Saturday night members of the original battalion held their annual reunion in memory of the battle of Fleurbaix, on July 19, 1916, where the battalion took part, with distinction, in its first major action. Speakers paid special tribute to the late Lieut.-Colonel B.			

<table>
<tr><td colspan="4">A. Wark, V.C., D.S.O., who recently died, and who was an original member of the old battalion.</td></tr>
<tr><td>30 July, 1941</td><td>Goulburn Evening Post</td><td>55th Battalion 15th Annual Reunion on July 19</td><td>Sargent's, Market Street, Sydney</td></tr>
<tr><td colspan="4">First they marched from Elizabeth Street to the Cenotaph, where Major E.W. Stutchbury placed a wreath in memory of those who had passed on in France and since. After "the last Post" followed by "Reveille", and the recitation of "In Flanders Fields the Poppies Grow", it was amazing to see the number of handkerchiefs being tucked away. In the few speeches made was stressed the need to be ready to rehabilitate the men who return from the 2nd A.I.F.</td></tr>
<tr><td>17 May, 1945</td><td>Sydney Morning Herald</td><td>56th Battalion Reunion at 8pm next Saturday</td><td>Legacy Club Rooms, George Street North</td></tr>
<tr><td>21 July 1945</td><td>The Newcastle Sun</td><td>30th Battalion Reunion dinner</td><td>Illoura Hall on 20 July.</td></tr>
<tr><td colspan="4">78 members held a reunion dinner last night. Also present was one number of the 2/30th. Colonel F. Street presided and members decided to hold these functions annually.</td></tr>
<tr><td>20 Oct. 1946</td><td>The Argus, Melbourne</td><td>57/60 Battalion Assn Reunion will be held Tuesday next at 8pm</td><td>11th Floor, Manchester Unity Building, City.</td></tr>
<tr><td>27 Oct. 1945</td><td>The Argus, Melbourne</td><td>60th Battalion Reunion on 5 November (Cup eve)</td><td>Soldiers Memorial Hall, 401 Church Street, Richmond.</td></tr>
<tr><td>2 July 1946</td><td>Sydney Morning Herald</td><td>55th Battalion Reunion on 13 July</td><td>Junior Red Cross Rooms, 188 George Street, Sydney</td></tr>
<tr><td colspan="4">Members will meet at the corner of Macquarie Street and Martin Place, at 7.15 p.m. for a ceremony of remembrance at the Cenotaph.</td></tr>
</table>

<table>
<tr><td>17 Sep. 1946</td><td>The Argus, Melbourne</td><td>59th Battalion on Show Eve, 25 Sep, at 8pm</td><td>RSL Hall, 201 Church Street, Richmond</td></tr>
<tr><td>14 Nov. 1946</td><td>Werribee Shire Banner</td><td>32nd Battalion smoke night reunion</td><td>Footscray Drill Hall on Friday 22 November.</td></tr>
<tr><td>21 Aug, 1947</td><td>Portland Guardian Vic.</td><td>29th Battalion Reunion</td><td>Geelong</td></tr>
<tr><td colspan="4">About 120 attended, many coming front other parts of the State. On the following morning about 50 members were present, when a beautiful wreath, with the battalion's colors predominating, was laid on the memorial stone in memory of departed comrades, by Col. Patterson.</td></tr>
<tr><td>19 Sep. 1947</td><td>The Age, Melbourne</td><td>59th Battalion Annual meeting and smoke night reunion</td><td>Next Tuesday night at Eton Hall, Commercial Road, Prahran.</td></tr>
<tr><td>26 June 1948</td><td>Sydney Morning Herald</td><td>53rd Battalion Assn annual Fromelles commemoration reunion</td><td>On July 17 at the Blue Tea Rooms, 39 Rowe Street, city</td></tr>
<tr><td colspan="4">Members will meet at Angel Place at 7.45 p.m. for a wreath laying ceremony.</td></tr>
<tr><td>5 Nov. 1948</td><td>Newcastle Morning Herald and Miners Advocate</td><td>55th Battalion, A.I.F., reunion</td><td>On Saturday 13 November at the Junior Red Cross Rooms, 188 George Street, Sydney</td></tr>
<tr><td colspan="4">Members will fall-in at the corner of Macquarie street and Martin Place at 7 p.m. for a ceremony at the Cenotaph.</td></tr>
<tr><td>21 Sep, 1949</td><td>The Argus, Melbourne</td><td>59th Battalion Reunion</td><td>29 September, at 8pm, Upper Room, Melbourne Town Hall</td></tr>
</table>

<table>
<tr><td>1 Oct. 1949</td><td>The Dubbo Liberal and Macquarie Advocate</td><td>54h Battalion Reunion</td><td>Proposed for Saturday 22 October, in the Drill Hall, Dubbo.</td></tr>
<tr><td>22 Oct, 1949</td><td>The Argus, Melbourne</td><td>60th Battalion Reunion</td><td>On 31 October, at 7.45pm, at the Masonic Hall, 254 Swan Street, Richmond</td></tr>
<tr><td>29 Oct. 1949</td><td>Sydney Morning Herald</td><td>55th Battalion Assn. 21st annual reunion</td><td>To be held on 12 November.</td></tr>
<tr><td>29 Nov. 1949</td><td>The Dubbo Liberal and Macquarie Advocate</td><td>54th Battalion first Reunion Dinner</td><td>Masonic Hall, Dubbo on Saturday night last.</td></tr>
<tr><td colspan="4">About 80 ex-members of the 54th Battalion, including representatives from both World Wars, attended the Battalion's first reunion to be held in Dubbo. Before the reunion commenced, members marched to the War Memorial and laid a wreath to the memory of members of the battalion who fell in both wars. The function was organised by a committee on behalf of the Dubbo and District 54th Re-union Association, and proved so popular that it is intended to conduct one annually.</td></tr>
<tr><td>12 Apr. 1950</td><td>The Argus, Melbourne</td><td>58/59th Battalion Association Reunion</td><td>On 24 April, at 8pm, at Coburg Football Pavilion.</td></tr>
<tr><td>15 Apr. 1950</td><td>Sporting Globe, Melbourne</td><td>The Get Together of 60th Battalion 1st AIF Assn</td><td>Memorial Hall, 401 Church Street, Richmond on Anzac Day, 25 April, 1950</td></tr>
<tr><td colspan="4">Members are advised the Battalion will fall In on Anzac Day in Russell Street, between Bourke and Collins Streets, on the South side of Bourke Street. The Get Together will be held immediately after the march.</td></tr>
<tr><td>4 Oct. 1950</td><td>Goulburn Evening Post</td><td>55th Battalion Reunion to be held on 28 October.</td><td></td></tr>
</table>

<table>
<tr><td>10 Oct. 1950</td><td>Sydney Morning Herald</td><td>55th Battalion 22nd annual reunion</td><td>Legacy Rooms, 188 George Street, city, on 28 Oct. at 8pm</td></tr>
<tr><td>14 Nov. 1950</td><td>Dubbo Liberal and Macquarie Advocate</td><td>54th Battalon Assn annual reunion.</td><td>Masonic Hall, Dubbo on Saturday night last.</td></tr>
<tr><td colspan="4">Several speakers referred in glowing terms to the achievements of the 54th Battalion especially in the first world war when the unit suffered heavy casualties. The work of Legacy was also mentioned. Support was requested for the good work done by this organisation in helping children of old comrades.</td></tr>
<tr><td>4 Apr. 1951</td><td>The Argus, Melbourne</td><td>60th Battalion annual reunion</td><td>Richmond Memorial Hall immediately after Anzac Day march.</td></tr>
<tr><td>15 Sep. 1951</td><td>Sporting Globe, Melbourne</td><td>60th Battalion annual reunion to be held.</td><td>Memorial hall, 401 Church Street, Richmond, on show eve, 26 September, at 7.45pm sharp.</td></tr>
<tr><td>29 Oct. 1951</td><td>Sydney Morning Herald</td><td>55th Battalion Assn. upcoming reunion</td><td>Millers Point Drill Hall on 16 November.</td></tr>
<tr><td colspan="4">Members will assemble at the corner of George and Barrack Streets at 7 p.m. for the Cenotaph remembrance ceremony.</td></tr>
<tr><td>10 Nov. 1952</td><td>Newcastle Morning Herald and Miners' Advocate</td><td>30th Battalion 27th Assn annual reunion held.</td><td>Soldiers' Memorial Institute, Perkin Street, Newcastle, Saturday 8 November.</td></tr>
<tr><td colspan="4">The 70 present included 18 visitors from Sydney and others from Woy Woy, Taree, Kurri Kurri and Cessnock. Representatives of the 2/30th Battalion were also present. The battalion's colours, purple and gold, were prominently displayed at the dinner and</td></tr>
</table>

<table>
<tr><td colspan="4">decorating the tables were the printed names of battles and engagements in which the battalion had fought.</td></tr>
<tr><td>1 July 1953</td><td>The Sydney Morning Herald</td><td>54th Battalion annual smoko reunion to be held</td><td>Winns Auditorium, on Thursday, 16 July, at 8pm.</td></tr>
<tr><td colspan="4">A wreath in memory of fallen comrades will be laid on the Cenotaph at 7.30 p.m.</td></tr>
<tr><td>19 Sep. 1953</td><td>The Age, Melbourne</td><td>59th Battalion reunion to be held in Show Week</td><td>Melbourne Town Hall, 8pm on 24 September</td></tr>
<tr><td>23 Oct. 1953</td><td>Murrumbidgee Irrigator, Leeton</td><td>56th Battalion reunion will be held</td><td>On 31 Saturday 31 October at Cootamundra</td></tr>
<tr><td>21 Oct. 1954</td><td>The Sydney Morning Herald</td><td>53rd Battalion (Whale Oil Guards) will hold a reunion</td><td>On Saturday 6 November, at 80m, at the Railways Institute, Devonshire Street, Sydney</td></tr>
<tr><td>18 Apr. 1955</td><td>The Argus, Melbourne</td><td>29th Battalion (A.I.F.) Association</td><td>Kensington Municipal Hall, opposite Newmarket Station, 8 p.m</td></tr>
<tr><td>13 Sep. 1955</td><td>The Argus, Melbourne</td><td>59th Battalion Assn Reunion to be held</td><td>Lower Melbourne Town Hall at 8pm on Thursday, 29 Sept.</td></tr>
<tr><td>8 Feb. 1956</td><td>The Argus, Melbourne</td><td>57/60th Battalion Assn Get Together to be held</td><td>Aspendale Palais, Sunday, beginning 10am</td></tr>
<tr><td>5 Sep. 1956</td><td>The Argus, Melbourne</td><td>57th Battalion AIF Assn smoke night reunion</td><td>Lower Melbourne Town Hall on Tuesday 18 September, at 7.45pm.</td></tr>
</table>

Many units appear to have held reunions in April, to coordinate with the annual ANZAC Day Parade. The 32nd, 54th and 55th Battalions all held their reunions in July. Those units which held reunions in Sydney can be seen to have done so in July.

Other units held reunions at different times. For example, the 30th Battalion in Newcastle held their reunion in November, on the anniversary of the date on which their unit sailed for the Great War.

The constant with all these reunions was reference to Fromelles.

The 19th of July men remembered, commemorated, and marched, on or as close to 19th July as they could do so. Right up until the last of those servicemen could do so no longer, in 1966.

LEST WE FORGET

---ooo000ooo---

Bibliography & Source Material

Material from Authors: -

AWMM: Australian War Memorial Museum: The Relics and Records of Australia's Effort In the Defence of The Empire, 1914-1918. Bookmark: http://trove.nla.gov.au/work/16827641.

Baker: Chris Baker - Was the Australian Official History More Truthful than the British? Originally an Essay submitted as part of the requirements for the degree of Master of Arts in the course of British First World War Studies at the University of Birmingham Centre for First World War Studies 2007.

Barton: Peter Barton - The Lost Legions of Fromelles, Allen & Unwin, 2014

Bean: Bean, C.E.W., Chapter XII - *Official History of Australia in the War of 1914-1918 Volume 3, The AIF in France 1916* Chapters 12 & 13 The Battle of Fromelles. Compiled by C.E.W. Bean, the history devotes almost 120 pages to the Battle of Fromelles (pp 328-447). This is available on line through the Australian War Memorial website www.awm.gov.au.

Beaumont: Joan Beaumont – Broken Nation: Australians In The Great War. 2013. Bookmark: http://trove.nla.gov.au/work/184016477.

Beaumont: Joan Beaumont – Why Australians Remember Fromelles. The Interpreter, Lowy Institute. 18 July, 2016.

Beckett & Corvi: Beckett, Dr Ian F; Corvi, Steven J (editors) (2006). Haig's Generals. London: Pen & Sword. ISBN 9781844158928

Birdwood: General Walter Riddell Birdwood: 'Khaki and Gown'

Browning: Neville Browning – Fromelles to Nauroy: 31st Battalion A.I.F. 1915-1919. Published Huntingdale W.A., N. Browning, 2013.

Burla: Bob Burla – Crossed Boomerangs, A History of All 31st Battalions.

Butcher: A.C. Butcher – An Examination of The Operational Influence of An Infantry Brigade Commander on the Western Front. Published 1999. Thesis (BA Hons) – University of New South Wales, Australian Defence Force Academy, Dept. of History 1999.

Carne: Lucy Carne: Heroes of Fromelles Set for Goodbye. The Advertiser, Adelaide 16 July, 2010 (NewsCorp)

Cassar: G. Cassar, (2004) Kitchener's War: British Strategy from 1914 to 1916. Washington: Brassy's ISBN 1-57488-708-4. P. 166

Cobb: Paul Cobb, Fromelles 1916. The History Press. ISBN 9780752456010

Collins: Collins Australian Pocket English Dictionary, William Collins Sons & Co, 1981, Reprinted 1986.

Cooksey: Jon Cooksey and Jerry Murland – The Battles of French Flanders: Neuve Chapelle, Aubers Ridge, Festubert, Loos and Fromelles. Pen & Sword Military, Bttle Lines: the Western Front 2015. Bookmark: http://trove.nla.gov.au/work/19393885.

Coombes: David Coombes – Anatomies of A Battle: Fromelles: Tactical Blunder or Mindless Slaughter. Thesis (B.A. (Hons)) University of Adelaide, Dept. of History 1993. Summary: http://digital.library.adelaide.edu.au/theses/09AR/09ARC7756.pdf

Corfield: Robin S. Corfield - Don't Forget Me, Cobber: The Battle of Fromelles. Carlton Vic.: MUP, - Miegunyah Press 2009.

Crawley: Walter Crawley – Diary: Private Walter Crawley, 58th Battalion, Wold War 1, 1916-1917. Part of Military Memorabilia Collection. Bookmark: http://museumvictoria.com.au/footer/-links/rights/

Dupuy: Dupuy, Trevor Nevitt (1992). Understanding war: History and Theory of Combat. London: Leo Cooper. ISBN 0-85052-293-5.

Edmonds: J. E. Edmonds, G.C. Wynne, and A.F. Becke, *The History of the Great War: Based on Official Document: Military Operations France and Belgium 1915*, Macmillan & Co, London, 1927, pps 63-64, 70, 84,

Elliott: Harold Edward Elliott - The Battle of Fleurbaix or Fromelles. Reprinted from The Duckboard, September 1, 1930. Available: http://handle.slv.vic.gov.au/10381/151930.

Ellis: Ellis, A.D. - The Story of the Fifth Australian Division, Being An Authoritative Account of the Division's Doings in Egypt, France and Belgium. First published 1920, now available in the Public Domain.

Enright: Michael Enright Ph D. – On the Western Front: Eyewitness Accounts from Australians In The Great War. Atglen, PA Schiffer Publishing 2015.

FitzSimons: Peter FitzSimons (sound recording) Battles of Fromelles and Pozieres; read by Richard Aspel. Melbourne Vic., ABC Audio, 2015. Available: World Wide Web.

Gatfield: John Gatfield – The RSL Book of World War 1. Edited by John Gatfield with Richard Landels. Bookmark: http://trove.nla.gov.au/work/191629743

Giles: Nicki Giles: "Yesterday, 17th July, 1916, The Battle of Bazentin Ridge Ended Published 18 July, 2014. See: - https://www.forces-war-records.co.uk/blog/2014/07/18/yesterday-17th-july-1916-the-battle-of-bazentin-ridge-ended Nicki Giles

Glasgow: University of Glasgow – Commemoration and Protection of Battlefields in Policy and Practice. Impact Case Study (REF3b).

Green: James Green – News From No-man's land. With introduction by Sir W.R. Birdwood. Published: London. Charles H. Kelly, 1917. Also available online: http://nla.gov.au/nla.obj-52799059

Hart: B.H. Liddell Hart - Foch (1937)

Hart: B.H. Liddell Hart – The Real War, first published 1930.

Holmes Peter Holmes – A Magnificent ANZAC. The Untold Story of Lieutenant Colonel Maurice Wilder Neligan

Jacobsen: Kris Jacobsen – Matilda's Son: A Life Lost At the Battle of Fromelles. Published Kambah A.C.T. Kris Jacobsen, 2012.

King: Jonathan King – The Western Front Diaries: The ANZACSs' own story, battle by battle. Bookmark: http://trove.nla.gov.au/work/34122440.

Kuring: Ian Kuring - Redcoats to Cams: A History of Australian Infantry 1788-2001

Laffin: John Laffin - Guide to Australian Battlefields of the Western Front, 1916-1918. Available: http://trove.nla.gov.au/work/11060053.

Associated article on Wikipedia: Peaceful Penetration.

Lee: Roger Lee - The Battle of Fromelles, 1916. Canberra A.C.T. Army History Unit – Australian Army campaigns series; 8. Published: Big Sky Publishing Pty Limited 2010. Bookmark: http://trove.nla.gov.au/work/37376389.

Lee: Roger Lee - British Battle Planning in 1916 and The Battle of Fromelles: A case Study of An Evolving Skill. Available: http://trove.nla.gov.au/work/192726228.

Lindsay: Patrick Lindsay - Fromelles. Bookmark: http://trove.nla.gov/work/25023760.

Lycett: Tim Lycett & Sandra Playle - Fromelles: The Final Chapters: How The Buried Diggers Were Identified and Their Lives Reclaimed. Melbourne Penguin Group, 2013.

McLachlan: Mat McLaughlin - Walking With The ANZACS: A Guide to Australian Battlefields on the Western Front. Available: http://trove.nla.gov.au/work/26296614.

McMullin: Ross McMullin – Disaster at Fromelles. Available: http://awm.gov.au/wartime/36/article.asp

McMullin: Ross McMullin – Pompey Elliott. Melbourne, Scribe Publications, 2008. Bookmark: http://trove.nla.gov.au/work/33183485.

McMullin: Ross McMullin – Farewell Dear People. Melbourne, Scribe Publications, 2012.

Middlebrook: M. Middlebrook - First Day on the Somme (1971)

Millar: John Dermott – The Soul of Anzac. General Sir William Birdwood and the AIF 1914-1918. 2015 Australian Scholarly Publishing.

Morrissey: Peter Morrissey – Anzac Battlefields of World War 1. [electronic resource] published: Peter Morrissey 2009. Available http://www.anzacbattlefields.com

Mueller: Robert J. Mueller – Fields of War: Fifty Key Battlefields in France and Belgium. Published: Arlington Heights [III] : French Battlefields c.2009.

Noonan: David Noonan – Those We Forget. Recounting Australian Casualties of the First World War. Melbourne University Press 2014.

Pedersen: Peter Pedersen - ANZACS on the Western Front: the Australian War Memorial Battlefield Guide. Available: http://trove.nla.gov.au/work/38424193.

Pedersen: Peter Pedersen with Chris Roberts - Battleground Europe: Fromelles French Flanders. Pen & Sword Books, first published in Great Britain in 2002.

Pollard: Tony Pollard – Pheasant Wood: Fromelles Evaluation of Possible Mass Graves. Published 2007 Royal United Services Institute of NSW. Open to the Public.

Rodger: Alexander Rodger – Battle Honours of the British Empire and Commonwealth Land Forces 1662 -1991. The Crowood Press, Ramsbury, Marlborough Wiltshire SN8 2HR. ISBN 1 86126 637 5

Rosenhain: Carol Rosenhain - The Man Who Carried The Nation's Grief.

Rule: Andrew Rule – Lost and Found in Fromelles; Sydney Morning Herald 20 July, 2010.

Senior: Michael Senior - Fromelles 1916, Pen & Sword Books, Barnsley England.

Senior: Michael Senior - *Fromelles, 19/20 July 1916 – A Success After All?*, Stand To: The Journal of the Western Front Association, Number 83, August / September 2008, pp 33-36.

Senior: Michael Senior: Haking - A Dutiful Soldier: Lieutenant General Sir Richard Haking: XI Corps Commander 1915-1918: A Study in Corps Command (2012) (hbk ed.) Barnsley: Pen & Sword Military. ISBN 978-1-84884-643-2.

Simpson: Andy Simpson - Directing Operations: British Corps Command on the Western Front 1914-18. (2006) Stroud: Spellmount. ISBN 978-1-86227-292-7.

Sloan: H. Sloan – The Purple and Gold: A History of the 30th Battalion. Bookmark: http://trove.nla.gov.au/work/9876991.

St. Claire: St. Claire, Ross - Our Gift To the Empire. 54th Australian Infantry Battalion, 1916-1919. Published: the Junction, NSW: R. St. Claire, 2006.

Summers: Julie Summers - Remembering Fromelles: A New Cemetery for a New Century. Commonwealth War Graves Commission, 2010.

Travers: Tim Travers: The Killing Ground. The British Army, the Western Front and the Emergence of Modern Warfare 1900-1918. Pen & Sword – reprinted 2009.

Van Os: Joanne Van Os – Ronan's Echo. A novel based on events at Fromelles, through women and forensic Anthropologists. MacMillan Press, 2014.

Wilkinson: Carole Wilkinson - Fromelles: Australia's Bloodiest Day At War. Fitzroy, Vic. Black Dog Books – the Drum. 2011.

Wikipedia: Wikipedia.org – Various items reproduced or sourced under the Wikipedia Creative Commons Attribution – Share Alike Licence.

---ooo000ooo---

Australian Unit Diaries on-line at the Australian War Memorial website www.awm.gov.au

1st Australian Division, General Staff, Headquarters War Diary July 1916, AWM4-1/42/18

1st Australian Division, Administrative Staff, Headquarters War Diary July 1916, AWM4-1/43/18

1st Infantry Brigade War Diary July 1916, AWM4-23/1/12

4th Australian Division, General Staff, Headquarters War Diary July 1916, AWM4-1/48/4

4th Australian Division, Administrative Staff, Headquarters War Diary July 1916, AWM4-1/49/4

5th Australian Division, General Staff, Headquarters War Diary July 1916, AWM4-1/50/5

5th Australian Division, Administrative Staff, Headquarters War Diary July 1916, AWM4-1/51/2

Military Orders and Instructions.

18.05.1922: Army Council Instruction 458 of 1922

4.09.1922: Army Order 338 of 1922

8.09.1922: Army Council Instruction 458 of 1922

16.12.1922: Army Order 470 of 1922

16.12.1922: Army Council Instruction 619 of 1922

1.08.1923: Military Order 336 of 1923

02.1924: Army Order 49 of 1924

10.05.1924: Military Order 167 of 1924

26.07.1924: Military Order 300 of 1924

1.11.1924: Military Order 480 of 1924

24.01.1925: Australian Army Order 37 of 1925

02.1925: Army Order 55 of 1925

9.03.1927: Australian Army Order 112 of 1927

Classified Records now Available for Examination: -

National Archives of Australia file 421/5/2

National Archives of Australia file 472/1/87

National Archives of Australia file 478/2/600

National Archives of Australia file 1926/525

(UK) War Office file 32/5921

Other Records Available Generally:

Australian War Memorial File 72/1/195

International Committee of The Red Cross – Grand Guerre files: https://grandeguerre.icrc.org/en/

Newspaper Reports referenced: -

1916.07.21 Evening Telegram, Newfoundland.

24.07.1916: The Argus, Melbourne, Where The Australians Attacked.

24.07.1916: The Argus, Melbourne: C.E.W. Bean - Australia's Share In Great Offensive

9.09.1916: Goulburn Evening Penny Post: Letter from Private Joseph Stephenson 14 MG CO.

27.03.1917: The Sun, Perth, WA: Trenches Won And Lost. Anzacs Lacked Support, in which Private A. Cohen described fighting at Fromelles, in which his division took part.

3.05.1917: The Sun, Sydney: letter by Private N. Batten to Mr. Butler, General Secretary of the Try Society, South Yarra.

6.09.1917: The Bendigonian, Bendigo Victoria, Captain H. M. Plowman, And Lieut. J. H. Sterling, M.C.

2.11.1917: Young Witness (Young NSW): A Returned Hero Who Won The D.C.M. At Fleurbaix.

3.08.1918: The Age, Melbourne on 3rd August, 1918: Australia's War Record. Four Years' Work Reviewed.

4.10.1918: The Argus, Melbourne: Repatriated Australians. Experiences Ln Germany.

18.10.1918: The Examiner, Launceston Tasmania: Private C.J. Mitchell Repatriated. Brutal Gaolers.

5.12.1918: The Daily Herald Adelaide: Heroes Of Fleurbaix. Australian Soldiers Returning

6.05.1919: The Moree Gwydir Examiner and General Advertiser: Starved And Ill-Treated.

17.07.1919: The Sydney Morning Herald, and the Western Star and Roma Advertiser, Toowoomba Qld, letter from "FLEURBAIX"

19.07.1919: The Sydney Morning Herald, Public Notices: Battle Of Fleurbaix.

22.07.1919: The Sydney Morning Herald, Sydney – letter to the Editor from Co.. Reuter E. Roth.

24.07.1919: The Sydney Morning Herald – Letter to the Editor from James Green.

26.07.1919: The Geelong Advertiser, Geelong Victoria: The Massacre At Fleurbaix Where Many Geelong Heroes Died

26.07.1919: The Sydney Morning Herald: Letter from Lt. Colonel C. MacLaurin, of the Army Medical Corps, AIF.

29.07.1919: The Sydney Morning Herald: Letter from 'One of the Fifth Division'.

30.07.1919: The Register, Adelaide - Commentary on the work of the Medical Officers, from J.R. Campbell, late Captain, A.I.F.

30.07.1919: The Sydney Morning Herald: Letter from J. Williams.

30.07.1919: The Sydney Morning Herald: Letter from Lt. Colonel Maclaurin.

20.03.1920: The Sydney Morning Herald: A Splendid Story, by F.M. Cutlack. A review of the book by Captain A.D. Ellis: The Fifth Division: - Its Great War Record.

10.04.1920: The Argus, Melbourne: Fromelles, 1916! A Glorious Failure. What Really Happened

20.07.1920: The Sydney Morning Herald: Fromelles. Fourth Anniversary.

22.07.1920: The Advocate, Melbourne: Battle Of Fromelles, A Hopeless Bungle.

23.07.1920: The Australian, Perth W.A.: The Fleurbaix Holocaust

31.07.1920: The Methodist, Sydney: Memorial Services. Fleurbaix And Pozieres

14.02.1922: House of Lords Debates Vol 49 cc126 [126]

19.07. 1924: The West Australian, Perth: Fleurbaix-Fromelles Battle.

19.07.1924: The West Australian, Perth: Fleurbaix-Fromelles Battle. First A.I.F. Engagement In France. Eighth Anniversary To-Day.

19.07.1933:	Newcastle Morning Herald and Miner's Advocate, NSW: Fleubaix Anniversary. Australians' Heavy Casualties
20.07.1933:	The Advertiser, Adelaide: Holocaust Of Fromelles. Australian Disaster In Early Days In France - Many Blunders.
30.11.1933:	The Western Mail, Perth: A Memory Of Fromelles.
24.04.1936:	Daily News Perth: First Interstate Unit Reunion
4.06.1936:	The Western Mail, Perth: In The Nursery. All Quiet At Fleurbaix by "Toxites" from Ballidu.
20.07.1936:	The Courier Mail, Brisbane: Famous Battle Recalled. Survivors Honour The Fallen.
26.09.1936:	The News, Adelaide: Gathering Of Diggers For 32nd Battalion Reunion.
28.04.1938:	The Western Mail, Perth WA: Chapter XVII. Fleurbaix. The article later became part of the *Story of the Eleventh Battalion A.I.F., by Captain Walter C. Belford M.A.*
28.04.1938:	The Western Mail, Perth WA: Fleurbaix Inferno.
12.05.1938:	The Western Mail, Perth WA: In Supports at Fleurbaix.
19.07.1939:	Goulburn Evening Penny Post: Fromelles - Australians In France.
17.03.1940:	Sunday Mail, Brisbane Qld: Heroes Of The Somme. V.C.'s - For Valour
28.07.1961:	The Canberra Times: France Honours Old Blood-Tie.
24.04.2010:	The Age, Melbourne – Bridie Smith: Push Grows To Have 'Fromelles' On Memorials
16.07.2010:	The Advertiser, Adelaide SA – Lucy Carne: Heroes Of Fromelles Set For Goodbye.
20.07.2010:	Sydney Morning Herald - Andrew Rule: Lost And Found In Fromelles.

About The Author

Geoffrey Benn grew up on a dairy farm outside Casino in northern NSW. He was surrounded by siblings in a family of 13 children. His mother's father always marched on Anzac Day but did not tell the family of the horrors he had witnessed at Gallipoli, in Egypt, and across Palestine in the Middle East. Nor did his grandfather talk about the death of his brother William.

Geoffrey's father served in the motorised regiments which remained in Australia in WWII.

And in turn Geoffrey served in the Australian Army Reserve in infantry and signals units.

After joining the workforce in the Local Courts of NSW, he took up legal studies as a mature age student, and graduated with a Bachelor of Laws. After 14 years of working in the courts, private practice beckoned, and he has both worked for others, and run his own business, from 1989 until retirement in 2022.

He is looking forward to travelling around Australia with his wife in the coming years.

---ooo000ooo---

Index

B

E

F

G

H

I

J

K

M

N

O

P

Q

R

S

T

U

V

www.ingramcontent.com/pod-product-compliance
Lightning Source LLC
LaVergne TN
LVHW071412210926
848486LV00004B/258